Handbook of Lipid Research 3

Sphingolipid Biochemistry

Handbook of Lipid Research

Editor: Donald J. Hanahan
The University of Texas Health Center at San Antonio
San Antonio, Texas

Handbook of Lipid Research 3

Sphingolipid Biochemistry

Julian N. Kanfer

University of Manitoba Faculty of Medicine
Winnipeg, Manitoba, Canada

and

Sen-itiroh Hakomori

Fred Hutchinson Cancer Research Center and
University of Washington
Seattle, Washington

Plenum Press · *New York and London*

ISBN 0-306-41092-3

Preface

Interest in and emphasis upon different aspects of the sphingolipids have, in general, followed the biochemical developments of the day. The early investigators were preoccupied principally with the isolation of "pure" compounds and structural elucidation. This historical perspective is found in the discussion presented in Chapter 1 (Section 1.1.2 and Table III). Still, the isolation and structural characterization of glycolipids are the basic foundation of all our knowledge of enzymology, immunology, and cell biology. Recent information obtained on structure has greatly affected the interpretation of various phenomena related to glycolipids. New structures suggest a new role of glycolipids as antigens and receptors. Ten years ago, only four neutral glycolipids and two gangliosides were known in human erythrocytes. We now know structures of at least twenty additional neutral glycolipids and ten additional gangliosides in human erythrocytes that are known to be important blood group, heterophil, and autoantigens. Erythrocytes are only one example of a cell type whose glycolipid profile has been extensively studied. Our defective knowledge in immunology and cell biology may be due to incomplete understanding of structural chemistry. Modern methodology based on methylation analysis, mass spectrometry, and enzymatic degradation has supplemented classical analysis based on clorimetry. Nuclear magnetic resonance spectroscopy is still in the development stage, but will eventually replace various chemical analyses. However, important future studies should be directed toward elucidating the organizational structure of glycolipids in membranes.

Knowledge of the biosynthesis of the sphingolipids has also followed a classical pattern of development. Soon after the availability of radioisotopes, early *in vivo* studies provided insight into the pathways of biosynthesis. Perhaps the most significant contribution was by Saul Roseman and his colleagues who demonstrated the logical sequential addition of carbohydrates from their nucleotide derivatives to produce these glycosphingolipids *in vitro*. A number of laboratories subsequently participated in providing the details of the individual reactions responsible for the production of the simplest compound, sphingosine, and eventually for that of the more complex gangliosides. Even though these pathways are accepted as dogma and appear in most general textbooks of biochemistry, the enzymes responsible for the individual reactions have not been purified. The data have been derived largely from studies employing particles, frequently of ill-defined character. Information about these reactions is presented in Chapter 2.

There is a wealth of information on the enzymatic degradation of the sphingolipids. Most of the enzymes catalyzing the hydrolysis of particular bonds have been purified, some to homogeneity. This contrasts with the lesser amount of information available on biosynthetic activity and can be readily ascribed to two circumstances: The hydrolytic enzymes do not appear to be hydrophobic proteins firmly embedded in a membrane matrix, and, therefore, are relatively simple to "solubilize," making them amenable to purification. A pioneer in providing descriptive information about the sphingolipid hydrolases is Shimon Gatt. Secondly, the existence of a number of sphingolipid storage diseases, caused by diminished hydrolytic activity, has provided the required "relevancy" to enhance the likelihood of grant support. A discussion of the hydrolytic activity and the sphingolipidoses is presented in Chapters 2 and 3. The early workers could not foresee the impact of studies on the individual hydrolases which became an important contribution to, and essential building block for, the fields of prenatal diagnosis, clinical genetics, and mass population screening, and which stimulated discussions on the floor of the United States Congress.

Investigations into the nature of tissue antigens revealed that many were lipid-soluble molecules. Early workers realistically preferred working with aqueous soluble compounds and devoted their efforts to the isolation and elucidation of the various blood group substances. Greater progress on the structural basis of various glycolipid antigens has been achieved with the availability of reliable immunological methods that can be applied to glycolipids, and by various modern methods in structural analysis (see Chapter 1). A summary of our knowledge on the structural concept of glycolipid specificity is discussed in Chapter 4. Of particular interest is the introduction of the monoclonal antibody approach and the discovery of tumor-associated glycolipid markers. We are at an early stage of development and Chapter 4 could only be an introduction to the coming great age of glycolipid immunology.

The realization that glycolipids are present on the external surface of cell membranes and that they possess antigenic activity has brought them into the modern-day arena of cell biology. Here again, their potential relevance to human disease has resulted in great activity on the sphingolipids. There are numerous reports of modifications in glycolipid composition and metabolism associated with oncogenic transformation. A clear-cut dependence on differentiation and development has also been demonstrated in more recent studies. There are indications that glycolipid changes are caused by activation of transforming genes. Studies in this area have inspired hope that the role of glycolipids in regulation of cell growth and cell recognition would eventually be realistic. Although our knowledge in this area is extremely premature and fragmentary, it is reviewed in Chapter 4.

The external aspect of the plasma membrane is also believed to contain the complement of cellular receptors for a diversity of agents. Since this coincides with the location of the glycosphingolipids, there has been a flurry of activity in the attempt to correlate receptors and glycolipids. This is described in the last Chapter.

It is difficult to predict the probable future developments, but it seems reasonable to predict that a component of future efforts will be a continuation of present ones. However, revolutionary developments can be expected in the application of monoclonal antibodies, physical instrumentation analysis, and gene cloning. A precise analysis of glycolipid organization in membranes, roles in cell growth regulation, cell adhesion and cell recognition, and their genetic and epigenetic control will be exciting developments for the near future. The knowledge of glycolipid function will be extremely useful in prevention and treatment of various human diseases.

There will probably not be any great effort expended on the biosynthesis of the sphingolipids, mainly because the individual steps have been accepted and are "ancient history." Some groups may think it important actually to purify and study the isolated enzymes responsible for the individual reactions. However, genetic control of glycosyltransferases at transcriptional and translational levels and epigenetic regulation of the assembly of glycosyltransferases in membranes will be the central theme of glycolipid studies in the future. Various diseases, including cancer, inflammation, and aging, may well be related to abnormalities of these processes.

There will be continued isolation and structural elucidation of glycolipids. The work will be focused on many uncharacterized species which are available only in small quantities in specific types of cells or tissues and are limited to a specific stage of development.

There has been a diminution of the previous flurry of activity on the sphingolipidoses largely because the enzyme deficiencies are understood, the patients can be identified, pedigrees can be screened, and reliable prenatal diagnosis can be offered. Perhaps the future efforts will reveal the pathophysiological basis for the characteristic disease process resulting from the accumulation of a particular sphingolipid. The techniques of recombinant DNA technology and monoclonal antibodies will undoubtably be utilized to reveal the primary genetic abnormalities responsible for the separate diseases. A good deal of current interest centers upon the anecdotal patient who presents uniquely to the physician. These patients may provide useful materials and challenges to accepted dogma so that more incisive investigations can be carried out.

The more difficult areas for prediction, but probably the most exciting, must be in cell biology. Perhaps some day we will be able to speak with some degree of certainty about the function of these intriguing molecules.

This book was written to fill a need for a single source of information in this field. It was not written for the sphingolipid specialist, although it is hoped that even the specialist will find some useful information. It was prepared for those who are curious to learn about these intriguing molecules. Immunologists, cell biologists, physicians, biochemists, educators, and students will find it useful for their various personal needs.

Inevitably, a few of our colleagues may have some degree of dissatisfaction for a number of possible reasons. We may have missed a contribution, we may disagree with them, we may misinterpret them, we may be incorrect in a statement, or our style may not coincide with theirs. Nonetheless, a text such as this inevitably reflects the authors' biases, experience, and particular research interests. We have attempted to be comprehensive, but not necessarily encyclopedic. We wish to thank the many researchers who have been actively engaged in the study of the sphingolipids for providing so much interesting data for us to read, understand, and digest.

S. Hakomori is greatly indebted to Mrs. Charlotte Pagni for reference arrangement and typing of the manuscript, to Drs. Kiyohiro Watanabe, William W. Young, and Reiji Kannagi for their collaboration in providing references and figures, and to Dr. Roger A. Laine for reading Chapter 1. Section 1.3.2 was based partially on a review written in Japanese by Drs. A. Hayashi and T. Matsubara. He is particularly grateful to a number of scientists and publishers who allowed him to cite published materials and to reproduce various figures.

Julian N. Kanfer
Sen-itiroh Hakomori

Contents

Chapter 2
Sphingolipid Metabolism
Julian N. Kanfer

Chapter 5
Glycolipid Antigens and Genetic Markers
Sen-itiroh Hakomori and William W. Young, Jr.

Chapter 6
Glycosphingolipids as Receptors
Julian N. Kanfer

Chapter 1

Chemistry of Glycosphingolipids

Sen-itiroh Hakomori

1.1. Introduction

1.1.1. Classification and Nomenclature of Glycosphingolipids

Glycosphingolipids are the glycosides of N-acylsphingosine, the trivial name of which is ceramide. A great deal of structural variation in fatty acids, sphingosines, and carbohydrates results in a great number of chemically distinct glycosphingolipids. Sphingosines are a group of related long-chain aliphatic 2-amino-1,3-diols [long-chain bases (LCBs)], of which D-*erythro*-1,3-dihydroxy-2-amino-4,5-*trans*-octadecene (sphingosine or sphingenine or sphing-4-enine) occurs most frequently in animal glycosphingolipids. The structure and molecular species of LCBs and ceramide are discussed in Section 1.4.

The simplest glycolipid, cerebroside, was classified in the past as cerebron, kerasin (or phrenosin), nervon, and oxynervon, according to the kind of fatty acid residues present in the ceramide moiety because the solubility of cerebrosides is greatly influenced by the kind of fatty acids present. With increasing complexity of the carbohydrate structure, however, solubility and other physical properties are more dependent on the latter.

Thus, classification of glycosphingolipids has been based on their carbohydrate structure rather than on the ceramide moieties (see Table I). The carbohydrate residue of glycosphingolipids is a definitive determinant in antigenicity and immunogenicity of cells (see Chapter 5) and may be involved in intercellular recognition and cell social activities (see Chapter 4).

Glycosphingolipids are classified as neutral glycolipids, sulfatides (sulfate-containing), and gangliosides* (sialic acid-containing), depending on the substituted groups of the carbohydrates (Table I).

---

* The term "ganglioside" is a generic name for the sialosylated glycolipid irrespective of its core structure, whereas the term "ganglio-series" or the prefix "ganglio-" signifies the glycolipid or oligosaccharide structure containing Galβ1→3GalNAcβ1→4Galβ1→4Glc or GalNAcβ1→4Galβ1→4Glc as the common core.

Table I. Classification and Nomenclature of Glycosphingolipids

A. Classification according to substitution
　1. Neutral glycolipids: glycolipids that contain hexoses, *N*-acetyl hexosamines, and methyl pentoses
　2. Sulfatides: glycolipids that contain sulfate esters
　3. Gangliosides (sialosylglycolipids): glycolipids that contain sialic acids
　4. Phosphoinositido-glycolipids: glycolipids that contain phosphoinositido-ceramide
B. Classification according to basic carbohydrate structures
　1. Simpler glycolipids (representing the basic structure) Cerebrosides and dihexosyl ceramides (see Table XX)
　2. Globo-series glycolipids
　　Globotriaosylceramide (ceramide trihexoside), globoside, Forssman glycolipid, and others (see Tables IIA and XXI)
　3. Lacto-series glycolipids
　　　Lactotriaosylceramide, paragloboside, sialosylparagloboside, H_1 glycolipid and many other analogs, and polyglycosylceramide (see Tables IIA and XXII)
　4. Muco-series glycolipids (see Tables IIA and XXIII)
　5. Simpler gangliosides and hematosides (sialosyl glycolipids without amino sugars) (see Table XXIV)
　6. Ganglio-series glycolipids (see Tables IIA,B and XXV)
C. Classification according to number of carbohydrate residues
　　Ceramide mono-, di-, tri-, tetra-, penta-, to polyglycosylceramide

On the other hand, glycosphingolipids can be classified according to the number of carbohydrate residues. Ceramide mono- to tetrakaidecasaccharides have been clearly identified up to the present time. Ceramides with a greater number of carbohydrate residues (20–50) have been isolated and partially characterized as "polyglycosylceramides,"[84,85] although these components have not been isolated in a pure state. Alternative classification depends on the backbone structure of the carbohydrate, i.e., the globo-, lacto-, muco-, ganglio-, and gala-series of glycolipids (see Table I, Section B). The "globoseries" contains globo-triaosylceramide (Galβ1→4 or 1→3Galβ1→4Glc→Cer), the "lacto-series" includes lactotriaosylceramide (GlcNAcβ1→ 3Galβ1→ 4Glc→Cer), the "muco-series" contains mucotriaosylceramide (Galβ1→4 or 1→3Galβ1→4Glcβ1→1Cer), the "ganglio-series" contains gangliotriaosylceramide (GlcNAcβ1→4Galβ1→4Glcβ1→1Cer), and the "gala-series" contains galabiosylceramide (Galβ1→4Galβ1→1Cer) as the common structure. Details of structural variation and distribution are discussed in Section 1.4, and the known structures are shown in Tables XX–XXV. A few important and basic features of each group of glycolipids can be described as follows: (1) The globo-series glycolipids are generally not substituted with fucose, galactose, or sialic acid; however, a few exceptions have been described recently.[379,548,589] (2) The lacto-series glycolipids show the most extensive substitution with (a) sialosyl, (b) fucosyl, (c) and α- and β-galactosyl moieties, and (d) extension and branching by repeating N-acetyllactosamine units, thus resulting in a great number of distinct molecular structures. The sialosylated species include the major ganglioside of extraneural tissue and the fucosylated species includes the major blood group antigens at the cell surface. (3) Gangliosides without hexosamine are called by the trivial name "hematoside," which constitutes the major extraneural ganglioside together with sialosyllactosamine glycolipids. (4) The ganglio-series structure is usually sialosylated; the nonsialosylated structure is rare, but present as specific cell-surface markers or antigens. A minor population of the ganglio-series is fucosylated (fucoganglioside) (see Table XXV, Section D).

All glycosphingolipids can be termed by naming the monosaccharide or oligosaccharide plus *osylceramide*, e.g., galactosylceramide and lactosylceramide. Systematic names of oligosaccharides are so cumbersome that they have been renamed as recommended by the IUPAC-IUB Commission on Biochemical Nomenclature* (Table IIA). Abbreviations of these glycolipid structures are also suggested in Table IIA. New symbols for each series of structures are *Gb* (or GbOse) for globo-series, *Lc* (or LcOse) for lacto-series, and *Gg* (GgOse) for ganglio-series plus arabic numbers indicating the number of carbohydrate residues, and the letters a, b, c, etc., to indicate the positional and steric isomers. For example, globoside is globotetraosylceramide, and its symbol would be Gb4a (or GbOse4a); cytolipin R is globo*neo*tetraosylceramide, and its symbol would be Gb4b (or GbOse4b); Forssman glycolipid is globopentaosylceramide, and its symbol would be Gb5 (or GbOse5); GM_{1a} ganglioside is sialosylgangliotetraosylceramide, and its symbol would be NeuAcGg4a; asialo GM_2 is gangliotriaosylceramide, and its symbol would be Gg3 (or GgOse3).

The location of sialosyl and fucosyl residues can be assigned by Roman numerals for the sequential number of sugar residues counting toward the nonreducing end from the glucose attached to ceramide. The position of sialic acid linkage can be expressed by the Arabic numeral. Thus, GM_{1a} would be II^3(NeuAc)GgOse4Cer, GD_{1b} would be II^3(NeuAc)$_2$GgOse4Cer, GT_{1a} would be IV^3(NeuAc)II^3(NeuAc)$_2$GgOse4Cer (see Table XXV), two sialosylparaglobosides with sialosyl2→3Gal and sialosyl2→6Gal would be IV^3(NeuAc)LcOse4Cer and IV^6(NeuAc)LcOse4Cer, fucosylated GM_1 would be IV^2(Fuc)II^3(NeuAc)GgOse4Cer, and sialosyllacto-N-*nor*hexaosylceramide would be VI^3(NeuAc)LcOse6Cer.

Structures belonging to the ganglio-series have been abbreviated most frequently according to the symbols proposed by Svennerholm,[57] which are listed in Table IIB.

---

* The recommendations were published under the title "The nomenclature of lipids" in: *The Journal of Lipid Research* **19**, 121–125 (1978) and *Lipids* **12**, 455–468 (1977).

Table IIA. Three Major and Two Minor Series of Basic Structures of Oligosaccharides Found in Glycosphingolipids and Their Symbols

Oligosaccharides	Trivial names	Proposed symbols for glycolipids
1. Globo-series (major)		
Galα1→4Galβ1→4Glc	Globotriaose	Gb3a or GbOse3a
Galα1→3Galβ1→4Glc	Globoneotriaose	Gb3b or GbOse3b
GalNAcβ1→3Galα1→4Galβ1→4Glc	Globotetraose	Gb4a or GbOse4a
GalNAcβ1→3Galα1→3Galβ1→4Glc	Globoneotetraose	Gb4b or GbOse4b
GalNAcα1→3GalNAcβ1→3Galα1→4Galβ1→4Glc	Globopentaose	Gb5 or GbOse5
2. Lacto-series (major)		
GlcNAcβ1→3Galβ1→4Glc	Lactotriaose	Lc3 or LcOse$_3$
Galβ1→3GlcNAcβ1→3Galβ1→4Glc	Lactotetraose	Lc4a or LcOse4a
Galβ1→4GlcNAcβ1→3Galβ→4Glc	Lactoneotetraose	Lc4b or LcOse4b
Galβ1→4GlcNAcβ1→3Galβ1→4GlcNAcβ1→3Galβ1→4Glc	Lactohexaose (or lactonorhexaose)	Lc6 or LcOse6
Galβ1→4GlcNAcβ1 → / ³₆Galβ1→4GlcNAcβ1→3Galβ1→4Glc / Galβ1→4GlcNAcβ1 ←	Lactoisooctaose	Lc8 or LcOse8
3. Ganglio-series (major)		
GalNAcβ1→4Galβ1→4Glc	Gangliotriaose	Gg3 or GgOse3
Galβ1→rGalβ1→3GalNAcβ1→4Galβ1→4Glc	Gangliotetraose	Gg4 or GgOse4
GalNAcβ1→3GalNAcβ1→4Galβ1→4Cer	Gangliopentaose	Gg5 or GgOse5
4. Muco-series (minor)		
Galβ1→4Galβ1→4Glc	Mucotriaose	Mc3 or McOse3
Galβ1→4Galβ1→4Galβ1→4Glc	Mucotetraose	Mc4 or McOse4
5. Gal-series (minor)		
Galα1→4Gal	Galabiose	Ga2 or GaOse2
Galα1→4Galα1→4Gal	Galatriaose	Ga3 or GaOse3

Table IIB. Ganglioside Symbols According to the Assignment of Svennerholm[57] *and Holmgren et al.*[559]

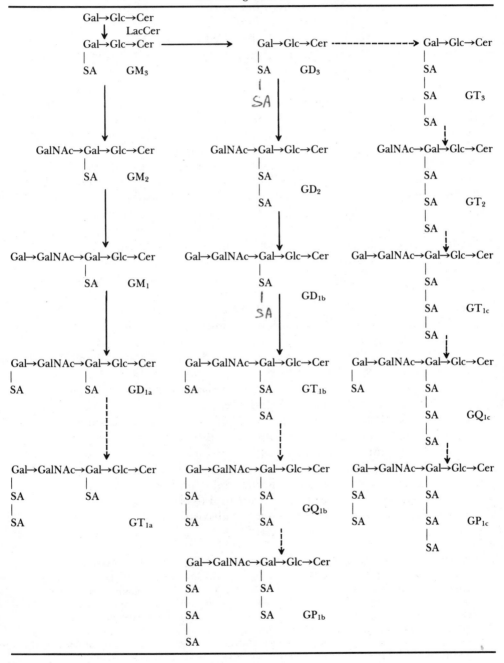

1.1.2. Brief History of the Chemistry of Glycosphingolipids

The chemistry of glycosphingolipids has been developed in association with two areas of research, one in the studies of the chemical composition of brain and nervous tissue and the other in the studies of the accumulating lipid in tissues of patients with "sphingolipidosis." Important events in the development of glycolipid chemistry are listed in Table III.

Historically, glycosphingolipids were discovered as the major lipid component of the brain. In 1865, Liebreich[1] described the presence of a unique homogeneous glycoside, soluble in hot ethanol and precipitable in cold. The composition of this glycoside was described as being invariable among various species, and he named it "protagon," signifying that it was the most important essential chemical entity of the brain. From 1874 to 1876, Thudichum[2] (see Section 1.6), a surgeon-chemist in London, described the chemical composition of the brain and the presence of cerebroside (or cerebral galactoside) and its exact chemical composition, including a unique aliphatic alkaloid called "sphingosine"—a description that is now indisputable. Results of his systematic studies were compiled in a monograph that is now regarded as a monumental classic of brain chemistry as well as sphingolipid chemistry, published first in English and subsequently in German.[3] Although Thudichum correctly described Liebreich's protagon as a mixture of phospholipid (sphingomyelin) and "cerebroside," his findings were not accepted by many influential biochemists at that time.* This resulted in the bitter and violent "protagon controversy" that was meticulously described in a biography of Thudichum's life by Drabkin[4] in 1958. The chemical compositions of cerebroside and sphingomyelin and the presence of sphingosine were fully confirmed in later days by O. Rosenheim[5] in London, H. Thierfelder and E. Klenk[7] in Tübingen, and P.A. Levene and W.A. Jacobs in New York[9] only after Thudichum's death in 1901. When the famous monograph *Die Chemie der Cerebroside und Phosphatide* was published in 1930 by Thierfelder and Klenk,[7] the protagon controversy was settled and use of the name "protagon" was discontinued. Cerebroside was classified into four groups according to the representative fatty acid residues: kerasin (containing 24 : 0 fatty acid), cerebron or phrenosin (24 : 0 α-hydroxy), nervon (24 : 1), and oxynervon (24 : 1 α-hydroxy). Galactosylceramide, or cerebroside, is found in the brain in unusually high concentrations, in striking contrast to extraneural tissues.

The most characteristic component of glycosphingolipids is the aliphatic

---

* Thudichum's discoveries concerning cerebroside and many other compounds were not accepted simply because he had an unusually wide range of capability. According to Hakomori[617]: "It is indeed amazing that Thudichum's finding on cerebroside and sphingosine and many of his other discoveries such as bile pigment, cruentin (hematoporphyrin), sarcosine, etc., were attacked as falsified results by his opponents, Liebreich, Hoppe-Seyler, Malley and Gamgee, mostly on the grounds that any man who had written a book on urine, another on gallstones, several on wines, a cook book, who had developed new surgical tools and procedures and who had interested himself in public health and the course of mange to name a few could possibly find things in the brain that no one else had ever bothered to look for previously. Thudichum's honor was not fully restored until a thorough re-investigation after his death in 1901."

Johann Ludwig Wilhelm Thudichum
(see biographical sketch, p. 145)

amino alcohol originally discovered by Thudichum[2] and named by him "sphingosine." He characterized the substance as a base with the empirical formula $C_{17}H_{35}NO_2$, but the carbon number was later corrected to 18 by Thierfelder and Klenk.[6] The presence of an olefinic double bond was discovered by Kitagaua and Thierfelder.[8] However, there was a long-lasting controversy as to the relative position of hydroxy groups and an amino group. The first structure assigned was 1-amino-2,3-dihydroxyheptadecene by Levene and co-workers (e.g., Levene and Jacobs,[9] Levene and West[10]), then corrected as 3-amino-1,2-dihydroxyoctadecene by Klenk and Diebold,[11] and finally determined as 2-amino-1,3-dihydroxyoctadecene by Carter *et al.*[12,13]

Herbert E. Carter
(see biographical sketch, p. 147)

Table III. Important Events in Glycosphingolipid Chemistry

Year	Event	Workers and references
1874	Discovery of sphingolipid (cerebroside and sphingomyelin) and sphingosine	Thudichum[2,3]
1924–1934	Accumulation of glucocerebroside in organs of patients with Gaucher's disease; first description of sphingolipidosis and glucosylceramide	Lieb,[27] Aghion[29]
1930	Isolation of α-hydroxy-fatty acid-containing cerebroside and classification of cerebrosides	Thierfelder and Klenk[7]
1933	Isolation and characterization of sulfatide	Blix[33]
1936–1938	Discovery of sialic acid and of the presence of N-acetylhexosamine in brain glycolipid	Blix[38,42]
1935–1942	Discovery, isolation, and characterization of brain ganglioside	Klenk[31,45]
1942	Isolation and characterization of dihexosyl-ceramide	Klenk and Rennkampf[36]
1942–1950	Determination of the structure of sphingosine as 2-amino-1,3-dihydroxy-octadecene and correct structure of cerebroside	Carter *et al.*,[12,13,63] Ohno[15,16]
1953–1958	Determination of the diastereomeric structure of sphingosine as *trans*-D-*erythro* ("*trans* 2S,3R")	Carter and co-workers, [64,65] Shapiro *et al.*[66]
1951	Isolation and characterization of hematoside from erythrocyte membranes, and species difference of glycolipid in erythrocytes	Yamakawa and Suzuki,[47] Yamakawa[51]
1952–1963	Structure of phytosphingosine as D-*ribo*-2-amino-1,3,4-trihydroxyoctadecane; discovery of phytosphingoglycolipids	Carter *et al.*,[21] Oda,[20] Carter and Hendrickson[70]
1958–1973	Total chemical synthesis of sphingosine, ceramide, and various glycosphingolipids	Shapiro and co-workers[66–69]
1960	Discovery of C_{20}-sphingosine in brain ganglioside	Proštenick and Majhofer-Oreščanin[25]
1962–1963	Determination of position of sulfate group in sulfatide	Yamakawa, *et al.*,[34] Stoffyn and Stoffyn[35]
1963	Isolation and structure of four major gangliosides of brain	Kuhn and Wiegandt,[60]
1963	Accumulation of a ceramide trihexoside in organs of a patient with Fabry's disease	Sweeley and Klionsky[32]
1964	Isolation and characterization of glucosamine-ganglioside (sialosyllacto-N-*neo*tetrasylceramide	Kuhn and Wiegandt[71]
1963–1965	Structure of globoside	Yamakawa *et al.*[72,73]
1963–1965	Structure of GM_2 ganglioside	Makita and Yamakawa,[61] Ledeen and Salsman[62]
1964	First isolation of fucoglycosphingolipid	Hakomori and Jeanloz[74]
1968	Isolation and characterization of pure blood group ABH active glycosphingolipid	Hakomori and Strycharz[75]
1970	Characterization of over 60 kinds of long-chain bases	Karlsson[76]
1971	Structure of globo-series glycolipids: anomeric structure of ceramide trihexosides and globoside and correct structure of Forssman hapten glycolipid	Hakomori *et al.*,[77] Siddiqui and Hakomori[78]

(continued)

Table III. (Continued)

Year	Event	Workers and references
1972–1973	Isolation of a glycolipid having two carbohydrate chains by branching: ceramide megalosaccharide	Hakomori et al.,[79] Wiegandt[80]
1973	Sialylgalactosylceramide as the major myelin ganglioside	Ledeen et al.[439]
1974–1976	Identifiction of P[k], P, P$_1$ antigens	Naiki and Marcus,[81,82] Marcus et al.[83]
1976	Discovery of "polyglycosylceramide"	Koscielak et al.,[84] Gardas[85]
1970–1978	Discovery and characterization of unusual glycolipids in invertebrates (e.g., those that contain sialic acid as internal sugar residue, those that contain mannose, xylose, and O-methyl-galactosamine)	Hori and co-workers,[510–518] Matsubara and Hayashi,[520,521] Kochetkov and Smirnova[508]
1973–1979	Isolation and characterization of new type of gangliosides	Svennerholm et al.,[86] Wiegandt,[80,87] Rauvala,[428] Watanabe et al.,[116,606] Iwamori and Nagai,[474,498] Chien et al.,[390]

(see Section 1.6), who observed that *N*-benzoyldihydrosphingosine was not oxidized by periodic acid,[12] whereas dihydrosphingosine was oxidized by sodium periodate to form equimolar amounts of formaldehyde, formic acid, ammonia, and palmital,[13] i.e., $CH_3(CH_2)_{14} \cdot CH(OH) \cdot CH(NH_2)CH_2OH \rightarrow CH_3(CH_2)_{14} \cdot CHO + HCHO + NH_3 + HCOOH$. According to this revised structure of sphingosine, the correct structure of cerebroside was established as 1-*O*-galactopyranosylceramide.[14] It is interesting to note that Ohno[15] independently threw doubt on the Klenk–Diebold structure and proposed 2-amino-1,3-dihydroxyoctadecene or octadecane on the basis of lead tetraacetate oxidation of sphingosine, dihydrosphingosine, and their *N*-benzoyl derivatives. Hexadecanal (palmital) and *(trans)*-hexadecene-(2,3)al were identified as thiosemicarbazone derivatives in the lead tetraacetate oxidation products of sphingosine and dihydrosphingosine, respectively.[15] *N*-Benzoyl derivatives of sphingosine and dihydrosphingosine were not oxidized under the same conditions.[16] Thus, the 2-amino-1,3-dihydroxy(*trans*-4,5)octadecene structure was independently proposed for sphingosine. On the basis of this structure, a revised structure of cerebroside was proposed by Nakayama.[17] Pröstenik[18] suggested that Seydel[19] had had good evidence for the 2-amino-1,3-dihydroxyalkene structure through his thesis work in 1941. No details were described, however. The diastereoisomeric structures of sphingosine and dihydrosphingosine as to C-2 and C-3 have been determined as D-*erythro*(2*S*,3*R*) by Carter and co-workers[64,65] (see Section 1.4.1).

For many years, only two bases, sphingosine (octadecasphingenine) and dihydrosphingosine (octadecasphinganine) were known until Oda[20] and Carter *et al.*[21] established the structure of phytosphingosine in plant and microbial sources* and Pröstenik and Majhofer-Oreščanin[25] and Stanaćev and Chargaff[26] found C_{20} sphingosine in brain gangliosides. Gas chromatography–mass spectrometry was introduced for analysis of sphingosine and ceramide, and their molecular species in various glycolipids have been analyzed in greater detail by Karlsson, Samuelsson, and Sweeley and their colleagues (see Sections 1.3.2 and 1.4.1).

The first sphingoglycolipidosis was reported in 1924; Lieb[27] described the accumulation of "galactocerebroside" in the spleen of a patient with Gaucher's disease. The carbohydrate moiety of the accumulated glycolipid was questioned and thought not to be galactose because it showed different optical rotation than galactose[28] and was finally identified as glucose by Aghion[29]† and then by Halliday *et al.*[30] Therefore, this disease is now regarded as a deficiency of β-glucosidase (see Chapter 3). A number of studies on various glycosphingolipidosis have been carried out and have greatly enriched our knowledge of the chemistry and metabolism of glycosphingolipids. In fact, Klenk's discovery of gangliosides[31] and Sweeley's discovery of ceramide trihexosides[32] were rooted in studies of sphingolipidosis.

The presence of sulfate-containing glycolipids was suggested by Thudichum,[2,3] but they were difficult to separate from phospholipids. In 1933, Blix[33] (see Section 1.6) in Uppsala was the first to isolate a pure compound containing fatty acid, sphingosine, galactose, and sulfate ester, and the structure was assigned as the 6-*O*-sulfate ester of galactosylcerebroside. The position of sulfate was later correctly established to be 3-*O*-galactose by Yamakawa *et al.*[34] and Stoffyn and Stoffyn.[35]

A glycolipid having more than one carbohydrate residue was first found in bovine kidney by Klenk and Rennkampf[36] in 1942. The sugar composition was eventually identified to be both glucose and galactose; this seemingly simple result required a tremendous amount of laborious work at that time, since the carbohydrate composition had to be identified by optical rotation, fermentability by yeast, and isolation and characterization of glucose and galactose as phenylhydrazone (osazone), and therefore required at least several grams of pure substance, starting from many kilograms of the tissue. A similar glycolipid with a lactosyl residue termed "cytolipin H" was claimed to

---

* The presence of a nitrogen-containing lipid substance in various mushrooms, molds, and yeasts has been known since Zellner described it in 1911 as "fungus cerebrin."[22] The yeast cerebrin was isolated and identified as an amide formed from 4-amino-1,3,5-trihydroxyeicosane ("cerebrin base") and fatty acids.[23,24] However, Oda[20] reinvestigated the structure of cerebrin base and concluded that it was 2-amino-1,3,4-trihydroxyoctadecane. Carter et al.[21] reported the presence of a base with the same structure as that reported by Oda. Because its structure showed a close similarity to that of animal sphingosine base and it was found in plants, the name phytosphingosine was proposed.

† Identified as glucosazon, which was distinguished from galactosazon. A copy of Aghion's thesis was made available through the kind courtesy of Dr. Bernard Zalc (Hôpital de la Salpêtrière, Paris), to whom the author is gratefully indebted. Dr. Aghion was a practicing physician in Paris until recent years; she is now retired.

Gunnar Blix
(see biographical sketch, p. 148)

be a tumor-specific lipid hapten in human epidermoid carcinoma by Rapport et al.[37] However, the lactosylceramide structure was suggested on the basis of the fact that lactose was the best hapten inhibitor in the complement-fixation reaction, not on the basis of chemical structural analysis. Permethylation was impossible to apply to the glycolipid class of compounds at that time, and the structural study of carbohydrate residues in glycolipids was extremely difficult. As will be discussed later, modern methylation analysis and identification of methylated carbohydrate residues in glycolipids were developed only after the mid-1960s.

The history of gangliosides (sialosylglycolipids) would require many pages for a complete recounting and will therefore be briefly summarized in the following few paragraphs: A crystalline polyhydroxyamino carboxylic acid was isolated from the hydrolysate of salivary gland mucine by Blix.[38] The compound was characterized by giving a purple color by Bial's orcinol–HCl reagent in contrast to the green color development for hexoses and pentoses. The compound was also characterized by the positive direct Ehrlich reaction in contrast to N-acetylhexosamines, which gave the indirect Ehrlich reaction (Morgan–Elson reaction). It was found to be widely distributed in various glycoproteins, and the name "sialic acid" was later given to it by Blix et al.[39] Two independent observations described the presence of lipid fractions showing a purple color development with Bial's orcinol–HCl reaction. Walz[40] in Tübingen found a lipid fraction of bovine spleen that showed a purple color development with Bial's orcinol reagent, and Levene and Landsteiner[41] in the Rockefeller Institute, New York, described a similar lipid fraction in horse kidney during the purification of Forssman antigen. These reports were the first recognition of present-day gangliosides.

Ernst Klenk
(see biographical sketch, p. 148)

In 1935, Klenk[31] in Cologne (see Section 1.6) described, in the brain of a patient who died of amaurotic idiocy (Tay–Sachs' disease), a large accumulation of a new type of lipid, called "Substanz X," that showed an intense purple color development with Bial's reaction. In 1938, Blix[42] discovered in normal bovine brain a lipid fraction that contained hexosamine and showed an indirect Ehrlich reaction with acetylacetone (Elson–Morgan color reaction). Subsequently, Klenk[43] isolated from the methanolysate of "Substanz X" a crystalline polyhydroxyamino carboxylic acid called "neuraminic acid." The neuraminic acid gave a purple color reaction with orcinol and gave the positive direct Ehrlich reaction, which was identical to that of Blix's substance containing "sialic acid." The concentration of the lipid-bound "neuraminic acid" was much higher in gray matter than in white matter,[44] and finally a "homogeneous" glycosphingolipid was isolated from bovine brain and spleen that had a fatty acids/sphingosine/hexose/neuraminic acid molar ratio of $1 : 1 : 3 : 1$, and it was given the name "ganglioside."[45] In this original paper, hexosamine was not considered as the component of ganglioside. The presence of hexosamine in brain glycolipid was previously found by Blix,[42] but was thought to be a component of "sialic acid" that Blix considered to be a disaccharide composed of hexouronic acid and hexosamine[42]; however, Klenk[45,46] was unable to detect hexosamine in his crystalline "neuraminic acid" obtained from a methanolysate of ganglioside. Blix *et al.*[39] identified the hexosamine of ganglioside as galactosamine, in an aqueous hydrolysate, using paper chromatography. Klenk[46] identified galactosamine as a component of ganglioside independent of sialic acid, thus finally establishing that brain gangliosides were composed of fatty acid, sphingosine, hexoses, galactosamine, and sialic acid.

At almost the same time that Klenk established the correct composition of brain gangliosides, a new type of sialic acid-containing glycolipid was iso-

Tamio Yamakama
(see biographical sketch, p. 149)

lated from horse erythrocyte stroma by Yamakawa and Suzuki[47] (see Section 1.6) and was named "hematoside." In contrast to the brain gangliosides, hematoside did not contain hexosamine. A crystalline polyhydroxyamino carboxylic acid was isolated from a methanolysate of hematoside and termed "hematamic acid." Consequently, "hematamic acid" was identified as the methylester-methylglycoside of Klenk's "neuraminic acid."[48] Discovery of "hematoside" firmly established the presence of sialic acid-containing glycolipid in extraneural tissues and suggested that gangliosides could be an important component of cell-surface membranes.[51]

Klenk and Lauenstein[49] noticed that the major glycolipid of human erythrocyte membranes contained galactosamine but lacked sialic acid. This finding was confirmed by Yamakawa and Suzuki,[50] and the glycolipid was named "globoside" because it was obtained as a birefringent "globular" precipitate. This work, together with the previous one on hematoside,[47] raised the possibility that the carbohydrate composition of the major glycolipids might vary according to species and organs. Species differences in the glycolipid composition of erythrocyte membranes was clearly demonstrated by a series of studies carried out by Yamakawa[51]: (1) cat, dog, and horse contained a hematoside (containing sialic acid, but no hexosamine) as the major component; (2) human, guinea pig, rabbit, and monkey contained "globoside" or a glycolipid without sialic acid; and (3) cow contained gangliosides with both sialic acid and hexosamine.

Ganglioside was viewed by some investigators as a macromolecular complex. The presence of a highly complex acidic macromolecule in brain and other tissue was claimed by Folch *et al.*[52] The compound was obtained simply by evaporation and drying of the dialyzed "Folch's upper phase" separated by "partition–dialysis." Because the fraction formed strands, it was called "strandin"[52]; it was soluble in both water and chloroform–methanol and had a carbohydrate and lipid composition similar to that of ganglioside.[53,54] Folch and co-workers regarded ganglioside as a degradation product of "strandin." This view was further supported by Rosenberg and Chargaff[53] and Chatagnon and Chatagnon.[54] Rosenberg and Chargaff named the fraction obtained by repeated partition–dialysis "mucolipid," which represents a high-molecular-weight complex including peptide and gangliosides.

The macromolecular theory for ganglioside was based on the following findings: (1) the fraction prepared by repeated partition–dialysis,[52] avoiding any drastic procedure, was homogeneous on electrophoresis and ultracentrifugation, contained a "bound polypeptide," and had a molecular weight of 180,000; (2) the polypeptide showed a consistent amino acid composition and demonstrated a definitive binding interaction with the lipid and carbohydrates; and (3) the chemical composition did not coincide with a simple integral molar ratio, and a methylation study indicated the presence of a number of nonreducing terminals and di-*O*-methyl or mono-*O*-methyl sugars (i.e., it indicated a highly branched structure). A possibility for micellar aggregation in aqueous solution of ganglioside became obvious later, and the fractions termed "strandin" and "mucolipid" were obviously heterogeneous in various chromatographic procedures and did not show macromolecular aggregation when molecular weight was determined in organic solvents with suitable polarity, such as dimethylformamide.[55] However, the possibility remained open that some ganglioside might be present that was associated with gangliophilic polypeptides (see Chapter 4).

Heterogeneity of gangliosides was first clearly noticed by Svennerholm[56] (see Section 1.6) in 1956 by chromatography of brain ganglioside on a powdered cellulose column in a chloroform–methanol–water solvent system. Ganglioside was further separated into discrete bands on thin-layer chromatography on silica gel plates, first performed by Weicker *et al.*[58] in 1960. In

Richard Kuhn
(see biographical sketch, p. 149)

subsequent studies, the heterogeneity of gangliosides was ascribed to differences in sialosyl substitution, since the "core structure" (desialylated structure) was rather homogeneous. The core structure of ganglioside was suggested by Svennerholm[59] as galactosyl-galactosaminyl-galactosyl-glucosyl-ceramide. Eventually, the definitive structure of brain ganglioside was established in a monumental paper by Kuhn and Wiegandt[60] (see Section 1.6) in 1963, in which the structure, "ganglio-*N*-tetraose" (Galβ1→3GalNAcβ1→4Galβ1→4Glc→Cer), was firmly established as the core and the position of substitution by sialosyl residues to the middle galactosyl and to the terminal galactosyl residue of ganglio-*N*-tetraose was clearly demonstrated. The structure of ganglioside accumulating in brains of patients with Tay–Sachs' familial amaurotic idiocy was later identified to be different from that of normal brain ganglioside. The structure was determined to be GalNAcβ1→ 4(NeuAc2→3)Galβ1→4Glc→Cer by Makita and Yamakawa[61] and Ledeen and Salsman.[62] Although structures of the major gangliosides in neural and extraneural tissue have been well established up to the present time, the presence of more than 30 minor gangliosides has been detected through a recent technique developed by Nagai and co-workers.[114,117,123]

Isolation and structural determination of a number of other glycolipids in erythrocyte membranes and other tissues of various species were performed by many investigators during the last decade with the use of improved techniques for isolation and chemical derivatization of lipids and carbohydrates. Our knowledge of the structural chemistry of glycolipids has been greatly enriched thereby, and the details are discussed in Section 1.4. The important events in the history of glycolipid chemistry are listed in chronological sequence in Table III.

In the Sections 1.2 and 1.3, methods of extraction, isolation, and characterization of glycosphingolipids are discussed with reference to Tables

Lars Svennerholm
(see biographical sketch, p. 150)

IV–XVIII. The established structure of glycolipids is summarized in Tables XX–XXVII.

1.2. Isolation of Glycosphingolipids

This section describes the principles of the major methodology currently used in the isolation and characterization of glycosphingolipids, although the reader should refer to the original references for the details of each procedure. The important analytical techniques developed in glycosphingolipid research are listed in Table IV.

1.2.1. Extraction

Glycosphingolipids are coextracted with other lipids and nonlipid materials by homogenizing the wet tissues or cells with at least 20 volumes (vol./wt.) of chloroform–methanol (2 : 1) in a Waring blender for 3 min according to Folch *et al.*[88] Polysialogangliosides are not quantitatively extracted under those conditions, and homogenization with increased methanol concentration such as chloroform–methanol–water (1 : 2 : 0.05) may be necessary. However, extraction with 100% methanol rather decreases the extractability of various glycolipids and is not recommended. Slomiany and Slomiany[89] observed that highly complex glycosphingolipids were not extractable from tissue by homogenization with chloroform–methanol (2 : 1), but they were extractable from tissue residues by stirring for 24 hr at room temperature with 0.4 M sodium acetate dissolved in methanol–chloroform–water (60 : 30 : 8).

Glycolipids are also quantitatively extracted by aqueous tetrahydrofuran according to a modified method of Tettamanti *et al.*[90] originally described by Trams and Lauter.[91] Tissue is first homogenized by 1 volume (vol./wt.) 0.01 M phosphate buffer (pH 6.8) followed by homogenizing with 8 volumes (vol./wt.) of tetrahydrofuran for 1 min.

When a large quantity (kilograms) of tissue is extracted for the purpose of preparation, the tissue can first be homogenized with 5–10 volumes of acetone in a Waring blender and filtered, and the residue reextracted with fresh acetone, refiltered, and dried. The acetone powder thus prepared can be successively extracted with 10 volumes (vol./wt.) of chloroform–methanol–water (2 : 1 : 0.05) and chloroform–methanol–water (1 : 2 : 0.1).[115] It should be noted, however, that a small portion of some glycolipids such as GM_3-GM_4 gangliosides and cerebroside is lost during the extraction with acetone. An alternative and cheaper method for treating a large amount of tissue is as follows: The tissue is homogenized with 10 volumes of absolute ethanol, and the homogenate heated in a water bath at 60°C for 15 min, and filtered on a prewarmed büchner funnel. The insoluble residue is homogenized in 10 volumes of 90% ethanol, and treated in the same way. The first and second filtrates are combined and evaporated in a rotary evaporator to dryness, and

Table IV. Important Analytical or Preparative Methods in Glycolipid Characterization

Partitioning of glycolipids in chloroform–methanol extract in aqueous methanol (upper phase) and chloroform–methanol phase (lower phase): handy preparation of gangliosides	Folch *et al.* (1951)[52]
Thin-layer chromatography (TLC) of gangliosides	Weicker *et al.* (1960)[58]
Permethylation with methylsulfinylcarbanion, methyliodide in dimethylsulfoxide	Hakomori (1964)[234]
Sugar analysis in glycolipids by gas chromatography (GC) as trimethylsilyl derivative	Sweeley and Walker (1964)[187]
Preferential hydrolysis of fatty acyl amide linkage to sphingosine by 1 N KOH in *n*-butanol	Taketomi and Yamakawa (1963)[626]
Release of oligosaccharides from glycosphingolipid through ozonolysis and alkaline degradation (Wiegandt's degradation)	Wiegandt and Baschang (1965)[198]
Methylation analysis by GC–mass spectrometry of neutral and amino sugar derivatives	Björndal *et al.* (1967),[235] Stellner *et al.* (1973)[237]
Sphingosine analysis by GC as *N*-acetyl-*O*-trimethylsilyl derivative	Carter and Gaver (1967)[138]
Direct-probe mass spectrometry of derivatized whole glycolipids	Sweeley and Dawson (1969),[254] Karlsson (1970),[76] Karlsson *et al.* (1974)[259]
Separation of total glycosphingolipid from lipid extract as acetylated derivatives	Saito and Hakomori (1971)[106]
Stepwise enzymatic degradation by exoglycosidase for determination of sugar sequence and anomeric linkage	Hakomori *et al.* (1971)[77]
Chromium trioxide oxidation for distinguishing α and β anomeric linkages	Hoffman *et al.* (1972),[247] Laine and Renkonen (1975)[249]
Oxidation of allylic hydrosyl of sphingosine to 3-ketosphingosine by 2,3-dichloro-5,6-dicyanobenzoquinone	Kishimoto and Mitry (1974)[203]
Ion-exchange chromatography of neutral glycolipids and ganglioside on diethylaminoethyl (DEAE)–Sephadex	Rouser *et al.* (1969),[108] Yu and Ledeen (1972)[109]
Ganglioside separation and mapping through DEAE–Sepharose chromatography and TLC combination	Nagai and co-workers (1976–1978)[114,117,123]
High-performance liquid chromatography (HPLC) of neutral glycolipids and gangliosides (analytical application)	McCluer and Jungalwala (1976)[125]
HPLC of neutral glycolipids (preparative application)	Watanabe and Arao (1981)[561]
Identification of lacto-series glycolipids by their general susceptibility to endo-β-galactosidase of *Escherichia freundii*	Fukuda *et al.* (1978)[232]
Trifluoroacetolysis for separation of oligosaccharides from glycosphingolipids	Svensson and co-workers (1978–1979)[205,208]
Application of nuclear magnetic resonance spectra in glycolipid structural analysis	Falk *et al.* (1979),[284–286] Dabrowski *et al.* (1980)[288,592]
Application of monoclonal antibodies in glycolipid analysis	Young *et al.* (1979–1981)[593,594]
TLC with immunostaining	Magnani *et al.* (1980)[595]

the residue is extracted with chloroform–methanol (2 : 1). The insoluble residue (nonlipid contaminant) is removed by filtration through "celite."[115]

Recently, a new method for extraction of glycolipids having more than 10 sugar residues (ceramide megalosaccharide) from human erythrocyte membranes has been used (Nudelman, E., and Hakomori, S., unpublished data). Erythrocyte membranes (ghosts) were homogenized with 5.5 volumes of isopropanol, 2 volumes of hexane and 1 volume of water in a Waring blender for 3 minutes, and stirred on a magnetic stirrer overnight, and filtered. The residue was homogenized with 5 volumes of isopropanol–hexane–water (55 : 20 : 25) (vol./vol./vol.), and treated in the same way. Two extracts were combined and evaporated in a rotary evaporator, and the dried residue was dissolved in chloroform–methanol (2 : 1). The insoluble residue was removed by filtration as described above and partitioned with 6 volumes of water according to the procedure of Folch *et al.*[52,88]

The conditions described above may not extract ceramides with many (over 30) sugar residues (polyglycosylceramide or ceramide polysaccharides), which have been described by Koscielak *et al.*[84] and Gardas.[85] In their method, erythrocyte stroma were first extracted three times with 83% ethanol, and the dried stroma powder was ground with 10 volumes of 0.02 M phosphate buffer, pH 7.5, with 10 volumes of butanol for 1 hr at 4°C. The homogenate was filtered and the filtrate left standing for 24 hr to allow separation of the upper butanol and the lower aqueous phase. Polyglycosylceramide was isolated from the aqueous phase by ultrafiltration, diethylaminoethyl (DEAE)–Sephadex chromatography, and chromatography as the acetylated derivative.[84,85] Dejter-Juszynski *et al.*[92] solubilized a similar "macroglycolipid" from membranes by agitation of 1 part of aqueous membrane suspension with 6 parts of chloroform and 3 parts of methanol, i.e., treatment by the method of Hamaguchi and Cleve[93] for extraction of "glycophorin." The upper-phase fraction was treated with aqueous butanol and DEAE–Sephadex chromatography in 1% Triton X. The method, however, may not clearly distinguish between ceramides with more than 10 sugar residues (megaloglycosyl and polyglycosylceramide).

1.2.2. Purification of Lipid Extract: Elimination of Nonlipid Contaminants

Nonlipid contaminants present in the lipid extract can be eliminated by chromatography on Sephadex G-25 in chloroform–methanol–water according to the method described by Wuthier[94] or on a Sephadex LH-20 column, originally described for lipid extract by Maxwell and Williams.[95] Total lipids have been separated from nonlipid contaminants on a 100-cm column of Sephadex LH-20 in a void volume of chloroform–methanol–water (2 : 1 : 0.05).[97]

Nonlipid contaminants in the Folch extract can be eliminated by partitioning with water, the well-known Folch partition procedure.[52,88] The nonlipid contaminants are in the aqueous upper phase. A modified procedure with addition of 0.2 volume of 0.88% KCl has often been practiced as described by Suzuki.[96] These methods are simple, but highly polar glycolipids and gangliosides are also eliminated in the upper phase; the methods are therefore useful for separating gangliosides and highly polar glycolipids from relatively nonpolar lipids. The highly polar glycolipids present in the "upper-phase fraction" can be separated from the nonlipid contaminants by gel filtration on LH-20 in chloroform–methanol–water (1 : 1 : 0.01)[97] or by dialysis. Most of the nonlipid contaminants with low molecular weights can be dialyzed from the "upper-phase fraction," leaving gangliosides and highly polar neutral glycolipids if some precaution is taken. Some of the glycolipids can be lost when the upper-phase fraction is directly dialyzed and when ganglioside concentration is lower than the critical micellar concentration, which is approximately 150 μg/ml.[98] The loss of glycolipids was negligible when the upper phase was evaporated and dissolved in water and dialyzed through a "Spectrapor dialysis tube" with pore size 3000 (Spectra Medical Industries, Inc., Los Angeles, California).[97]

1.2.3. Separation of Gangliosides and Long-Chain Neutral Glycolipids

1.2.3.1. Separation of Gangliosides and Long-Chain Neutral Glycolipids from Other Lipid Classes by Folch's Partition Procedure

The highly polar glycolipids (gangliosides and neutral glycolipids with long carbohydrate chains) can be separated from other lipid classes by Folch's partition procedure[88] in the presence of 0.73% NaCl or 0.88% KCl[96]; the presence of these salts prevents extraction of acidic phospholipid and sulfatides into the upper phase. Repeated partition (5–10 times) with chloroform–methanol–0.1% sodium chloride (1 : 10 : 10) according to the modification described by Svennerholm[99] extracted about 90% of GM_3 (hematoside) and all other higher gangliosides. About 10% of GM_3 and essentially all of GM_4 (sialosylgalactosylceramide) remained in the lower phase. The partition of neutral glycolipid into "upper" and "lower" phases with Folch's procedure is not clear-cut. If the method is applied to human erythrocyte glycolipids, the upper phase contains 90% of the ceramides with more than five carbohydrate residues and higher gangliosides; the lower phase contains 90% of ceramides with fewer than four carbohydrate residues. However, this partition pattern of glycolipids in the upper and lower phases varies greatly depending on the coexisting phospholipid, other lipids, and the concentration of ions present in the lipid extract. The well-documented factor for the extractability of ganglioside in Folch's upper phase is the concentration of Ca^{2+}: gangliosides were not extracted in the upper phase in the range of 0.005–0.2 M concentration of $CaCl_2$. At a $CaCl_2$ concentration below 0.002 M or above 0.16 M, essentially all of the gangliosides were found in the upper phase.[100]

Two alternate schemes of glycolipid fractionation are presented diagrammatically in Fig. 1.

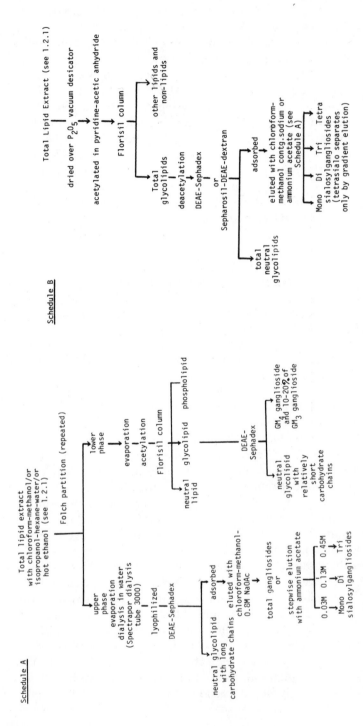

Figure 1. Scheme of glycolipid fractionation. The total lipid extract can be prepared either by direct extraction of tissue with chloroform–methanol (2 : 1) or isopropanol–hexane–water/or hot ethanol (see 1.2.1) or by hot ethanol. In Schedule A, the total extract, dissolved in chloroform–methanol (2 : 1) was partitioned repeatedly according to the Folch method or its modifications. The neutral long-chain glycolipids (ceramide hepta- to octasaccharide or larger) and gangliosides were separated from the upper phase by DEAE–Sephadex chromatography. The short-chain glycolipids (ceramide tetrasaccharide or shorter) and GM$_4$ ganglioside were separated from the lower phase after acetylation followed by chromatography on a Florisil column. Ceramide hexa- to tetrasaccharides and GM$_3$ ganglioside are distributed in both the upper and the lower phase. This procedure is recommended for preparation of a relatively large amount of gangliosides and neutral glycolipids, but is not suitable for quantitative analysis. Schedule B is the direct application of the acetylation procedure (see Section 1.2.3.3 and Table V) to the total lipid extract, followed by separation of glycolipids into neutral glycolipid and gangliosides. Gangliosides can be further separated into classes according to the amount of sialic acids, i.e., mono-, di-, tri-, and tetrasialosylgangliosides.

1.2.3.2. Separation of Glycolipids from Folch's Lower Phase

Ceramide monohexoside and sulfatide (sulfate ester of cerebroside) can be separated from phospholipids and other lipid classes by chromatography on a Florisil column in chloroform–methanol (2 : 1) according to the method described by Radin *et al.*[101]

Alkaline degradation of phospholipids followed by separation of glycolipids has been often used[102]; however, sphingomyelin and plasmalogens were not degraded and were still present in the glycolipid fraction. Glycerol-phospholipids can be degraded according to Dawson's procedure; namely, lipids were dissolved in 10 volumes (vol./wt.) of chloroform, and an equal volume of 1 N KOH in methanol was added. This was kept at 37°C for 16 hr, followed by neutralization with HCl and dialysis. The glycolipid can be separated from the alkali-degraded material according to the procedure described by Vanier *et al.*[103] Chromatography was carried out on a silica gel H column; eluates with chloroform and chloroform–acetic acid (95 : 5, vol./vol.) were discarded. Chloroform–methanol (9 : 1, vol./vol.) eluted glucosylceramide and cerebrosides, chloroform–methanol–water (65 : 24 : 4, vol./vol./ vol.) eluted neutral glycolipids, sulfatides, sphingomyelin, and GM_3, and chloroform–methanol–water (60 : 30 : 6 or 60 : 40 : 10, vol./vol./vol.) eluted other gangliosides.

Sulfatide can be separated from Folch's lower phase by a "linked distribution method"[104] or by DEAE–cellulose chromatography. Sulfatide was eluted with chloroform–methanol (2 : 1) containing 5% of 0.5 M lithium chloride.[105]*

1.2.3.3. Separation of Total Glycolipids from Other Lipid Classes

All glycolipids can be quantitatively isolated as acetylated derivatives from the total lipid extract freed from nonlipid contaminants.[106] The total lipid fraction was evaporated to dryness and repeatedly dehydrated by evaporation with toluene and absolute ethanol. The residue was further dried in a vacuum desiccator over phosphorus pentoxide, and the dried residue was dissolved in dried pyridine–acetic anhydride (2 : 1). The solution was acetylated in pyridine and acetic anhydride, followed by evaporation *in vacuo*. The dry residue, containing the acetylated glycolipid, was subjected to column chromatography on Florisil, which was eluted successively with 3 column volumes of the following solvents: hexane–1,2-dichloroethane (DCE) (1 : 4), DCE alone, DCE–acetone (1 : 1) or DCE–methanol (9 : 1), and DCE–methanol–water (2 : 8 : 1). The acetylated glycolipids were completely eluted with DCE–acetone (1 : 1) or DCE–methanol (9 : 1), leaving all the phospholipids on the column. The solution of acetylated glycolipids was evaporated to dryness *in vacuo*,

---

* A small amount of sulfatide is also present in the Folch's upper phase that is adsorbed on DEAE–Sepharose or DEAE–Sephadex. The adsorbed sulfatide was eluted with about 0.1 M ammonium acetate (overlapped with GD_{1a} or GD_{1b} ganglioside) (see Momoi *et al.*[114] and Fig. 2).

Table V. Separation of Total Glycosphingolipids[a]

Solvent for elution	Substances eluted (alcoholic hydroxy group in acetylated state)	Recovery
DCE	Cholesterol	Not measured
DCE–acetone (1 : 1) fraction	Ceramide, cerebroside, lactosylceramide, ceramide trihexoside, globoside, ceramide pentahexoside (Le^a-active), ceramide octasaccharide (Le^b-active), hematoside, monosialoganglioside, disialosylhematoside, trisialoganglioside	95–100%
DCE–methanol–water (2 : 8 : 1) fraction	Cardiolipin, lecithin, phosphatidylethanolamine, phosphatidylserine, sphingomyelin, phosphatidic acid	80–85% Not measured

[a]A mixture of 17 lipids as above was treated by acetylation procedures; glycolipids were about 30–50 μg for each substance, and phospholipids and cholesterol were about 100–200 μg for each.

followed by deacetylation in 0.1% sodium methoxide in chloroform–methanol. The separation of each lipid class is shown in Table V.

The method is generally applicable to separate total glycolipids from phospholipids and cholesterol in any quantity of lipid extract; however, it cannot be applied to glycolipids possessing an *O*-acetyl or *O*-acyl group or to ester-cerebrosides. The fraction of glycolipid eluted from a Florisil column with DCE–acetone (1 : 1) or DCE–methanol (9 : 1) sometimes contains fast-migrating nonpolar lipids that are neither glycolipids nor phospholipids. This material has not yet been identified (unpublished observation). Glycolipids can be separated from phospholipids by their capability of complex formation with phenylboronic acid covalently linked to the nonpolar polystyrene matrix. A method was proposed by Krohn *et al.*[107] in which all phospholipids were eluted from the phenylboronic acid column with chloroform–methanol (10 : 2, vol./vol.), whereas glycolipids and gangliosides were eluted with addition of water to this solvent.

1.2.4. Separation of Individual Glycolipids

1.2.4.1. Separation of Gangliosides from Neutral Glycolipids

Total neutral glycolipids can be separated from total gangliosides by chromatography on DEAE–cellulose, originally described by Rouser *et al.*,[108] or on DEAE–Sephadex according to the method described by Yu and Ledeen.[109] The latter method is easier to perform. DEAE–Sephadex A-25 was washed with methanol containing 0.8 M sodium acetate followed by preparation of the column in chloroform–methanol–water (30 : 60 : 8), washed with the same solvent extensively. The lipid solution in the same solvent was applied on the column and washed with the same solvent. The washings contained total neutral glycolipids, whereas all the gangliosides were retained on the column, which was subsequently eluted with methanol containing 0.8 M sodium acetate or with chloroform–methanol–0.8 M sodium acetate (30 : 60 : 8). The ganglioside fraction was evaporated and dialyzed in water. Recently, Kundu and Roy[110] proposed using DEAE–silica gel for rapid separation of ganglioside from neutral glycolipids. DEAE–silica gel was prepared by reacting silica gel with γ-glycidoxy propyltrimethoxysilane and *N,N'*-diethylethanolamine. Kawamura and Taketomi[111] proposed separating ganglioside adsorbed on silica gel G–"Kieselguhr G" (1 : 1, wt./wt.) in chloroform–methanol–28% ammonia (32 : 8 : 1, vol./vol./vol.). Neutral glycolipids were not adsorbed, and the adsorbed gangliosides were eluted with neutral solvent (chloroform–methanol–water, 60 : 35 : 8, vol./vol./vol.).[111]

1.2.4.2. Separation of Individual Gangliosides

Gangliosides were separated by a long column of "Anasil S" [Analab, Connecticut (a mixture of activated silica gel and magnesium oxide)] in a chloroform–methanol–water system according to the procedure by Penick *et al.*[112] or on columns of silica gel G or silica gel H in a chloroform–methanol–water system.[113] The flow rate on these columns was too slow to be practical.

Recently, an improved method based on a combination of DEAE–Sephadex chromatography with gradient elution and chromatography on a long column of porous, perfectly spherical silica gel ["Iatro beads" (Iatron Lab., Inc., Tokyo 110, Japan)] with gradient elution has been developed by Momoi *et al.*[114] A ganglioside mixture was first applied on a DEAE–Sephadex A-25 column activated with chloroform–methanol–0.8 M sodium acetate (30 : 60 : 8), and the column was washed with 10 volumes of chloroform–methanol–water (30 : 60 : 8) followed by application of ganglioside solution in the same solvent, followed by washing with methanol and a gradient elution with 0.03–0.3 M ammonium acetate. As shown in Fig. 2, three major groups of gangliosides (A, B, C) corresponding to mono-, di-, and trisialoganglioside, respectively, and other minor fractions were separated. Each fraction was concentrated, dialyzed, and lyophilized. Each group of gangliosides was further separated into components by silica gel chromatography (see Section 1.2.4.3). More recently, Fredman *et al.*[135] developed new porous glass beads covalently linked to DEAE–dextran as a chromatography absorbent that could separate mono-, di-, tri-, and tetrasialosylgangliosides.

1.2.4.3. Separation of Neutral Glycolipids and Gangliosides on a Silica Gel Column

The total neutral glycolipid fraction or the subgroup of gangliosides (mono-, di-, and trisialosyl groups), separated by DEAE–Sephadex chromatography, can be further separated into individual components by chromatography on "Biosil A"[115,116] or on "Iatrobead" columns[117,118] in chloroform–methanol systems. Gradient elution is recommended as seen in the Fig. 3 caption. An example of separation of neutral glycolipids of human erythrocyte membranes is shown in Fig. 3.

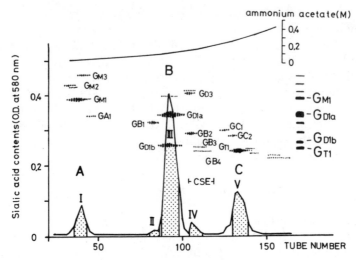

Figure 2. Elution profile of gangliosides from a DEAE–Sephadex column. Each fraction eluted from DEAE–Sephadex with an ammonium acetate gradient in methanol was analyzed on thin-layer chromatography. Three fractions represent mono- (A), di- (B), and trisialogangliosides (C). Fraction A-I contains GM_3, GM_2, GM_1, and GA_1. Fraction B-II contains GB_1; fraction B-III contains GD_{1a} and GD_{1b}; fraction B-IV contains GD_3, GB_2, GB_3, and GB_4. Fraction C-V contains GC_1, GC_2, GT_1, and other minor components. GA_1: GM_1 with NeuGlyc; GB_1: GalNAc→4(NeuAc2→3)Gal1→3GalNAc→4(NeuAc2→3)Gal1→4Glc→Cer; GB_2: GD_{1a} with NeuGly. The elution position for sulfatide is indicated by CSE, which overlaps fraction IV. GB_3, GB_4, GC_1, and GC_2 are uncharacterized components. (From Ref. 114, courtesy of Dr. Y. Nagai.)

1.2.4.4. Separation of Gangliosides and Neutral Glycolipids as Acetylated Derivatives

Fractions eluted from Iatrobeads or Biosil A columns, while showing a homogeneous band on thin-layer chromatography (TLC), may still contain multiple components with similar migration properties. Further resolution into individual components can be carried out only after acetylation, and the acetylated glycolipids can be separated on silica gel G or H plates. Separation of H_1 glycolipid (L-Fucα1→2Galβ1→4GlcNAcβ1→3Galβ1→4Glcβ1→1Cer) and galactosylparagloboside (Galβ1→3Galβ1→4GlcNAcβ1→3Galβ1→4Glc β1→1Cer) was seen only after acetylation and TLC on silica gel G with DCE–methanol (9 : 1).[119] Similarly, H_3 glycolipid (a branched-chain ceramide decasaccharide) and other components can be separated only after acetylation.[120] Forssman glycolipids can be readily separated from other ceramide pentasaccharides after acetylation and TLC with DCE-acetone (55 : 45).[121]

Recently, Kato et al.[122] separated the P_1 antigen glycolipid from human erythrocyte membranes as the acetylated compound by chromatography on a column of Florisil and Kiselguhr 60-HR (1 : 1, wt./wt.) with various concentrations of acetone in DCE. Gangliosides with I activity and others were separated as acetylated compounds on TLC developed with 1,2-dichloro-

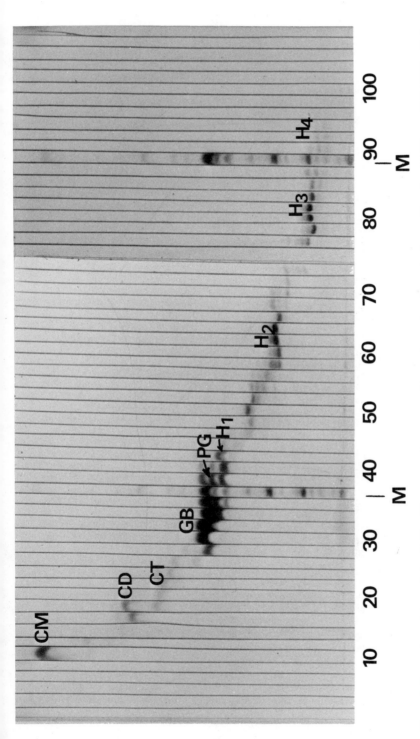

Figure 3. Separation of neutral glycolipids of human erythrocyte membranes through a silica gel column, "Iatrobeads" (60 μm, 6RS-8060) with low-pressure operation. The upper-phase neutral glycolipid fraction of human O-erythrocyte membranes prepared according to Schedule A in Fig. 1 was dissolved in the starting solvent and placed on the stainless column (0.9 × 100 cm) packed with Iatrobeads 6RS-8060 (60 μm diameter) prepared in the starting solvent (chloroform–methanol–water, 75 : 25 : 3, vol./vol./vol.). Gradient elution was performed between 200 ml of the starting solvent and 200 ml of chloroform–methanol–water (20 : 80 : 15, vol./vol./vol.) within 250 min. The elution speed was 2 ml/min. Aliquots of 4 ml (2 min) were separated by a fraction collector. Each fraction was analyzed by thin-layer chromatography with solvent (chloroform–methanol–water, 60 : 35 : 8). CM, ceramide monohexoside; CD, ceramide dihexoside; CT, ceramide trihexoside; GB, globoside; PG, paragloboside. H₁, H₂, H₃, H₄ are variant H-active glycolipids (for H variants, see Hakomori and Watanabe[115] and Hakomori et al.[310,407]); M, mixture of glycolipids before separation. This figure is taken from the author's own experimental data.[562] Further separation has been achieved; see Kannagi et al.[603]

A

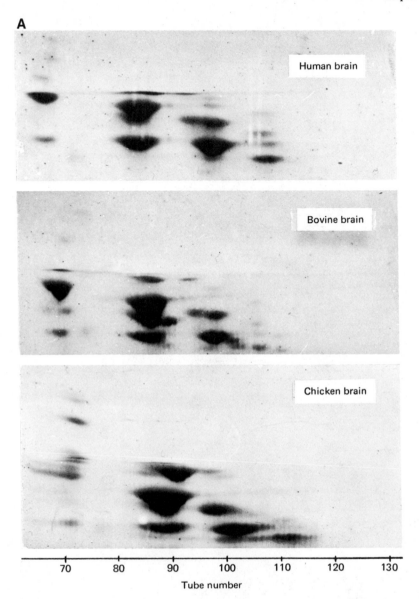

Figure 4. Ganglioside mapping patterns of various species of brain. Total gangliosides were separated on DEAE–Sepharose in chloroform–methanol with ammonium acetate gradient (0.05–0.45 M in methanol) in a three-chamber gradient system (outer chamber 0.05 M, middle chamber 0.15 M, inner chamber 0.45 M).[114] Each fraction was analyzed with TLC with chloroform–methanol–2.5 M ammonia (60 : 40 : 9, vol./vol.) and stained with resorcinol reagent. Figure 4A demonstrates the mapping pattern of human, bovine, and chicken brain as indicated. The relative position of migration of each ganglioside in the ganglioside mapping procedure is shown in Fig. 4B. The positions of known ganglioside are shown by solid outlines and those of unknown gangliosides by dotted outlines. GD_2 was located at the same position as GD_{1a}. (Data from Ref. 123, courtesy of Dr. Y. Nagai.)

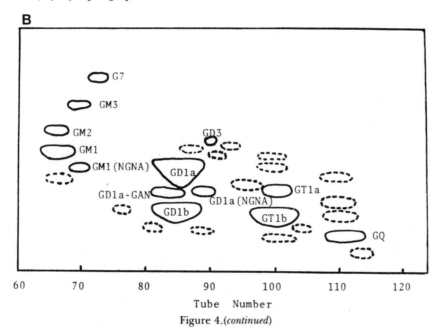

Figure 4.(*continued*)

ethane/acetone–water (20 : 30 : 1 or 4 : 16 :1).[118] Glycolipids with blood group A and H activities were finally purified as acetylated compounds with various solvent systems as shown in Table VI (solvents 12–14) and in Fig. 7.[115]

1.2.4.5. Ganglioside Mapping

As many as 25 brain gangliosides and 30 erythrocyte gangliosides were separated by the combination of DEAE–Sephadex or Sepharose chromatography and TLC of the DEAE-eluate fractions.[118,123] The pattern of separation is characteristic for the source of gangliosides and can be called "ganglioside mapping." The TLC pattern of the neutral glycolipid fraction eluted from an "Iatrobead" column is also characteristic for the cells and tissue. These mapping patterns are shown in Fig. 4.

1.2.4.6. Separation of Glycosphingolipids by High-Pressure or High-Performance Liquid Chromatography

The application of high-performance liquid chromatography (HPLC) for the separation and analysis of glycosphingolipids has been pioneered by McCluer and associates.[124,125] In their studies, fully benzoylated glycosphingolipids were separated, and the pattern was recorded by UV absorption. A system for the separation of neutral glycosphingolipids by HPLC was described by Evans and McCluer.[124] The glycolipids were benzoylated and separated on dry-packed Z pax (Instrumental Product Division, E.I. DuPont de Nemours & Co., Wilmington, Delaware) with a UV detector system. Ceramide mono-

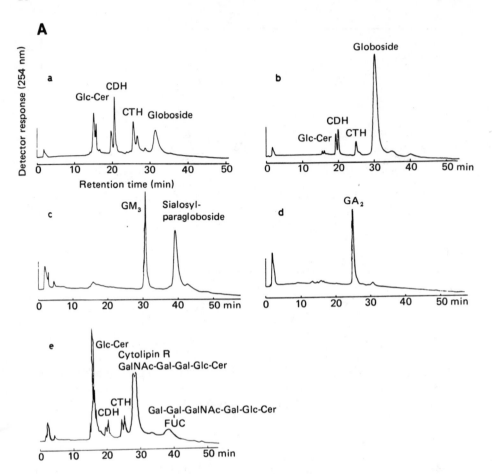

Figure 5. HPLC of glycosphingolipids. Glycolipids are separated and detected as benzoyl or *p*-nitrobenzoyl derivatives, but three different derivatizations have been used. A, Separation of *O*-acetylated *N*-*p*-nitrobenzoylated derivative. Total glycolipid fraction was isolated as acetate (see Saito and Hakomori[106] and Table V), then *N*-*p*-nitrobenzoylated with *p*-nitrobenzoylchloride and pyridine and separated in a Zorpax Sil column [silica gel, 2.1 mm × 25 cm (E.I. du Pont de Nemours, Inc., Wilmington, Delaware) and eluted with a linear gradient of 0.5–7% isopropanol in hexane–chloroform (2 : 1, vol./vol.) at a flow rate of 0.5 ml/min for 40 min. a, Standard sample mixture; b, human erythrocyte neutral glycolipids; c, gangliosides of human erythrocytes; d, glycolipid of guinea pig erythrocytes (GA_2 = asialo GM_2); e, neutral glycolipid of rat granuloma. Data from Ref. 405.) B, HPLC separation as *O*-benzoylated derivative. Glycolipids were *O*-benzoylated by benzoylanhydride and DMAP in pyridine. Separation was performed through a Zilpax column (2.1 mm × 50 cm) and eluted with a 13-min linear gradient of 2.5–25% dioxane in hexane with detection at 230 nm. a, Standard glycosphingolipid, *O*-benzoylated; b, plasma glycolipids, *O*-benzoylated; c, plasma glycolipids, *N,O*-perbenzoylated by benzoylchloride. Peaks 1, 2, 3, and 4 were *O*-benzoylated GlcCer, LacCer, GbOse₃Cer, and GbOse₄Cer; peaks 1', 2', 3', and 4' were *N,O*-benzoylated derivatives of the same glycolipids as above. Peak a is unknown. Peak b is GalCer containing α-hydroxy fatty acid. Data from Ref. 354.) C, HPLC separation of perbenzoylated gangliosides on a 25-cm × 2.1-mm Micro-pak SI-10 column (Varian) with detection at 262 nm. Eluted with 1–30% acetic acid and in hexane pumped at 1.5 ml/min. (Reproduced with permission from the authors and publishers.)

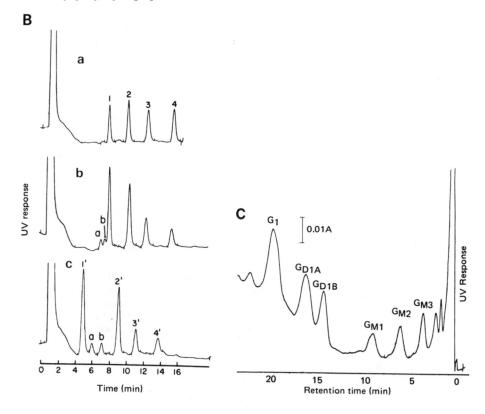

Figure 5. (*continued*)

to tetrahexosides were completely resolved within 25 min, and 1.0 nmole of glycolipid was recorded as a full-scale recorder response. Further extensive studies for the separation of gangliosides have been reported by McCluer and Jungalwala.[125] Six common gangliosides of brain were well separated as benzoylated derivatives on a 25 cm × 2.1 mm Micro Pak SI-10 (Varian) column with detection at 262 nm. Gradient elution of 1–30% acetic acid in hexane was used as an example (Fig. 5C). A drawback of their method is the use of fully benzoylated derivatives, which includes *N*-benzoylation. The amide nitrogens of sphingosine, hexosamines, and sialic acid were all *N*-benzoylated in addition to *O*-benzoylation, and the compounds were not recoverable. However, *N*-benzoylation can be avoided by reacting glycolipids with benzoylan-hydride in pyridine in the presence of *N,N'*-dimethyl-4-aminopyridine (DMAP) as catalyst. The *O*-benzoylated globoside obtained using this method can be recovered completely by treating with 0.5 N methanolic sodium hydroxide for 1 hr at 37°C. The method therefore has great potential value for both analytical and preparative purposes[354] (Fig. 5B). Tjaden *et al.*[126] described

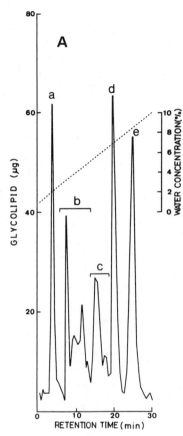

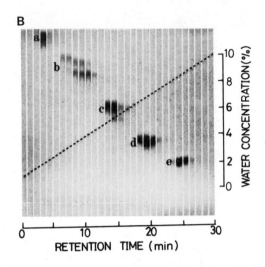

Figure 6. Preparative HPLC of glycosphingolipids without derivatization through an Iatrobead (10 μm, 6RS-8010) column. Glycolipids were separated through a 4-mm or 8-mm × 50-cm column of Iatrobeads 6RS8010. Elution was performed by a linear gradient, isopropanol–hexane–water (50 : 49 : 1 to 55 : 20 : 25), or isopropanol–hexane–water (55 : 40 : 5 to 55 : 30 : 15). Separation pattern of GlcCer (a), LacCer (b), GbOse₃Cer (c), GbOse₄Cer (d), and GbOse₅Cer (e) within 30 min. Isopropanol–hexane–water (55 : 44 : 1 to 55 : 35 : 10) was passed through the column at a flow rate of 2.0 ml/min, and eluates were collected every 0.5 min by a fraction collector. The change of water content is indicated. Panel A shows the intensity of orcinol reaction applied to aliquots of each fraction; panel B shows the TLC pattern of each fraction.

a complete separation of the known brain gangliosides with Silica SI-60 (E. Merck, Darmstadt, Federal Republic of Germany) with a particle size of 9 ± 1.5 μm. The elution pattern was detected by a moving-wire system equipped with a flame-ionization detector. Suzuki *et al.*[127] and Yamazaki *et al.*[405] separated various glycolipids after *O*-acetylation followed by *N*-benzoylation. Quantitative analysis has been performed for erythrocyte glycolipids of various species through a silica gel column. Zorbax Sil (E.I. DuPont de Nemours, Wilmington, Delaware) with a gradient elution of 0.5–7.0% isopropanol to hexane–chloroform (2 : 1) and reversed-phase chromatography on "μ-Bondapak C18" [phenylalkyl silica gel (Waters Assoc.)] with 3% isopropanal–3% chloroform in hexane were used. More recently, Watanabe and Arao[561] successfully separated various neutral glycolipids without derivatization through the "Iatrobead 6RS-8010" column (10 μm porous silica gel, Iatron Chemical Co., 1-11-4 Higashi-Kanda, Chiyoda-Ku, Tokyo) in the solvent system isopropanol–hexane–water (see Fig. 6A,B). The method has been applied to

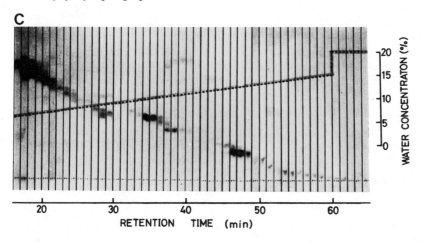

C

RETENTION TIME (min)

WATER CONCENTRATON (%)

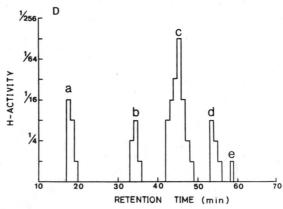

D

H-ACTIVITY

RETENTION TIME (min)

Figure 6.(*continued*) C, Separation pattern of various neutral glcyolipids present in the upper phase of human blood group O-erythrocyte membranes. Elution was performed with isopropanol–hexane–water (55 : 42 : 3 to 55 : 30 : 15, by volume) at a flow rate of 2.0 ml/min, programmed for 60 min, then with isopropanol–hexane–water (55 : 25 : 20). Each fraction was analyzed by TLC. D, Distribution of H activity in each fraction, separated as above, and determined by inhibition of hemagglutination caused by *Ulex europeus* lectin. The figures are reproduced partially from Watanabe and Arao,[561] and Watanabe and Hakomori.[562]

separation of blood group H glycolipids (see also Fig. 6C). The procedure has been slightly modified and successfully applied in separation of a number of neutral glycolipids and gangliosides from erythrocytes and from human cancer tissue.[596,602–604] However, separation of glycolipids in the HPLC systems is not particularly greater than that with conventional silica gel columns or with TLC of acetylated compounds, although the time required for analysis is much shorter and the results are highly reproducible. Separation between glycolipids with the same number of sugars is difficult even on HPLC. These separations are possible, however, by using acetylated compounds on activated silica gel plates.[119–121]

1.2.4.7. Preparative Separation of Glycolipids by Thin-Layer Chromatography

Several milligrams of glycolipids can be separated on a 20 × 20 cm thin-layer plate under the same conditions as the analytical procedure (see Section

1.3.1); however, the following points are important: (1) Glycolipid bands separated on TLC shoud be immediately extracted. If TLC plates are dried extensively after development, no glycolipids are recovered from them, even with repeated extraction and sonication. (2) The recommended procedure is as follows: Glycolipids are placed as a line on the full middle portion and a narrow "guide area" of a prescored, "crackable" TLC plate. Glycolipids are allowed to separate in a suitable solvent system in a tank, and the "guide area" is removed by cracking in the developing tank. Glycolipid bands in the "guide area," revealed by orcinol reaction, and the corresponding bands in the major middle portion of TLC plate are extracted immediately. (3) When less polar glycolipids (up to ceramide tetrasaccharide) or glycolipid acetates are separated on a TLC plate, the plate is briefly dried in air (not more than a few minutes!) and sprayed with 0.01% "Primerline," and glycolipid spots are detected under UV light.[583] Glycolipid bands are immediately extracted with chloroform–methanol–water (1 : 1 : 0.1) in an ultrasonic bath.

1.3. Characterization of Glycosphingolipids

If glycolipids were separated by various procedures as described in Section 1.2, their homogeneity should be determined by thin-layer chromatography (TLC) in various solvent systems (see Section 1.3.1), and if possible, homogeneity of the oligosaccharides liberated from the glycolipid should be examined (see Section 1.3.4). The migration behavior of glycolipids on TLC in various solvents as compared to reference standards is a good diagnostic method for estimating numbers of carbohydrate residues (Section 1.3.1). A number of reliable methods are now available for analysis of ceramides, long-chain bases (LCBs), and fatty acids (see Section 1.3.2) and for analysis of carbohydrate composition (see Section 1.3.3). The sequence of carbohydrates in classical studies was determined by partial degradation with weak acid, but it is now analyzed by step-by-step hydrolysis with the purified exoglycosidases in the presence of a suitable detergent (see Section 1.3.5). A new endo-β-galactosidase that specifically hydrolyzes "lacto-series" glycolipids (lactoglycosylceramides) is particularly useful for structural analysis and identification of this class of glycolipids (see Section 1.3.5). These procedures with enzyme degradation will yield information not only on sugar sequences but also on anomeric linkage of sugars. The position of the glycosidic linkage will be analyzed by methylation analysis. Great technical advances have been made during the past 10 years due to the development of a new methylation and mass spectrometry of partially *O*-methylated hexoses and partially *O*-methylated and *N*-methylated hexosamines (see Section 1.3.6). Direct-probe mass spectrometry of a whole or an enzyme-degraded glycolipid after trimethylsilylation or methylation offers useful information on both sequences of carbohydrates and ceramide structures, although the positions of the linkage, kinds of hexoses and amino sugars, and anomeric linkages cannot yet be completely determined (see Section 1.3.7). Complementary to mass spectrometry, anomeric linkages in glycolipids can be determined by nuclear magnetic resonance (NMR) spectroscopy (see Section 1.3.9.3) and by chromium trioxide oxidation (see Section 1.3.7). In addition, the antigenic properties of glyco-lipids will yield useful information on the structure of glycolipids (see Chapter 5).

1.3.1. Analysis by Thin-Layer Chromatography

The solvent systems shown in Table VI and the reagents shown in Table VII have been used in many studies. Mixtures of chloroform–methanol–water in varying ratios (Table VI, solvents 1–4) are often used, and those with addition of calcium (solvent 6) and ammonia (solvent 7) are particularly useful for separation of gangliosides. Solvent systems with tetrahydrofuran (solvents 9 and 10) and propanol with or without ammonia (solvent 7) have been used for separation of gangliosides and neutral glycolipids. Glycolipid acetate have been successfully separated in mixtures of 1,2-dichloroethane(DCE)–meth-anol–water or butylacetate–acetone–water. The mobility of glycolipids on TLC roughly indicates the number of sugar residues present in glycolipids. The choice of solvents and spray reagents depends on cases, as indicated in Table VII. For preliminary identification of glycolipids, the mobility and the reac-tivity of the spots on TLC are compared with those of the known reference compounds. The most useful reagents are resorcinol–HCl spray for ganglio-sides and orcinol–H_2SO_4 spray for neutral glycolipid and gangliosides. Re-cently, separation of glycolipids on TLC plates has been greatly improved by a new type of TLC plate. These high-performance (HP) TLC plates have been used with uniform superfine silica gel, which has a mean particle size of 5 μm and a mean pore diameter of 80 nm. HPTLC separates 10–15 gangliosides and neutral glycolipids on one-dimensional development. Such plates are commercially available through Whatman Co., Ltd., and E. Merck Co., Ltd. A TLC plate with a "preadsorbent" spotting area, called a "linear-K plate," showed better separation of various glycolipids than ordinary TLC plates (available through Whatman Co., Ltd).

Homogeneity of glycolipids is often tested on TLC in various solvents. Drawbacks of this approach are: (1) Many glycolipids with the same number of carbohydrates are difficult to separate (see below). (2) Some glycolipids having the same carbohydrate structure are separated into two or multiple spots due to differences in the ceramide moiety. Therefore, homogeneity should be tested by the following methods: (1) TLC of acetylated derivatives using specific solvent systems (see Table IV and Fig. 7). Multiple spots due to the difference of ceramides will often become a single spot as the acetate; on the other hand, a single spot will be separated into multiple components differing in carbohydrate structure. (2) TLC on borate-impregnated silica gel will separate some glycolipids that are not separated on ordinary silica gel plates.[128] (3) Paper chromatography or gas–liquid chromatography (GLC) of oligosaccharides released from some glycolipids will be separated better than original glycolipids.

Glycolipids with longer carbohydrate chains are very difficult to separate

Table VI. Solvent Systems for Separation of Glycosphingolipids

Solvent No.	Silica gel TLC plates	Solvents and proportions (by volume)	Glycosphingolipids
1	G or H	Chloroform–methanol–water (65 : 25 : 4)	Neutral glycolipids with shorter carbohydrate chains and hematosides
2	G or H	Chloroform–methanol–water (60 : 35 : 8)	Gangliosides and neutral glycolipids having relatively long carbohydrate chains
3	G	Chloroform–methanol–water (55 : 45 : 10)	Neutral glycolipids and gangliosides with long carbohydrate chains
4	G or H	Chloroform–methanol–water (85 : 15 : 1.5)	Ester-cerebrosides
5	G or H	Chloroform–methanol–15 M ammonia–water (60 : 35 : 1 : 7)	Gangliosides
6	G or H	Chloroform–methanol–0.02% CaCl$_2$ · 2H$_2$O (60 : 40 : 9)	Gangliosides
7	G	n-Propanol–15 M ammonia (7 : 3)	Gangliosides
8	G	Chloroform–methanol–90% formic acid (90 : 18 : 12)	Separation of cerebroside into molecular species
9	G	Tetrahydrofuran–diisobutylketone–water (45 : 5 : 6)	Gangliosides
10	G or H	Tetrahydrofuran-2-butanone–water (8 : 2 : 1)	Neutral glycolipids
11	G or H	DCE–methanol–water (90 : 20 : 0.5)	Glucosylceramide, galactosylceramide, ceramide with normal fatty acid, and ceramide with hydroxy fatty acid
12	G or H	DCE–methanol–water (88 : 12 : 0.1)	Fully acetylated glycolipids with shorter chains
13	G	DCE–acetone–water (55 : 45 : 5)	Fully acetylated glycolipids with longer chains (10–15 sugar residues)
14	G	Butylacetate–acetone–water (25 : 8 : 3)	Same as above

Table VII. Reagents for Detection of Glycosphingolipids on Thin-Layer Chromatographic Plates and the Choice of Solvent and Reagents

Reagent	Use	Compounds detected
Iodine vapor	Expose briefly to iodine vapors.	Natural glycolipid $+ +$, phospholipids $+ + +$, and gangliosides $+$
Water spray or water–methanol (1 : 1)	Spray plate heavily with water.	Neutral glycolipid $+$, phospholipids $+$, and gangliosides $+ +$ (acetylated glycolipids were not detectable)
Orcinol–H_2SO_4	Orcinol (0.5 g) in 100 ml 3 M H_2SO_4, spray lightly, heat at 130°C for 2–5 min.	Neutral glycolipids and gangliosides
Resorcinol–HCl	Resorcinol (200 mg) in 10 ml H_2O, 80 ml conc. HCl, and 0.25 ml 0.1 M $CuSO_4$, total volume made up to 100 ml with H_2O. Spray the reagent, cover with glass plate, clamp tightly, and heat at 150°C for 10 min.	Gangliosides
Diphenylamine	Stock solution: 10 g diphenylamine dissolved in 100 ml acetone. Spray mixture of 2.5 ml stock solution, 1.25 ml H_3PO_4, 0.125 ml aniline, and 10 ml acetone. Plates heated at 120°C for 25–30 min. Blue-gray spots for sugars.	Neutral glycolipid, phospholipids, gangliosides
Molybdenum trioxide spray (Dittmer–Lester reagent)	MoO3 (4.01 g) dissolved in 100 ml 25 N H_2SO_4 on heating (solution A); 0.178 g metallic molybdenum dissolved in 50 ml solution A on boiling (solution B). Spray a mixture of A, B, and water (1 : 1 : 2, vol./vol./vol.).	Phospholipids
Hypochloride–benzidine	(1)"Chlorox" brand bleaching solution 5 ml, benzene 50 ml, and glacial acetic acid 5 ml. (2)Benzidine 0.5 g, KI 0.1 g, 50 ml ethanol, and filtered. Spray (1), aerate, and spray (2). Blue spots for all amides.	Ceramides and all other glycosylceramides

under ordinary conditions. Several ceramide pentasaccharides, such as glycolipids with H_1, Le^a, and Le^x specificities, galactosylparagloboside, and Forssman glycolipid, are not separated well on silica gel G plates in chloroform–methanol systems. Some of them are separated well as acetates in a DCE–methanol or butylacetate–acetone system.[115]

Ceramides with a larger number of carbohydrate residues, e.g., ceramide deca-, dodeca-, and eicosasaccharides, occasionally present in tissues and erythrocytes are difficult to separate according to their carbohydrate differ-

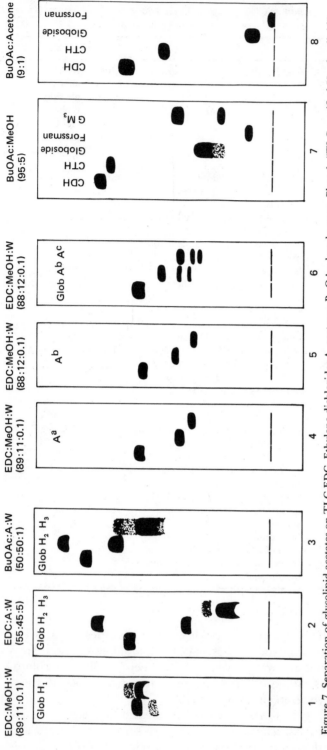

Figure 7. Separation of glycolipid acetates on TLC EDC, Ethylene dichloride; A, acetone; BuOAc, butylacetate. Plate 1: "TLC-purified H_1 glycolipid" gave two bands after acetylation on TLC. The H_1 glycolipid acetate runs slightly slower than globoside acetate. Plates 2 and 3: "TLC-purified H_2 and H_3 glycolipid" runs as acetate. H_2 acetate gave two bands with equal intensity. One runs faster than globoside acetate and another runs slower than globoside acetate, in EDC–acetone–water (for plate 2) and BuOAc–MeOH–water (for plate 3). The lower band was identified as H_2 glycolipid. The upper band was not identified. H_3 glycolipid gave essentially a single spot in both solvents, although a small amount of contaminant was detectable. Plates 4–6: "TLC-purified A^a, A^b, A^c glycolipid" usually consisted of three components. Three bands of acetates were separated from the A^a fraction, and each component is demonstrated in plate 4; the lower two bands were A-active. Similarly, three bands of A^b acetates were separated into each component as shown in plate 5. The lower two bands were A-active; the upper band was inactive. Three A^c-acetate bands are shown in the third

ences on regular TLC. However, they could be separated as acetylated derivatives.[84,85,115,120,269]

Galactosyl- and glucosylceramides were best separated on silica gel impregnated with sodium tetraborate; namely, 30 g of silica gel G is suspended in 75 ml of 1% sodium tetraborate, applied as a 0.25-mm layer to glass plates, dried, and activated at 120°C. On this plate, glucosylceramide ran much faster than galactosylceramide developed with chloroform–methanol–water–15 M ammonia (280 : 70 : 6 : 1).[128] Two ceramide tetrasaccharides with the structures $Gal\beta1\rightarrow4GlcNAc\beta1\rightarrow3Gal\beta1\rightarrow4Glc\beta1\rightarrow1Cer$ and $GalNAc\alpha1\rightarrow3Gal NAc\beta1\rightarrow3Gal\alpha1\rightarrow4Gal\rightarrow Cer$ of NILpy cells, which comigrated on a regular silica gel TLC plate, were separated on a borate-impregnated silica gel plate.[129]

Cerebrosides with different fatty acid residues have been separated on silica gel G with chloroform–methanol–90% formic acid (70 : 18 : 12).[130] Cerebrosides and lactosylceramide often gave double spots with solvent 1 or 3 (Table VI). The upper spot contains nonhydroxylated fatty acid (e.g., cerebronic acid).[130] In extraneural tissue, the upper spot contains longer fatty acid residues such as C_{22} and C_{24} and the lower spot contains C_{18} and C_{16}.[131,132,164] Ceramide tri-, tetra-, and pentasaccharides usually gave a single spot, irrespective of the heterogeneity of the ceramide moiety, although they occasionally showed double or triple spots. For example, globoside and Forssman glycolipid of hamster NIL cells and Forssman glycolipid of sheep erythrocytes show double spots. The upper spot of Forssman glycolipid isolated from sheep erythrocytes contained octadecasphingenine (C_{18}-sphingosine), whereas the lower spot contained octadecasphinganine (C_{18}-dihydrosphingosine).[133] Globoside of human and pig erythrocytes and Forssman glycolipid of goat erythrocytes each showed a single spot, whereas gangliotriosylceramide of mice lymphoma cells showed triple spots.[164] Each, upper, middle, and lower band was identified as having the ceramide containing C24 : 0 or C24 : 1 fatty acid, C16 : 0 fatty acid, and C16 : 0 α-hydroxy fatty acid, respectively.[287]

1.3.2. Characterization of Ceramide Structure, Fatty Acids, and Long-Chain Bases*

The structure of ceramides can be determined by three procedures: (1) Fatty acids and LCBs are analyzed by gas chromatography–mass spectrometry (GC-MS) after glycolipids are methanolyzed. (2) Ceramides can be released from cerebroside and other glycosphingolipids by periodate oxidation in chloroform–methanol followed by reduction and acid hydrolysis (Smith degradation). The liberated ceramides can be directly analyzed by GC-MS or by their fatty acid and LCB profiles. (3) The trimethylsilylated or methylated glycosphingolipid can be analyzed by direct-probe mass spectrometry. The method is, however, limited to glycolipids with relatively short carbohydrate chains.

---

* A part of this section was prepared on the basis of a Japanese review of sphingolipid chemistry by Hayashi and Matsubara.[134]

The methodology and some problems in these analyses are described in the following sections.

1.3.2.1. Conditions for Methanolysis

It has been well known that various artifacts from LCBs are produced during the methanolysis or hydrolysis of glycosphingolipids. This is essentially due to acid-catalyzed rearrangement and substitution at an allylic group; namely, C-3 and C-5 of an allylic LCB will form a carbonium ion due to a resonance hydrid [-CH=CH-C- = -CH-CH=CH-]; thus, nucleophilic addition of hydroxyl ions (from water) or methoxyl ion (from methanol) at the 3 or 5 position will yield a variety of positional steric isomers. The mechanism of production of various artifacts has been extensively discussed in reviews by Karlsson,[76] Taketomi and Kawamura,[136] and Gaver and Sweeley.[137] The conditions for methanolysis of glycosphingolipid yielding minimum amounts of by-products were described by Gaver and Sweeley[137] and Carter and Gaver,[138,139] Namely, 1 N HCl–methanol containing 10 M (18%) of water, heated at 70°C for 18 hr[137] or at 80°C for 16 hr.[138] Weiss and Stiller[140] recommended methanolysis in concentrated HCl–methanol–water (3 : 29 : 4) at 78°C for 18 hr, and Morrisson and Hay[141] used 10% aqueous barium hydroxide in dioxane at 110°C for 24 hr. After methanolysis, fatty acid methyl esters present in the methanolysate were extracted with hexane or petroleum ether; the lower-phase methanolic HCl layer that contained hydrochloride of LCBs was extracted with ethyl ether or chloroform after the methanolic HCl was neutralized and slightly excess amounts of potassium or sodium hydroxide were added. Alternatively, the lower methanol–HCl phase was evaporated to dryness and the dried residue was extracted with chloroform and put on a silicic acid column ["Biosil A," 1 × 4 cm (Biorad Chemical Co., Richmond, California)]. Fatty acids were eluted with chloroform, and LCBs were eluted with methanol.[137,157,184]

1.3.2.2. Analysis of Fatty Acids

The fatty acid methyl ester extracted from the methanolysate by hexane or the fraction separated by silicic acid chromatography can be analyzed by gas chromatography (GC). Extensive reviews of fatty acid analysis on GLC have been published.[142–144] For the detection and separation of hydroxy fatty acids, TLC on silica gel H or G in hexane–ether (85 : 15 to 70 : 30, vol./vol.) is most convenient.[145,146] The α-hydroxy fatty acid methyl esters can be separated from normal fatty acid methyl esters by chromatography on a "Florisil" column[147,148] and then analyzed by GLC.[149,150] Unsaturated fatty acids can be separated on 10% silver nitrate-impregnated silica gel G ("argentation chromatography") developed with hexane–ether mixtures (9 : 1 to 4 : 6, vol./vol.); they are separated depending on the number, position, and steric configuration (*cis* or *trans*) of double bonds.[151] Saturated and unsaturated α-hydroxy fatty acids were separated by argentation chromatography on 5%

silver nitrate silica gel G.[146] Recently, trimethylsilylated α-hydroxy fatty acids present in complex mixtures of fatty acid methyl esters have been identified by mass chromatography with mass number 73 and $[M-59]^+$.[152] Mass spectrometry of α-hydroxy fatty acids in glycosphingolipids[155] and unsaturated poly-en fatty acids[156] has been described (for monographs, see also Budzkiewitz *et al.*[153] and Frigerio and Castagnoli.[154]) The fatty acid profile can also be obtained by direct-probe mass spectrometry of permethylated glycolipids[287]; (see also Section 1.3.8 and Tables XVII-IA, IIA).

1.3.2.3. Quantitation, Isolation and Identification of Long-Chain Bases by Thin-Layer Chromatography

Quantitative determination of LCBs in sphingolipids and in total lipids has been performed based on the determination of the chloroform-soluble nitrogen or amino group present in the acid hydrolysate of lipids. A number of methods have been proposed (cited in Naoi *et al.*,[366] but the most sensitive and reliable method is the fluorimetric fluorescamine assay of the organic-solvent-soluble amino nitrogen of a 6 N HCl hydrolysate as described by Naoi *et al.*[366]

Identification of LCBs is more difficult than that of fatty acids because of artifact formation as described previously. The LCB fraction obtained by aqueous methanolysis[137,138] was purified by silica gel chromatography as previously described.[137,157] Separation of LCBs into subgroups (e.g., saturated, unsaturated, dihydroxy, trihydroxy) can be carried out by TLC as free bases[158] or as *N*-dinitrophenyl (DNP) derivatives.[159] Three groups, saturated dihydroxy, unsaturated dihydroxy, and trihydroxy bases, were separated by TLC with hexane–chloroform–methanol (50 : 50 : 15) on borate-impregnated silica gel G plates. Mono-en and di-en compounds were separated by argentation chromatography on silver nitrate-impregnated silica gel G plates.[159] Saturated and unsaturated mono-en bases were separated on silica gel G plates with chloroform–methanol–ammonia (80 : 20 : 2).[160]

1.3.2.4. Gas Chromatography–Mass Spectrometry of Long-Chain-Base Derivatives

LCBs can be *O*-trimethylsilylated by pyridine–hexamethyldisilazane (HMDS)–trimethylchlorosilane (TMCS) (10 : 2 : 1) at 65°C for 3–5 min, or at room temperature for 15 min, under which condition *N*-trimethylsilylation does not take place. The *O*-trimethylsilylated LCBs can be separated on a 1–3% SE-30 or OV-1 column.[161] Carter and Gaver[138,139] recommended *N*-acetylation followed by *O*-trimethylsilylation with HMDS–pyridine–TMCS (2.6 : 2.0 : 1.6), and then analysis of the reaction product on a 3% SE-30 column. Under these conditions, *erythro* and *threo* derivatives were separated. The retention time was determined as dihydrosphingosine equal to 18, which was called equivalent chain lengths (ECL). The value was useful for identification of various LCBs by GLC.[139] However, LCBs with different number and location of double bands are not separated well on GC. Polito *et al.*[157] identified hexadecasphing-4-enine, heptadecasphing-4-enine, sphinga-4,14-

Table VIII. Gas–Liquid
Chromatography of Sphingosines[a]

Sphingosines	Relative retention times
Erythro-C_{18}	1.00
Threo-C_{18}	0.91
Phyto	1.51
Erythro-C_{20}	1.89
Threo-C_{20}	1.74

[a] Retention times relative to C_{18}-*erythro*-sphin-
gosine chromatographed as *N*-acetyl-*O*-tri-
methylsilyl derivatives on a column of 3% OV
17 (2m × 3mm) at 210°C, according to Gaver
and Sweeley[137] and Carter and Gaver.[138]

dienine, and other compounds by GC-MS as *N*-acetyl-*O*-trimethylsilyl derivatives before and after permanganate-periodate oxidation. The retention times
of a few major LCBs are shown in Table VIII.

Mass spectra of trimethylsilylated LCBs have been determined for dihydroxy bases[76,161] and for trihydroxy bases.[162,163] *N*-acetylated, *O*-trime-
thylsilylated LCBs have been analyzed by GC-MS.[132,157,161,165,166] The common ions yielded from dihydroxy bases were *m/e* 174, 157, 247, and 243.
Although the molecular ion was not recorded, peaks for M-15, M-103, M-
105, and M-174 were diagnostic for molecular weight and gave details of
aliphatic chain structure. Hara and Matsushima[167] described a successful
simultaneous *N*-acetylation and *O*-trimethylsilylation for amino sugars, and
the method can be applied for analysis of LCBs.[134] White *et al.*[165] described
analysis of the branched LCBs by mass spectrometry. Mendershausen and
Sweeley[168] described the mass spectra of *N*-carbobenzoxy derivatives of
sphinganine and ketosphinganine, whereas Karlsson[76] reported mass spectra
of *N*-DNP-*O*-trimethylsilyl LCBs.

For determination of the position of the double bond, GC-MS is quite
useful. Polito *et al.*[166] and Hayashi and Matsubara[161] oxidized *N*-acetyl LCBs
with osmic acid, thereby converting the double bond to a diol. GC-MS of the
trimethylsilyl (TMS) derivative of the diol derivative was diagnostic for determining the position of the double bond. A more complicated procedure
using *N*-DNP LCBs was used after group classification of LCBs on TLC. Each
group was oxidized with osmic acid, purified and trimethylsilylated, and the
fragmentation pattern was determined by mass spectrometry.[76] Determination of the double-bond position of *N*-acetyl LCBs employed permanganate–periodate oxidation (Rudoloff oxidation) followed by GLC.[157,161] Alternatively, triacetyl LCB was ozonolyzed, and the resulting aldehyde was
analyzed on GLC.[169]

1.3.2.5. Characterization of Ceramides Liberated from Sphingolipids

Ceramides can be released from cerebroside and other glycosphingolipids
by the Smith degradation according to the procedure described by Carter *et*

al.[170] Ceramides of gangliosides were also released by the same procedure.[171] The molecular species of ceramides (the kinds of combinations of fatty acids and LCBs) can be determined by examining the released and separated ceramides. The ceramide can be identified by GC-MS according to the procedures described by Samuelsson and Samuelsson[172–174] or Hammerström and co-workers.[175–177] However, the method is applicable only to dihydroxy bases, since trihydroxy bases containing ceramide were degraded by periodate oxidation.

1.3.2.6. Determination of Steric Isomers of Long-Chain Bases

NMR, circular dichroism (CD), optical rotation, infrared (IR) spectrometry, and synthesis of stereospecific compounds have been used for determination of the steric configuration. Steric isomers at C-2 and C-3 (*threo* or *erythro*) have been determined by retention time on GLC[157] or mobility on TLC.[76,158,178] NMR spectra of 4-sphinganine and sphinganine have been determined, and IR spectra of the *trans* double bond (at 970 cm^{-1}, or 10.3 nm) of LCBs have been recorded.[179] Martin-Lomas and Chapman[180] described NMR spectra of triacetylsphingosine. The dihydroxy base should have the *erythro* configuration if the optical rotation of the ethanol solution of *N*-DNP-derivative shows dextrorotatory properties.[181] Similarly, the configuration of the trihydroxy base should be D-*ribo* if the DNP derivative shows dextrorotation.[76]

1.3.2.7. Identification of Ceramide Structures in Glycosphingolipids by Direct-Probe Mass Spectrometry of Derivatized Glycosphingolipids

Successful analysis of ceramide structure has been carried out by direct-probe mass spectrometry of intact cerebroside as the TMS derivative or the methylated derivative.[172–177] TMS cerebroside can be separated by gas chromatography on a short (1-ft.) column of 2% OV-1 at 300–320°C. The major useful fragment ions and their assignments are shown in Table IX. Through this technique, the ions for ceramide, fatty acid, and LCB can be distinguished, and these ions are useful for estimating the molecular species of ceramides. Direct-probe mass spectrometry applied to TLC-separated permethylated glycolipids offers good information not only on sugar sequences (see Section 1.3.8) but also on the ceramide structure. Ions in the mass range from *m/e* 500 to 650 signal the structure of the ceramide moiety (see Table XVII). As an example, three major neutral glycolipids (bands a, b, and c) of mouse lymphoma L5178 were separated on TLC and analyzed by direct-probe mass spectrometry after permethylation. The mass spectrum of "band a" contained prominent ions at *m/e* 661 and 659, which would result from a ceramide containing octadecasphinganine and nervonic acid (*m/e* 659) or lignoceric acid (*m/e* 661). The spectrum of "band b" contained an ion at *m/e* 549, which could result from a ceramide containing octadecasphingenine and palmitic acid. The spectrum of "band c" showed an ion at *m/e* 579, which would result from a ceramide containing octadecasphingenine and α-hydroxy palmitic acid. The

Table IX. Major Fragment Ions Useful for Mass Spectrometric Analysis of Trimethylsilyl (TMS) Cerebroside and Ceramide*

LCB ion[b]	Fatty acid ion[c]		Ceramide ion[a]
R_1—CH =CH—CH $\parallel$ TMS—O$^+$	CH—CH$_2$—O—R$^+$ $\parallel$ NH $\mid$ CH$_2$ $\mid$ R$_2$	CH—CH$_2$—O—R$_3{}^+$ $\parallel$ NH $\mid$ CO $\mid$ HC—OTMS $\mid$ R$_2$	M − d + 102
			M − d + 74
. . . .			M − d + 102 − 90
R_1—CH$_2$—CH $\parallel$ TMS—O$^+$	CH—CH$_2$—O$^+$ $\mid$ H N—TMS $\mid$ CH$_2$ $\mid$ R$_2$	CH—CH$_2$—O$^+$ $\mid$ H N—TMS $\mid$ H C—OTMS $\mid$ R$_2$ [R$_2$—CHOTMS]$^+$	M − d − 16
			M − d − 16 − 90

*For fragment ions from permethylated derivatives, see Table XVII, sections I-A and II-A.

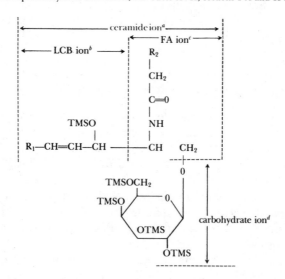

ion at *m/e* 253 was consistently present in each spectrum, indicating the presence of octadecasphingenine as the major LCB of these glycolipids. (D. L. Urdal and S. Hakomori.[287]

1.3.2.8. Example of the Determination of Fatty Acids and Long-Chain Bases in Glycolipids

A slightly modified combination of the methods of Gaver and Sweeley[137] and Carter and Gaver[138,139] has been used for determination of fatty acids and sphingosine bases in glycolipids. Aqueous methanolic HCl (1 ml) is added to a sample of 3–500 μg of glycolipid in a screw-capped test tube (Teflon-lined cap) and heated overnight at 70°C in a heating block. The reaction mixture is cooled to room temperature and evaporated to dryness under a stream of nitrogen, and the residue is dissolved in a small volume of chloroform. The solution is applied to a small column (1 × 4 cm) of Biosil A (Biorad, Inc., Richmond, California) prepared in chloroform. Fatty acids and methylesters are eluted with 20 ml of chloroform and sphingosine bases with 20 ml of methanol.

 a. Sphingosine Analysis. The methanol eluate prepared as described above is evaporated to dryness under nitrogen, the residue is dissolved in 1 ml of chloroform and shaken with 1 ml of 0.1 N NaOH, and the mixture is centrifuged. The chloroform layer is evaporated under nitrogen, dissolved in a methanol–acetic anhydride mixture (4 : 1), and kept overnight at room temperature. This is followed by evaporation under a nitrogen stream. The residue is dissolved in chloroform and transferred to a small conical tube to which is added 20 μl of a freshly prepared trimethylsilylating reagent. After 20 min, a 1-μl aliquot of the solution is taken for injection per assay, which is performed on a 6-ft. column of 3% SE-30 or OV-17, coated on 80–100 mesh Diatoport S or Gas-Chrom Q, and chromatographed at 210°C. Retention times are shown in Table VIII.

 b. Fatty Acid Analysis. The chloroform eluate from a Biosil A column (see above) or the hexane or petroleum ether extracts of a methanolysate (Section 1.3.3.2) are evaporated to dryness under nitrogen and treated with diazomethane. Alternatively, the residue is combined with 1 ml of 0.5 N anhydrous methanolic HCl and heated for a few hours at 70°C. The fatty acid methyl esters were analyzed (after being dried and redissolved in hexane) by GC through a 2 mm × 5 ft. column of 3% SE-30 coated on 80–100 mesh "Supelcoport," with temperature programmed from 140 to 240°C/min. The presence of α-hydroxy fatty acid was determined after trimethylsilylation. The trimethylsilylated methyl esters of α-hydroxy fatty acids yield distinctive mass spectra through GC-MS,[152] as previously mentioned. The method is simpler and more accurate than the analysis of the α-hydroxy fatty acid fraction separated from normal fatty acid through "Florisil" chromatography.

1.3.3. Determination of Carbohydrate Composition

 The carbohydrate composition of glycolipids cannot be determined by a single method. Methanolysis in anhydrous methanolic HCl, followed by N-acetylation of amino sugars and sialic acid, and determination of the methyl glycosides of sugars by GLC as TMS derivatives[182,183] have been used most frequently. The retention times of major sugars present in glycolipids are shown in Table X. The method is also useful in determination of the ratio of hexoses to sphingosines. However, the recovery of amino sugars by this method is not always quantitative, and results have often been inconsistent. The second method with hydrolysis of glycolipids in 90–95% acetic acid containing 0.3–0.5 N sulfuric acid followed by reducing hexoses and hexosamines

Table X. Retention Times of Various Monosaccharides in Gas–Liquid Chromatography

Sugar	As trimethylsilyl derivatives of methyl glycosides[a]		As alditol acetates[b]
	Major peaks	Minor peaks	
Fucose	0.335, 0.365, 0.400	—	0.16
Galactose	0.690, 0.760, 0.820	—	0.92
Glucose	0.865, 0.910	—	1.00
Mannitol	1.00	—	0.68
N-Acetylglucosamine	1.19	1.05, 1.08	6.25
N-Acetylgalactosamine	1.27	1.07, 1.16	7.05
N-Acetylneuraminic acid methyl ester	1.90	1.67, 1.83, 2.00	—

[a] Retention times are relative to hexatrimethylsilyl mannitol for pertrimethylsilyl methyl glycosides. Chromatography was performed on a 2-m × 3-mm column of 3% SE-30 on 100–120 Supelcoport in an F&M Hewlett Packard 402 with flame-ionization detectors; flash heater and detector temperatures, 280°C; carrier gas, nitrogen at 35 ml/min. The column oven was programmed at 3°C/min for 150–250°C.
[b] Chromatography was performed on a 4-ft × 0.4-inch 3% ECNSS-M glass column coated on Gas-Chrom Q, 170°C.

by borohydride and analysis of hexitol and hexosaminitol acetates gives satisfactory recovery for neutral and amino sugars, but causes destruction of sialic acid.[184]

1.3.3.1. Example of the Determination of the Molar Ratios of Methylpentose, Hexose, and Hexosamine

A chloroform–methanol solution containing 50–200 µg of glycolipid is pipetted into a Pyrex glass tube with Teflon-lined screw cap and evaporated to dryness under a nitrogen stream. Three-tenths of a milliliter of 0.5 N sulfuric acid in 90% acetic acid is added to the residue, which is placed in a heating block at 80°C for 16 hr. Hydrolysis is continued for 5 hr with the addition of 0.3 ml water. This condition has been found satisfactory for glycolipid with relatively short carbohydrate residues up to about six. If the number of carbohydrate residues increases, more of the sugar residues at the peripheral region tend to be destroyed than those sugar residues proximal to ceramide. We recently found that a milder condition is better for quantitative recovery of sugars: hydrolysis at 80°C in 90% acetic acid containing 0.3 N sulfuric acid for 6 hr, omitting the water addition and the additional heating.[185] A similar condition was used for hydrolysis of bacterial lipopolysaccharides.[186] The hydrolysate is filtered through 200 mg of dried Dowex 1X8 in acetate form. Methanol (2 ml) is used to wash the resin. The combined filtrate is evaporated to complete dryness under nitrogen and left in a KOH desiccator for several hours. The residue is dissolved in 0.3 ml of water and reduced by the addition of 10 mg of sodium borohydride and left for 3 hr at room temperature. The excess borohydride is destroyed by a drop of glacial acetic acid. Chloroform (0.3 ml) is added, mixed, and centrifuged to remove fatty material released. This is repeated three times. The water layer is evaporated to dryness under a nitrogen stream, and the residue is further evaporated four times with 3 ml of methanol containing 1 drop of glacial acetic acid and finally dried completely over phosphorus pentoxide in a vacuum desiccator. The residue is dissolved in 0.5 ml of acetic anhydride and heated at 100°C for 2 hr. The acetic anhydride is evaporated under a nitrogen stream with the addition of excess toluene. The residue is partitioned between equal volumes of water and chloroform. The chloroform layer is transferred to a small conical test tube by a capillary pipette and evaporated to dryness under nitrogen.

The residue is dissolved in 20–50 μl of acetone; 1–2 μl of the solution is taken for injection on a 4-ft. × 1/16-inch 3% ECNSS-M glass column coated on Gas-Chrom Q (Applied Science Laboratories, State College, Pennsylvania). The temperature is set at 170°C. The retention times of various sugars relative to that of glucitol equal to 1.00 are as follows: fucitol, 0.16; mannitol, 0.68; galactitol, 0.92; glucitol, 1.00; glucosaminitol, 6.25; and galactosaminitol, 7.05.

1.3.3.2. Example of the Determination of the Molar Ratios of Fucose, Hexoses, Hexosamines, and Sialic Acid

Sugars are released from pure glycosphingolipids by methanolysis in 1 N methanolic HCl.[182] Glycosphingolipid (50–100 μg) is dissolved in 0.5–1 ml of this reagent, and the sealed tube is kept at 80°C for 18 hr. Mannitol can be added as an internal standard. After cooling, the fatty acid methyl esters are removed by three successive extractions, each with a 3-ml portion of hexane or petroleum ether. A few milligrams of solid silver carbonate is added to the lower methanol–HCl phase until neutral pH is obtained. Then 0.05 ml of acetic anhydride is added and left for at least 6 hr at 25°C, and the mixture is finally centrifuged or filtered to remove silver carbonate. The filtrate is evaporated in a small conical test tube under a stream of nitrogen, and the residue is dissolved in 50 μl of a mixture of pyridine–HMDS–TMCS (5 : 2 : 1)[187] and allowed to react for 15 min at 25°C. The precipitate formed does not interfere with the GLC analysis. A 1- to 6-μl aliquot is injected into a 2-m × 3-mm column of 3% SE-30 or OV-1 (on Supelcoport, Gas-Chrom Q, Chromosorb W, or similar acid-washed, silanized diatomaceous support) at 150°C, and the column oven is programmed at 3°C/min to 250°C. Flash heater and detectors are maintained at 280°C. The nitrogen carrier flow rate is 35 ml/min. Relative retention times of the derivatives of fucose, hexoses, hexosamines, and sialic acid, with respect to hexatrimethylsilyl mannitol, have been reported.[187]

The sugar mixture with a predetermined molar ratio including mannitol is analyzed to determine the flame-detector response for each sugar. The hexosamines should be added to the standard mixture as *N*-acetyl derivatives, and usually exhibit 0.7 mass response when compared with glucose. Sialic acid gives a response of 1.2 vs. an expected 1.6.

1.3.3.3. Determination of the Ratio of Hexoses to Sphingosines

For the determination of the ratio of hexoses to sphingosines, a combination of the GLC methods for hexoses (Sections 1.3.3.1 and 1.3.3.2) and sphingosines (Section 1.3.2.8a) can be used. The detector response of each component can vary depending on column and GLC equipment Therefore, it should be determined with authentic mannitol, glucose, galactose, and sphingosine, mixed in a certain molar ratio and carried through either the anhydrous or the aqueous methanolysis procedure, followed by N-acetylation and O-trimethylsilylation, and chromatographed on 3% SE-30 with programming from 150 to 275°C at 3°C/min. An unknown sample with mannitol as internal standard is methanolyzed under the same conditions, and the peak areas are compared to those obtained for the standards. A known standard glycosphingolipid, such as human erythrocyte globoside, can be used in addition to the free sugars and sphingosines. Methylglycosides of hexoses, hexosamines, and sphingosines in the methanolysate of glycolipids can also be determined as trifluoroacetyl derivatives.[188]

1.3.3.4. Determination of N-Acetyl, N-Glycolyl, and N,O-Substituted Neuraminic Acid

With 0.05 N methanolic HCl, sialic acid is liberated from gangliosides as a methyl β-ketoside of N-acetylneuraminic acid methyl ester or as a methyl β-ketoside of N-glycolylneuraminic acid methyl ester which were determined by GLC as trimethylsilyl derivatives.[189] The dried ganglioside samples were treated with 1–2 ml of 0.05 N methanolic HCl, freshly prepared by addition of 1 part of concentrated HCl (12 N) to 240 parts of distilled methanol. Samples were heated in a Teflon-lined screw-capped test tube for 1 hr at 80°C. The cooled solutions were extracted three times with 3 ml of hexane to remove the small quantity of liberated fatty acid esters that may obstruct GLC analysis. The methanolic solutions were evaporated to dryness with a slow stream of N_2. Internal standard (phenyl β-glucosaminide) was added in a small volume of methanol.

The mixture was evaporated with N_2 and finally dried in a vacuum desiccator for 10–15 min. TMS derivatives were formed according to the method described by Carter and Gaver[139] and analyzed by GC on 3% OV-1 or 3% OV-225. N-Glycolyl-(4-O-acetyl)neuraminic acid present in a specific fraction of horse hematoside[190] was analyzed by the same method except that milder methanolysis conditions (0.02 N methanolic HCl, 80°C, 1 hr) were used.[189] The retention times of various sialic acids present in glycolipids are shown in Table XI.[189] Free sialic acid, liberated by aqueous hydrolysis, can be analyzed by GLC. The free sialic acid was heated with *bis*-(trimethylsilyl)trifluoroacetamide in acetonitrile to form the TMS ether-ester.[191] A similar analysis of sialic acid was made by heating sialic acids with N-trimethyl silylimidazole.[192]

Table XI. *Relative Retention Times of Trimethylsilylated and Methylated Sialic Acids*

Column	Column	
	OV-1	OV-225
I. TMS derivatives[a]		
NeuAc: α-methyl ketoside methyl ester	0.71	0.41
NeuAc: β-methyl ketoside methyl ester	0.89	0.45
NeuGlyc: α-methyl ketoside methyl ester	1.50	—
NeuGlyc: β-methyl ketoside methyl ester	1.69	0.53
NeuGlyc: O-acetyl methyl ketoside methyl ester	2.08	1.18
Phenyl N-acetyl-α-D-glucosaminide[b]	1.00	1.00
II. Methylketoside methylester of permethylated sialic acid[c]	SE-30 (2.2%)	
NeuAc: 4,7,8,9-tetra-O-methyl	1.00	
NeuAc: 4,7,9 -tri-O-methyl-, 8-O-acetyl	1.30	
NeuGlyc: 4,7,8,9-tetra-O-methyl	1.50	

[a] Relative retention times were measured relative to the internal standard, the absolute values of which were 10.0 min on OV-1 and 20.4 min on OV-255. Column conditions were: OV-1 3% on 100–120 Chromosorb W HP, 6 ft., 205°C; OV-255 3% on 100–120 Supelcoport, 4 ft., 190°C. Helium carrier gas was employed at a flow rate of 70 ml/min. All samples were chromatographed as TMS derivatives.
[b] Internal standard.
[c] Relative retention time is expressed in comparison with that of 4,7,8,9,-tetra-O-methyl NeuAc as 1.00.

The most reliable quantitative analysis of N-glycolyl and N-acetyl neuraminic acid is based on methylation followed by methanolysis and GC analysis of the methyl ketoside methyl esters of 4,7,8,9-tetra-O-methyl, and N'-methyl-derivatives of N-acetyl, and N-glycolyl neuraminic acids, which are separated well on GC. If O-acetylation is included after methanolysis, 3-O-acetyl-4,7,9-tri-O-methyl sialic acid can be detected, which has diagnostic value for the presence of the sialosyl2→8sialosyl structure. Rauvala and Kärkkäinen[193] elaborated the condition of the methylation analysis for sialic acid and determined the mass spectra of 4,7,8,9-tetra-O-methyl and 8-O-acetyl-4,7,9-tri-O-methyl derivatives of N-acetyl or N-glycolyl neuraminic acid. HCl methanol (0.5 N) at 80°C for 3 hr was recommended for the methanolysis of methylated sialic acid. The retention times are shown in Table XI.

Recently, 8–12 different kinds of sialic acids having various O-substitutions have been isolated from glycoproteins and identified. These sialic acids have been separated by TLC on cellulose plates and silica gel plates and by GLC (3% OV-17 or 2.8% SE-30); the conditions of the analyses have been elaborated in studies by Schauer and associates[194–196] (see Table XII). The O-substituted sialic acid has not often been detected in glycolipids. One kind of hematoside isolated from horse erythrocytes was identified as having 4-O-methyl-N-glycolyl neuraminic acid.[190,196] Recently, a brain ganglioside with GT_{1b} structure was found to contain 9-O-acetyl neuraminic acid.[197] Thus, gangliosides with O-substituted sialic acid may be found more often in the future by applying a suitable procedure to detect O-substituted sialic acid. Adoption of the methods of Schauer and associates[194,195] for ganglioside analysis is certainly warranted.

Table XII. Chromatographic Behaviors and Fragment Ions Obtained on Mass Spectrometry[a]

Sialic acid	Dowex 2 × 8 HCOOH(M) for elution	TLC on cellulose[b]	TLC on silica gel[c]	GLC on SE-30[d]	Fragment ions (m/z, see Ref. 196)						
					A	B	C	D	E	F	G
NeuAc	0.05–0.40	0.57	0.39	1.00	668	624	478	298	317	205	173
4-O-Ac-NeuAc	—	0.76	0.61	1.18	638	594	448	298	—	205	143
7-O-Ac-NeuAc	0.05–0.20	0.76	0.61	1.04	638	594	—	—	317	205	173
9-O-Ac-NeuAc	0.05–0.45	0.76	0.61	1.13	638	594	478	298	317	175	173
4,9-Di-O-Ac-NeuAc	—	0.83	0.73	1.31	608	564	448	298	—	175	143
7,9-Di-O-Ac-NeuAc	0.10–0.25	0.83	0.73	1.14	608	564	—	—	317	175	173
8,9,-Di-O-Ac-NeuAc	0.25–0.45	—	—	1.19	608	564	478	298	317	—	173
7,8,9-Tri-O-Ac-NeuAc	0.15–0.25	—	—	1.07	578	534	—	—	317	—	173
9-O-L-Lac-NeuAc	—	0.70	0.61	2.55	740	696	478	298	317	277	173
4-O-Ac-9-O-Lac-NeuAc	—	—	—	3.01	710	666	448	298	—	277	143
NeuGl	0.10–0.45	0.48	0.39	1.81	756	712	566	386	317	205	261
4-O-Ac-NeuGl	—	0.65	0.61	2.02	726	682	536	386	—	205	231
7-O-Ac-NeuGl	—	—	—	1.83	726	682	—	—	317	205	201
9-O-Ac-NeuGl	0.20–0.45	0.65	0.61	2.04	726	682	566	386	317	175	261
7,9-Di-O-Ac-NeuGl	0.35–0.45	—	—	2.01	696	652	—	—	317	175	261
8,9-Di-O-Ac-NeuGl	—	—	—	1.99	696	652	566	386	317	—	261
7,8,9-Tri-O-Ac-NeuGl	—	—	—	1.93	666	622	—	—	317	—	261

[a] Data from Refs. 194–196.
[b] In n-butanol–n-propanol–0.1 N HCl (1 : 2 : 1, vol./vol./vol.).
[c] In n-propanol–water (7 : 3, vol./vol.).
[d] Methyl esters of O-TMS derivative; 3.8% SE-30 on chromosorb W/AW-DMCS, HP, 80–100 mesh; glass column, 4.0 mm × 2 ft.; per-O-TMS neuraminic acid methyl ester as 1.00.

1.3.4. Release of Oligosaccharides from Glycosphingolipids

1.3.4.1. Release of Oligosaccharides by Endo-β-galactosidase

This reaction is discussed in Section 1.3.5.3.

1.3.4.2. Release of Oligosaccharides by Oxidation of the Olefinic Double Bond followed by Alkaline-Induced Degradation

Oligosaccharides of glycosphingolipids can be liberated by alkaline degradation after the double bond of sphingosine has been converted to an aldehyde by ozonolysis[198,199] or by osmium periodate.[200]

According to the method described by Wiegandt and Baschang,[198] 10–100 mg of glycolipids were dissolved in hexane–methanol or in methanol or tertiary butanol, and the sample was ozonolyzed by passing ozone for 30 min. The solvent was evaporated *in vacuo,* the residue dissolved in water, and the pH adjusted at 10.5–11 with 0.2 N sodium carbonate and allowed to remain at room temperature for 12–16 hr. During the incubation period, the pH should be kept as indicated above. The solution was centrifuged at 10,000 rpm for 20 min, and the supernatant was lyophilized after neutralization with Dowex 50 (H$^+$). This method has been adapted to work with milligram quantities of glycolipid.[199]

Oligosaccharides of glycosphingolipids can also be liberated by osmium-catalyzed periodate oxidation.[200] The method has been improved recently by replacing $NaIO_4$ by HIO_4 during the oxidation, which gave quantitative yields of the oligosaccharides.[201] Glycolipid acetate derived from 1–3 mg of starting material was dissolved in 0.5 ml of methanol, and was added to 0.2 ml of freshly prepared 0.2 M HIO_4 in methanol followed by 25 μl of freshly prepared 5% osmium tetroxide in ether, and the mixture was left at 5°C overnight. A drop of glycerol, 6 ml of chloroform–methanol (2 : 1), and 1.5 ml of water were added to separate two layers. The lower layer was washed four times with 2.5-ml portions of chloroform–methanol–water (1 : 10 : 10). The organic phase, dried in a stream of nitrogen, gave a glycolipid-aldehyde. The material was dissolved in 0.3 ml of methanol, and 50 μl of 0.2 M sodium methoxide in methanol was added and left at room temperature for 1 hr, followed by addition of a drop of water, and left an additional hour at room temperature and neutralized with acetic acid. The reaction mixture was evaporated to dryness under nitrogen, dissolved in 0.5 ml of chloroform–methanol–water (1 : 10 : 10), and partitioned repeatedly with 0.5-ml portions of chloroform–methanol–water (60 : 35 : 8). The aqueous layer contained essentially pure oligosaccharides. These liberated oligosaccharides can be identified by paper chromatography or by GC through a short (1-ft.) column of OV-17 according to the method described by Ohashi and Yamakawa.[202] The oligosaccharide patterns liberated from erythrocyte glycolipids of human, horse, dog, guinea pig, and goat were compared by this method.[202]

The liberation of oligosaccharides proceeds through double-bond transposition, similar to the epimerization of sugar (Lobry-de-Bryn/Alberta van

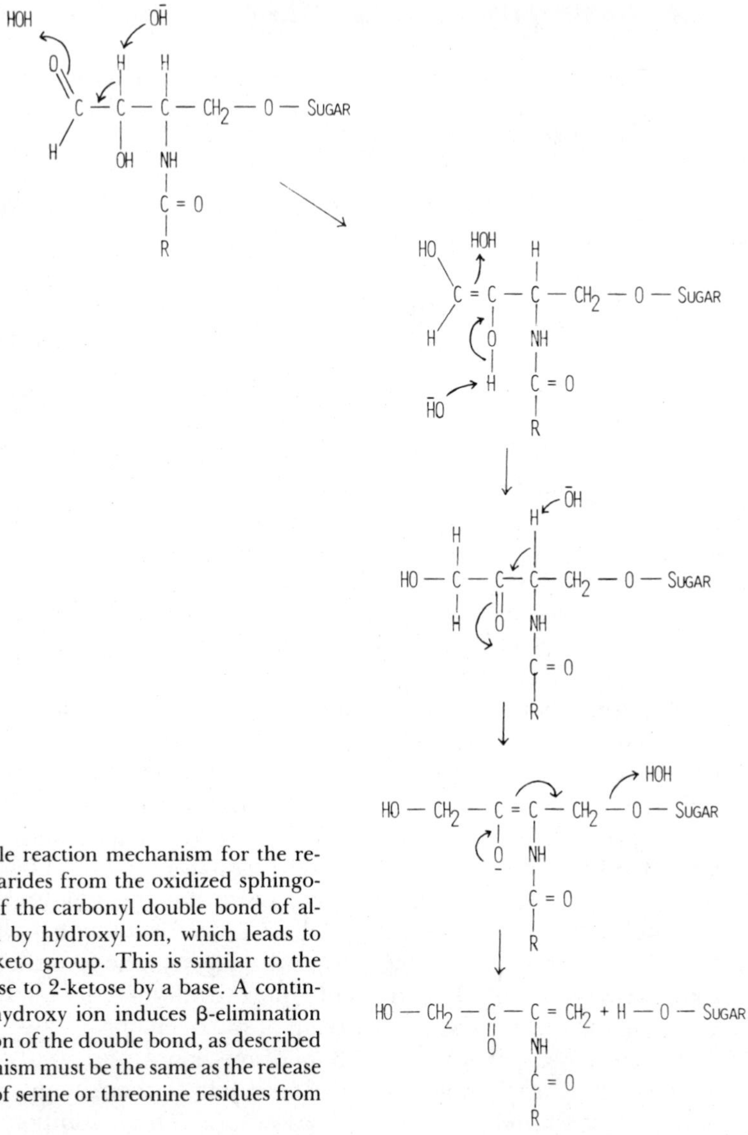

Figure 8. A possible reaction mechanism for the release of oligosaccharides from the oxidized sphingolipids. Migration of the carbonyl double bond of aldehyde is induced by hydroxyl ion, which leads to formation of a 3-keto group. This is similar to the conversion of aldose to 2-ketose by a base. A continuous attack with hydroxy ion induces β-elimination due to the migration of the double bond, as described above. The mechanism must be the same as the release of the *O*-glycoside of serine or threonine residues from a peptide.

Eckenstein transformation, i.e., conversion of aldose to 2-ketose). The carbohydrates attached to C-1 of 3-keto-intermediate (II) can be readily released by a β-elimination reaction induced by hydroxy ion (see Fig. 8). This mechanism has been supported by the fact that 3-keto-sphingosine containing glycolipid, which was formed by oxidation of glycosphingolipid through the Kishimoto oxidation with 2,3-dichloro-5,6-dicyanobenzoquinone,[203] was readily degraded by β-elimination in a weaker base solution (0.005 M Na_2CO_3 in chloroform–methanol, 2 : 1) than required for ozonolysis or osmium tetroxide–periodate oxidation product.[204]

1.3.4.3. Release of Oligosaccharides by Trifluoroacetolysis

When glycoproteins and glycosphingolipids were treated with a mixture of trifluoroacetic acid (TFA) and trifluoroacetic anhydride (TFAA), all hydroxy groups were rapidly trifluoroacetylated, since the reagent TFAA–TFA should contain the ions $[CF_3CO]^+$ and $[CF_3COO]^-$. The N-acetyl groups of hexosamines were exchanged with N-trifluoroacetyl function. A strong electron-attracting property of the O-trifluoroacetyl groups renders the glycosidic acetal group electron-deficient. The O-glycosidic linkages were therefore all stable during trifluoroacetolysis, and no anomerization was evident. However, peptides, ceramides, and sialic acids were extensively degraded, leaving sialic acid-free intact oligosaccharides liberated from peptides and ceramides. The reaction was first described by Nilsson and Svensson[205] as an N-deacetylation method for N-acetyl amino sugars and subsequently applied to oligosaccharide liberation from glycosphingolipids and glycoproteins.[206,207] The same method was used for a specific cleavage of the O-glycosidic bond linked to serine and threonine.[208] Recently, the method was used for analysis of total oligosaccharides liberated from total glycolipids of erythrocytes with various blood group status.[451] Oligosaccharides liberated were all N-trifluoroacetylated and were suitable for GC-MS identification after methylation.[588] Oligosaccharides were not liberated from glycolipids containing dihydrosphingosine, which is resistant to the conditions described (Nilsson, B., and Zopf, D., personal communication).

1.3.5. Determination of Carbohydrate Sequence

1.3.5.1. Sequential Degradation by Weak Acid Hydrolysis and Controlled Smith Degradation

Repeated hydrolysis of glycolipids with mild acid (for example, 0.1 N HCl for 30 min), followed by dialysis of liberated sugars, and examination of the resulting glycolipids by TLC was used for determination of the carbohydrate sequences in globoside[72] and in Le^a-active glycolipid.[209] The method, however, requires a relatively large amount of substance and has not been used often in recent structural studies.

Sialosyl residues of glycolipids were readily hydrolyzed by 1% aqueous acetic acid at 100°C for 1 hr or by 0.05 N sulfuric acid at 80°C for 1 hr. No other sugar residues can be hydrolyzed under such conditions. The hydrolysis product, the asialo core, can be examined by TLC.

Fucosyl residues in glycolipids were not readily hydrolyzed by weak acid as compared with these residues in oligosaccharides and glycoproteins. For example, the fucosyl residue of H$_1$ glycolipid was not readily hydrolyzed by 0.1 N HCl at 100°C for 1 hr, whereas it was hydrolyzed in 0.1 N trichloroacetic acid at 100°C for 2 hr.[119]

A micellar solution or aqueous emulsion of glycosphingolipids can be oxidized by periodate only at their nonreducing terminus when the concentration of periodate is lower than 0.05 M. A sequential degradation is therefore possible by a repeated periodate oxidation, reduction, and mild acid hydrolysis (Smith degradation).[210] Such a method has been used for the elimination of the terminal N-acetylgalactosamine of globoside,[77] for the elimination of the terminal carbohydrate residues in A-active, B-active, and Le^a-active glyco-lipids,[209] and for the conversion of paragloboside to lacto-N-triaosylceramide.[381]

1.3.5.2. Simultaneous Determination of the Sequence of Carbohydrate and Anomeric Linkages by Enzymatic Hydrolysis

The most specific and useful method for sequential degradation of carbohydrate is the use of specific exoglycosidases that can hydrolyze only a specific nonreducing residue of the carbohydrate chain. This method will simultaneously determine anomeric linkages and requires less than 100 μg of glycolipid. The following enzymes have been used: *Turbo cornutus* α-L-fucosidase,[211]* jack bean β-galactosidase,[213,215] fig α-galactosidase,[77,215] jack bean β-N-acetylhexosaminidase,[214] hog liver α-N-acetylgalactosamini-dase,[216] beef brain β-glucosidase,[217] and sialidase from various sources. Hydrolysis is usually carried out in 0.05 M citrate or acetate buffers containing sodium deoxytaurocholate (1 mg/ml); pHs are adjusted according to the op-timal range of the enzyme.[77] The conditions of hydrolysis, particularly the detergent concentration, have been studied extensively.[218]

A successful hydrolysis of a specific sugar residue greatly depends on selection of a hydrolase suitable for certain types of glycolipids or oligosac-charides. The terminal β-Gal residue in type 2 chain (Galβ1→4GlcNAcβ1→3Galβ1→R) is readily hyrolyzed by jack bean β-galactosidase, but the same Gal residue in type 1 chain (Galβ1→3GlcNAcβ1→3Galβl→R) was less sus-ceptible. In contrast, β-Gal residue of Le^a or Le^x structure was completely insensitive to various β-galactosidases (S. Hakomori, unpublished observa-tions).[622] An α-N-acetylgalactosaminidase from the limpet, even after exten-sive purification, still possesses α-galactosidase activity, and the enzyme may have a dual specificity.[219] Similarly, jack bean β-N-acetylhexosaminidase has a dual specificity to β-N-acetylglucosamine and β-N-acetylgalactosamine. The sialosyl residue linked to the terminal Gal of gangliotetraosylceramide was readily hydrolyzed, but that linked to the internal Gal residue of gangliote-traosyl or gangliotriosylceramide (GM_1 or GM_2) was not hydrolyzed by siali-dase from various sources in the presence or absence of nonionic deter-gent.[220–224] The sialidase-resistant sialic acid in GM_2 or GM_1 ganglioside was later found to be hydrolyzed by *Clostridium perfringens* sialidase in the presence of bile salt[225,487] or in a monodisperse solution below the critical micelle concentration.[226] More recently, a new type of sialidase was isolated and characterized from *Arthrobacter ureafaciens* culture filtrate,[227] which can hy-drolyze sialidase-resistant sialosyl residue of GM_1 ganglioside in the presence of nonionic detergent.[228–229] Sialidase from influenza virus preferentially

---

* The enzyme was prepared by Seikegaku Kogyo Ltd. and is available from Miles Laboratories, but required further purification by affinity chromatography.[212]

hydrolyzes sialosylα2→3Gal linkage rather than α2→6Gal linkage,[490] and this tendency is quite remarkable in New Castle disease virus sialidase and in fowel plaque virus.[624,625] The latter did not hydrolyze sialosyl α2→8 sialosyl linkage in polysialogangliosides, but hydrolyzed the terminal sialosyl 2→3Gal residue in GD_{1a} or GT_{1b}.[489] The sialosyl linkage of 4-*O*-acetylsialic acid was completely resistant to sialidase of any source, and GM_3 hematoside containing this sialic acid was not hydrolyzed by sialidase unless the ganglioside was treated with base.[454]

Sequential degradation of glycolipids by various exoglycosidases can be conveniently followed by TLC analysis. An example of the experimental conditions is as follows: Glycolipid (100 μg) is dissolved in 100 μl of the buffer solution containing sodium taurocholate (100 μg/100 μl) in a small conical test tube. The glycolipids are readily dissolved by warming, followed by agitation with a Vortex mixer and immersion in an ultrasonic bath. The enzyme solution (2–50 μl) is added (total activity 0.5–2 units), and the mixture is incubated at 37°C for 18 hr. After incubation, the reaction mixture is shaken with 6 volumes of chloroform–methanol (2 : 1) and then centrifuged. The lower layer is removed by capillary pipette, evaporated to dryness, and analyzed by TLC on a silica gel H plate. Solvent 1 or solvent 2 (see Table VI) is used for development, and the plate is sprayed with orcinol–sulfuric acid. The conversion of a given glycolipid to its derivative with one less sugar moiety occurs when the non-reducing sugar is removed by enzymatic hydrolysis, and the glycosidases that cause this conversion indicate the kind of terminal sugar residue. Glycolipid with one less sugar residue, produced by enzymatic hydrolysis, ran on TLC faster than before enzymatic treatment. However, there were a few exceptions: a highly complex ganglioside showed slower migration on TLC after sialidase treatment.[116,185] The resulting glycolipid after one enzyme treatment is again dissolved in a suitable buffer containing sodium taurocholate. Enzymatic hydrolysis is repeated according to the data of sugar analysis and the molar ratio of carbohydrates. Typical examples are shown in Fig. 9.

1.3.5.3. Degradation by Endo-β-galactosidase

A keratan sulfate-degrading enzyme of *Escherichia freundii* induced by addition of whale nasal cartilage keratan sulfate in the culture medium was reported by Kitamikado and Ueno.[230] The enzyme was purified by Fukuda and Matsumura[231] and was identified as an endo-β-galactosidase hydrolyzing the β-galactosyl linkage of R→3Galβ1→4GlcNAc→R; thus, lacto-*N*-*neo*tetraitol (Galβ1→4GlcNAcβ1→3Galβ1→4 sorbitol) is converted to a trisaccharide (Galβ1→4GlcNAcβ1→3Gal) and sorbitol. The lacto-*N*-glycosylceramide series having the common structure R→GlcNAcβ1→3Galβ1→4Glc(or GlcNAc) were hydrolyzed at the β-galactosyl linkage to Glc or GlcNAc.[232] The β-Gal linkages in the globo-series or the ganglio-series of glycolipids were not hydrolyzed.[232] The kinetics of hydrolysis of various substituents of paragloboside and H_2 and H_3 glycolipids were studied. The following findings were of particular interest: (1) Sialosyl substitution at the terminal galactosyl residue

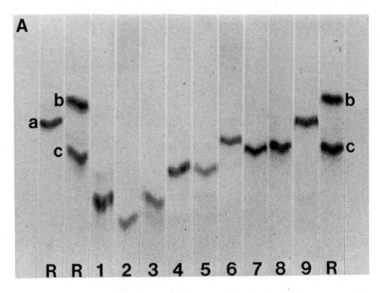

Structure I

Galα1→3Galβ1→4GlcNAcβ1
                              ↘
                                 §Galβ1→4GlcNAcβ1→3Galβ1→4Glcβ→Cer
                              ↗
NeuAcα2→3Galβ1→4GlcNAcβ1

Structure II

Galα1→3Galβ1→4GlcNAcβ1
                              ↘
                                 §Galβ1→4GlcNAcβ1→3Galβ1→4Glcβ→Cer
                              ↗
Galβ1→4GlcNAcβ1

Structure III

Galβ1→4GlcNAcβ1
                         ↘
                            §Galβ1→4GlcNAcβ1→3Galβ1→4Glcβ→Cer
                         ↗
Galβ1→4GlcNAcβ1

Structure IV

Galβ1→4GlcNAcβ1
                         ↘
                            §Galβ1→4GlcNAcβ1→3Galβ1→4Glcβ→Cer
                         ↗
GlcNAcβ1

Structure V

GlcNAcβ1
                    ↘
                       §Galβ1→4GlcNAcβ1→3Galβ1→4Glcβ→Cer
                    ↗
Galβ1→4GlcNAcβ1

Structure VI

GlcNAcβ1
                ↘
                   §Galβ1→4GlcNAcβ1→3Galβ1→4Glcβ→Cer
                ↗
GlcNAcβ1

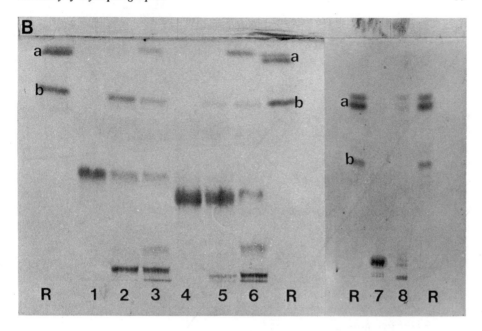

Structure VII   Galβ1→4GlcNAcβ1→6Galβ1→4GlcNAcβ1→3Galβ1→4Glcβ→Cer

Structure VIII   Galβ1→4GlcNAcβ1→3Galβ1→4GlcNAcβ1→3Galβ1→4Glcβ→Cer

Structure IX   GlcNAcβ1→6Galβ1→4GlcNAcβ1→3Galβ1→4Glcβ→Cer

Figure 9. TLC patterns of glycolipid degradation by exoglycosidases (A) and by endo-β-galactosidase of *Escherichia freundii* (B). A—lanes: R, references: a, sialosylparagloboside; b, paragloboside; c, lacto-*N-nor*hexaosylceramide; 1, I-active ganglioside of bovine erythrocytes [Structure I (see below)]; 2, Structure II (Structure I treated with sialidase); 3, Structure III (Structure II treated with α-galactosidase); 4, Structure IV (Structure I treated with sialidase, β-galactosidase, and α-galactosidase); 5, Structure V (Structure I treated with α-galactosidase, β-galactosidase, and sialidase); 6, Structure VI (Structure III treated with β-galactosidase); 7, Structure VII (Structure II treated with β-galactosidase and β-*N*-acetylhexosaminidase); 8, Structure VIII, lacto-*N-nor*hexaosylceramide (sialosyl-lacto-*N-nor*hexaosylceramide treated with sialidase); 9, Structure IX, lacto-*N-nor*hexaosylceramide treated with β-galactosidase. B—lanes: R, references: a, ceramide monohexoside; b, ceramide trihexoside; 1, H_2 glycolipid; 2, H_2 glycolipid incubated for 2 hr with 2.5 munits of endo-β-galactosidase (condition 1[232]); 3, H_2 glycolipid incubated with 25 mU of the enzyme for 18 hr (condition 2[232]); 4, A^b glycolipid incubated for 18 hr without enzyme; 5, A^b glycolipid incubated for 2 hr under "condition 1"; 6, A^b glycolipid incubated for 18 hr under "condition 2"; 7, H_3 glycolipid; 8, H_3 glycolipid incubated for 18 hr under "condition 2." Note that "condition 1" liberated a ceramide trisaccharide and "condition 2" liberated a ceramide monosaccharide from H_2 and A^b glycolipid. H_3 glycolipid (with branched structure) was not degraded under "condition 1," but was degraded under "condition 2" to give a ceramide monohexoside and a large oligosaccharide (which stayed at the origin). The data is taken from Watanabe *et al.*[185] and from Fukuda *et al.*[232]

of the lacto-*N*-*neo*tetraosyl structure greatly enhances the hydrolyzability of the internal Galβ1→4Glc linkage, thus producing a high yield of sialosyltetrasaccharide (NeuAcα2→3Galβ1→4GlcNAcβ1→3Gal). (2) The Galβ1→4GlcNAc linkage located in the middle of the repeating Galβ1→4GlcNAc (or Glc) unit as seen in H₂ or Aᵇ glycolipid (R→Galβ1→4GlcNAcβ1→3Galβ1→4GlcNAcβ1→3Galβ1→4Glc→Cer) was preferentially hydrolyzed relative to the Galβ1→4Glc linkage directly attached to ceramide and the oligosaccharide R→Galβ1→4GlcNAcβ1→3Gal was liberated. (3) The branched structure as found in H₃ glycolipid greatly reduced the hydrolyzability. The H₃ glycolipid was hydrolyzed at a higher concentration of enzyme, resulting in the liberation of a branched nonasaccharide, Fucα1→2Galβ1→4GlcNAcβ1→3(Fucα1→2Galβ1→4GlcNAcβ1→6)Galβ1→4GlcNAcβ1→3Gal, which was in turn degraded into a branched heptasaccharide and a disaccharide. The substrate specificity and kinetics of the enzyme action are shown in Fig. 10. Testing the susceptibility of glycolipid to this enzyme is a useful tool for the identification of an unknown glycolipid as belonging to the lacto-*N*-glycosyl series.

1.3.5.4. Sequence Determination of Carbohydrates by Direct-Probe Mass Spectrometry of Permethylated Glycolipids and Permethylated Oligosaccharides

This determination is discussed in Section 1.3.8.

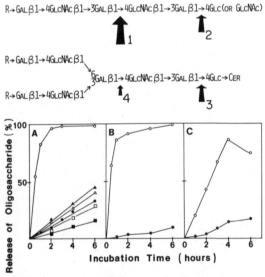

Figure 10. Susceptibility of various β-galactosidic linkages in glycolipids to endo-β-galactosidase of *E. freundii*. *Top:* The order of the susceptibility of β-galactosidic linkages is indicated by the relative sizes of the arrows: 1 ≫ 2 = 3 ≫ 4. The susceptibility of linkage 1 is further enhanced by the presence of the sialosyl residue at the terminus. *Bottom:* Kinetic analysis of various glycolipid structures of the lacto-*N*-glycosyl series. A, Oligosaccharide release from sialosylparagloboside (○), α-galactosylparagloboside (▲), β-galactosylparagloboside (△), paragloboside (●), H₁ glycolipid (□), and lacto-*N*triosylceramide (■). B, Release of Fucα1→2Galβ1→4GlcNAcβ1→3Gal from H₂ glycolipid (○) and release of GlcNAcβ1→3Gal from H₂ glycolipid (●). C, Release of a branched nonasaccharide from H₃ (○) and release of GlcNAcβ1→3Gal from the H₃ (●). In A and B, "condition 1" was used; in C, "condition 2" was used. This figure is based on data described in Fukuda *et al.*[232]

1.3.6. Determination of the Position of Glycosyl Linkages

Methylation study is the most useful confirmative method for determination of the position of glycosyl linkages. The classical technique has been greatly advanced by discovery of effective methylation methods[233,234] and improved technology for identification of the partially methylated sugars by GC-MS.[235,236] The permethylation condition originally described by Hakomori[234] was elaborated and improved by Sanford and Conrad.[239] The application of the methylation analysis for amino sugar derivatives[237,238] and for sialic acid derivatives[193,240,244] has been elaborated.

1.3.6.1. General Procedure for Methylation and Subsequent Hydrolysis

Glycolipid or oligosaccharide (0.3–0.5 mg) is placed in a Pyrex glass tube with a Teflon-lined screw cap. Anhydrous dimethylsulfoxide (DMSO) (0.1–0.5 ml) is added to the tube, and the contents are then stirred with a magnetic stirring bar. Methylsulfinyl carbanion (0.1–0.5 ml)* is added to the tube, which is then flushed with nitrogen, and the tube is screw-capped. The mixture is stirred for 3 hr at room temperature, followed by the addition of 0.5 ml methyl iodide. After 2 hr, the reaction mixture is applied to a small column (1 × 30 cm) of Sephadex LH20 (Pharmacia Fine Chemicals) prepared in chloroform–methanol (1 : 1, vol./vol.). The column is then eluted with the same solvent, and fractions (1 ml each) are collected with a fraction collector. About 5 µl of each fraction is spotted on a silica gel G TLC plate, which is then charred with orcinol–sulfuric acid. The fractions shown to contain sugars are combined and evaporated under nitrogen. The degree of methylation of glycolipid is checked by IR spectrometry using Irtran plates. The degree of methylation can be monitored by the concentration of remaining carbanion with a simple test using triphenylmethane.[245]

The permethylated glycolipid is degraded by either of the following two methods: (1) The permethylated glycolipid is dissolved in 90% formic acid and kept at 100°C for 2 hr. The formic acid is removed by vacuum distillation (or under nitrogen) at 40°C with toluene, followed by hydrolysis in 1 ml of 0.25 N H_2SO_4 at 100°C overnight, according to the method of Björndal *et al.*[235] (2) The permethylated glycolipid is dissolved in 0.3 ml of 95% acetic acid containing 0.5 N sulfuric acid at 80°C overnight, and 0.3 ml water is added and heating continued at 80°C for another 5 hr, according to the method of Stellner *et al.*[237] (3) For amino sugar analysis, the second method is necessary, while either method can be applied to analysis of neutral sugars.

---

* NaH–mineral oil (Lg), after removal of the oil with petroleum ether, is stirred with 20 ml of DMSO at 60°C until a clear green solution is obtained (about 1 hr). Aliquots of 1 ml can be stored under nitrogen at −20°C and can be used for at least 6 months.

The sulfate ion in the hydrolysate of either method is eliminated by passage through a small column of 200 mg of AG3-X4A acetate form (Biorad, Richmond, California). The column is washed with methanol (2 ml), and the hydrolysate and the methanol washings are evaporated under nitrogen. The residue is dissolved in 0.3 ml of water and reduced by the addition of 10 mg of sodium borohydride for at least 3 hr at room temperature. A drop of glacial acetic acid is added, and the mixture is evaporated under nitrogen. Several milliliters of methanol containing a few drops of glacial acetic acid are added and again evaporated under nitrogen. A white residue (sodium acetate) left in the tube is further dried in a vacuum desiccator over phosphorus pentoxide. The residue is mixed with 0.5 ml of distilled acetic anhydride and heated at 100°C for 2 hr. The reaction mixture is evaporated with toluene in a rotary evaporator. The residue in the flask is dissolved in chloroform and shaken with water. The chloroform layer is evaporated under a gentle stream of nitrogen in a small conical tube. The residue left on the tube is dissolved in 20–50 μl of acetone, and 1- to 2-μl aliquots are injected into GC or GC-MS equipment for analysis. All procedures can be scaled down to one-fifth of the amount of glycolipid and the reagents.

1.3.6.2. Identification of Partially O-Methylated Neutral Hexitols

Identification of partially O-methylated neutral hexitols is most conveniently performed on a 6-ft. column containing 3% ECNSS-M on Gas-Chrom Q or 3% ECNSS-M on Gas-Chrom W (nonsilanized absorbent) for separation of 3,4,6-tri-O-methylgalactose and 2,3,6-tri-O-methylglucose (Stellner and Hakomori, unpublished data), or on 3% OV-225-coated Supelcoport (Supelco Inc., Bellafonte, Pennsylvania). Separation of other partially methylated hexitols is quite satisfactory with 3% ECNSS-M on Gas-Chrom Q. The recommended temperature is 160°C. Retention times of O-methylated sugars have been extensively studied by Björndal et al.,[235] and their data are reproduced in Table XIII. Since mannose is absent in glycosphingolipids, separation of mannose derivatives and other derivatives has not been a serious problem for characterization of glycolipids except for glycosphingolipids of sea animals and plants (Sections 1.4.8 and 1.4.9). Because contaminating nonsugar peaks occur in these chromatograms, confirmative identification should be based not only on the retention time but also on mass spectra. For this purpose, the GC-MS combination is highly desirable. Mass spectra of all kinds of partially O-methylated neutral sugars have been examined by Björndal et al.,[235,236] and their data are reproduced in Table XIV.

Recently, chemical ionization mass spectrometry combined with GC through a capillary column has been used for analysis of the partially methylated hexitol acetates or TMS glycosides by McNeil and Albersheim[567] and by Laine.[568,587] The method is more sensitive than the established procedure with electron impact mass spectrometry and regular GC.

Table XIII. Retention Times (T Values) on 3%
ECNSS-M of Partially Methylated Sugars, in the
Form of Their Alditol Acetates, Relative to 1,5-Di-
O-acetyl-2,3,4,6-tetra-O-methyl-D-glucitol

Position of OCH$_3$	Gal	Glc	Fuc
2	8.1	7.9	1.67
3	11.1	9.6	2.05
4	11.1	11.5	—
6	5.10	5.62	—
2, 3	5.68	5.39	1.18
2, 4	6.35	5.10	1.12
2, 6	3.65	3.83	—
3, 4	6.93	5.27	—
3, 6	4.35	4.40	—
4, 6	3.64	4.02	—
2, 3, 4	3.41	2.49	0.65
2, 3, 6	2.42	2.50	•—
2, 4, 6	2.28	1.95	—
3, 4, 6	2.50	1.98	—
2, 3, 4, 6	1.25	1.00	—

Table XIV. Primary Fragments in the Mass Spectra of Partially Methylated Sugars in the Form
of Their Alditol Acetates

Position of CH$_3$	Mass numbers (m/z) of fragments									
	45	117	131	161	175	189	203	205	233	261
Hexoses										
2		X								
3 (4)						X				X
6	X									
2, 3		X								X
2, 4		X				X				
2, 6	X	X								
3, 4						X				
3, 6	X					X			X	
4, 6	X									X
2, 3, 4		X		X		X			X	
2, 3, 6	X	X							X	
2, 4, 6	X	X		X					X	
3, 4, 6	X	X		X						
2, 3, 4, 6	X	X		X				X		
6-Deoxy-hexoses										
2		X								
3						X	X			
4			X							X
2, 3		X					X			
2, 4		X	X							
3, 4			X			X				
2, 3, 4		X	X	X	X					

1.3.6.3. Identification of Partially O-methylated Amino Sugars and Sialic Acid by Gas Chromatography and Mass Spectrometry

The N-acyl groups of N-acetylhexosamines and sialic acids are converted to N-methyl-acetoimido groups of the corresponding sugars by the permethylation procedure described by Hakomori.[234] Therefore, partially O-methylated amino sugars and sialic acid should be identified as 2-deoxy-2-N-methyl-2-acetamidohexoses or 5-deoxy-5-N-methyl-5-acetamido-nonulosonic acid. Syntheses, determination of mass spectra, and retention times on GC of various partially O-methylated-2-deoxy-2-N-methyl-2-acetamindohexoses have been published recently by Stellner *et al.*,[237] and their data are reproduced in Tables XV and XVI. According to Tai *et al.*,[238] the recovery of partially O-methylated 2-N-methyl-2-acetamido-2-deoxyhexoses is better on an OV-17 column than on an ECNSS-M column. A possible O-demethylation at the reducing-terminal N-acetylhexosamine has been discussed recently.*

---

* If the acetolysis–hydrolysis procedure[237] is applied for cleavage of permethylated oligosaccharide that has a terminal hexosaminitol derivative derived from reducing terminal N-acetylhexosamine, O-demethylation of C-1 of the substituted, terminal hexosaminitol derivative occurs (route I below).[241] N-demethylation of the C-2 N-methyl-acetamido group was previously claimed to occur during the hydrolysis procedure (route II).[242,243] Caroff and Szabo[241] presented evidence based on mass spectra of authentic synthetic product and claimed that the final acetylation product is the 1-O-acetyl-2-(N-methylacetamido)-2-deoxyhexitol derivative (see Structure A below) rather than the 2-N-acetyl-acetamido-2-deoxyhexitol derivative (see Structure B). In glycolipid analysis, the demethylation problem is not serious, since all oligosaccharide chains have no reducing terminal and are protected by ceramide. However, the problem will arise when treating oligosaccharide released from glycolipid.

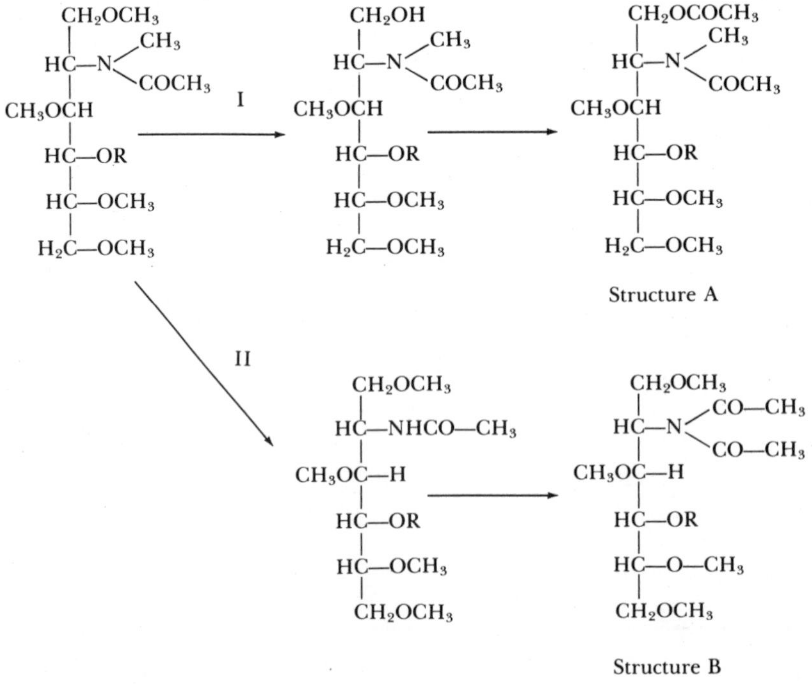

Structure A

Structure B

Table XV. Retention Times (T Values) of Partially Methylated
Amino Sugars, in the Form of Their Alditol Acetates, Relative to
1,5-Di-O-acetyl-3,4,6-tri-O-methyl-2-deoxy-2-N-
methylacetamidoglucitol[a]

Position of CH₃	Glc—N—COCH₃ | CH₃	Cal—N—COCH₃ | CH₃
3, 4, 6	1.00[b]	1.32
3, 6	1.71[b]	1.82
4, 6	2.32[b]	2.48
3, 4	2.33[b]	3.15
6	3.01[b]	2.73
3	3.75[b]	4.07

[a] Generally, 6-ft. glass columns containing 3% ECNSS-M on Chromosorb Q, 100–200 mesh, were used.
[b] 1,5-Di-O-acetyl-2,3,4,6-tetra-O-methyl-D-glucitol = 0.18.

Sialic acids (N-glycolyl- or N-acetylneuraminic acid) are present at the terminal position and often form a sialosyl2→8sialosyl residue in gangliosides. The sialosyl residues can be readily identified by mass spectrometry after permethylation, or after permethylation and methanolysis in 0.05 N HCl in methanol. Rauvala and Kärkkäinen[193] recently introduced methylation analysis of sialic acid and identified 8-O-acetyl-4,6,7-tri-O-methyl-N-methyl-N-acetylneuraminic acid by mass spectrometry, and the technique was applied to a structural study of brain ganglioside and brain glycoproteins.[240,244] For further information on methylation analysis of complex carbohydrates with mass spectrometry, readers should refer to excellent reviews by Lindberg and Lönngren[636] and Rauvala *et al.*[641]

1.3.7. Determination of Anomeric Configuration (α or β) in Glycolipids by Chromium Trioxide Oxidation

Angyal and James[246] described a preferential oxidation of equatorial glycosidic linkages of acetylated glucopyranoside to 5-ketohexulosonate by

Table XVI. Primary Fragments in the Mass Spectra of Partially Methylated Amino
Sugars in the Form of Their Alditol Acetates

Position of CH₃	Mass numbers (m/z) of fragments								
	45	158	161	189	202	205	233	261	274
3		X			X			X	
4		X		X					X
6	X	X							
3, 4		X		X	X				
3, 6	X	X			X		X		
4, 6	X	X	X						X
3, 4, 6	X	X	X		X	X			

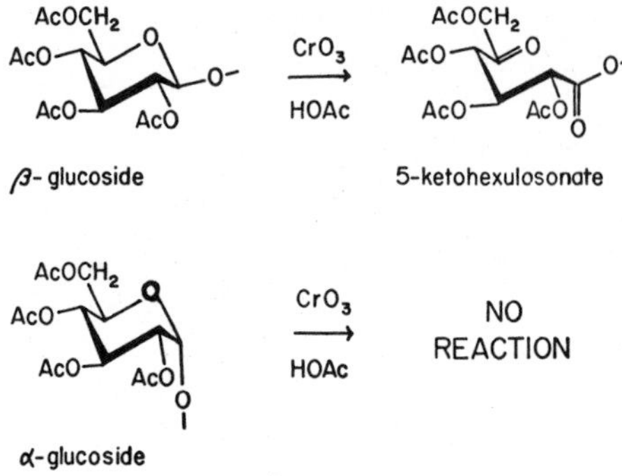

AcOCH₂ ... AcO ... AcO ... AcO ... O⁻

β- glucoside

CrO₃ / HOAc →

AcOCH₂ ... AcO ... AcO ... AcO ... O⁻ ... O

5-ketohexulosonate

AcOCH₂ ... AcO ... AcO ... AcO ... O

α-glucoside

CrO₃ / HOAc →

NO REACTION

Figure 11. Differential oxidation of axial and equatoral glycosidic linkages by chromium trioxide (Reproduced with permission from Ref. 248.)

chromium trioxide in acetic acid, in which condition axial glycosidic linkages were not oxidized (see Fig. 11). The possibility of using this differential reactivity of axial and equatorial glycosidic linkages for the analysis of anomeric configuration of glycosides in lipopolysaccharide has been studied by Hoffman et al.[247] and in various glycosphingolipids by Laine and Renkonen.[248,249] However, the α-galactofuranosyl residue was oxidized much more readily than the β-galactofuranosyl residue. Therefore, the oxidizability of the anomeric linkage in furanoside by chromium trioxide is quite different from that in pyranosides.[250] The method was applied to the analysis of the anomeric linkage of galactofuranoside-containing glycolipid of thermophilic bacteria[250] and to the analysis of mannose-containing glycolipid of freshwater bivalves (corbicula).[251] The glycolipids (100–300 μg), together with myoinositol as an internal standard, were acetylated in 0.1 ml of pyridine–acetic anhydride (1 : 1, vol./vol.) at 100°C for 15 min. An aliquot (50–100 μg) of the acetylated samples was evaporated to dryness, 0.5 ml of acetic acid and 50 mg of CrO_3 were added, and the mixture was treated for 15 min at 40°C in an ultrasonic bath. The reaction mixture was diluted with 1 ml of water and extracted with 6 ml of chloroform–methanol (2 : 1, vol./vol.) and then twice extracted with 2 ml of chloroform. The monosaccharide composition in the original and in the oxidized sample was analyzed by acetolysis followed by hydrolysis according to the method of Yang and Hakomori.[184] However, careful control is necessary to evaluate the results of their analysis, since not all the β-glycosides were degraded and not all the α-glycosides were resistant to oxidative degradation (personal communication from Dr. T. Hori, and our own experience).

1.3.8. Direct-Probe Mass Spectrometry of Glycolipids

The first complete mass spectra including molecular ions of trimethylsilylated cerebroside and ceramides were presented in 1970 by Karlsson.[76] Consequently, glycolipids with more than one carbohydrate have been analyzed as TMS or acetyl derivative.[252,253] Sweeley and Dawson described mass spectra of trimethylsilylated globoside and GM_1 ganglioside.[254,255] In recent studies, however, fully methylated derivatives have been used frequently, rather than acetylated, trimethylsilylated, or trifluoroacetylated derivatives, since methylated derivatives are more stable and the mass increase is only 14 units for each point of substitution compared with 42 for acetyl, 72 for TMS, and 96 for trifluoroacetyl derivatives.[256,259] Permethylated glycolipids, prepared by the Hakomori methylation,[209] have methyl groups not only at all alcoholic hydroxyls and sialic acid carboxyls, but also at amido nitrogens of amino sugars and of sialic acid.[237,260] Such derivatives are stable and volatile and show good mass spectra. In addition, a second derivative was obtained by reduction with LiAlH$_4$, converting all amide groups to the corresponding amines, thus converting all N-methyl-acetamido groups of hexosamines and sialic acids to N-methyl-N-ethyl groups as shown below and all carboxyl methyl ester to primary alcohol that could then be further trimethylsilylated.[256–259]

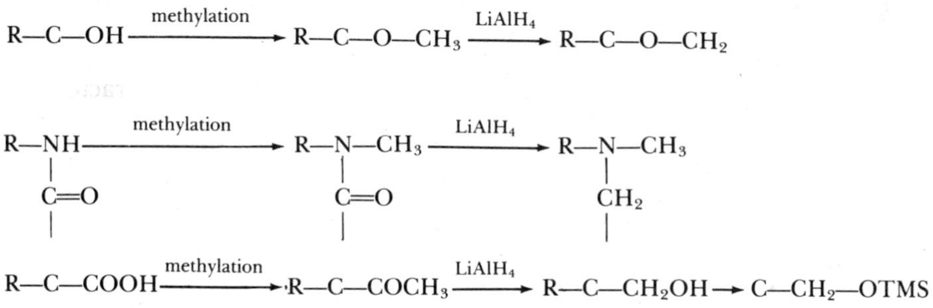

$$R{-}C{-}OH \xrightarrow{\text{methylation}} R{-}C{-}O{-}CH_3 \xrightarrow{\text{LiAlH}_4} R{-}C{-}O{-}CH_2$$

$$
\begin{array}{ccc}
R{-}NH & R{-}N{-}CH_3 & R{-}N{-}CH_3 \\
| & | & | \\
C{=}O & C{=}O & CH_2 \\
| & | & |
\end{array}
$$

methylation ⟶ ; LiAlH$_4$ ⟶

$$R{-}C{-}COOH \xrightarrow{\text{methylation}} R{-}C{-}COCH_3 \xrightarrow{\text{LiAlH}_4} R{-}C{-}CH_2OH \rightarrow C{-}CH_2{-}OTMS$$

The derivative thus prepared is highly suitable for mass spectrometric analysis, and especially suitable for gangliosides. Some gangliosides do not give ions for sialosyl residues unless reduced and trimethylsilylated.[261]

Forssman hapten glycolipid, GM$_3$ ganglioside (hematoside) with N-acetylneuraminic acid[257,259] and with N-glycolylneuraminic acid,[261] disialosylhematoside (GD$_3$ ganglioside),[262] disialoganglioside of brain,[258] and blood group H, A, and Lewis glycolipids of dog and human intestines[256,263,264] have been analyzed by this procedure.

The method was also applied to structural determination of blood group glycolipids with branched structure. The method combined with enzymatic degradation is successfully applied for determining the branching point of two "type 2" chains[415] carrying H determinants in H$_3$ glycolipid.[115,120] In this study, the permethylated H$_3$ glycolipid was reduced in sodium bis-(2-methoxy-ethoxy)aluminum hydride in benzene and separated the TLC before analysis by mass spectrometry.

Ledeen et al.[265] have analyzed various permethylated glycolipids by mass spectrometry without reduction and have obtained consistent reasonable mass spectra except those for sialic acid. A similar approach was used for structural studies of blood group B and H glycolipids by Hanfland and Egge,[266,267] and 11 glycosphingolipids including a novel fucoganglioside (see Section 1.3.9) of hog adipose tissue have been studied with this method by Ohashi and

Yamakawa.[268] More recently, the method has been extensively utilized not only for structural analysis of various new types of human erythrocyte gangliosides,[185,269,270] but also for analysis of oligosaccharides derived from glycolipids[232,269] and from glycoproteins.[271,272] Essential mass numbers derived from permethylated or permethylated–reduced glycolipids are useful for diagnosis of glycolipid structure and are listed in Table XVII.

Recently, Oshima *et al.*[273] determined the chemical ionization mass spectra of trimethylsilylated glucosylsphingosine, ceramides, and glucosylceramides in isobutane, methane, and ammonia as reactant gases. Eleven molecular species of ceramides and cerebrosides were separated. Each was characterized

Table XVII. Assignment for Fragment Ions Derived from Methylated or Methylated–Reduced Glycolipids

I.   Fragments from methylated–reduced glycosphingolipids
     A.   Ceramide fragments
         $CH_3(CH_2)_{12}CH{=}CH\text{-}CH\text{-}CH\text{-}CH_2\text{-}O\text{-}Glc\text{-}$

	Fatty acid	*m/e*		− MeOH
A:	C_{24}	851	(850 + 1)	819
	C_{22}	823	(822 + 1)	791
	C_{20}	795	(794 + 1)	763
	C_{18}	767	(766 + 1)	734
	C_{16}	739	(738 + 1)	706
B:	C_{24}	646		614
	C_{22}	618		586
	C_{20}	590		558
	C_{18}	562		530
	C_{16}	534		502
C:	C_{24}	392	(393 − 1)	
	C_{22}	364	(365 − 1)	
	C_{20}	336	(337 − 1)	
	C_{18}	308	(309 − 1)	
	C_{16}	280	(281 − 1)	

Table XVII. (Continued)

B. Sugar fragments

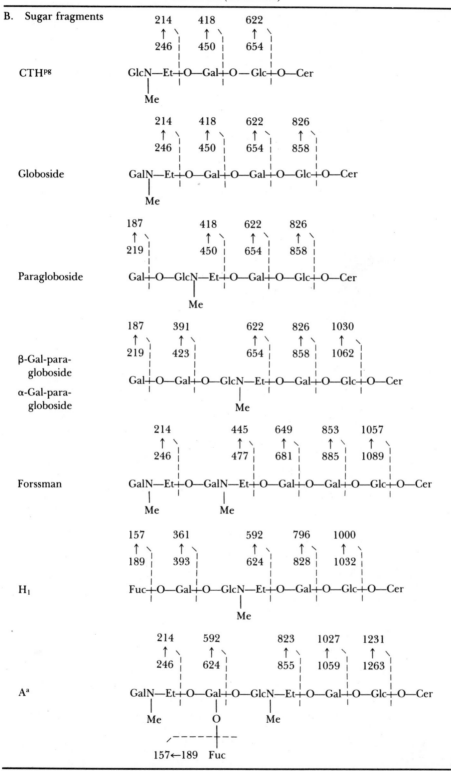

CTH[pg]

```
        214          418          622
         ↑ \          ↑ \          ↑ \
        246 |        450 |        654 |
                                            |
GlcN—Et+O—Gal+O—Glc+O—Cer
  |
 Me
```

Globoside

```
        214          418          622          826
         ↑ \          ↑ \          ↑ \          ↑ \
        246 |        450 |        654 |        858 |
                                                        
GalN—Et+O—Gal+O—Gal+O—Glc+O—Cer
  |
 Me
```

Paragloboside

```
 187                 418          622          826
  ↑ \                 ↑ \          ↑ \          ↑ \
 219 |               450 |        654 |        858 |

Gal+O—GlcN—Et+O—Gal+O—Glc+O—Cer
              |
             Me
```

β-Gal-para-globoside
α-Gal-para-globoside

```
 187      391                 622          826         1030
  ↑ \      ↑ \                 ↑ \          ↑ \          ↑ \
 219 |    423 |               654 |        858 |        1062 |

Gal+O—Gal+O—GlcN—Et+O—Gal+O—Glc+O—Cer
                       |
                      Me
```

Forssman

```
        214                 445          649          853         1057
         ↑ \                 ↑ \          ↑ \          ↑ \          ↑ \
        246 |               477 |        681 |        885 |        1089 |

GalN—Et+O—GalN—Et+O—Gal+O—Gal+O—Glc+O—Cer
  |              |
 Me             Me
```

H₁

```
 157      361                 592          796         1000
  ↑ \      ↑ \                 ↑ \          ↑ \          ↑ \
 189 |    393 |               624 |        828 |        1032 |

Fuc+O—Gal+O—GlcN—Et+O—Gal+O—Glc+O—Cer
                     |
                    Me
```

A[a]

```
        214          592                 823         1027         1231
         ↑ \          ↑ \                 ↑ \          ↑ \          ↑ \
        246 |        624 |               855 |        1059 |        1263 |

GalN—Et+O—Gal+O—GlcN—Et+O—Gal+O—Glc+O—Cer
  |          |              |
 Me          O             Me
         /------+---
      157←189  Fuc
```

(continued)

Table XVII. (Continued)

II. Fragments from methylated glycosphingolipids
   A. Ceramide fragments
      Ch₃(CH₂)₁₂CH=CH-CH-CH-CH₂-O-Glc-O-

$$CH_3(CH_2)_{12}CH=CH-CH+CH-CH_2+O-Glc+O-$$

with labels D, B, A above, MeO below first CH, N—Me / C=O / R below, and ......C...... at bottom.

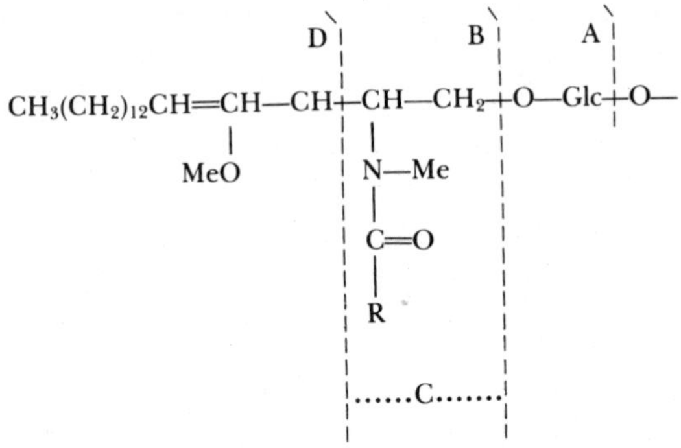

	Fatty acid	m/e		− MeOH
A:	C₂₄	865	(864 + 1)	833
	C₂₂	837	(836 + 1)	805
	C₂₀	809	(808 + 1)	777
	C₁₈	781	(780 + 1)	748
	C₁₆	753	(752 + 1)	720
B:	C₂₄	660*		628
	C₂₂	632		600
	C₂₀	604		572
	C₁₈	576		544
	C₁₆	548		516

* + 2 (662, C24): Acquisition of two hydrogens by the olefin-
            containing ceramide during fragmentation
* − 2 (658, C24:1): Unsaturated fatty acid

C:	C₂₄	406*	(407 − 1)
	C₂₂	378	(379 − 1)
	C₂₀	350	(351 − 1)
	C₁₈	322	(323 − 1)
	C₁₀	294	(295 − 1)

* − 2 (404, C24:1, 407-1-2): Unsaturated fatty acid

Sugar fragments

187
↑
219

CMH   Gal+O—Cer

157
↑
189

Fuc+O—Cer

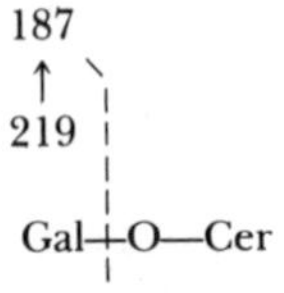

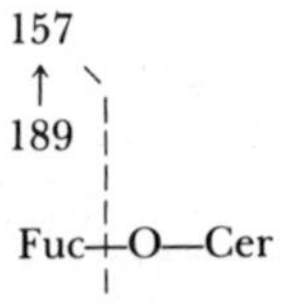

Table XVII. (Continued)

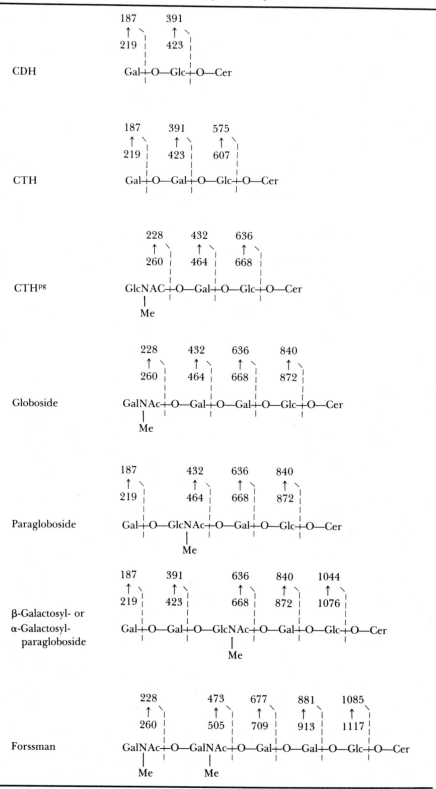

CDH

```
187        391
 ↑  \       ↑  \
219  |     423  |
     |         |
Gal─┼─O──Glc─┼─O──Cer
     |         |
```

CTH

```
187        391        575
 ↑  \       ↑  \       ↑  \
219  |     423  |     607  |
     |         |         |
Gal─┼─O──Gal─┼─O──Glc─┼─O──Cer
     |         |         |
```

CTH^pg

```
   228        432        636
    ↑  \       ↑  \       ↑  \
   260  |     464  |     668  |
        |         |         |
GlcNAC─┼─O──Gal─┼─O──Glc─┼─O──Cer
   |    |         |         |
   Me
```

Globoside

```
   228        432        636        840
    ↑  \       ↑  \       ↑  \       ↑  \
   260  |     464  |     668  |     872  |
        |         |         |         |
GalNAc─┼─O──Gal─┼─O──Gal─┼─O──Glc─┼─O──Cer
   |    |         |         |         |
   Me
```

Paragloboside

```
187           432        636        840
 ↑  \          ↑  \       ↑  \       ↑  \
219  |        464  |     668  |     872  |
     |            |         |         |
Gal─┼─O──GlcNAc─┼─O──Gal─┼─O──Glc─┼─O──Cer
     |      |     |         |         |
            Me
```

β-Galactosyl- or
α-Galactosyl-
paragloboside

```
187        391           636        840       1044
 ↑  \       ↑  \           ↑  \       ↑  \       ↑  \
219  |     423  |         668  |     872  |    1076  |
     |         |             |         |         |
Gal─┼─O──Gal─┼─O──GlcNAc─┼─O──Gal─┼─O──Glc─┼─O──Cer
     |         |      |     |         |         |
                      Me
```

Forssman

```
   228           473        677        881       1085
    ↑  \          ↑  \       ↑  \       ↑  \       ↑  \
   260  |        505  |     709  |     913  |    1117  |
        |            |         |         |         |
GalNAc─┼─O──GalNAc─┼─O──Gal─┼─O──Gal─┼─O──Glc─┼─O──Cer
   |    |      |     |         |         |         |
   Me         Me
```

(continued)

Table XVII. (Continued)

H₁

$$157 \quad 361 \quad\quad 606 \quad 810 \quad 1014$$
$$\uparrow\ \diagdown\quad \uparrow\ \diagdown\quad\quad \uparrow\ \diagdown\quad \uparrow\ \diagdown\quad \uparrow\ \diagdown$$
$$189\ |\quad 393\ |\quad\quad 638\ |\quad 842\ |\quad 1046\ |$$

$$\text{Fuc} + \text{O} - \text{Gal} + \text{O} - \text{GlcNAc} + \text{O} - \text{Gal} + \text{O} - \text{Glc} + \text{O} - \text{Cer}$$
$$\qquad\qquad\qquad\qquad\ \ |$$
$$\qquad\qquad\qquad\qquad\ \ \text{Me}$$

Aᵃ

$$228 \quad 606 \quad\quad 851 \quad 1055 \quad 1259$$
$$\uparrow\ \diagdown\quad \uparrow\ \diagdown\quad\quad \uparrow\ \diagdown\quad \uparrow\ \diagdown\quad \uparrow\ \diagdown$$
$$260\ |\quad 638\ |\quad\quad 883\ |\quad 1087\ |\quad 1291\ |$$

$$\text{GalNAc} + \text{O} - \text{Gal} + \text{O} - \text{GlcNAc} + \text{O} - \text{Gal} + \text{O} - \text{Glc} + \text{O} - \text{Cer}$$
$$\quad\ \ |\qquad\quad\ |\qquad\qquad\quad |$$
$$\quad\ \ \text{Me}\qquad\quad \text{O}\qquad\qquad\ \text{Me}$$
$$\qquad\qquad\qquad\quad |$$
$$\qquad\qquad\diagup\ -\ -\ +\ -\ -$$
$$157 \leftarrow 189 \quad \text{Fuc}$$

GM₃
(N-acetyl)

$$856 + 1$$
$$857$$

$$344 \quad\ 548 \quad\ 752$$
$$\uparrow\ \diagdown\quad \uparrow\ \diagdown\quad \uparrow\ \diagdown$$
$$376\ |\quad 580\ |\quad 784\ |$$
$$\qquad\qquad\qquad\qquad\qquad \text{CH}_3$$
$$\qquad\qquad\qquad\qquad\qquad |$$
$$\qquad\qquad\qquad\qquad\quad \text{N} - \text{CO} - (\text{CH}_2)_{22} - \text{CH}_3$$
$$\qquad\qquad\qquad\qquad\qquad |$$
$$\text{NeuNAc} + \text{O} - \text{Hex} + \text{O} - \text{Hex} + \text{O} - \text{CH}_2 - \text{CH} + \text{CH} - \text{CH} = \text{CH} - (\text{CH}_2)_{12} - \text{CH}_3$$
$$\quad\ \ |\qquad\qquad\qquad\qquad\qquad\qquad 660\ |\ \text{O} - \text{CH}_3$$
$$\quad\ \ \text{CH}_3\ |$$

GM₃
(N-glycolyl)

$$374 \quad\ 568 \quad\ 782 \quad\ 887$$
$$\uparrow\ \diagdown\quad \uparrow\ \diagdown\quad \uparrow\ \diagdown$$
$$406\ |\quad 610\ |\quad 814\ |$$
$$\qquad\qquad\qquad\qquad\qquad \text{CH}_3$$
$$\qquad\qquad\qquad\qquad\qquad |$$
$$\qquad\qquad\qquad\qquad\quad \text{N} - \text{CO} - (\text{CH}_2)_{22} - \text{CH}_3$$
$$\qquad\qquad\qquad\qquad\qquad |$$
$$\text{NeuNGl} + \text{O} - \text{Hex} + \text{O} - \text{Hex} + \text{O} - \text{CH}_2 - \text{CH} + \text{CH} - \text{CH} = \text{CH} - (\text{CH}_2)_{12} - \text{CH}_3$$
$$\quad\ \ |\qquad\qquad\qquad\qquad\qquad\qquad 660\ |\ \text{O} - \text{CH}_3$$
$$\quad\ \ \text{CH}_3\ |$$

GM₁

$$187 \quad\ 432 \quad\ 1045 \quad\ 1201$$
$$\uparrow\ \diagdown\quad \uparrow\ \diagdown\quad\quad\quad \uparrow\ \diagdown$$
$$219\ |\quad 464\ |\qquad\quad 1233\ |$$

$$\text{Hex} + \text{O} - \text{HexN} + \text{O} - \text{Hex} + \text{O} - \text{O} + \text{Cer}$$
$$\qquad\qquad\qquad\qquad\quad |$$
$$\qquad\qquad\qquad\qquad\quad \text{O}$$
$$\qquad\qquad\qquad\qquad\quad |$$
$$\quad\ -\ -\ -\ -\ +\ -\ -$$
$$344 \leftarrow 376 \quad \text{NeuNAc}$$
$$\qquad\qquad\qquad\quad |$$
$$\qquad\qquad\qquad\quad \text{CH}_3$$

Table XVII. (Continued)

GD$_{1a}$

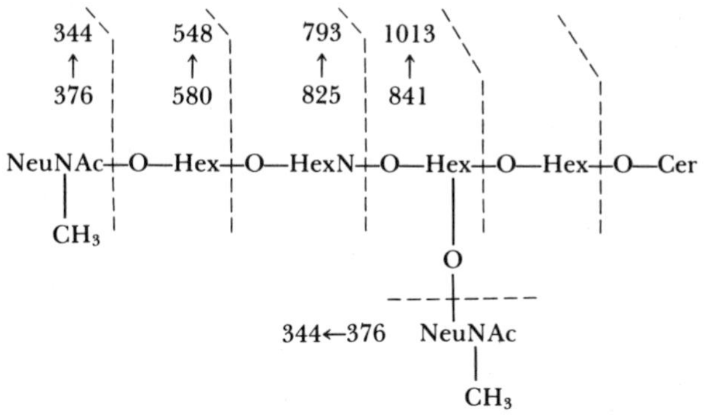

$$344 \quad 548 \quad 793 \quad 1013$$
$$\uparrow \quad\quad \uparrow \quad\quad \uparrow \quad\quad \uparrow$$
$$376 \quad 580 \quad 825 \quad 841$$

NeuNAc—O—Hex—O—HexN—O—Hex—O—Hex—O—Cer

CH$_3$

O

344←376   NeuNAc

CH$_3$

Fuco-GM$_1$

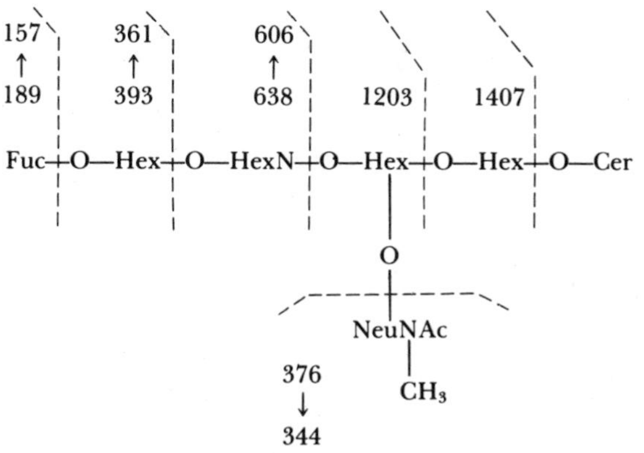

$$157 \quad 361 \quad 606$$
$$\uparrow \quad\quad \uparrow \quad\quad \uparrow$$
$$189 \quad 393 \quad 638 \quad 1203 \quad 1407$$

Fuc—O—Hex—O—HexN—O—Hex—O—Hex—O—Cer

O

NeuNAc

376   CH$_3$
↓
344

^{a} Fa, fatty acid.

by a characteristic quasi-molecular ion (QM$^+$, M$^+$ + 1), molecular ion minus one or two trimethylsilanol (MH-90, MH-90-90). Similar patterns were obtained irrespective of reagent gas. Thus, the method is extremely useful to determine the precise composition of molecular species (i.e., combination of sugar, fatty acid, and sphingosines).

Ando *et al.*[274] determined the chemical ionization mass spectra of oligosaccharides derived from ceramide trihexoside, globoside, paragloboside, Forssman glycolipid, asialo GM$_2$ (ganglio-*N*-triosylceramide), and asialo GM$_1$ (ganglio-*N*-tetraosylceramide). The spectra were all characterized by QM$^+$ or [M-59]$^+$ or both in all of the peracetylated oligosaccharides. The chemical ionization spectra with ammonia gas were characterized by [MH]$^+$ and [M + MH$_4$]$^+$. It is, however, unlikely that chemical ionization is as useful as electron impact (EI) procedure for the determination of the sugar sequence of whole glycosphingolipids.

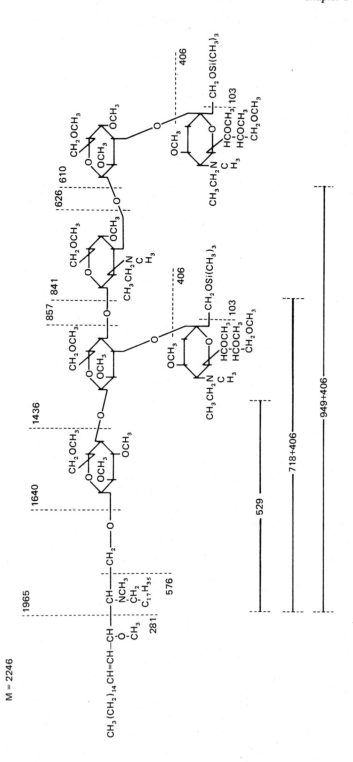

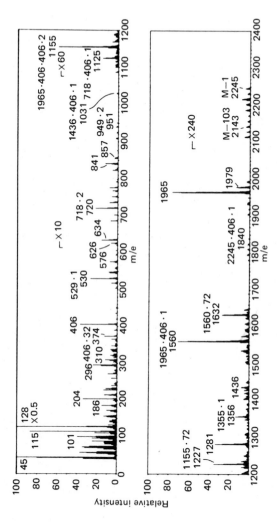

Figure 12. Direct-probe mass spectra of various permethylated or permethylated–reduced glycolipids and of oligosaccharides derived therefrom. A, Chemical formula of the TMS derivative of permethylated–reduced disialoganglioside (GD$_{1a}$) of brain. Note that all *N*-acetyl groups were converted to *N*-methylethyl groups and the alcohol group of the reduced sialic acid was trimethylsilylated. The mass spectrum of the TMS derivatives of methylated and reduced GD$_{1b}$ ganglioside is shown in the lower panel. The conditions of the analyses were: electron energy, 30 eV; acceleration voltage, 3.2 kV; trapped current, 500 mA; probe temperature, 320°C; ion source temperature, 290°C. B, Chemical formula and mass spectrum of permethylated disialoganglioside (GD$_{1a}$) of brain. The conditions of the analyses were: electron energy, 35 eV; acceleration voltage, 5 kV; trapped current, 500 mA; probe temperature, 350°C; ion source temperature, 310°C. The formula for the interpretation is shown in the figure with indication of primary fragment. C, Mass spectra of permethylated–reduced fucoganglioside (G9) of human erythrocyte membranes. Permethylated fucoganglioside was aluminum hydride reduced and trimethylsilylated according to the procedure of Karlsson *et al.,*[259] as modified by Watanabe *et al.*[120] D, Chemical formula and mass spectrum of oligosaccharide liberated from the fucoganglioside by endo-β-galactosidase of *E. freundii* according to "condition 1" described by Fukuda *et al.*[232] The oligosaccharide was reduced with sodium borohydride, permethylated, and purified on TLC. The conditions for mass spectrometry were as follows: electron energy, 30 V; ion energy programmed from +3.5 V; extractor, +8.0 V; lens, 35 V; emission 0.5 mA; electron multiplier, 2220 V; sensitivity, 10^{-7} A/V (Reproduced with permission from the authors and publishers.)

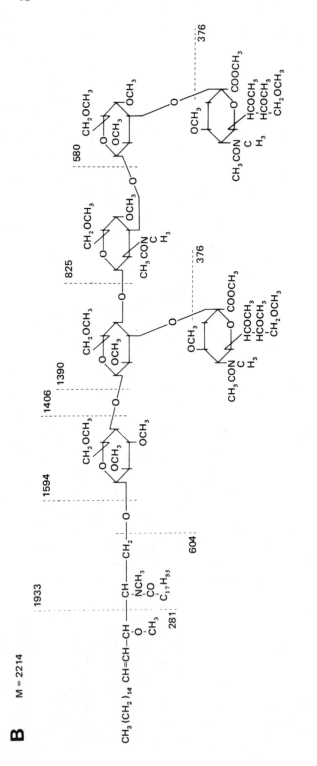

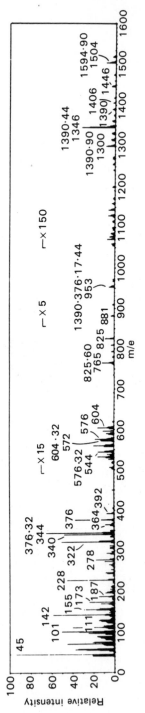

Figure 12. *(continued)*

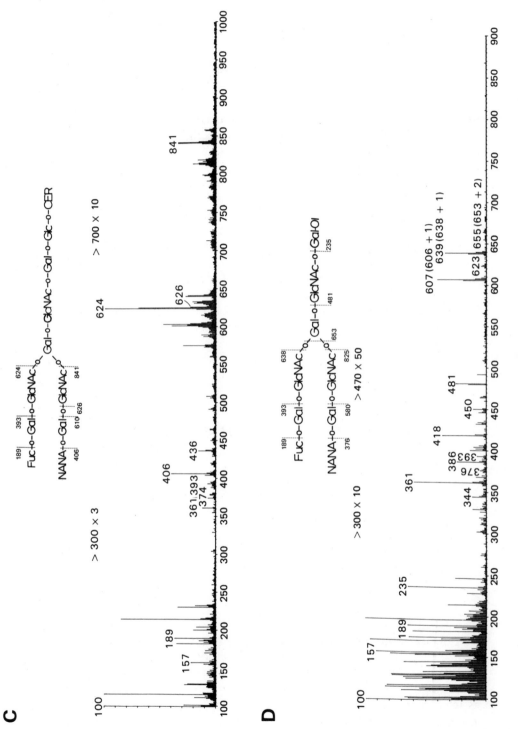

Figure 12. (*continued*)

1.3.8.1. Direct-Probe Mass Spectra of Methylated Glycolipid Mixtures with Selected Ion Monitoring and with Programmed Temperature Increase

Direct-probe mass spectrometry of permethylated or permethy-lated–reduced (LiAlH$_4$) derivatives as described above can be performed for a mixture of glycolipids if the sample placed in a probe cuvette is heated slowly (1–5°C/min, from 150 to 350°C) and the mass numbers important for diagnosis of glycolipid structures are monitored with the change of temperature. Thus, a fractional evaporation of glycolipid components is effected according to their number of sugar residues. A difference in fatty acid chain length of eight carbon atoms gave a separation effect similar to that given by a difference of one sugar residue. Thus, separation between hexaglycosyl-ceramide (peak at *m/z* 1647) and the trihexosylceramide (*m/z* 1066) appeared to be complete (see Fig. 13). A complete or partial separation of compounds during the fractional evaporation was useful for resolving ambiguity in the interpretation of fragments with identical mass from different compounds. The method was applied for analysis of neutral glycolipids obtained from the small intestine of one single rabbit[275] and for analysis of neutral glycolipids of cat small intestine.[276] In these analyses, however, the small fragment that is most useful for identification of glycolipid component was the so-called "F" ion that is derived from the cleavage between C-2 and C-3 of sphingosine:

$$
\begin{array}{c}
\text{F} \qquad\qquad \text{OMe} \\
\text{sugar—O—CH}_2\text{—CH—CH—CH—C}_{14}\text{H}_{29} \\
\text{CH}_3\text{—N} \\
\text{CH}_2 \\
\text{CH}_3\text{—OCH} \\
(\text{CH}_2)_n \\
\text{CH}_3
\end{array}
$$

The F ion for ceramide mono, di-, and trihexoside and B-active hexasaccharide was 658, 862, 1066, and 1646, respectively, if the fatty acid is α-hydroxy C$_{24}$. For identification with this method, it is essential to use a mass spectrometer with the capability to analyze a mass range up to 2000 or preferably higher. Nine neutral glycolipids from rabbit intestine (Fig. 13) as well as from cat intestine have been analyzed.[275,276] For further information on applications of mass spectrometry with both electron impact and chemical ionization procedures, readers should refer to a recent review by McNeil *et al.*[640]

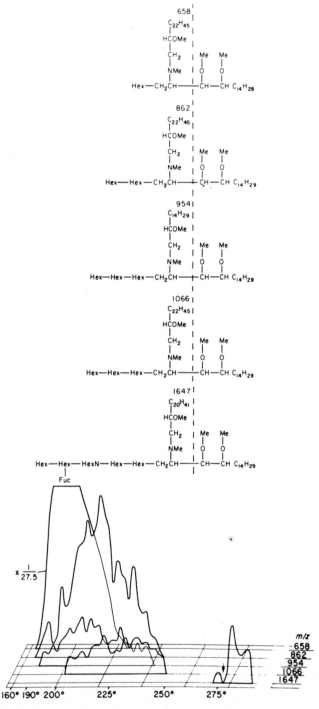

Figure 13. Selective ion recording of the neutral glycolipids from rabbit small intestine and simplified formulas for interpretation. In this analysis, 100 μg of the permethylated–reduced glycolipid mixture was used. The curves represent relative intensity of selective ions as a function of probe temperature. The direct-probe cuvette was heated 5°C/min from 160 to 190°C and 1°C/min from 190 to 300°C. Mass spectra were recorded every 2 min. Data from Breimer *et al.* [275] (Reproduced courtesy of Dr. K.-A. Karlsson, with permission from Heyden and Sons, Ltd.)

1.3.9. Spectrometric Analysis of Glycosphingolipids: Infrared (IR), Nuclear Magnetic Resonance, and Electron Spin Resonance Spectra

1.3.9.1. Infrared Spectra

Glycosphingolipids are characterized by a few common IR absorption spectra: (1) the absorption bands for amide at 1650 and 1550 cm^{-1}; (2) strong absorption at 3600–3000 cm^{-1} due to OH-stretching of carbohydrate; (3) CH-stretching absorption at 2970–2850 cm^{-1} and at 1450 cm^{-1}; and (4) absorption at 980–970 cm^{-1} due to *trans* double bond (Fig. 14 a–e). Sulfatides are characterized by sulfate absorption at 1245 cm^{-1} (due to symmetric stretching vibration of SO$_2$) and at 820 cm^{-1} (due to equatorial substitution of C-O-H (Fig. 14f). If the sample is contaminated by glyceride and glycerolphospholipid, the spectra show an absorption at 1750 cm^{-1} due to the ester carbonyl. Ester-cerebroside and glycosyldiglycerides are characterized by a strong absorption at 1750 cm^{-1}. The IR spectra of glycosphingolipids have been used in various classical studies by Yamakawa's group.[34,51,72,73] The spectra are useful to detect (1) contamination of phospholipids, (2) sulfatides, and (3) degree of methylation. The permethylated glycolipids do not show an absorption band due to OH (2900–2800) or to NH-CO- (1550 cm^{-1}).[234]

1.3.9.2. Proton Magnetic Resonance Spectra*

The proton magnetic resonance (PMR) spectra of monosaccharides and oligosaccharides have been extensively studied; the assignment of spectra, the chemical shift value (expressed in δ ppm) or τ value (10 ppm–δppm) and the spin–spin coupling constant (cps or J in H$_z$) showed a characteristic value for each proton (H$_1$ to H$_6$) of pyranose or furanose ring structure (for reviews see Hall,[277] Lemieux and Stevens,[278] Inch,[279] Bathbone and Stephen,[280] and Rowe and Rowe.[281] These spectral signals vary depending on steric interactions. The following are rules for anomeric proton spectra (Lemieux's empirical rule): (1) The axial anomeric protons of β-glucosides and β-galactosides showed a higher coupling constant (6.5–9.0 H$_z$) than that of equatorial anomeric protons of α-glucosides or α-galactosides (1–4 H$_z$). (2) The equatorial anomeric protons (protons linked to an α-glycoside carbon) gave signals at lower fields than that of axial anomeric protons (protons linked to a β-glycoside carbon). Both the chemical shift and the coupling constant offer useful information for the status of anomeric protons (α or β) and ring conformation (C-1 or 1-C) of carbohydrates.[277–280]

As little as five years ago, information available from NMR of glycosphingolipids was limited by lack of sensitivity, resolution, and experience to

---

* This section was written by Steve Levery and S. Hakomori.

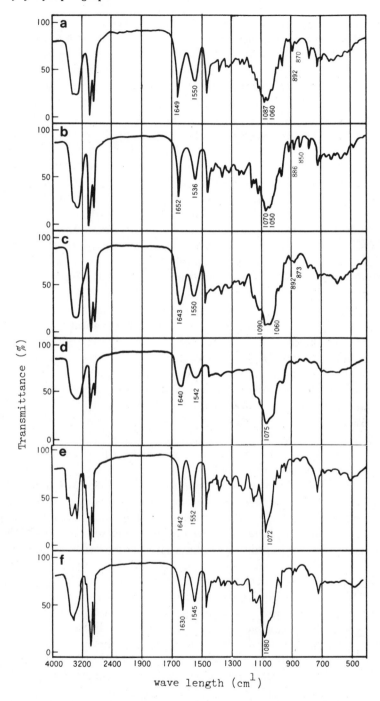

Figure 14. IR spectra of glycosphingolipids. a, Lactosylceramide; b, globotriaosylceramide (ceramide trisaccharide); c, globotetraosylceramide (globoside); d, blood group B-active ceramide hexasaccharide; e, cerebroside (kerasin); f, cerebroside (phrenosin). Data through the courtesy of Dr. S. Ando (with permission from the Japanese Biochemical Society).

assignment of anomeric protons of a few simple compounds, available in large quantity (1–10 mg). However, the development and proliferation of Fourier Transform (FT), high field superconducting magnets, and sophisticated data acquisition techniques has greatly increased the scope and sensitivity of the NMR method.[305] Since the simpler PMR investigations initiated by Kawanami,[282,283] the store of information has been greatly enriched through recent studies by Falk *et al.*[284–286] and Dabrowski *et al.*[288,631]

PMR spectra of glycosphingolipids vary greatly depending on a number of factors, such as the kind of derivative, solvent, temperature, condition, and type of instrument; therefore, results should be cautiously interpreted. Three approaches have been used: (1) the oligosaccharide liberated from glycolipid was analyzed in D_2O; (2) methylated or trimethylsilylated derivatives were analyzed in benzene, chloroform, or carbon tetrachloride; and (3) glycolipids were directly analyzed without derivatization and dissolved in pyridine-d_5 or in DMSO-d_6. The first method was ued only in the past; with high-resolution equipment (such as 360 M Hz), the anomeric proton was very well separated from other protons. The second or third approach is often performed in current studies.

a. The PMR Spectra of Oligosaccharides Released from Glycolipids. An oligosaccharide of a ceramide trihexoside (Gal-Gal-Glc ceramide) isolated from Nakahara–Fukuoka sarcoma was analyzed by Kawanami, and a signal for the equatorial anomeric proton (δ = 4.98 ppm, J = 3.5) was clearly detected.[283] This was the first clear evidence indicating the presence of α-galactosidic linkage in ceramide trihexoside, although the glycolipid was obtained from a special tumor tissue. The NMR spectra of the reduced oligosaccharides liberated from ceramide trihexoside and rat renal globoside were reinvestigated by Kawanami and Tsuji[283]; they demonstrated a very clear doublet at τ = 5.02, δ = 4.98 ppm with small coupling constant (J = 3.5 Hz) due to an equatorial anomeric proton in both oligosaccharides (see Fig. 15A).

b. PMR Spectra of Derivatized Glycosphingolipids. The PMR spectra of ceramide trihexoside and globoside of human erythrocytes were studied as TMS derivatives dissolved in carbon tetrachloride. The spectrum indicating the presence of an equatorial anomeric proton was detected.[77] The PMR spectrum of acetylated glycolipid was studied, and the anomeric proton was well defined.[180] The PMR spectrum of permethylated or permethylated–reduced glycolipids with mono- to octasaccharides were systematically studied by Falk *et al.*[284–286] In general, the signals for anomeric protons (H_1) are all associated with the δ ppm values of 5.2–4.5 and are well separated from other proton signals. As a typical example, the spectra of permethylated globoside and Forssman glycolipid are shown in Fig. 15G, H, and the spectra of permethylated H-active pentaglycosylceramide with type 2 and type 1 chains are shown in Fig. 15 I, J, respectively. The spectrum of permethylated globoside is, as expected, almost identical to that of globotriosylceramide; ie., δ = 4.20 ppm (J = 7.7 Hz) for β-Glc, δ = 4.34 ppm (J = 7.4 Hz) for β-Gal, δ = 4.95 ppm (J = 2.8 Hz) for α-Gal, and an additional doublet at δ = 4.63 ppm (J = 7.8 Hz) for the β-proton of GalNAc. The high chemical shift of the GalNAc β-

anomeric proton is probably caused by a deshielding from the acetylated amino group (Fig. 15G). Permethylated ceramide pentasaccharide (Forssman glycolipid) produced a spectrum sharing with that of globoside; δ = 4.20 ppm (J = 8.0 Hz) for the H_1 β-proton of Glc, δ = 4.32 ppm (J = 7.4 Hz) for the β-proton of Gal, δ = 4.95 ppm (J = 2.9 Hz) for the α-proton of Gal, and δ = 4.64 ppm (J = 7.0 Hz) for the β-proton of GalNAc. In addition, a doublet at δ = 5.0–5.1 ppm is overlapping with the proton signals associated with the olefinic double bond of sphingosine, which was eliminated by hydrogenation. After reduction with $LiAlH_4$, a distinctive doublet at δ = 5.02 ppm (J = 3.2 Hz) was observed (Fig. 15H). The spectrum of permethylated H_1-glycolipid [lacto-N-fucopentaosyl(IV)ceramide] was characterized by the presence of a sharp doublet at δ = 5.43 ppm with a small coupling constant (J = 4.1 Hz). All other spectra are approximately the same as that of paragloboside (Fig. 15I). The spectrum of an isomeric H-active glycolipid, lacto-N-fucopentaosyl(I)ceramide, with the type 1 carbohydrate chain (galactosyl linkage to GlcNAc is 1→3) showed a strikingly different chemical shift value of the β-anomeric proton after reduction, i.e., moved downfield about 0.8 ppm to δ = 5.13 ppm (J = 8.0 Hz) (Fig. 15J). This may be due to a deshielding effect from C-2 of $GlcNH_2$. Thus, the spectra after $LiAlH_4$ reduction are useful to distinguish type 1 from type 2 chain.

c. PMR Spectra of Intact Glycosphingolipids. The NMR spectra of intact ceramide trihexoside molecules accumulated in the kidney of a patient with Fabry's disease were determined in pyridine-d_5 at various temperatures.[289] The anomeric protons of β-D-glucopyranosyl, (attached to ceramide), an internal β-D-galactopyranosyl, and a terminal α-galactopyranosyl residue were characterized, respectively by δ = 4.74 ppm (J = 7.6 Hz); δ = 4.94 (J = 7.5 Hz); and δ = 5.45 ppm (J = 4.1 Hz) (see Fig. 15B).[289] With a similar approach, the PMR spectra of lacto-N-triaosylceramide and paragloboside were determined in pyridine-d_5 at high temperature; various anomeric protons were well separated[381] (see Fig. 15C).

Recently, very clear spectra of various glycolipids were demonstrated by Dabrowski and associates[592,631] with the deuterium-exchanged samples dissolved in DMSO-d_6 containing 2% D_2O using high-resolution proton NMR spectroscopy (360 MHz and 500 MHz) (see Fig. 15D–F). The δ-values and the coupling constants of each H_1 proton of various glycolipids are also shown in the legend for Fig. 15 D–F.

Recent work (since 1979) has included complete assignments of all ceramide and sugar resonances in some simple glycolipids, i.e., glucosylceramide[633] and galactosylceramide,[634] and most of the sugar protons of some other, i.e., globoside,[592] Lewis-active glycolipids,[635] and an I-active ceramide decasaccharide.[590] Useful spectra can now be obtained on as little as 100 μg of sample with 500-MHz instruments and signal averaging techniques. Dabrowski, Hanfland and Egge have recently extensively reviewed their own work in this field and interested readers should refer to their excellent discussion for more detailed information.[631]

The most recent investigations have made unambiguous assignments with

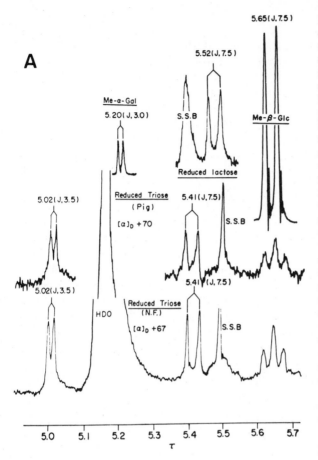

A

Me-α-Gal
5.20(J,3.0)

S.S.B

5.52(J,7.5)

5.65(J,7.5)

Me-β-Glc

Reduced lactose

Reduced Triose
(Pig)

[α]_D +70

5.02(J,3.5)

5.41(J,7.5)

S.S.B

5.02(J,3.5)

HDO

Reduced Triose
(N.F.)

[α]_D +67

5.41(J,7.5)

S.S.B

5.0   5.1   5.2   5.3   5.4   5.5   5.6   5.7
τ

Figure 15. PMR spectra of gly-colipids and oligosaccharides derived therefrom. (A) PMR spectra of the reduced oligosaccharide isolated from pig erythrocyte glycolipids and that from Naka-hara–Fukuoka sarcoma as compared with various oligosaccharides with known structure. These two reduced oligosaccharides showed a clear doublet at $\tau = 5.02$ with $J = 3.5$ Hz, indicating α-anomeric proton. (Data from Ref. 283.) (B) *Left:* Partial PMR spectra of galactosylceramide (CMH) and galactosylgalactosylglucosylceramide (CTH). The spectra were recorded at 220 MHz on 75 mg of CTH and 50 mg of CMH in 0.5–1 ml of pyridine-d$_5$ containing a trace of D$_2$O at 30°C with a sweep width of 500 Hz. Note that a clear doublet with $\delta = 5.45$ ppm ($J = 4.1$ Hz) is separated, indicating the presence of α-anomeric proton. *Right:* Partial PMR spectra of various glycolipids and derivatives. Galactosylceramide (CMH), hydrogenated galactosylceramide [CMH (H$_2$)], lactosylceramide (CDH), and CTH were recorded at 100 MHz in 0.5–1 ml of pyridine-d$_5$ at 30°C. Doublet e in CTH indicates the presence of α-anomeric proton. Reproduced from Clark *et al.*[289] (C) PMR spectra of lacto-*N*-triaosylceramide (a) and paragloboside (b). Glycolipid (10 mg) was dissolved in 0.3 ml of pyridine-d$_5$ containing 6 μl of D$_2$O and measured at 90°C with a Varian XL-100 spectrometer. Reproduced from Ando *et al.*[381] (D) PMR spectrum of globotriaosylceramide in DMSO-d$_6$ with 360 MHz operating frequency. The deuterium-exchanged samples were dissolved in DMSO-d$_6$ containing 2% D$_2$O. Sample concentration: 0.2%. The assignments for doublets a–c are as follows: a, $\delta = 4.81$ ppm ($J = 4.0$ Hz) for Galα1→4; b, $\delta = 4.28$ ppm ($J = 7.7$ Hz) for Galβ1→4; c, $\delta = 4.17$ ppm ($J = 8.1$ Hz) for Glcβ1→4. Reproduced from Dabrowski *et al.*[592] (E) PMR spectrum of globoside in DMSO-d$_6$ with the same conditions as above. The assignments for doublets a–d are as follows: a, $\delta = 4.83$ ppm ($J = 3.6$ Hz) for Galα1→4; b, $\delta = 4.54$ ppm ($J = 8.1$ Hz) for GalNAcβ1→3; c, $\delta = 4.28$ ppm ($J = 7.7$ Hz) for Galβ1→4, d, $\delta = 4.17$ ppm ($J = 8.1$ Hz) for Glcβ1→1 (namely, GalNAcβ1→3Galα1→4 Galβ1→4Glcβ1→1Cer). Reproduced from Dabrowski *et al.*[592] (F) PMR spectra of the following ceramide decasaccharide. The chemical shift value (in δ ppm) of each anomeric proton is shown under each assignment (for doublets a–e) and the coupling constant (in J Hz) is shown in parentheses. Reproduced from Hanfland *et al.*[590]

(c) 4.42   (8.2)

Galα1→3Galβ1→4GlcNAcβ1
   (a)   (d)   ↘
            §Gal β1→4GlcNAcβ1→3Galβ1→4Glcβ1→1Ceramide
         ↗   (d)   (b)   (e)
Galα1→3Galβ1→4GlcNAcβ1   4.30   4.67   4.17
   (a)   (d)   (b)   (7.3)   (8.5)   (7.7)
δ = 4.84   4.34   4.67
(J = 3.6)   (7.3)   (8.5)

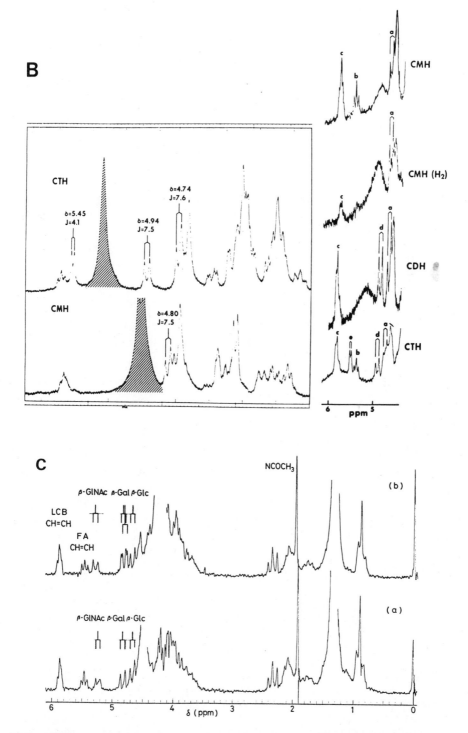

(G) *Top:* PMR spectrum of permethylated globo-*N*-tetraosylceramide (globoside), 10 mg in 0.5 ml of chloroform, 60 pulses at 40°C. The spectrum was taken in the 270-MHz NMR spectra on a Bruker WH-270 spectrometer, operating in the Fourier-transform mode. Chemical shift values were determined using the residual solvent peaks as internal standard, but are given as δ values relative to tetramethylsilane. *Bottom:* PMR spectrum of permethylated–reduced globo-*N*-tetrao-

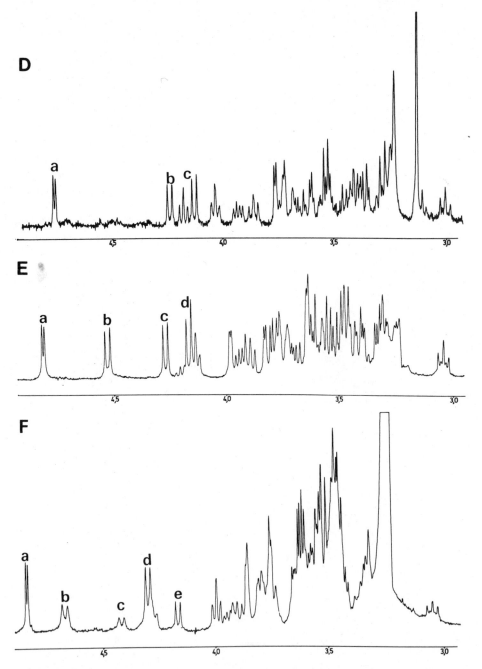

sylceramide with LiAlH₄, 6 mg in 0.5 ml of chloroform, 262 pulses at 25°C. The signal for the α-anomeric proton of the Gal residue shifted slightly downfield, whereas that for the β-anomeric proton of GalNAc shifted upfield; however, the signal for the β-anomeric proton of Gal or Glc was unchanged by reduction. The complex signals above 5 ppm due to the *trans*-double bond of sphingosine disappeared on reduction. (H) PMR spectra of permethylated Forssman glycolipid. *Top:* Spectrum of permethylated and Pt-hydrogenated Forssman glycolipid, 1.75 mg in 0.5 ml of chloroform, 560 pulses at 40°C. *Bottom:* Spectrum of permethylated Forssman glycolipid, catalytically hydrogenated with Pt, and further reduced with LiAlH₄, 1 mg in 0.5 ml of chloroform, 1869 pulses at 42°C. Signals for the α-anomeric proton of α-GalNAc and of α-Gal were

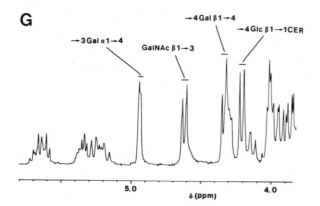

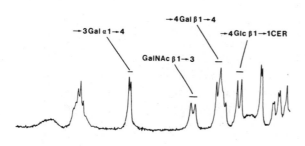

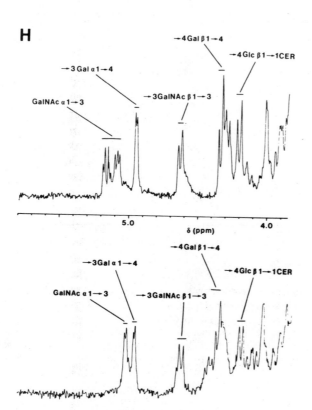

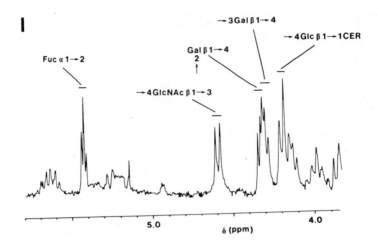

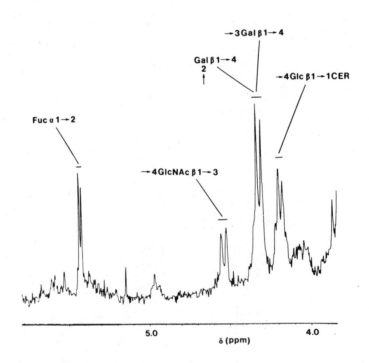

separated from other signals after LiAlH$_4$ reduction. (I)*Top:* PMR spectrum of permethylated H-active pentaglycosylceramide (H$_1$ glycolipid) with a type 2 carbohydrate chain (Fucα1→2Galβ1→4GlcNAcβ1 →3Galβ1→4Glcβ1→1Cer), 2 mg in 0.5 ml of chloroform, 730 pulses at 40°C. *Bottom:* PMR spectrum of permethylated–reduced H$_1$ glycolipid with a type 2 carbohydrate chain, 1 mg in 0.5 ml of chloroform, 1800 pulses at 40°C. (J)*Top:* PMR spectrum of permethylated H-active pentaglycosylceramide with a type 1 carbohydrate chain

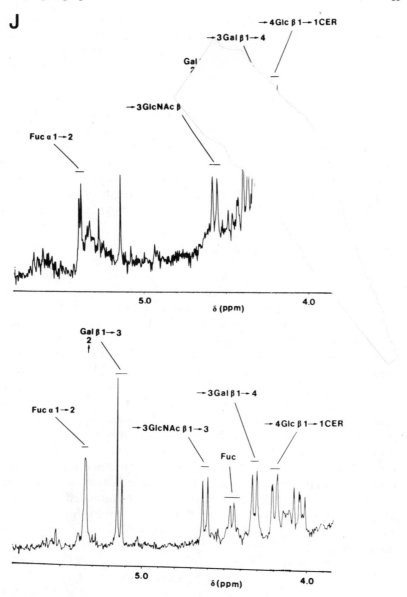

(Fucα1→2Galβ1→3GlcNAcβ1 →3Galβ1→4Glcβ1→1Cer), 2.5 mg in 0.5 ml of chloroform, 600 pulses at 40°C. *Bottom:* PMR spectrum of permethylated–reduced H-active pentaglycosylceramide with a type 1 carbohydrate chain, 2.5 mg in 0.5 ml of chloroform, 600 pulses at 40°C. The signals for the β-anomeric proton of galactose linked to *N*-acetylglucosamine shifted extensively downfield after LiAlH$_4$ reduction. Such a shift of the signal was not observed in type 2 chain.

extensive use of the spin-decoupling and the nuclear overhauser effect (NOE) experiments. Application of the most advanced data acquisition and transformation software already available may soon make it possible to perform complete structure determinations in one or two experiments. Although, up to now, a majority of studies have been on known structures, many more unknown structures will routinely be done by NMR in the future, aided by the older established techniques of enzyme degradation and GC/MS.

It should be pointed out that the major limitation of NMR analysis may be size rather than complexity *per se*. Because larger molecules tumble in a spinning NMR sample tube at a slower rate, and slower tumbling results in broadening of the resonance lines, the loss of resolution for individual peaks of a large molecule could become severe enough to make assignments too difficult; also line-broadening can result in a great loss of sensitivity.

1.3.9.3. Carbon-13 Nuclear Magnetic Resonance Spectra

The natural abundance of carbon-13 Fourier-Transform NMR ($[^{13}C]$-NMR) spectra offers an elaborate status of the description of each carbon atom such as the conformational status, interaction with the neighboring groups, involvement with hydrogen bonding, and the nature of metal binding, and other properties[290–292] (for a textbook, see Abraham and Loftus[643]. A much larger range of ^{13}C chemical shift as compared with ^1H chemical shift makes it easier to evaluate and identify each spectrum. The information of a spectrum is therefore suitable to highly complex molecules such as glycosphingolipids. Despite these merits, the applicability of the method is limited because a large amount of pure sample (300–1000 mg) is necessary to obtain reliable spectra. Therefore, the method has only been applied for the elaborate structural analysis of abundantly available compounds the primary structures of which were already established. The method is not suitable for analyzing the structures of unknown compounds that are available in small quantities. $[^{13}C]$-NMR data have been accumulated for monosaccharides,[293–295] oligosaccharides,[296–298] phospho- and sphingolipids[299] and for analysis of the linkages of oligosaccharides[300] and the conformation of the glycosidic bond.[301] More recently, a conformational structure and the status of hydrogen bonding of sialic acid were elaborated. Three hydrogen bonds are present to stabilize the conformation of sialic acid, as can be seen in Fig. 16.[302] Proceeding from these basic data, Sillerud *et al.*[303] and Koerner *et al.*[304] studied $[^{13}C]$-NMR spectra of cerebroside, and GM_1 ganglioside. The chemical shift of each carbon atom has been assigned. Interestingly, most of the oligosaccharide resonance occurs in the range from $\delta = 60$–105 ppm, while most of the ceramide carbons resonate $\delta = 10$–56 ppm. The GM_1 spectrum showed about 40 resonances from approximately 50 carbon types that are expected to yield distinct peaks. To determine the details of the interaction of GM_1 with metal ions, the influence of paramagnetic europium on the $[^{13}C]$-NMR spectrum was examined. The alteration of resonances (change in chemical shift, line-broadening) was computed and the results indicate that the oxygen atoms on C-2 and C-3 of the terminal galactose; C-12 and C-13

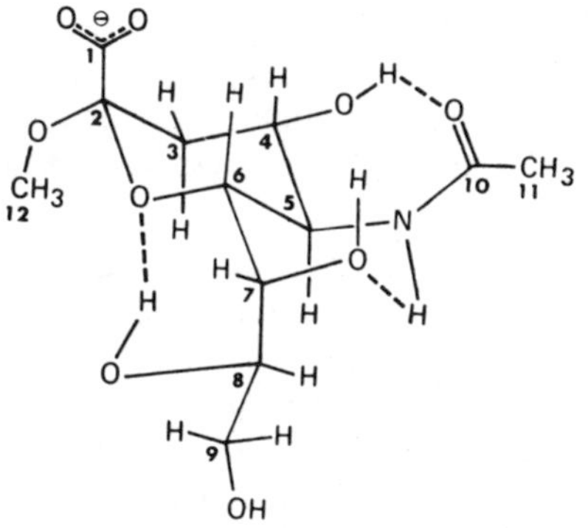

Figure 16. Structure of sialic acids as revealed by [^{13}C]-NMR spectrometry. [^{13}C]-NMR spectrometry indicates the presence of at least three hydrogen bonds. That between the C-4 hydroxy group and the carbonyl group of the *N*-acetyl residue (C-10), that between the C-7 hydroxy O and NH group, and that between the C-8 hydroxy group and the pyranose O atom are shown here only for the α-anomer. This conformation is independent of anomeric configuration. (Data from Ref. 302, courtesy of Drs. Czarniecki and Thornton.)

(*N*-acetylcarbonyl) of the penultimate GalNAc; and C-21 (carboxylate), C-30 (*N*-acetylcarbonyl) and C-28 (glyceryl side chain) of the α-sialic acid residue all participate in the strong binding of europium, as shown in Fig. 17. This is particularly significant in view of the fact that 2-*O*-methyl-α-NeuAc binds metal cations such as Ca^{2+} only very weakly; although the β-anomer is relatively a better ligand, it is not the natural isomer. Thus gangliosides would not bind Ca^{2+} very well if sialic acid were the only coordination moiety available.

Another PMR-[^{13}C]-NMR study by Jaques *et al.*[638] using similar techniques, established that *N*-glycolylneuraminic acid binds Ca^{2+} more strongly than *N*-acetylneuraminic acid due to participation of the hydroxyl of the *N*-glycolyl group, and confirmed that the glyceryl side chain is a necessary participant in binding by studying the effect of its removal via periodate oxidation on the binding constant. Because these experiments were done on the β-anomers, however, the physiological significance is somewhat lessened.

Most recently, a [^{13}C]-NMR study by Daman and Dill[639] on the binding of Mn^{2+} and Gd^{3+} to α-NeuAc showed that (1) Mn^{2+} and Gd^{3+} both bind significantly; (2) their binding sites are slightly different, and surprisingly; (3) the carboxyl group is not necessary for metal-ion binding.

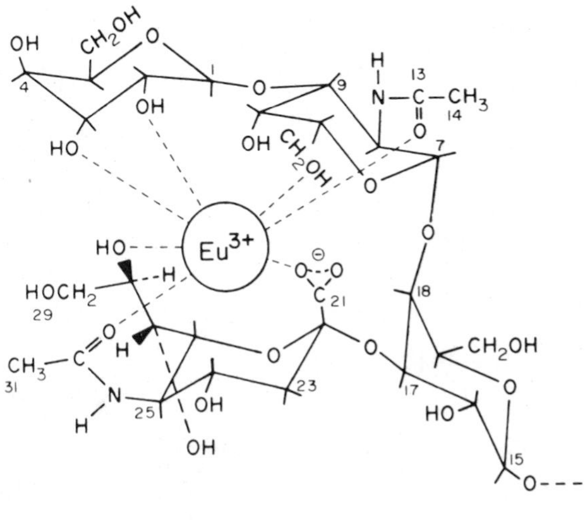

Figure 17. Conformational structure of a part of the carbohydrate chain of the GM$_1$ ganglioside–Eu complex. Proposed structure of the metal cation-binding site on GM$_1$ based on the spectral perturbations produced by paramagnetic europium in the [^{13}C]-NMR spectrum as described Sillerud *et al.*[303] (Courtesy of Dr. Sillerud.)

These and other studies on some sialic acid-containing natural substances seem to make it clear that there will be significant differences in results depending on which metal and which carbohydrate source is used and generalizations should be viewed with caution. The method calls for work with well-defined systems.

1.3.9.4. Electron Spin Resonance or Electron Paramagnetic Resonance Spectra

The electron spin resonance (ESR) spectrum with the spin-label technique has proven useful in membrane research, giving the basic dynamic status of the fluid bilayer of membranes, slow flip-flop,[360] and rapid lateral diffusion.[316,408] Sharom and Grant[307,308] have synthesized spin-labeled derivatives of two glycolipids: galactosylcerebroside with the spin-labeled steric acid and GM_1 ganglioside with a spin label at the terminal galactosyl residue (Fig. 18). The typical electron paramagnetic resonance (EPR) spectrum of spin-labeled ganglioside is shown in Fig. 19. These components were incorporated into liposomes, and their ESR spectra were studied. The mobility and behavior

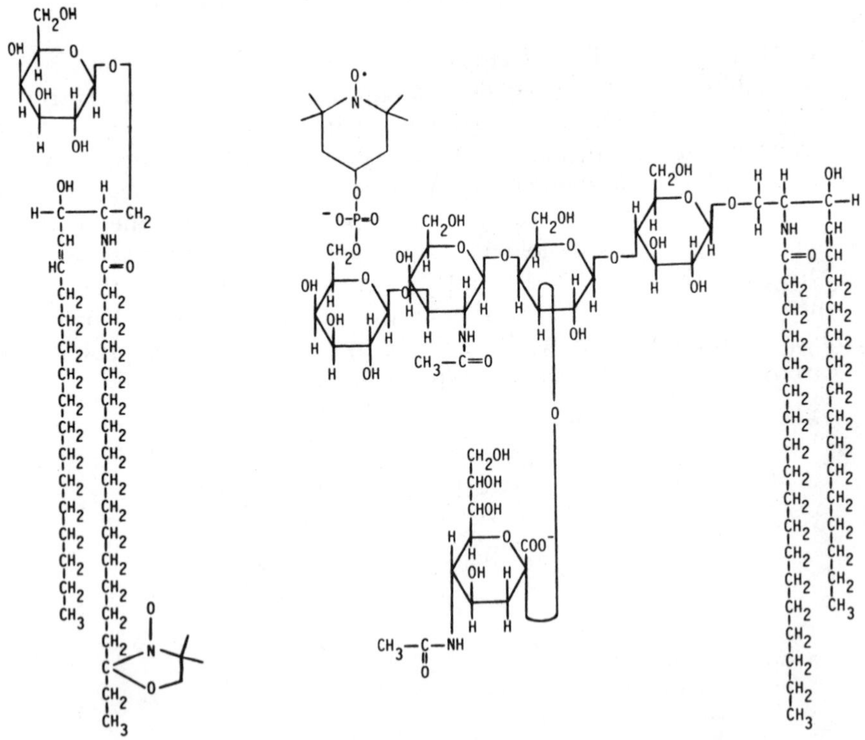

SPIN-LABELED
GALACTOSYL CERAMIDE

SPIN-LABELED GANGLIOSIDE

Figure 18. Structure of spin-labeled glycosphingolipids.

Figure 19. Typical EPR spectrum of spin-labeled GM$_1$ ganglioside with 0.45 spin-label per ganglioside at 3.3 mole% in egg lecithin bilayer. A spectral parameter related to the correlation time, τ_c, is used to express the effect on ganglioside head-group mobility, which is derived by $\tau_c = 6.5 \times 10^{-10} W_o[(h_o/h_{-1})^{1/2}-1]$. W_o is the linewidth of the midfield line, and h_o and h_{-1} are indicated in the figure. Since the W_o value is practically unchanged, the fraction $[(h_o/h_{-1})^{1/2}-1]$ is used to express the ganglioside mobility (see Chapter 4, Fig. 3). Data from Ref. 308, courtesy of Dr. C. W. M. Grant.

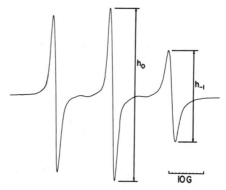

of galactosylcerebrosides in the lipid bilayer were similar to those of phosphatidylcholine: negligible flip-flop rate, a flexibility gradient in the acyl chains, and exclusion from phosphatidylserine. However, glycolipids seem to have the capacity to increase rigidity in lipid bilayers. Studies with spin-labeled ganglioside indicate that the head-group oligosaccharides are in rapid, random motion. This mobility of the head group significantly decreased with increasing bilayer concentration of unlabeled ganglioside. This is perhaps caused by a cooperative head-group interaction. Thus, it is most likely that glycosphingolipid may regulate the membrane fluidity and rigidity.

1.4. Structure of Glycosphingolipids

Exact three-dimensional structures of glycosphingolipids are not known, although a recent study on the X-ray diffraction pattern of crystalline cerebroside indicated that the axis of the galactopyranosyl plane was oriented almost perpendicular to the axis of two aliphatic chains of ceramide, forming a "shovel" shape[309] (see Fig. 20). Glycosphingolipids have been obtained in amorphous form except cerebroside; therefore, conformational analyses of glycolipids based on X-ray diffraction have been limited. However, according to their chemical structure, molecular models can be readily constructed based on an arbitrary choice of the conformation of carbohydrates. Figure 21 shows several models of glycosphingolipids including those with blood group A and H specificities.[310] The two aliphatic chains of ceramide form a "fork" or "tail" through which the glycolipid molecules are closely associated with membrane components. It is generally assumed that glycolipids are present in the outer leaflet of the lipid bilayer, although evidence for such organization is limited to erythrocyte membranes. The carbohydrate moiety of a glycolipid forms a hydrophilic "head" that protrudes from the lipid bilayer to the outer environment. Organization of glycolipids in cell-surface membranes is discussed in Chapter 4. In this section, however, the major variations of the chemical structures in the ceramide and carbohydrate moieties of glycosphingolipids are discussed.

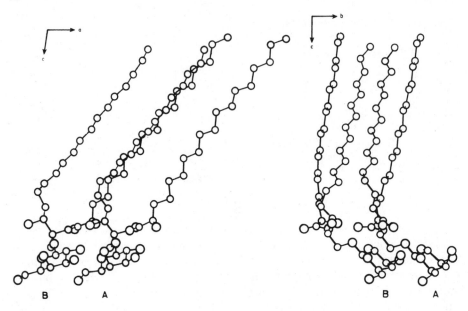

Figure 20. Conformation of two cerebroside molecules (A and B) viewed along two different axis (a and b). The data are based on X-ray crystallography of the single-crystal analysis of cerebroside [D-galactopyranosylβ1→1-*N*-(2-D-hydroxyoctadecanoyl)-D-dihydrosphingosine $(C_{42}H_{83}O_9N \cdot \frac{1}{2}C_2H_5OH)$]. The molecules are packed in a typical bilayer arrangement with adjacent double layers separated by ethanol molecules of crystallization. The planes of the sugar links are turned almost parallel to the layer interface, which gives the molecule a "shovel" shape. Together with the polar ceramide part, the galactose head group forms an intensive layer network of hydrogen bonds within the polar region of each layer. The chains tilt at an angle of 49° toward the polar boundary. (Data from Ref. 309, courtesy of Dr. I. Pascher.)

1.4.1. Structural Variation in Ceramides

A great number of molecular species result from various combinations of sphingosine analogs [long-chain bases (LCBs)] and fatty acids. As discussed in Section 1.3.2, the analytical methods for characterization of ceramides, LCBs, and fatty acids have been greatly advanced, and the molecular species of ceramides can be determined relatively easily through gas chromatography–mass spectrometry. More than 60 kinds of sphingosines are listed in the table presented in the review article by Karlsson[76] in 1970, although most of them are very minor components. Sphingosine and dihydrosphingosine are most widely distributed among animal tissue sphingolipids. C_{20}-sphingosine and C_{20}-dihydrosphingosine have been found as components of brain ganglioside since Proštenick and Majhofer-Oreščanin[25] and Stanaćev and Chargaff[26] described the presence in the ganglioside. Phytosphingosines constitute the major base component of higher plants,[21] protozoa,[311] yeasts[23,24] (also see the first footnote on page 10), and fungi.[22] Karlsson[312] and Michalec and Kolman[313] found C_{18}- and C_{20}-phytosphingosines in human kidney cerebrosides. Yang and Hakomori[184] described a colonic tumor glycolipid that had C_{18}-phytosphingosine as the major base component. Thus, phytosphingosine, despite its name, is widely distributed in animal sphingolipids as well. A few types of branched LCBs (e.g., $\Delta 4$-16-methyl-C_{17}-sphingosine) have been found in protista[314,315] and in bovine kidney and blood.[316] It was suggested that a branched sphingosine of bovine blood and tissue may be derived from their rumen protista.[316] The major classes of LCBs are shown in Table XVIII and their steric structures in Table XIX.

Two possible geometric isomers as to the double bond, *cis* or *trans,* and four stereoisomers as to the two asymetric carbons, C-2 and C-3, have been considered. The *trans* configuration of the double bond of sphingosine was determined by Ohno,[16] who identified *trans*-hexadecene(2,3)al as semicarbazone after the lead tetraacetate oxidation and *trans*-hexadecanoic acid after the permanganate oxidation of sphingosine. The result was supported by infrared spectra of sphingosine[317] and sphingomyelin[318] by an absorption at 10.3 μm, characteristic for a *trans* double bond. As to C-3 configuration relative to C-1, D- and L-isomers are possible, and as to C-2 configuration relative to C-3, *erythro-* and *threo*-diastereoisomers are possible, as follows:

C-1	CH_2OH	CH_2OH	CH_2OH	$CH_2—OH$
C-2	H—C—NH$_2$	H$_2$N—C—H	H$_2$N—C—H	H—C—NH$_2$
C-3	H—C—OH	HO—C—H	H—C—OH	HO—C—H
	R	R	R	R
	D	L	D	L
	erythro	*erythro*	*threo*	*threo*
	(2S,3R)	(2R,3S)	(2R,3R)	(2S,3R)

The natural dihydrosphingosine was identified as having D-*erythro*(2S,3R) configuration based on the melting point, and the optical rotation of the synthetic D-*erythro*-dihydrosphingosine was identical to that of the natural compound.[64,65,319,320] The *SR* nomenclature described above according to Cahn *et al.*[321] is unequivocal as compared to the conventional D–L and *threo–erythro* designations. In phytosphingosine, the stereoisomeric structure as to C-2, C-3, and C-4, was determined as D-*ribo*, or 2S,3R,4R, by Carter and Hendrickson[70] (see Tables XVIII and XIX). The chemical synthesis of sphingosine, ceramides, and glycosphingolipids is discussed in Section 1.5.

Sphingosines and long-chain aliphatic bases do not exist in cells and in tissues as the free bases, but are always bound to fatty acid, to carbohydrate, to phosphorylcholine, or to phosphorylinositol. A rare exception as to the form of ceramide has been reported: Kotchetkov *et al.*[322,323] found a ceramide

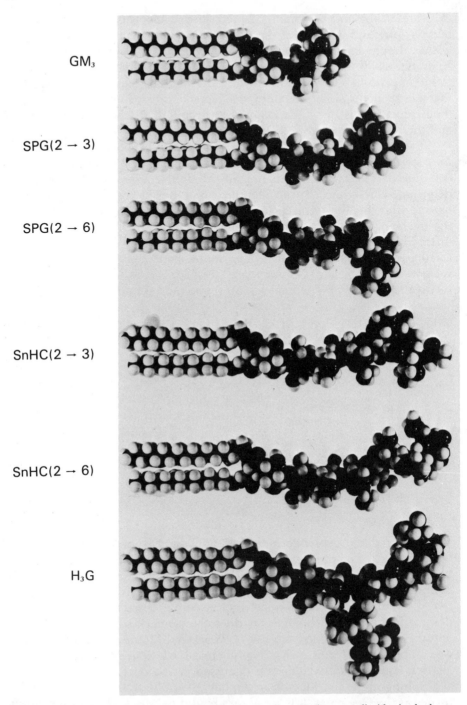

GM₃

SPG(2 → 3)

SPG(2 → 6)

SnHC(2 → 3)

SnHC(2 → 6)

H₃G

Figure 21. Molecular models of glycosphingolipids. *First line*—Various gangliosides in the lacto-*N*-glycosyl series: SPG(2→3), sialosylparagloboside with NeuAc2→3Gal structure; SPG(2→6), sialosylparagloboside with NeuAc2→6Gal structure; SnHC(2→3), sialosyllacto-*N-nor*hexaosyl-ceramide with NeuAc2→3Gal structure; SnHC(2→6), sialosyllacto-*N-nor*hexaosylceramide with

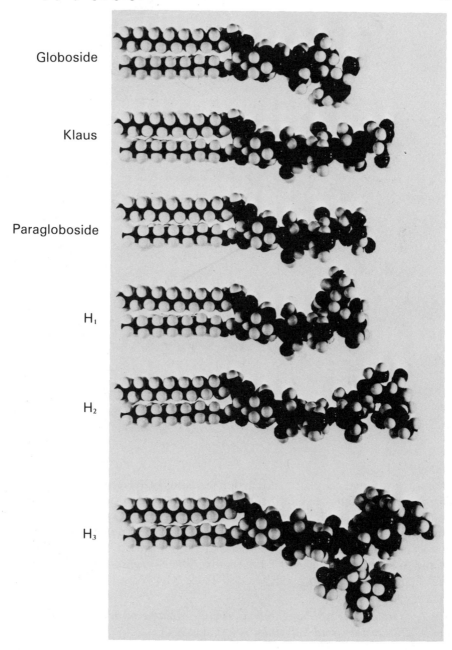

Globoside

Klaus

Paragloboside

H_1

H_2

H_3

NeuAc2→6Gal structure; H_3G, ganglioside G9 of human erythrocytes.[116] For the chemical formulas of these gangliosides, see Table XXII.[116,606] *Second line*—Some neutral glycolipids: Klaus, a β-galactosylparagloboside the structure of which was determined by Dr. Klaus Stellner[388]; H_1, H_2, H_3, and (next page) A^a, A^b, A^c, three variants of blood group determinants isolated from human erythrocytes.[120,310,407,596] (Models were prepared by Dr. Kiyohiro Watanabe.)

(continued)

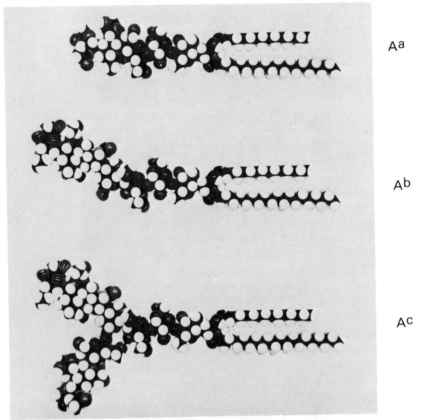

Figure 21.(*continued*)

bound to plasmal (aliphatic aldehyde) through 3-*O*-alkenyl (vinyl ether linkage) or to aliphatic alcohol through 3-*O*-alkyl. Cerebrosides having such 3-*O*-substituted ceremides should be less polar and run on thin-layer chromatography (TLC) as a fast-migrating component in less polar solvents. Galactosylcerebroside with 3-*O*-fatty acylceramide was reported[324,325] and was later found to be indistinguishable from regular "ester-cerebrosides,"[326,327] which are discussed in Section 1.4.2 (see also Table XX).

1.4.2. Simpler Glycosphingolipids: Ceramide Monohexosides, Ceramide Dihexosides, and Sulfatides (see Table XX)

Galactosyl- and glucosylceramide are the major monohexosylceramides. A new class of cerebrosides containing a fatty acid ester group and termed "ester cerebrosides" has been isolated from brain. Since Kotchetkov, Norton, and their associates[619,620] demonstrated the new type of glycolipid having a higher TLC mobility than a regular cerebroside, many studies have been focused on its chemical characterization. Klenk and Löhr[324] isolated five fractions of fast-migrating cerebroside from human brain; these were iden-

Table XVIII. Structural Variation and Classification of Major Long-Chain Bases

Number of carbon atoms	Abbreviated formula	Systematic name	References
I. Saturated, straight-chain, dihydroxy			
18.	d18:0	D-*Erythro*-1,3-dihydroxy-2-amino-octadecane (sphinganine or dihydrosphingosine)	63–65, 309
20.	d20:0	1,3-Dihydroxy-2-amino-eicosane	25, 26, 158, 171, 341
II. Monounsaturated, straight-chain, dihydroxy			
14	d14:1^4	D-*Erythro*-1,3-dihydroxy-2-amino-*trans*-4-tetradecene	342–347
16	d16:1^4	D-*Erythro*-1,3-dihydroxy-2-amino-*trans*-4-hexadecene	137, 157, 342–345, 347, 349
18	d18:1^4	D-*Erythro*-1,3-dihydroxy-2-amino-*trans*-4-octadecene (sphingosine)	12, 13–15, 350–352
20	d20:1^4	D-*Erythro*-1,3-dihydroxy-2-amino-*trans*-4-eicosene (eicosasphingosine)	25, 26, 158, 171, 341, 353
22	d22:1^9	*Erythro*-1,3-dihydroxy-2-amino-*cis*-9-docorene	356
22	d22:1^{13}	*Erythro*-1,3-dihydroxy-2-amino-*cis*-13-docorene	356
III. Diunsaturated, straight-chain dihydroxy			
18	d18:2^{4,14}	D-*Erythro*-1,3-dihydroxy-2-amino-*trans*-4, *cis*-14-octadecadiene	356, 157, 178, 355
18	d18:2^{4,8}	D-*Erythro*-1,3-dihydroxy-2-amino-*trans*-4,*trans*-8-octadecadiene	356, 357
22	d22:2^{4,9}	*Erythro*-1,3-dihydroxy-2-amino-*trans*-4,*cis*-9-docosadiene	356
IV. Saturated, straight-chain, trihydroxy			
18	t18:0	D-*Ribo*-1,3,4-trihydroxy-2-amino-octadecane (phytosphingosine)	135, 352, 358
20	t20:0	D-*Ribo*-1,3,4-trihydroxy-2-amino-eicosane	162
V. Unsaturated, straight-chain, trihydroxy			
18	t18:1^8	D-*Ribo*-1,3,4-trihydroxy-2-amino-*trans*-8-octadecene (dehydrophytosphingosine)	358
VI. Saturated, branched-chain, dihydroxy			
19	Anteiso d19:0	1,3-Dihydroxy-2-amino-16-methyloctadecane	135, 342
VII. Unsaturated, branched-chain, dihydroxy			
18	Iso d18:1^4	1,3-Dihydroxy-2-amino-16-methyl-4-heptadecene	135, 138, 316, 332, 359, 360
VIII. Saturated, branched-chain, trihydroxy			
19	Anteiso t19:O	1,3,4-Trihydroxy-2-amino-16-methyloctadecane	135, 316

Table XIX. Fischer Projection Formulas of Major Long-Chain Bases

CH_2OH	CH_2OH	CH_2OH	CH_2OH	CH_2OH	CH_2OH	CH_2OH
$HC\text{-}NH_2$	$HC\text{-}NH_2$	$HC^a\text{-}NH_2$	$HC^a\text{-}NH_2$	$HC\text{-}NH_2$	$HC\text{-}NH_2$	$HC\text{-}NH_2$
$HC\text{-}OH$	$HC\text{-}OH$	$HC\text{-}OH$	$HC\text{-}OH$	$HC\text{-}OH$	$HC\text{-}OH$	$HC\text{-}OH$
$HC=CH$	$-CH_2$	$HC=CH$	$-CH_2$	$HC\text{-}OH$	$HC\text{-}OH$	$HC\text{-}OH$
$-CH_2$	$-CH_2$	$-CH_2$	$-CH_2$	$-CH_2$	$-(CH_2)_3$	$-CH_2$
$-CH_2$	$-CH_2$	$-CH_2$	$-CH_2$	$-CH_2$	$HC=CH$	$-CH_2$
$-(CH_2)_{10}$	$-CH_2$	$-(CH_2)_{12}$	$-CH_2$	$-CH_2$	$-(CH_2)_8$	$-CH_2$
$-CH_3$	$-(CH_2)_{10}$	$-CH_3$	$-(CH_2)_{10}$	$-(CH_2)_{10}$	$-CH_3$	$-(CH_2)_{12}$
	$-CH_3$		$-CH_3$	$-CH_3$		$-CH_3$
Sphingosine	Dihydrosphingosine	C_{20}-Sphingosine (eicosasphingosine)	C_{20}-Dihydrosphingosine (eicosasphinganine)	Phytosphingosine	Dehydrophytosphingosine	C_{20}-Phytosphingosine
D-*Erythro*-1,3-dihydroxy-2-amino-*trans*-4-octadecene	D-*Erythro*-1,3-dihydroxy-2-amino-octadecane	D-*Erythro*-1,3-dihydroxy-2-amino-*trans*-eicosene	D-*Erythro*-1,3-dihydroxy-2-aminoeicosane[a]	D-*Ribo*-1,3,4-trihydroxy-2-amino-octadecane	D-*Ribo*-1,3,4-trihydroxy-2-amino-*trans*-8-octadecene	D-*Ribo*-1,3,4-trihydroxy-2-amino-eicosane
2S,3R-2-Amino-1,3-dihydroxy octadecene	2S,3R-2-Amino-1,3-dihydroxy-octadecane	2S,3R-2-Amino-1,3-dihydroxy-eicosene	2S,3R-2-Amino-1,3-dihydroxy-eicosane	2S,3R,4R-2-Amino-1,3,4-trihydroxy-octadecane	2S,3R-2-Amino-1,3,4-trihydroxy-8-octadecene	—

[a] D-Erythro configuration is tentative, not yet conclusively identified through comparison with synthetic compounds.

tified as a cerebroside containing 3-*O*-fatty acylsphingosine, 6-*O*-fatty acylgalactose, and 3-*O*-fatty acylgalactose. Kishimoto *et al.*[325] isolated a fast-migrating cerebroside from pig brain in which the 6-hydroxy position of galactose was acylated with fatty acid. Tamai *et al.*[326] isolated a similar fraction containing ester cerebroside, and finally identified it as a mixture of 3-*O*-fatty acylgalactosylceramide and 6-*O*-fatty acylgalactosylceramide; however, a later study by Tamai did not find any unequivocal evidence for the presence of a cerebroside having an ester linkage at the 3-hydroxy group of sphingosine.[327]

Recently, a novel xylosylceramide was isolated from the salt gland of sea gull by Karlsson *et al.*,[328] and L-α-fucosylceramide was isolated from human colonic carcinoma by Watanabe *et al.*[329] The ceramide moiety of fucosylceramide consisted of a palmitoyl residue as the major fatty acid, and the LCB contained eicosasphingenine in addition to octadecasphingenine. The quantity of fucosylceramide in metastatic deposits was higher than in the original tumors.

Two kinds of ceramide dihexosides can be distinguished, as can be seen in Table XX. The most common ceramide dihexoside is lactosylceramide, widely distributed in various organs and tissues except brain and nervous tissue. The distribution of the second dihexosylceramide, α-galactosyl1→4βgalactosylceramide ("galabiosylceramide"), has not been extensively studied. Since galabiosylceramide is not easily separated from lactosylceramide on TLC or by other chromatographic techniques, systematic analysis of the dihexosylceramide fraction of various tissues and cells is necessary. The presence of galabiosylceramide in the kidneys of various mice strains (BALB/c, C_3H, C57/BL) was noticed by Gray,[330] who found not only that the quantities were higher in male than in female mice but also that the level was controlled by testosterone. The glycolipid level in the kidneys of castrated male mice or in female mice increased severalfold when these mice were administered testosterone.[330,331] The same glycolipid accumulated in organs of patients with Fabry's disease, and the anomeric linkage of the terminal galactosyl was determined as α.[332,333] The same glycolipid was found in NIL cells, and the same structure was found in one of the Forssman glycolipids isolated from the same cells.[129] Recently, about 20% of dihexosylceramide isolated from human neutrophil leukocytes was found to be this glycolipid.[509]

Sulfatides. The presence of sulfur in brain lipid fraction has been known since it was described by Thudichum, but the compound identified as "sulfated cerebroside" was very difficult to separate from phospholipids before the technique of chromatography was introduced. Even after Blix[33] isolated and characterized "sulfatide" from brain, the presence of a sulfate-containing phospholipid, termed "phospho-sulfolipid," was described.[104] The correct structure was established as 3-*O*-sulfated galactopyranosylceramide by methylation in 1964.[34,35] The compound has been found in relatively large quantities in white matter of brain.[574] Consequently, it was identified as the major component of myelin sheath membrane[575,576] and astroglia.[577] A relative abundance of sulfated glycolipid in kidney, gastrointestinal epithelia, and in a few other tissues that display a high level of Na^+ and K^+-Na^+ transport

Table XX. Simpler Glycosphingolipids: Ceramide Monohexosides and Dihexosides and Sulfatides

1. Monohexosylceramides

Galβ1→1Cer	Isolation: Thudicum[2,3]; structure: Carter et al.[63];
Glcβ1→1Cer	Occurrence: Gaucher spleen, Aghion,[29] Halliday et al.,[30] Klenk[361]; liver, spleen, serum, Svennerholm[362]
Ester cerebroside	Occurrence and isolation: from white matter, Klenk and Löhr,[324] Tamai and co-workers[326,327]
Xylβ1→1Cer	Occurrence and isolation: from sea gull salt gland, Karlsson et al.[328]
L-Fucα1→1Cer	Occurrence and isolation: from human adenocarcinoma, Watanabe et al.[329]

2. Dihexosylceramide (CDH)

Galβ1→4Glcβ1→1Cer (lactosylceramide)	Occurrence: Klenk and Rennkampf[36]; as human tumor hapten ("cytolipin H"), Rapport et al.[37]; identity of cytolipin H with ceramide dihexoside of bovine kidney, Rapport et al.[363]; Isolation: from erythrocytes, Yamakawa et al.[364]; from kidney, Makita et al.[365]; structure: Yamakawa et al.[34]
Galα1→4Galβ1→1Cer	In mouse kidney showing sex hormone response, Gray[330]; in Fabry kidney, Sweeley and Klionsky[32]; structure: Handa et al.,[332] Li et al.[333]; in hamster NIL cells, Gahmberg and Hakomori[129]

3. Sulfatides

H$_2$SO$_3$→3Galβ1→1Cer	Occurrence and isolation: Blix;[33] structure: Yamakawa et al.[34]; Stoffyn and Stoffyn[35]
H$_2$SO$_3$→3Galβ1→4Glcβ1→1Cer	Occurrence and isolation from kidney: Martensson[335]; structure: Stoffyn et al.[336]; in dog intestine: McKibbin[572]; in trout testis: Levine et al.[573]
H$_2$SO$_3$6Glcα1→6Glcα1→6Glcα1→3Glyceride	Occurrence and isolation: Slomiany et al.[334]
H$_2$SO$_3$→Galβ1→4Galβ1→4Glcβ1→1Cer	Occurrence in hog gastric mucosa: Slomiany et al.[570]
Galβ1→4GlcNacβ1→3Galβ1→4Glcβ1→1Cer	Occurrence in hog gastric mucosa: Slomiany and Slomiany[571]

Galβ1→4GlcNacβ1→3Galβ1→4Glcβ1→1Cer
$$6$$
$$\uparrow$$
$$H_2SO_3$$

GalNAcβ1→4Galβ1→4Glcβ1→1Cer   Occurrence and isolation from rat kidney: Tadano and Ishizuka[435]
$$3$$
$$\uparrow$$
$$H_2SO3$$

activity led to speculation that this compound is functionally related to the K$^+$-Na$^+$ transport activity of the cell membrane, particularly K$^+$-Na$^+$-dependent ATPase activity[578] (see also Chapter 4). In fact, sulfatide content became significantly higher in the remaining hyperfunctional kidney after unilateral nephrectomy.[579] However, exact location of sulfatide, revealed by immunofluorescence with antisulfatide antibodies, was not coincident with the location of K$^+$-Na$^+$-dependent ATPase.[580]

A sulfated lactosylceramide was first isolated and characterized from rat kidney by Mårtensson,[335] and its structure was established by Stoffyn *et al.*[336] This glycolipid represented as much as 30% of the total sulfatide of rat kidney. Similar to sulfated cerebroside, gastrointestinal mucosae[572] also contain this glycolipid. Sulfate esters of a few other glycolipids have been reported by Slomiany and co-workers[334,570,571] to occur in gastrointestinal mucosae. Recently, a third sulfated glycosphingolipid was isolated and unequivocally characterized. Tadano and Ishizuka[435] isolated and characterized a new type of sulfated GM_2 ganglioside from rat kidney, and it is believed to be a component of renal tubules in kidney of various species.[615]

Another type of interesting glycolipid-containing sulfate is "seminolipid," or sulfate ester of galactosyldiglyceride (1-*O*-alkyl-2-*O*-acyl-3-3'-*O*-sulfogalactopyranosyl glycerol). Its presence is mainly confined to germinal cells (spermatozoa and testis)[337–340] and brain.[558]

1.4.3. Globo-Series Glycolipids (see Table XXI)

The major glycosphingolipid of human erythrocytes was found to be a neutral glycolipid containing hexoses and galactosamine and was termed "globoside," in striking contrast to the fact that gangliosides were the major glycolipid component of horse and bovine erythrocytes[47,49–51] (see Section 1.1.2). The correct carbohydrate composition and sequence of globoside were shown by Yamakawa *et al.*[72] in 1963, although the position of terminal GalNAc initially proposed as 1→6 was corrected to 1→3 by subsequent studies.[73] In this structure, a ceramide trihexoside was assigned for the first time as the internal moiety, which was identical to that isolated from organs of a patient with Fabry's disease[32] and that isolated from human erythrocytes.[367] The complete anomeric structure of ceramide trihexoside and globoside has been determined in subsequent studies.[77] An important feature of the globo-series structure is the presence of an α-galactosyl residue at the terminus of ceramide trihexoside and at the penultimate sugar residue of globoside.*[77,332,368] Kawanami[282] first found the presence of the α-galactosyl linkage in a ceramide trihexoside present in Nakahara–Fukuoka sarcoma on the basis of nuclear magnetic resonance spectra. Subsequently, several studies including enzymatic degradation supported this linkage in globoside and in ceramide

---

* The presence of α-Gal residue in the penultimate (or next to the penultimate) residue of any carbohydrate chain of higher animals and man is not known except for this globo-series case. The positive optical rotation of globoside from human erythrocytes ($[\alpha]_D$ = +19.5°), globoside from pig erythrocytes (+19.7°), a ceramide trihexoside from Fabry's disease (+24.2°), and a ceramide trihexoside isolated from a partial hydrolyzate of human erythrocyte globoside (+23.0°), was reported by Yamakawa,[55] Miyatake *et al.*,[627] and Miyatake.[628] These classical findings strongly indicated that these globo-series glycolipids should have an α anomeric linkage within their carbohydrate chains. Other glycolipids reported by these authors showed negative optical rotation, including lactosyl ceramide of human erythrocytes ($[\alpha]_D$ = 9.7°), and that isolated from partial hydrolyzate of human globoside (−14.0°). The suggestion was finally proven by enzymatic degradation with fig α-galactosidase, that only this α-galactosidase was capable of hydrolyzing the α-Gal residue of glycolipids;[77] other α-galactosidases were virtually inactive for this substrate.

Table XXI. Glycolipids with Globo-Series Structure

Structure	Occurrence / References
Galα1→4Galβ1→4Glcβ1→1Cer (GbOse₃Cer, CTH)	Occurrence: in Fabry kidney, Sweeley and Klionsky,[32]; erythrocytes, Vance and Sweeley,[367]; structure: Hakomori et al.,[77] Li and Li,[368] Handa et al.[332]; identification as p^k antigen: Naiki and Marcus,[81,82] Marcus et al.[83]
Galα1→3Galβ1→4Glcβ1→1Cer (isoGbOse₃Cer, rat CTH)	Occurrence: in rat kidney, Arita and Kawanami[369]
GalNAcβ1→3Galα1→4Galβ1→4Glc→Cer (GbOse₄Cer, globoside)	Occurrence: Klenk and Lauenstein,[49] Yamakawa and Suzuki[50]; structure: Yamakawa et al.,[34,73] Hakomori et al.[77]
GalNAcβ1→3Galα1→3Galβ1→4Glc→Cer (isoGbOse₄Cer, cytolipin R)	Occurrence and isolation: Rapport et al.[370]; structure: Laine et al.[371] Siddiqui et al.,[372] Kawanami[373]
GalNAcα1→3GalNAcβ1→3Galα1→4Galβ1→4Glc→Cer (GbOse₅Cer, Forssman antigen)	Occurrence: Brunius,[374] Papiermeister and Mallette,[375] Yamakawa et al.[364]; isolation: Makita et al.[365]; structure: Siddiqui and Hakomori,[78] Sung et al.,[376] Taketomi et al.,[377] Ziolkowski et al.[378]
Fucα1→2Galα1→4Galβ1→4Glcβ1→1Cer (III²FucGbOse₃Cer)	Isolation and structure: from hog gastric mucosa, Slomiany et al.[379]
GalNAcα1→3GalNAcβ1→3Galα1→4Galβ1→4Glcβ1→1Cer 　　　　　　　　　　4 　　　　　　　　　　↑ 　　Galβ1→3GalNAcβ1 (branched Forssman antigen)	Isolation and structure: from dog gastric mucosa, Slomiany and Slomiany[380,597]
GalNAcβ1→3GalNAcβ1→3Galα1→4Galβ1→4Glcβ1→1Cer ("para-Forssman") (IV³βGalNAcGbOse₄Cer)	Isolation and structure: from human erythrocytes, Kon et al.,[598] Ando et al.[599]
Galβ1→3GalNAcβ1→3Galα1→4Galβ1→4Glcβ1→1Cer (IV³βGalGbOse₄Cer)	Isolation and structure: from green monkey kidney cells, Blomberg et al.[589]; from human teratocarcinoma cells, Kannagi et al.[605]
NeuAcα2→3Galβ1→xGalNAcβ1→3Galα1→4Galβ1→4Glcβ1→1Cer (IVNeuAcα2→3GalGbOse₄Cer)	Isolation and structure: from chicken muscle cells, Chien and Hogan[548]; from human teratocarcinoma cells, Kannagi et al.[605]

trihexoside of erythrocytes[77] and that accumulated in tissues of patients with Fabry's disease.[289,332,368] Globoside of human erythrocytes, human kidney, pig erythrocytes, and pig kidney have been identified as having the same structure[77,372]; however, "cytolipin R,"[370] a globosidelike glycolipid of rat sarcoma[370] and in rat kidney,[373] has been identified as GalNAcβ1→3Galα1→3Galβ1→4Glc→ceramide.[371,372] This structure has sequence and anomeric configuration of carbohydrate residues identical to those of globoside; the only difference is in the positional linkage of the α-Gal residue.

In the early studies, Forssman hapten glycolipid was regarded as a ceramide tetrasaccharide having the same carbohydrate structure as globoside except that the terminal GalNAc in Forssman hapten glycolipid was assigned as α in contrast to β in globoside.[288,365] However, more recent studies showed that Forssman hapten glycolipid can be converted to globoside by hydrolysis with α-N-acetylgalactosaminidase, and all other results by carbohydrate analysis and methylation studies indicated that the Forssman hapten glycolipid must have two N-acetylgalactosamine residues at the terminus with GalNacα1→3GalNAcβ1→4Gal sequence.[78] This structural assignment has been confirmed by various investigators with samples isolated from various species of animals.[237,257,259,376–378] Thus, the globo-series of glycolipids has been established as presented in Table XXI. It should be noted that no sialosyl or fucosyl substitutions occur in this series of glycolipids, except one glycolipid reported by Slomiany *et al.*[379] that has an L-α-fucosyl substitution at the α-galactosyl residue of ceramide trihexoside and another one reported by Chien and Hogan[548] that has a sialosyl-galactosyl substitution in globoside. More recently, a galactosyl substitution in globoside was isolated and partially characterized from green monkey kidney cells. The anomeric structure of the terminal Gal was tentatively assigned as β because the glycolipid was incapable of inhibiting blood group B or P_1 activity.[589] A similar glycolipid was isolated from human teratocarcinoma and unequivocally characterized on the basis of methylation analysis, direct-probe mass spectrometry, enzyme degradation, and NMR-spectroscopy.[605]

The globo-series of glycolipids are now identified by Naiki and Marcus[81–83] as the important isoantigens; ceramide trihexoside and globoside represent P^k and P antigen, respectively (see Chapter 5).

1.4.4. Lacto-Series Glycolipids (see Table XXII)

The first clear evidence for the existence of "lacto-series" glycolipids was the isolation and characterization of sialosyl-lacto-N-neotetraosylceramide by Kuhn and Wiegandt[71] in 1964. Subsequently, another lacto-series of glycolipids substituted with a fucose residue was found to accumulate in some human adenocarcinomas.[74,395] The structure was identified as lacto-N-fucopentaose(III) ceramide.[184] An extensively purified glycolipid fraction with blood group A, B, and H activities was found to comprise lacto-series glycolipids containing N-acetyl glucosamine and fucose.[75] A ceramide tetrasaccharide called "paragloboside" was identified as lacto-N-neotetraosylceram-

Table XXII. Glycolipids with Lacto-Series Structure

I. Basic structure

Lacto-N-triosylceramide (LcOse₃Cer)	GlcNAcβ1→3Galβ1→4Glcβ1→1Cer	Occurs in trace amounts in human erythrocytes. Isolation and structure: Ando et al.[381]; accumulation in HEMPAS erythrocytes, Joseph and Gockerman;[382] accumulation in adenocarcinoma of colon, Watanabe and Hakomori[383]; in melanoma, Karlsson[384]; in human neutrophil leukocytes, Macher and Klock[509]
Lacto-N-*neo*tetraosylceramide (nLcOse₄Cer)	Galβ1→4GlcNAcβ1→3Galβ1→4Glc1→1Cer (paragloboside)	Isolation and structure: from human erythrocytes, Siddiqui and Hakomori,[385] Ando and Yamakawa[386]; identification as NILpy tumor-specific antigen, Sundsmo and Hakomori[387]; from neutrophil leukocytes, Macher and Klock[509]
Lacto-N-tetraosylceramide (LcOse₄Cer)	Galβ1→3GlcNAcβ1→3Galβ1→4Glcβ1→1Cer	Isolation and structure: from human meconium, Karlsson and Larson[515]

Lacto-*N-nor*hexaosylceramide (nLcOse₆Cer)	Galβ1→4GlcNAcβ1→3Galβ1→4GlcNAcβ1→3Galβ1→4Glcβ1→1Cer	As i antigen, Neiman *et al.*[392]
Lacto-*N-nor*octaosylceramide (nLcOse₈Cer)	Galβ1→4GlcNAcβ1→3Galβ1→4GlcNAcβ1→3Galβ1 →4GlcNAcβ1→3Galβ1→4Glcβ1→1Cer	Possible i antigen, present as a core structure of the corresponding ganglioside and fucosyl derivatives, Kannagi *et al.*[602]
Lacto-*N-iso*octaosylceramide (iLcOse₈Cer)	Galβ1→4GlcNAcβ1 $\searrow$ $\overset{3}{\underset{6}{}}$ Galβ1→4GlcNAcβ1→3Galβ1→4Glcβ1→1Cer $\nearrow$ Galβ1→4GlcNAcβ1	Possible I antigen present as a core structure of the corresponding gangliosides and fucosyl derivatives, Watanabe *et al.*[120,116,606]

II. Derivatives of paragloboside (lacto-*N-neo*tetraosylceramide, nLcOse₄Cer)

a. Galactosyl substitution

	Galβ1→3Galβ1→4GlcNAcβ1→3Galβ1→4Glcβ1→1Cer (β-galactosylparagloboside, IV³βGalnLcOse₄Cer)	Isolation and structure: from human erythrocytes, Stellner and Hakomori[388]
	Galα1→3Galβ1→4GlcNAcβ1→3Galβ1→4Glc1→1Cer (α-galactosylparagloboside, IV³αGalnLcOse₄Cer)	Occurrence and isolation: from rabbit erythrocytes, Eto *et al.*[389]; structure: Stellner *et al.*[237] occurrence: in bovine erythrocytes, Chien *et al.*[390]
	Galα1→4Galβ1→4GlcNAcβ1→3Galβ1→4Glcβ1→1Cer (p₁ antigen, IV⁴αGalnLcOse₄Cer)	Occurrence: Marcus[391]; isolation and structure: Naiki and Marcus[82]

(continued)

Table XXII. (Continued)

b. Fucosyl substitution of paragloboside	Fucα1→2Galβ1→4GlcNAcβ1→3Galβ1→4Glcβ1→1Cer (H₁ glycolipid, type 2 chain H glycolipid, IV²FucnLcOse₄Cer)	Isolation and structure: from human erythrocytes, Stellner et al.,[119] Koscielak et al.,[393] Smith et al.[420]
	Fucα1→2Galβ1→4GlcNAcβ1→3Galβ1→4Glcβ1→1Cer 　　　　　　　　　　3 　　　　　　　　　　↑ 　　　　　　　　αGalNAc1 (Aᵃ glycolipid, type 2 chain A glycolipid, IV²FucIV³αGalNAcnLcOse₄Cer)	Isolation and structure: from human erythrocytes, Hakomori et al.,[79] Ando and Yamakawa[386]; from hog gastric mucosa, Slomiany et al.[418]; isolation and characterization: from dog and human intestine mucosa, McKibbin et al.,[421] mass spectrometric identification, Smith et al.[422]
	Fucα1→2Galβ1→4GlcNAcβ1→3Galβ1→4Glcβ1→1Cer 　　　　　　　　　　3 　　　　　　　　　　↑ 　　　　　　　　αGal1 (B₁ glycolipid, type 2 chain B glycolipid, IV²FucIV³ αGalnLcOse₄Cer)	Isolation and characterization: from human erythrocytes, Koscielak et al.,[393] Hanfland and Egge,[266,267] Hanfland et al.[394]; from pancreas, Wherret and Hakomori[398]
	Galβ1→4GlcNAcβ1→3Galβ1→4Glcβ1→1Cer 　　　　3 　　　　↑ 　　　αFuc1 (Leˣ glycolipid; III³FucnLcOse₄Cer)	Isolation and structure: from human adenocarcinoma, Yang and Hakomori[184]; from hog gastric mucosa, Slomiany et al.[418]
	Fucα1→2Galβ1→4GlcNAcβ1→3Galβ1→4Glcβ1→1Cer 　　　　　　　　　3　　　3 　　　　　　　　　↑　　　↑ 　　　　　　αGalNAcl　　αFuc1 (Difucosylated A glycolipid; III³FucIV²FucIV³αGalNAcnLcOse₄Cer)	Isolation and structure: from hog gastric mucosa, Slomiany and Slomiany[419]; mass spectrometric detection: from rabbit small intestine, Breimer et al.[275]; from dog small intestine, Smith et al.[424]

c. Sialosyl substitution of LcOse₃Cer and paragloboside (nLcOse₄Cer)

NeuAcα2→GlcNAcβ1→3Galβ1→4Glcβ1→1Cer

Occurrence and partial characterization in human lymphocytes and neutrophil: Macher et al.[632]

NeuAc2→3Galβ1→4GlcNAcβ1→3Galβ1→4Glcβ1→1Cer
(sialosylparagloboside, IV³NeuAcnLcOse₄Cer)

Occurrence in human erythrocytes, Wherrett and Brown[399]; in leucocytes, Wherrett[401]; in human peripheral nerve, MacMillan and Wherrett,[400] Svennerholm[402]; in brain and peripheral nerve, Li et al.[403]

NeuGly2→3Galβ1→4GlcNAcβ1→3Galβ1→4Glc1→1Cer
(IV³NeuGlynLcOse₄Cer)

Occurrence, isolation, and structure: Kuhn and Wiegandt,[71] Wiegandt,[80] Wintzer and Uhlenbruck [404]

NeuAc2→6Galβ1→4GlcNAcβ1→3Galβ1→4Glc1→1Cer
(IV⁶NeuAcnLcOse₄Cer)

Occurrence, isolation, and structure from human erythrocytes, Watanabe et al.[606]; from meconium Nilsson et al.[207]

NeuAc2→8NeuAc2→3Galβ1→4GlcNAcβ1→3Galβ1→4Glc1→1Cer
(IV³ (NeuAc)₂nLcOse₄Cer)

Occurrence, isolation, and structure: Rauvala et al.[406]

d. Sialosylfucosyl substitution of paragloboside

NeuAc2→3Galβ1→4GlcNAcβ1→3Galβ1→4Glc→Cer
```
                          3
                          ↑
                        αFuc1
```
(III³FucIV³NeuAcnLcOse₄Cer)

Occurrence, isolation, and structure: Rauvala[428]

(continued)

Table XXII. (Continued)

e. GalNAc substitution of paragloboside

GalNAcβ1→3Galβ1→4GlcNAcβ1→3Galβ1→4Glcβ1→1Cer
(IV³GalNAcnLcOse₄Cer)

Isolation and structure: from human erythrocytes, Kannagi et al.[607]; from cancer tissue of blood group pp (Tjaˉ) individual, Kannagi et al.[608]

f. Sialosyl-2→3GalNAc substitution of paragloboside

NeuAc2→3GalNAcβ1→3Galβ1→4GlcNAcβ1→3Galβ1→4Glcβ1→1Cer
(IV³NeuAc2→3GalNAcnLcOse₄Cer)

Isolation and structure: from human erythrocytes, Watanabe, and Hakomori[415]

III. Derivatives of lacto-N-tetraosylceramide

a. Fucosyl substitution of lacto-N-tetraosylceramide ("type 1 chain")

Galβ1→3GlcNAβ1→3Galβ1→4Glcβ1→1Cer
　　　　4
　　　　↑
　　　αFuc1

(Leᵃ glycolipid, III⁴FucLeOse₄Cer)

Occurrence and isolation: Hakomori et al.[395]; characterization from human serum, Marcus and Cass[396]; multiplicity: Hanfland[397]; isolation and characterization: from human intestine, Smith et al.[423]

Fucα1→2Galβ1→3GlcNAcβ1→3Galβ1→4Glcβ1→1Cer
　　　　　　　　　4
　　　　　　　　　↑
　　　　　　　　αFuc1

(Leᵇ glycolipid, III⁴FucIV²FucLcOse₄Cer)

Occurrence and isolation: Marcus and Cass[396]; multiplicity: Handfland[397]; Isolation and structure. from human tumor, Hakomori and Andrews[414]

Fucα1→2Galβ1→3GlcNAcβ1→3Galβ1→4Glcβ1→1Cer
　　　　　　　　　3
　　　　　　　　　↑
　　　　　　　　αGalNAc1

(A active type 1 chain glycolipid, IV²FucIV³αGalNAcLcOse₄Cer)

Occurrence, Isolation, and possible structure: from human small intestine mucosa, McKibbin et al.[421]; mass spectrometric analysis: Smith et al.[422]

Fucα1→2Galβ1→3GlcNAcβ1→3Galβ1→4Glcβ1→1Cer
$$3$$
$$↑$$
$$αGal1$$
(IV²FucIV³αGalLcOse₄Cer)

Isolation and structure: from Fabry's pancreas, Wherrett and Hakomori[398]

Galβ1→3GlcNAcβ1→3Galβ1→4Glcβ1→1Cer
$$3\qquad4$$
$$↑\qquad↑$$
$$NeuAcα2\quad Fucα1$$
(III³FucIV³NeuAcLcOse₄Cer)

Isolation and characterization as human cancer-associated antigen defined by the monoclonal antibody NS-19-9, Magnani et al.[609,610]

IV. Derivatives of lacto-N-*nor*hexaosylceramide
a. αGalactosyl substitution

Galα1→3Galβ1→4GlcNAcβ1→3Galβ1→4Glcβ1→1Cer
(VI⁶GalnLcOse₆Cer)

Isolation and structure: from bovine erythrocytes, Chien et al.[390]

b. Fucosyl substitution

Fucα1→2Galβ1→4GlcNAcβ1→3Galβ1→4Glcβ1→1Cer
(H₂-glycolipid, VI²FucnLcOse₆Cer)

Watanabe et al.[120]

Fucα1→2Galβ1→4GlcNAcβ1→3Galβ1→4Glcβ1→1Cer
$$3$$
$$↑$$
$$αGalNAc1$$
(Aᵇ-glycolipid, VI²FucVI³GalNAcnLcOse₆Cer)

Hakomori et al.[79,310]

Fucα1→2Galβ1→4GlcNAcβ1→3Galβ1→4Glcβ1→1Cer
$$3$$
$$↑$$
$$αGal1$$
(BII-glycolipid, VI²FucVI³GalnLcOse₆Cer)

Hanfland and Egge,[266,267] Hanfland et al.[394]

(continued)

Table XXII. (Continued)

Galβ1→4GlcNAcβ1→3Galβ1→4GlcNAcβ1→3Galβ1→4Glcβ1→1Cer

            3
            ↑
         Fucα1

(Y_2 glycolipid, V^3FucLcOse$_6$Cer)

Isolation and structure: from human erythrocytes, Kannagi *et al.*[602]

Galβ1→4GlcNAcβ1→3Galβ14→GlcNAcβ1→3Galβ1→4Glcβ1→1Cer

          3   3
          ↑   ↑
      Fucα1   Fucα1

(difucosyllacto-N-*nor*hexaosylceramide, III^{3}FucV^{3}FucLcOse$_6$Cer)

Isolation and structure: from human adenocarcinoma, Hakomori *et al.*[604]

c. Sialosyl substitution

NeuAc/or NeuGly2→3Galβ1→4GlcNAcβ1→3Galβ1→4GlcNAcβ1

→3Galβ1→4Glcβ1→1Cer

(VI^{3}NeuAcnLcOse$_6$Cer)

From human spleen, Wiegandt[87] from bovine erythrocytes, Chien *et al.*[390]; Neimann *et al.*[392]; from human erythrocytes

NeuAcα2→6Galβ1→4GlcNAcβ1→3Galβ1→4GlcNAcβ1→3Galβ1→4Glcβ1→1Cer

(VI^{6}NeuAcnLcOse$_6$Cer)

Isolation and structure from human erythrocytes, Watanabe *et al.*[606]

V. Derivatives of lacto-N-*nor*octaosylceramide

a. Fucosyl substitution

Galβ1→4GlcNAcβ1→3Galβ1→4GlcNAcβ1→3Galβ1→4GlcNAcβ1→3Galβ1→4Glcβ1→1Cer

            3
            ↑
         Fucα1

("Z$_1$ glycolipid," VII^{3}FucLcOse$_8$Cer)

Isolation and structure: from human erythrocytes, Kannagi *et al.*[602]

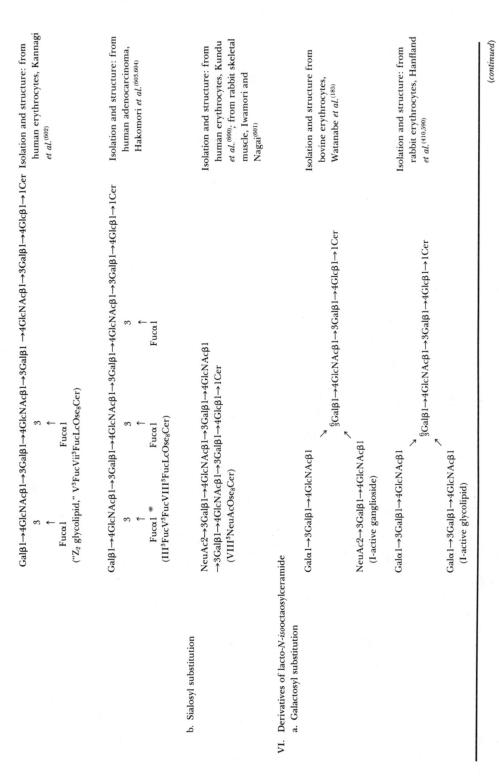

Galβ1→4GlcNAcβ1→3Galβ1 →4GlcNAcβ1→3Galβ1→4Glcβ1→1Cer Isolation and structure: from
   3   3   human erythrocytes, Kannagi
  ↑   ↑   et al.[602]
Fucα1 Fucα1
("Z₂ glycolipid," V³FucVii³FucLcOse₈Cer)

Galβ1→4GlcNAcβ1→3Galβ1→4GlcNAcβ1→3Galβ1→4Glcβ1→1Cer Isolation and structure: from
   3   3   3   human adenocarcinoma,
  ↑   ↑   ↑   Hakomori et al.[603,604]
Fucα1 Fucα1 Fucα1
(III³FucV³FucVIII³FucLcOse₈Cer)

b. Sialosyl substitution

NeuAc2→3Galβ1→4GlcNAcβ1→3Galβ1→4GlcNAcβ1 Isolation and structure: from
→3Galβ1→4GlcNAcβ1→3Galβ1→4Glcβ1→1Cer human erythrocytes, Kundu
(VIII³NeuAcOse₈Cer) et al.[600], from rabbit skeletal
                              muscle, Iwamori and
                              Nagai[601]

VI. Derivatives of lacto-N-*iso*octaosylceramide
a. Galactosyl substitution

Galα1→3Galβ1→4GlcNAcβ1
                   ↘
                    ⁶Galβ1→4GlcNAcβ1→3Galβ1→4Glcβ1→1Cer Isolation and structure from
                   ₃   bovine erythrocytes,
                   ↗   Watanabe et al.[185]
NeuAc2→3Galβ1→4GlcNAcβ1
(I-active ganglioside)

Galα1→3Galβ1→4GlcNAcβ1
                   ↘
                    ⁶Galβ1→4GlcNAcβ1→3Galβ1→4Glcβ1→1Cer Isolation and structure: from
                   ₃   rabbit erythrocytes, Hanfland
                   ↗   et al.[410,590]
Galα1→3Galβ1→4GlcNAcβ1
(I-active glycolipid)

(continued)

Table XXII. (Continued)

b. Fucosyl substitution

Fucα1→2Galβ1→4GlcNAcβ1
³⁶Galβ1→4GlcNAcβ1→3Galβ1→4Glcβ1→1Cer
Fucα1→2Galβ1→4GlcNAcβ1
(H₃ glycolipid)

Isolation and structure: from human erythrocytes, Watanabe et al.[120]

αGalNac1
↓
3
Fucα1→2Galβ1→4GlcNAcβ1
³⁶Galβ1→4GlcNAcβ1→3Galβ1→4Glcβ1→1Cer
Fucα1→2Galβ1→4GlcNAcβ1
3
↑
αGalNAc1
(Aᶜ glycolipid)

Isolation and structure: from human erythrocytes, Hakamori et al.[310,407], Fukuda and Hakomori,[596] mass spectrometric analysis of a sample from rat intestine, Karlsson[409]

Galβ1→4GlcNAcβ1
⁶⁶Galβ1→4[GlcNAc1→3Gal]₇-GlcNAcβ1→3Galβ1→4Glcβ1→1Cer
Galβ1→4GlcNAcβ1

Isolation: from human erythrocytes, Gardas[85]

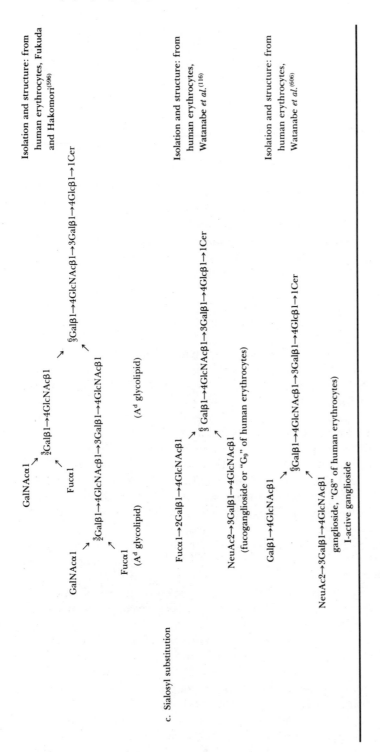

GalNAcα1
⟍
$^3_2$Galβ1→4GlcNAcβ1
⟋
Fucα1 $^6_3$Galβ1→4GlcNAcβ1→3Galβ1→4Glcβ1→1Cer
⟋   ⟋
GalNAcα1   ⟍   ↑
⟍ $^3_2$Galβ1→4GlcNAcβ1→3Galβ1→4GlcNAcβ1
↑   (A^d glycolipid)
Fucα1
(A^d glycolipid)

Isolation and structure: from human erythrocytes, Fukuda and Hakomori[596]

c. Sialosyl substitution

Fucα1→2Galβ1→4GlcNAcβ1
⟍
$^6_3$ Galβ1→4GlcNAcβ1→3Galβ1→4Glcβ1→1Cer
⟋
NeuAc2→3Galβ1→4GlcNAcβ1
(fucoganglioside or "C$_9$" of human erythrocytes)

Isolation and structure: from human erythrocytes, Watanabe et al.[116]

Galβ1→4GlcNAcβ1
⟍
$^6_3$Galβ1→4GlcNAcβ1→3Galβ1→4Glcβ1→1Cer
⟋
NeuAc2→3Galβ1→4GlcNAcβ1
ganglioside, "G8" of human erythrocytes)
I-active ganglioside

Isolation and structure: from human erythrocytes, Watanabe et al.[606]

ide.[385] Other lacto-series compounds were then found, including ceramide pentasaccharide from rabbit erythrocytes[389] and fucose- and glucosamine-containing glycolipid of gastrointestinal mucosae.[411,412] Consequently, a number of glycolipids have been isolated and characterized as blood groups ABH, Lewis, Ii, and P_1 antigens all of which possess lacto-series structures, as is discussed extensively in Chapter 5. Lacto-series constitute the major glycolipid component of various tissues and organs and its sialosyl derivative represents the major ganglioside of extraneural tissues. The lacto-series glycolipid has been considered to be absent in brain. Recently, however, a small quantity of lacto-N-fucopentaosyl(III)ceramide and its sialosyl derivative were found in brain.[546]

In striking contrast to the globo-series, the carbohydrate chains of lacto-series glycolipids are extensively substituted with fucosyl, sialosyl, and α or β-galactosyl, and with a repeating Galβ1→4GlcNAc structure. Combinations of these substitutions also occur, resulting in a large variety of glycolipid structures, as can be seen in Table XXII. Substitution patterns and the resulting variation of lacto-series glycolipids are described in the following sections.

1.4.4.1. Substitution of Paragloboside (Lacto-N-neotetraosylceramide) or Lacto-N-tetraosylceramide

Four types of substitutions have been known: (1) sialosyl, (2) fucosyl, (3) galactosyl, and (4) N-acetyllactosaminyl substitutions.

1. Sialosyl substitution resulted in four types of sialosylparaglobosides: (a) sialosylparagloboside with an NeuAc2→3 or NeuGly2→3Gal residue[71,80,385,386,399–403]; (b) sialosylparagloboside with an NeuAc2→6Gal residue[80,606]; (c) sialosylparagloboside with an NeuAc2→8NeuAc2→3Gal residue[406]; (d) substitution with an NeuAc2→3GalNAcβ1→3Gal residue.[413]

2. Fucosyl substitution of paragloboside resulted in H_1 glycolipid[119] and Le^x glycolipid,[184] and that of lacto-N-tetraosylceramide gave Le^a- and Le^b-active glycolipid.[209,396,397,414] Thus, fucosyl substitution of glycolipid with type 1 and type 2 chain[415,416] gave, respectively, the structures with greatly different immunological specificity.

A^a glycolipid[79,310,386] of human erythrocyte membranes is the derivative of H_1 glycolipid[119,393]; similarly, B_1 glycolipid[267,393,394] of human erythrocyte membranes is the derivative of H_1 glycolipid. The majority of B-active glycolipids accumulating in the pancreas of a patient with Fabry's disease contained the type 1 chain,[398] suggesting that the major glycolipid of pancreas must be the lacto-series glycolipids with type 1 chain.

3. Galactosyl substitution of paragloboside showed three different derivatives: (a) substitution with a Galβ1→3Gal residue represents a minor component of human erythrocyte membranes[388]; (b) substitution with a Galα1→3Gal residue represents the major glycolipids of rabbit erythrocyte membranes[237,389] and bovine erythrocyte membranes[390]; (c) substitution with a Galα1→4Gal residue has been assigned as the determinant for blood group P_1 antigen of human erythrocyte membranes.[82]

4. The extension of the carbohydrate chain through substitution with an

N-acetyllactosamine unit resulted in a linear nonbranched lacto-N-norhexaosylceramide[392] or a branched lacto-N-octaosylceramide.[185]

1.4.4.2. Substitution of Lacto-N-norhexaosylceramide (Table XXII, section III)

The lacto-series derivatives of nonbranched repeating N-acetyllactosamine structure have been isolated and characterized. These are: (1) fucosyl substitution resulted in H_2 glycolipids[120] and A^b and B_{II} glycolipids[266,267,310,407]; (2) sialosyl substitution resulted in two gangliosides with core structure recently isolated from bovine[390,392] and human erythrocytes[405] and human and bovine spleen and kidney[408]; (3) Gal substitution resulted in a blood group B-active heptagalactosylceramide recently found in bovine erythrocyte membranes by Chien *et al.*[390]

1.4.4.3. Substitution of Lacto-N-noroctaosylceramide

Two unbranched lacto-series glycolipids with three repeating N-acetyllactosamine units have recently been isolated and characterized. These are: (1) fucosyl substitution in the penultimate GlcNAc residue, called Z_1 glycolipid, showing a remarkable reactivity with stage-specific embryonic antigen-1 (SSEA-1) monoclonal antibody[602] (see Chapter 4); (2) sialosyl substitution at the terminal Gal.[600,601]

1.4.4.4. Substitution of Lacto-N-isooctaosylceramide

1. Fucosyl substitution at the terminal Gal of both the 1→3 and the 1→6 side chain resulted in H_3 glycolipid.[120]
2. A^c glycolipid was derived from H_3 glycolipid by addition of two α-GalNAc residues at two H-active determinants.[310,407,409]
3. Fucosyl substitution at the 1→3 side chain and sialyl substitution at the 1→6 side chain resulted in fucoganglioside (or "G-9" ganglioside) of human erythrocytes.
4. Sialylation at the Gal terminal of the 1→6 linked chain resulted in I-active ganglioside (or "G-8" ganglioside) of human erythrocytes.
5. In bovine erythrocyte ganglioside, the 1→6 side chain was α-galactosylated, which showed a moderate I activity.[185] A large variety of glycolipids with different structural units have been prepared from this glycolipid to determine which structural moiety is responsible for a large variety of Ii activities (see Chapter 5).[185,417]

The lacto-series glycolipids are present mainly in extraneural tissues and organs. They are also the major gangliosides of peripheral nerve,[400,403] although their location in peripheral nerve is unknown. Their presence in a small quantity in the central nervous tissue has been reported. It is interesting to note that this class of glycolipid represents the major antigens with various blood group determinants such as ABH, Lewis, Ii, and P_1 of erythrocytes and epithelial cells as well as the major ganglioside of blood cells and other tissues.

The majority of fucosylated glycolipids (fucolipids) belong to this class and are the principal glycolipid of mucous membrane of gastrointestinal tract, and carry the major blood group antigens of mucous membrane components.[310]

A ceramide trisaccharide of this series (lacto-triaosyl ceramide) is present in very low concentrations in normal erythrocytes and tissues,[381,383] whereas human neutrophil leukocytes,[509] some human adenocarcinomas of the gastrointestinal tract,[383] and human melanoma tissue,[384] and erythrocytes of a hereditary human disorder of erythropoiesis, Hereditary Erythroblastic Multinuclearity; Positive Acidified Serum test (HEMPAS)[382] contain relatively large quantities of this glycolipid. Di- and trifucosylated polylactosaminolipids[602] and sialosyl-Le[610] are unique for certain human cancers and may represent human cancer-associated antigens (see Chapter 4).[170]

Recently, lacto-series glycolipids were shown to be susceptible to endo-β-galactosidase of *E. freundii in vitro* as well as *in situ*.[230,231] The susceptibility to this enzyme is a convenient diagnostic value to distinguish the lacto-series glycolipids from the other classes (see Section 1.3.5.3). A remarkable difference in the composition of this class of glycolipid between fetal and adult erythrocytes has been found[383]; namely, a straight-chain repeating N-acetyllactosamine structure is predominant in fetal erythrocytes that is converted to a branched structure as found in adult erythrocytes during ontogenic development.[383,434] Such a difference is now implicated as the basis of the blood group i and I difference.[184,417]

1.4.4.5. Polyglycosylceramide (Megaloglycolipids)

Glycosphingolipids with complex structures having as many as 30–50 carbohydrate residues have been isolated from human erythrocyte membranes and partially characterized as the carrier of blood group ABHI determinants.[84,85,505] A preliminary structural study indicates that this class of compounds has a poly-N-acetyllactosamine structure analogous to the "lacto-series" structure discussed in this section. The presence of such compounds was originally claimed by Gardas and Koscielak,[505,506] who found that the major blood group-active determinant of erythrocyte membranes was recovered in an aqueous phase of the n-butanol–water extract of stroma and that the purified substance on diethylaminoethyl (DEAE)–Sephadex chromatography consisted of 90% carbohydrate, less than 1% amino acids, and 2% sphingosine. Further purification including acetylation in formamide–pyridine–acetic anhydride (10 : 5 : 4) followed by dialysis in water resulted in an insoluble polyglycosylceramide acetate that was fractionated on silica gel chromatography. Based on the chemical composition and methylation analysis of the purified sample, Koscielak *et al.*[84] claimed the following general composition:

$$(\text{Fuc})_{3-4}(\text{Gal})_n(\text{GlcNAc})_{n-2}(\text{Glc})_1(\text{sphingosine})_1(\text{GalNAc})_{2-3}$$

1.4.5. Muco-Series Glycolipids: Glycolipids with Digalactosyl to Hexagalactosyl Core Structure

The presence of a series of glycolipids containing di- to hexagalactosyl glycosides linked to ceramides in gastrointestinal mucosae has been described by Slomiany and co-workers and more recently by Karlsson and associates. The former authors described glycolipids in hog gastrointestinal mucosa with the following common core structures:

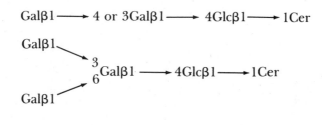

The first indication for the presence of this glycolipid series stems from the work by Slomiany and Horowitz[425] showing that blood group A glycolipid of hog gastric mucosa contains excess galactose, and these investigators established the exact structure in the following year.[426] In subsequent studies, they found the unique branching core structure Galβ1→ 3(Galβ1→ 6)Galβ1→4Glc in a large variety of glycolipids isolated from hog gastric mucosa.[429,430] They are the carriers of blood group A and H determinants (see Table XXIII). In a striking contrast, Karlsson and associates found in the nonepithelial tissue of rat intestinal mucosae the following series of poly-α-galactosyl glycolipids:

$$Gal\alpha1\rightarrow3Gal\alpha1\rightarrow3Gal\alpha1\rightarrow4Gal\alpha1\rightarrow4Glc\beta1\rightarrow1Cer$$

$$Gal\alpha1\rightarrow3Gal\alpha1\rightarrow3Gal\alpha1\rightarrow3Gal\alpha1\rightarrow3Gal\alpha1\rightarrow4Glc\beta1\rightarrow1Cer$$

$$Gal\alpha1\rightarrow3Gal\alpha1\rightarrow3Gal\alpha1\rightarrow3Gal\alpha1\rightarrow3Gal\alpha1\rightarrow4Gal\beta1\rightarrow1Cer$$

The claim for this novel structure was based on the presence of the dominant multiple signs for α-Gal protons as compared to β-Gal or β-Glc anomeric protons in proton magnetic resonance spectra (Section 1.3.9.2)[644] as well as on the expected mass numbers in mass spectra, particularly the "F" ion of the permethylated–reduced glycolipid fraction (see Section 1.3.8 and Fig. 13).[275,276,409] The presence of a ceramide tetrasaccharide (Hex-Hex-Hex-

Table XXIII. Muco-Series Glycosphingolipids

Structure	Reference
GalNAcα1→3Galβ1→3 or 4GlcNAcβ1→4Galβ1→1Cer 　　　　　　　2 　　　　　　　↑ 　　　　　　αFuc1	Isolation: from hog stomach mucosa, Slomiany and Horowitz[425]; structure: Slomiany et al.[426]
GalNAcα1→3Galβ1→3Galβ1→4Galβ1→Glcβ1→1Cer 　　　　　　　2 　　　　　　　↑ 　　　　　　αFuc1	Isolation and structure: from hog gastric mucosa, Slomiany et al.[427]
GalNAcα1→3Gal 1→3Gal 1→4Glcβ1→1Cer 　　　　　　2 　　　　　　↑ 　　　　　αFuc1	Isolation and structure: from hog gastric mucosa, Slomiany et al.[418]
GalNAcα1→3[Fucα1→2]Galβ1→4Galβ1 　　　　　　　　　　　　　　R₁ ⎫ 　　GalNAcα1→3Galβ1　　　　⎬ ³⁄₆Galβ1→4Glcβ1→1Cer 　　　　　　R₂ ⎫ 　　R₁ Galβ1 ⎬ 　　R₂ Galβ1	Isolation and structure: Slomiany and Slomiany[429]
R₁ Galβ1 　　　↗ ³⁄₆Galβ1→4Glcβ1→1Cer 　Galβ1	Slomiany and Slomiany[430]
Hex-Hex-Hex-Cer Hex-Hex-Hex-Hex-Cer HexN-Hex-Hex-Hex-Cer 　　　| 　　Fuc	Breimer et al.[275,276]
Galα1→3Galα1→3Galα1→3Galβ1→4Glc→Cer GalNAcβ1→3Galα1→3Galα1→3Galβ1→4Glc→Cer	Breimer et al.[569]

Hex-Cer), a ceramide pentahexoside (Hex-Hex-Hex-Hex-Hex-Cer), and a ceramide hexasaccharide with an A determinant linked to a ceramide trihexoside (HexN-[Fuc-]Hex-Hex-Hex-Cer) has been detected in cat small intestine glycolipids.[275,276] Although no enzymatic degradation or further chemical analysis of each component has been performed to establish these structures, these structures detected by direct-probe mass spectrometry must be similar to a muco-series analog. The evidence for claiming such a novel structure is quite sound, but further enzymatic studies to establish the poly-α-galactoside structure remain to be made. Contradiction between Slomiany's poly-β-galactosides and Karlsson's poly-α-galactosides could be due to a species difference.

The ganglioside with sialosyldigalactosylglucosylceramide reported by Feldman *et al.*[458] to be present in bovine lens and by Vance *et al.*[471] to be present in dog intestine could be a similar analog.

The polygalactosyl structure, Galβ→Galβ→Galβ, not only is limited to glycolipids but also was reported in glycoprotein. Slomiany and Meyer[431] described the core structure, Galβ1→3(Galβ1→6)Gal, in blood group glycoproteins of hog gastric mucosa. The polygalactosyl structure has been found in the peripheral region of glycolipids from human erythrocytes,[388] from frog fat ganglioside,[432] and from calf thymocyte membrane glycoproteins.[433]

1.4.6. Simpler Gangliosides and Hematosides: Sialosyl Glycolipids without Hexosamines (see Table XXIV)

1.4.6.1. Sialosylcerebroside (GM₄, G₇)

The simplest form of ganglioside is a sialosylcerebroside that has been found as a minor component of ganglioside. Kuhn and Wiegandt[71] and Wiegandt and Baschang[198] isolated a minor component of brain ganglioside that yielded sialosyl2→3galactose upon ozonolysis and alkaline degradation. Additional structural features of this glycolipid, particularly in the ceramide moiety, were studied by Klenk and Georgias,[436] Gielen,[437] and Siddiqui and McCluer,[438] and it has been termed GM_4, G_7, or Ggal. The ceramide of this glycolipid was characterized by the presenece of α-hydroxy fatty acids and the absence of C_{20}-sphingosine.[438] Thus, the ceramide moiety of this glycolipid is similar to that of cerebroside but dissimilar from that of normal brain gangliosides. It is conceivable that this glycolipid was synthesized from cerebroside, and this biosynthetic reaction has recently been demonstrated.[440] The glycolipid is much less polar than other gangliosides; therefore, it was not well extracted by Folch's partition. Recently, Ledeen *et al.*[439] found a relatively large quantity of this glycolipid in white matter of human brain by analysis using chromatography on DEAE–Sephadex without the Folch partition. The quantity of this glycolipid in lower-animal brain was found to be much less than that in human brain. Chimpanzee and monkey brain showed an intermediate value.[441,442] Thus, the glycolipid has been regarded as a characteristic component of myelin membranes,[443] and its level

Table XXIV. *Simpler Gangliosides and Hematosides (Sialosyl Glycolipids without Amino Sugars)*

Symbol or abbreviation	Structure	References
GM$_4^*$, G$_7$, Ggal	NeuAc2→3Galβ1→1Cer	Occurrence and isolation: from brain, Kuhn and Weigandt,[71] Klenk and Georgias,[436] Gielen[437]; structure: Siddiqui and McCleur[438]; relative abundance in human white matter and localization in myelin, Ledeen et al.[439]; occurrence: in rat kidney, Tadano and Ishizuka[435]; in mouse erythrocytes, Hamanaka et al.[549]
GM$_4$(Glc)*	NeuAc2→3Glc1→1Cer	Occurrence and isolation from sea urchin sperm: see Table XXVI
GM$_4^*$	NeuAc2→8NeuAc2→3Glc1→1Cer	
GM$_3$, G$_6$, G$_{Lac}$	NeuAc2→3Galβ1→4Glcβ1→1Cer	Occurrence and isolation: from brain, Kuhn et al.,[453] Svennerholm[455]; from human spleen, Svennerholm[362]; from dog erythrocytes, Klenk and Heuer[452]; structure: Handa and Yamakawa,[456] Puro[463]
	NeuGly2→3Galβ1→4Glcβ1→1Cer	Occurrence and isolation: from erythrocytes, Yamakawa and Suzuki,[47] Klenk and Wolter[48]; structure: Klenk and Padberg,[449] Handa and Yamakawa[456]

	4-OAc-NeuGly2→3Galβ1→4Glcβ1→1Cer	Occurrence and isolation: from bovine erythrocytes, Hakomori and Saito[454]; structure: Veh et al.[457]
	NeuAc→Gal→Gal→Glc→Cer (sequence of sugar arbitrary)	Occurrence and isolation: from lens, Feldman et al.[458]; from dog intestine, Vance et al.[471]
GD₃, G₃ₐ, G_{Lac}^2 II³(NeuGly)₂LacCer	NeuGly2→8NeuGly2→3Galβ1→4Glcβ1→1Cer	Isolation and structure: from cat erythrocytes, Handa and Handa[461]
II³(NeuAc)₂LacCer	NeuAc2→8NeuAc2→3Galβ1→4Glcβ1→1Cer	Isolation and structure: from bovine retina, Handa, Handa and Burton[462]; from bovine kidney, Puro[463]; from human, bovine, and rabbit retina, Holm et al.[464] Holm and Mansson[465]; mass spectrometric analysis, Holm et al.[467]; from elasmobranches brain, Avrova et al.[473]
II³NeuGly-NeuAcLacCer	NeuGly2→8NeuAc2→3Galβ1→4Glcβ1→1Cer	Isolation and structure: from rabbit thymus, Iwamori and Nagai[474]
GT₃, G_{Lac}^3 II³(NeuAc)₃LacCer	NeuAc→NeuAc→NeuAc2→3Galβ1→4Glcβ1→1Cer	Occurrence: in elasmobranches brain, Tettamanti et al.[475] Avrova[472]; structure: Avrova et al.[473]

increased with myelination.[439] These interesting studies carried out by Yu and Ledeen indicated a specific association of this ganglioside with a specific function of myelin membrane. However, a wider distribution of this glycolipid in various tissues other than myelin sheath has been noticed recently, the glycolipid having been found in mouse erythrocytes,[549] rat kidney,[435] and chicken egg yolk.[581]

Other simpler sialosylcerebrosides have been bound in sea urchin spermatozoa, and two of them have been identified by Nagai and associates as sialosyl2→6glucosylceramide and sialosyl2→8sialosyl2→6glucosylceramide.[444,445] Antibodies directed against these glycolipids were found to agglutinate various sea urchin spermatozoa, and the topological distribution of the glycolipids within the spermatozoa surface was studied.[446] Similar types of sialosylglucosyl glycolipids with different structures have been isolated from different strains of sea urchin spermatozoa by Kochetkov et al.,[447,448] as discussed in Section 1.4.8.

1.4.6.2. Sialosyllactosylceramide (Hematoside; GM₃, G₆, GLac, and Disialohematoside; GD₃)

Sialosyllactosylceramide was first isolated and characterized as the major glycolipid of horse erythrocytes by Yamakawa and Suzuki[47] and termed "hematoside." This was the first recognition of the presence of ganglioside in extraneural tissues and cells (see Section 1.1.2). The structure of horse hematoside was established as N-glycolylneuraminosyl2→3galactosylβ1→4glucosylβ1→1ceramide.[449,450,456] A glycolipid with the same carbohydrate composition, but having N-acetylneuraminic acid, was isolated from dog erythrocytes as the major component.[452] Later, a faster-migrating monosialoganglioside component of brain ganglioside was separated by chromatography and termed GM_3,[455] G_6, or GLac.[60,453] Hematoside having N-acetylneuraminic acid was found to be identical to GM_3 ganglioside of brain. The hematoside of horse erythrocytes includes a molecular species that has 4-O-acetyl-N-glycolylneuraminic acid and was resistant to hydrolysis by neuraminidase.[454,457]

Hematosides are widely distributed in almost all tissues and in most cultured cells, although their quantities vary significantly. Dramatic changes of this glycolipid associated with oncogenic transformation, cell contact, and cell cycle have been observed in some systems, suggesting a role of this glycolipid in maintenance of cell social activity and a possible function in cell growth regulation (see Chapter 4). While hematoside with an N-acetylneuraminic acid residue is a minor component of brain, it is the major component of astroglial cells in culture as described by Robert et al.[459] Although N-acetylhematoside is a minor component in neural tissue, it is the major component of various tissues and cells. The most convenient source for preparation of this glycolipid is dog erythrocytes, in which this glycolipid is the major component,[452] although some dog species contain N-glycolylhematoside.[450,456] Recently, a strain-specific distribution of N-glycolyl- and N-acetylhematoside among various dogs was found, and their genetic background has been studied[560,563] (see also Chapter 5). *N*-Glycolylhematoside has never been found in human tissues, whereas many other species contain this glycolipid as the major component. Interestingly, the serum of a patient who received an injection of animal serum for serum therapy contained antibodies (Hanganutziu–Deicher antibody) that are directed to *N*-glycolylhematoside[460,564] (see also Chapter 5). Horse erythrocytes are the most convenient source for preparation of *N*-glycolylhematoside.[47]

Disialosyllactosylceramide is also a very minor component in brain, but is present abundantly in some extraneural tissues and cells. The hematoside with a di-*N*-glycolylneuraminosyl residue was first discovered as the major glycolipid of cat erythrocytes by Handa and Handa.[461] Hematoside with a di-*N*-acetyl-neurominosyl residue, GD$_3$, was found as the major ganglioside of mammalian retinal tissue by Handa and Burton[462] and Holm and co-workers.[464–466] The composition of the ceramide moiety and the metabolism of this major retinal ganglioside have been studied.[465,466] and its structure was determined by mass spectrometry after methylation and reduction.[467] Despite a close relation to retina, optic nerve did not contain this glycolipid.[466] Relatively large quantities of this glycolipid with *N*-acetylneuraminic acid have been found in bovine kidney by Puro[463] and in chicken egg yolk by Li *et al.*[581] Probably these are the best sources for preparation of this glycolipid. On the other hand, the same glycolipid has been detected in relative abundance in human melanoma tissue,[468] chicken fibroblasts,[469] and human skin fibroblasts 8166.[470] The glycolipid is highly sensitive to cell contact and is deleted on cellular transformation in some isolated cell systems (see Chapter 4) and is regarded as the intermediate for biosynthesis of GD$_{1b}$ ganglioside (see Chapter 3).

Feldman *et al.*[458] described a hematoside type of ganglioside present in human lens that contained 2 moles of galactose and 1 mole each of glucose and sialic acid. A ganglioside with a similar carbohydrate composition was reported to be present in small intestine,[471] although the exact structures of these glycolipids remain to be elucidated. This may be related to muco-series glycosphingolipids (see Table XXIII).

1.4.7. Ganglio-Series Glycolipids: Gangliosides with Ganglio-N-triose, Ganglio-N-tetraose, and Ganglio-N-pentaose Structure
(see Table XXV)

1.4.7.1. Sialosylgangliotriaosylceramide (GM$_2$, G5, II^{3}NeuAcGgOse$_3$Cer) and Gangliotriaosylceramide (Asialo GM$_2$, GgOse$_3$Cer)

GM$_2$ ganglioside is a minor component of normal brain,[453,455] but is relatively abundant in myelin sheath membranes.[441] The glycolipid is well known as being accumulated in the brain of patients with Tay–Sachs' disease (Chapter 3). Identification of the accumulating material was the starting point of ganglioside study begun by Klenk about 40 years ago. This ganglioside was purified and characterized 30 years later by Ledeen and Salzman[62] and by Makita and Yamakawa[61] (see Table III). It is now known that GM$_2$ is dis-

Table XXV. Gangliosides with Ganglio-N-triose, Ganglio-N-tetraose, and Ganglio-N-pentaose Structure (Ganglio-series Gangliosides)

Symbol or abbreviation	Structure	Reference
A. Ganglio-N-triosyl derivatives Asialo GM_2 Ganglio-N-triosylceramide ($GgOse_3Cer$) "Tay–Sachs' globoside"	$GalNAc\beta1\rightarrow4Gal\beta1\rightarrow4Glc\beta1\rightarrow1Cer$	Occurrence—in Tay–Sachs' brain: Svennerholm and Raal,[476] Makita and Yamakawa,[61] Gatt and Berman[566]; structure: Makita and Yamakawa[61]; isolation and structure: from guinea pig erythrocytes, Seyama and Yamakawa[477], from mouse Kirsten tumor, Rosenfelder et al.[478]; from rat hepatoma, Hirabayashi et al.[540]
GM_2, G_5, G_{gtri}^{1} (II^3 NeuAcGgOse$_3$Cer)	$GalNAc\beta1\rightarrow4Gal\beta1\rightarrow4Glc1\rightarrow1Cer$ $\quad\quad\quad\quad\quad\quad\quad\quad\quad 3$ $\quad\quad\quad\quad\quad\quad\quad\quad\quad \uparrow$ $\quad\quad\quad\quad\quad\quad\quad\quad\alpha NeuAc2$	Accumulation in Tay–Sachs' brain, Klenk[31]; structure: Makita and Yamakawa,[61] Ledeen and Salsman[62]
GD_2 [II^3 (NeuAc)$_2$GgOse$_3$Cer]	$GalNAc\beta1\rightarrow4Gal\beta1\rightarrow4Glc\beta1\rightarrow1Cer$ $\quad\quad\quad\quad\quad\quad\quad\quad\quad 3$ $\quad\quad\quad\quad\quad\quad\quad\quad\quad \uparrow$ $\quad\quad\quad\quad\quad\quad\quad\quad\alpha NeuAc2$ $\quad\quad\quad\quad\quad\quad\quad\quad\quad 8$ $\quad\quad\quad\quad\quad\quad\quad\quad\quad \uparrow$ $\quad\quad\quad\quad\quad\quad\quad\quad\alpha NeuAc2$	Occurrence in human brain in trace amount and characterization: Kuhn and Wiegandt,[71] Klenk and Naoi[497]; occurrence in elasmobranches brain: Tettamanti et al.,[475] Avrova[472], structure: Avrova et al.[473]

B. Ganglio-*N*-tetraosyl derivatives

Asialo GM$_1$
(GgOseCer)

Galβ1→3Galβ1→4Glcβ1→1Cer

Occurrence: as rat erythrocyte antigen, Rapport et al.[504]; as mouse NK cell marker, Young et al.,[479] Kasai et al.[557]

Fucosyl-asialo GM$_1$
(IV^{2}FucGgOse$_4$Cer)

Fucα1→2Galβ1→3Galβ1→4Galβ1→4Glcβ1→1Cer

Occurrence and structure: in rat hepatoma, Matsumoto and Taki[539]

GM$_{1a}$, G$_4$, G$_{gtet}^1$
(II^{3}NeuAcGgOse$_4$Cer)

Galβ1→3GalNAcβ1→4Galβ1→4Glcβ1→1Cer
$$\underset{\displaystyle \alpha NeuAc2}{\overset{3}{\uparrow}}$$

Isolation and structure: from brain, Kuhn and Wiegandt,[60] Kuhn and Egge,[487] Svennerholm[59,455]

GM$_{1b}$
(IV^{3}NeuAcGgOse$_4$Cer)

NeuAc2→3Galβ1→3GalNAcβ1→4Galβ1→4Glcβ1→1Cer

Occurrence and structure: of biosynthesized product, Stoffyn et al.,[491] Yip[492]; in human erythrocytes, Watanabe et al.[606]; in rat hepatoma, Koizumi et al.[545]

GD$_{1a}$, G$_3$, G$_{gtet}^{2a}$
(IV^{3}NeuAcII^{3}NeuAcGgOse$_4$Cer)

Galβ1→3GalNAcβ1→4Galβ1→1Cer
$$\underset{\displaystyle \alpha NeuAc2}{\overset{3}{\uparrow}} \qquad \underset{\displaystyle \alpha NeuAc2}{\overset{3}{\uparrow}}$$

Isolation and structure: from brain, Kuhn et al.,[453,486] Kuhn and Wiegandt,[60] Kuhn and Egge,[233] Svennerholm[59,455]

GD$_{1b}$, G$_2$, G$_{gtet}^{2b}$
[II^3(NeuAc)$_2$GgOse$_4$Cer]

Galβ1→3GalNAcβ1→4Galβ1→4Glc→Cer
$$\underset{\displaystyle \underset{\displaystyle \alpha NeuAc2}{\overset{8}{\uparrow}}}{\overset{3}{\underset{\displaystyle \alpha NeuAc2}{}}}$$

Isolation and structure from brain, Kuhn and Wiegandt,[60] Kuhn et al.[453,486]

(continued)

Table XXV (Continued)

Symbol or abbreviation	Structure	Reference
GD$_{1a}$ with NeuAc and NeuGlyc	Galβ1→3GalNAcβ1→4Galβ1→4Glc1→1Cer 　　　　　　　3　　　　　　　　　3 　　　　　　　↑　　　　　　　　　↑ 　αNeuAc2　　αNeuGly2	Isolation and structure: Price et al.[493]
GT$_{1b}$, G$_1$, G$_{gtet}$^{3a} [IV³NeuAcII (NeuAc)$_2$GgOse$_4$Cer]	Galβ1→3GalNAcβ1→4Galβ1→4Glc→1Cer 　　　　　　　3　　　　　　　　　3 　　　　　　　↑　　　　　　　　　↑ 　αNeuAc2　　αNeuAc2 　　　　　　　　　　　　　　　　8 　　　　　　　　　　　　　　　　↑ 　　　　　　　　　　　　　　αNeuAc2	Isolation and structure: from brain, Kuhn et al.,[453,486] Kuhn and Wiegandt[60]
GT$_{1a}$ (IV³NeuAc$_2$ GgOse$_4$Cer)	Galβ1→3GalNAcβ1→4Galβ1→4Glc→1Cer 　　　　　　　3　　　　　　　　　3 　　　　　　　↑　　　　　　　　　↑ 　αNeuAc2　　αNeuAc2 　　　　　　　8 　　　　　　　↑ 　　　NeuAcα2	Isolation and structure: from brain, Ando and Yu[494]
GT$_{1b}$-O-Ac (9-O-acetylneuraminic acid containing GT$_{1b}$) [IV³NeuAcII³NeuAc (9-OAc)NeuAcGgOse$_4$Cer]	Galβ1→4GalNAcβ1→4Galβ1→4Glc1→1Cer 　　　　　　　3　　　　　　　　　3 　　　　　　　↑　　　　　　　　　↑ 　αNeuAc2　　αNeuAc2 　　　　　　　　　　　　　　　　8 　　　　　　　　　　　　　　　　↑ 　　　　　　　9-O-Ac-NeuAc2	Isolation and structure: from mouse brain, Ghidoni et al.[197]
GT$_{1c}$ [II³(NeuAc)$_3$GgOse$_4$Cer]	Galβ1→3GalNAcβ1→4Galβ1→4Glc1→1Cer 　　　　　　　3 　　　　　　　↑ 　αNeuAc2 　　　　　　　8 　　　　　　　↑ 　αNeuAc2	Occurrence suggested: Penick and McCluer[495] (structure has not been conclusively determined)

GQ₁b
(IV³NeuAc₂II³NeuAc₂GgOse₄Cer)

```
Galβ1→3GalNAcβ1→4Galβ1→4Glc1→1Cer
   3                    3
   ↑                    ↑
αNeuAc2              αNeuAc2
   8                    8
   ↑                    ↑
NeuAcα2              αNeuAc
```

Isolation and structure: from fish brain, Ishizuka and Wiegandt[489]; from cod brain, Ando and Yu[496]

GQ₁c
(IV³NeuAcII³NeuAc₃GgOse₄Cer)

```
Galβ1→3GalNAcβ1→4Galβ1→4Glc1→1Cer
   3                    3
   ↑                    ↑
αNeuAc2              αNeuAc2
   8                    8
   ↑                    ↑
αNeuAc2              αNeuAc2
   → 8αNeuAc2
```

Isolation and structure: from human, bovine, and chicken brain, Ando and Yu[496]

GP₁
(IV³NeuAc₂II³NeuAc₃GgOse₄Cer)

```
Galβ1→3GalNAcβ1→4Galβ1→4Glc1→1Cer
   3                    3
   ↑                    ↑
αNeuAc2              αNeuAc2
   8                    8
   ↑                    ↑
αNeuAc2              αNeuAc2
   8
   ↑
αNeuAc2
```

Isolation and structure: Ishizuka and Wiegandt[489]

C. Ganglio-*N*-pentaosyl derivatives
GD₁ₐGalNAc
(IV³NeuAcII³NeuAcGgOse₅Cer)

```
GalNAcβ1→4Galβ1→3GalNAcβ1→4Galβ1→4Glc1→1Cer
          3                    3
          ↑                    ↑
       αNeuAc2              αNeuAc2
```

Isolation and structure: Svennerholm et al.[86]

GM₁ₐGalNAc
(II³NeuAcGgOse₃Cer)

```
GalNAcβ1→4Galβ1→3GalNAcβ1→4Galβ1→4Glcβ1→1Cer
                               3
                               ↑
                            αNeuAc2
```

Isolation and structure: from human brain, Iwamori and Nagai[498]

(continued)

Table XXV (Continued)

Symbol or abbreviation	Structure	Reference
D. Ganglio-N-fucopentaosyl derivatives	L-Fucα1→2Galβ1→3GalNAcβ1→4Galβ1→4Glc1→1Cer 　　　　　　　　　　　3 　　　　　　　　　　　↑ 　　　　　　　　αNeuAc2 　(IV²FucII³NeuAcGgOse₄Cer)	Occurrence and characterization of the oligosaccharide: from bovine liver, Wiegandt[80]; from boar testis, Suzuki et al.[499]; from bovine brain, Ghidoni et al.[500]; from thyroid glands, Macher et al.[551]
	L-Fucα1→3Galβ1→3GalNAcβ1→4Galβ1→4Glc1→1Cer 　　　　　　　　　　　3 　　　　　　　　　　　↑ 　　　　　　　　αNeuGly2 　(IV³FucII³NeuGlyGgOse₄Cer)	Isolation and characterization: from pig adipose tissue, Ohashi and Yamakawa[501]
	L-Fucα1→2Galβ1→3GalNAcβ1→4Galβ1→4Glc1→1Cer 　　　　　　　　　　　3 　　　　　　　　　　　↑ 　　　　　　　　αNeuAc2 　　　　　　　　　　　8 　　　　　　　　　　　↑ 　　　　　　　　αNeuAc2 　(IV²FucII³NeuAc₂GgOse₄Cer)	Isolation and characterization: from pig cerebellum, Sonnino et al.[582]; from human brain, Ando and Yu[502]
E. Unusual polygalactosyl substitution	Galα1→3Galβ1→3Galα1→3GalNAcβ1→3Galβ1→4Galβ1→4Glcβ1→1Cer 　　　　　　　　　　　　　　　　　　　　　　　3 　　　　　　　　　　　　　　　　　　　　　　　↑ 　　　　　　　　　　　　　　　　　NeuAcα2	Isolation and characterization: from frog nervous tissue, Ohashi[591]

tributed widely in various extraneural tissues and cultured cells. Its presence has been reported in rat liver and hepatomas,[480] mouse fibroblasts,[481] and mouse tumors.[482]

A glycolipid termed asialo GM_2 ($GgOse_3Cer$, or ganglio-N-triaosylceramide) is absent or present only in trace amounts in normal human tissues, but accumulates in large amounts in the brains of patients with Tay–Sachs' disease,[483] and particularly in generalized gangliosidosis (Sandhoff's disease)[484,485] (see Chapter 3). However, a glycolipid with the same structure has been found to be the major component of guinea pig erythrocyte membranes, and its structure has been well established.[477] In various tissues of mice (BALB/c or DBA-2), the glycolipid was absent, or present only in very small quantities, but it is present abundantly in 3T3 cells transformed with the Kirsten strain of murine sarcoma virus and the tumors induced by inoculating such transformed cells[478] and in mouse lymphoma L5178.[642] Therefore, these glycolipids could be the tumor-associated markers (see Chapter 5).

1.4.7.2. Major Ganglioside of Brain (Sialosylgangliotetraosylceramide)

Since heterogeneity of gangliosides was demonstrated by cellulose chromatography[56] or by TLC[58] in a chloroform–methanol–water solvent system, a large-scale fractionation of brain gangliosides on a silicic acid column was carried out by Kuhn et al.,[453,486] who established the molecular species and finally found the common basic structure of brain ganglioside, G_I or GM_1. Its structure was established in the work of Kuhn and Wiegandt.[60] One monosialoganglioside (G_I or GM_1), two disialogangliosides (G_{II} and G_{III} or GD_{1a} and GD_{1b}), and one trisialoganglioside (G_{IV} or GT_{1b}) were separated from both human and calf brain. The correct chemical composition and structure of the four major gangliosides, G_I (GM_1), G_{II} (GD_{1a}), G_{III} (GD_{1b}), and G_{IV} (GT_1) were elucidated by Kuhn ad Wiegandt.[60] Independently, the correct carbohydrate sequence of the asialo core of gangliosides was suggested by Svennerholm[59]: (1) The carbohydrate composition of various ganglioside fractions indicated that the major gangliosides (G_{I-IV}) contain the same "asialo core," ganglio-N-tetraosylceramide.[60,233,455] (2) By enzymatic hydrolysis with sialidase, disialoganglioside G_{II} (GD_{1a}) and the trisialoganglioside G_{IV} (GT_{1b}) were all converted to monosialoganglioside G_I (GM_1).[60,455,483] (3) The carbohydrate sequence of ganglio-N-tetraose (asialo core structure) was assumed to be Gal→GalNAc→Gal→Glc→Cer by stepwise acid degradation.[59] (4) The definitive structure of ganglio-N-tetraose was determined by isolation and characterization of five oligosaccharides from the acetolysis product of a large amount of GM_1 ganglioside; the oligosaccharides characterized were (a) ganglio-N-triaose I (Galβ1→3GalNAcβ1→4Gal), (b) ganglio-N-triaose II (GalNAcβ1→4Galβ1→4Glc), (c) ganglio-N-biose I (Galβ1→3GalNAc), (d) ganglio-N-biose II (GalNAcβ1→4Gal), and (e) lactose (Galβ1→4Glc).[60] (5) A crucial experiment to establish the structure of GM_1 ganglioside was the isolation of various sialosyl oligosaccharides from the acetolysis product of GM_1 ganglioside.[60] Through acetolysis, the sialosyl residue was not readily released as compared with hydrolysis. The sialosyl oligosaccharides released and identified were: sialosyllactose (NAcNeu2→3Galβ1→4Glc), sialosylganglio-*N*-biose II [GalNAcβ1→4(NeuAc2→3)Gal], sialosylganglio-*N*-triaose I [Galβ1→3Gal-NAcβ1→4(NeuAc2→3)Gal], sialosylganglio-*N*-triaose II [GalNAcβ1→ 4(Neu-c2→3)Galβ1→4Glc], and sialosylganglio-*N*-tetraose [Galβ1→ 3GalNAcβ1→ 4(NeuAc2→3)Galβ1→4Glc]. Kuhn and Wiegandt[60] also discovered that the sialic acid of sialosyllactose was easily hydrolyzed by sialidase, but sialosyl residues of all other sialosyloligosaccharides were resistant to the action of sialidase. On the basis of these findings, the sialosyl residue of GM_1 ganglioside was definitively established as being linked to the middle galactose of ganglio-*N*-tetraose.[60] The difference in structure between G_I (GM_1) and other major gangliosides is due to the difference of sialosyl substitution.[60] Finally, these structures were proved by methylation analysis by Kuhn and Egge.[233]

These four gangliosides, G_{I-IV} or GM_1, GD_{1a}, GD_{1b}, GT_{1b}, were found in higher vertebrate brain as the major components, whereas they were only minor components in fish and frog brain in general, which contain highly sialylated gangliosides such as tetrasialo (GQ) and pentasialo (GP) gangliosides as seen in Table XXV. The presence of such highly sialylated gangliosides in brain of various fish (carp, codfish, haddock, and halibut) was first noted by Ishizuka *et al.*,[488] and their structure was established as ganglio-*N*-tetraosyl-ceramide with a NeuAc(α2→8)NeuAc(α2→8)NeuAc residue linked to the C-3 hydroxy of the nonterminal galactose as a common structure.[489] It is interesting to note that a new neuraminidase from fowl plague virus, which does not hydrolyze the NeuAc(α2→8)NeuAc linkage but specifically hydrolyzes the NeuAc(α2→3)Gal linkage,[490] was successfully used in this study.[489] A study by Avrova[472] also found relatively large quantities of tetra- and pentasialogangliosides and a relatively low concentration of monosialoganglioside in brain of carp, perch, and frog. However, Avrova observed that ray and lamprey brain contained a large quantity of monosialoganglioside; therefore, it is difficult to find a simple correlation between the brain ganglioside patterns and phylogenetic classification of brain. In a recent study by Avrova *et al.*,[473] a large quantity of GD_3 and GD_2 gangliosides was found in elasmobranches (cartilaginous fish) brain.

1.4.7.3. Minor Ganglioside Components and Fucoganglioside

Various minor gangliosides have been isolated and characterized from brain. Svennerholm *et al.*[86] isolated a new type of disialoganglioside with a pentasaccharide core that contained an additional GalNAcβ1→4 linkage to the terminal Gal of the known ganglio-*N*-tetraosylceramide. This new ganglioside was designated as GD_{1a}-GalNAc[86] or II^3,IV^3(NeuAc)$_2$GgOse$_5$Cer. Recently, great advances in systematic fractionation of gangliosides have been introduced by Nagai and associates.[114,123] A combination of DEAE–Sepharose chromatography with gradient elution and chromatography on the homogeneous porous silica gel column or silica gel TLC makes it possible to separate as many as 30 components from brain ["ganglioside mapping" (see Section

1.2.4.5)]. Through these new procedures, new minor gangliosides have been separated and characterized: GT_{1a} ganglioside,[494] CQ_{1b} ganglioside,[496] GQ_{1c} ganglioside[496] (see Table XXV, section B), a new GM_{1a}-GalNAc (GP) ganglioside[498] (see Table XXV, section C), a new type of GD_3,[474] and GD_{1b} ganglioside (containing a fucose residue)[502] (see Table XXV, section D). Two isomeric trisialogangliosides, GT_{1a} and GT_{1b}, have been isolated and well characterized. However, GT_{1c}, though its presence was suggested by the periodate oxidation study of Penick and McCluer,[495] has not been isolated or characterized.

Weigandt[80] first assumed the presence of a fucosylated GM_1 ganglioside in bovine liver when he obtained an oligosaccharide with the structure $Fuc\alpha1\rightarrow2Gal\beta1\rightarrow3GalNAc\beta1\rightarrow4(NeuAc2\rightarrow3)Gal\beta1\rightarrow4Glc$ released from the total ganglioside mixture of bovine liver. A ganglioside with the same structure has been isolated from bovine brain,[500] boar testis,[499] and hog adipose tissue.[501] The same fucoganglioside was isolated as one of the major gangliosides of thyroid glands[551] and of hepatoma H-35 cells.[97] Interestingly, one of the fucogangliosides isolated from hog adipose tissue contained an $Fuc\alpha1\rightarrow3Gal$ residue at the nonreducing terminal (see Table XXV, section D).

A GD_{1a}-type ganglioside containing both N-glycolylneuraminic acid and N-acetylneuraminic acid was isolated from bovine adrenal medulla[493] and from pig adipose tissue.[501] Similarly, a ganglioside with GD_3 structure, containing both N-glycolyl- and N-acetylneuraminic acid, has been isolated from rabbit thymus.[474]

Ganglio-N-tetraosylceramide (asialo GM_1, $GgOse_4Cer$) is normally absent in tissue except in brain of a patient with generalized gangliosidosis.[503] However, a novel lipid hapten of rat spleen and erythrocytes (cytolipin S) has been identified as showing a property similar to that of this glycolipid.[504] Antibodies directed to asialo GM_1 have been used to stain acute lymphatic leukemic cells with immunofluorescence[565] and to eliminate natural killer cell activity from lymphocyte populations in the presence of complement.[479,557]

1.4.8. Glycosphingolipids in Water-Living Invertebrates (see Table XXVI)

Glycosphingolipids of invertebrate species have been studied by several investigators. Although the studies have been limited to a few species, glycolipids with unusual carbohydrate composition and structure compared to higher animal glycolipids have been isolated. For example: (1) Mannose has never been found in glycolipids of higher animals, although it is found in plants,[248] but the majority of glycolipids in freshwater bivalves are mannosyl glycolipids, some having mannose 6-phosphate. (2) Fucose and sialic acids are usually present at the nonreducing external region of the carbohydrate chain in higher-animal glycosphingolipids. However, sea urchin eggs and sperm and starfish hepatopancreas contain sialic acid residues in an unusual internal position of the carbohydrate chain. Similarly, some of the neutral glycolipids of freshwater bivalves contain fucose residues at the internal position of the carbohydrate chain. (3) 3-O-Methyl or 4-O-methyl sugar residues are present

Table XXVI. Glycosphingolipids of Water-Living Invertebrates

Source	Structure	References
A. Gangliosides of sea urchin and starfish		
Sea urchin		
Pseudocentrotus depressus, Hemicentrotus pulcherrimus and Anthocideris, crassispina gametes	NeuAc2→6Glc1→1Cer NeuAc2→8NeuAc2→6Glc1→1Cer	Nagai and Hoshi,[444] Hoshi and Nagai[445]
Strongylocentrotus intermedius gonads	NeuAc2→6Glc1→8NeuAc2→6Glc→Cer NeuAc2→8NeuAc2→6Glc1→6Glc→Cer	Kotchetkov et al.[447] Kotchetkov et al.[448]
Echinocardium cordatum gonads	HSO_3→8NeuNGlyα2→6Glcβ1→1Cer	Prokazova et al.[507]
Strongylocentrotus intermedius eggs	NeuGlyα2→6Glcβ1→8NeuGlyα2→6Glcβ1→1Cer ↑ (±sulfate)	
Star fish		
Patiria pectinifera hepatopancreas	L-Arab1→3(&6)D-Gal1→3D-Gal1→4NeuAc2→Galβ1→4Glcβ1→1Cer	Kotchetkov and Smirnova[508]
Asterina pectinifera whole tissue	$Arab_p$1→6Gal_p1→3(8-0-Me)NeuNGlycα2→3Galβ1→4Glcβ1→1Cer $Arab_p$1→6Gal_p1→3NeuNGlycα2→3Galβ1→4Glcβ1→1Cer Gal_p1 ↑$^6_3$NeuNGlycα2→3Galβ1→4Glcβ1→1Cer $Arab_p$1→6Gal_p1	Isolation and partial characterization: Sugita and Hori[510]; structure: Sugita[511,512]

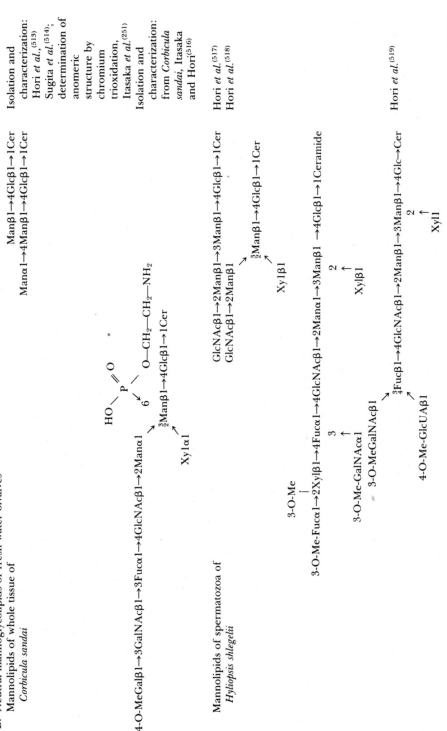

B. Neutral mannoglycolipids of fresh water bivalves
Mannolipids of whole tissue of *Corbicula sandai*

Manβ1→4Glcβ1→1Cer
Manα1→4Manβ1→4Glcβ1→1Cer

Isolation and characterization: Hori et al.,[513] Sugita et al.[514]; determination of anomeric structure by chromium trioxidation, Itasaka et al.[251]

4-O-MeGalβ1→3GalNAcβ1→3Fucα1→4GlcNAcβ1→2Manα1

$$\overset{3}{\underset{2}{\text{Man}}}\beta1\rightarrow4\text{Glc}\beta1\rightarrow1\text{Cer}$$

Xylα1

$$\text{HO}\diagdown\underset{\diagup}{\overset{\parallel}{P}}=O$$
$$\overset{6}{\diagup}\text{O}-\text{CH}_2-\text{CH}_2-\text{NH}_2$$
$$\overset{3}{\underset{2}{\text{Man}}}\beta1\rightarrow4\text{Glc}\beta1\rightarrow1\text{Cer}$$

Xylα1

Isolation and characterization: from *Corbicula sandai*, Itasaka and Hori[516]

Mannolipids of spermatozoa of *Hyliopsis shlegelii*

GlcNAcβ1→2Manβ1→3Manβ1→4Glcβ1→1Cer
GlcNAcβ1→2Manβ1

Hori et al.[517]
Hori et al.[518]

$$\overset{3}{\underset{2}{\text{Man}}}\beta1\rightarrow4\text{Glc}\beta1\rightarrow1\text{Cer}$$

Xylβ1

3-O-Me
|
3-O-Me-Fucα1→2Xylβ1→4Fucα1→4GlcNAcβ1→2Manα1→3Manβ1 →4Glcβ1→1Ceramide

3
↑
2
↑
Xylβ1

3-O-Me-GalNAcα1
3-O-MeGalNAcβ1

$$\overset{3}{\underset{4}{\text{Fuc}}}\beta1\rightarrow4\text{GlcNAc}\beta1\rightarrow2\text{Man}\beta1\rightarrow3\text{Man}\beta1\rightarrow4\text{Glc}\rightarrow\text{Cer}$$

2
↑
Xyl1

4-O-Me-GlcUAβ1

Hori et al.[519]

(continued)

Table XXVI (Continued)

Source	Structure	References
C. Glycolipids of sea bivalves Oyster *Osteria gigas*	3-O-MeFuc1 $\downarrow$ 2 GlcNAc1→3Gal1→4GalNAc1 $\searrow$ 3Gal1→3Gal1→4Glc→Cer 2 $\uparrow$ 3-O-MeGalNAc1→3Gal1 2 $\uparrow$ αFuc1	Matsubara and Hayashi,[520] Hayashi *et al.*[521]
Turbo cornutus muscle and viscera	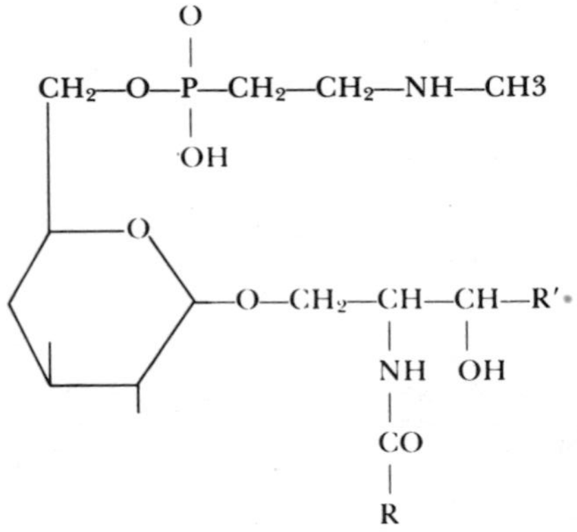	Matsuura[522]

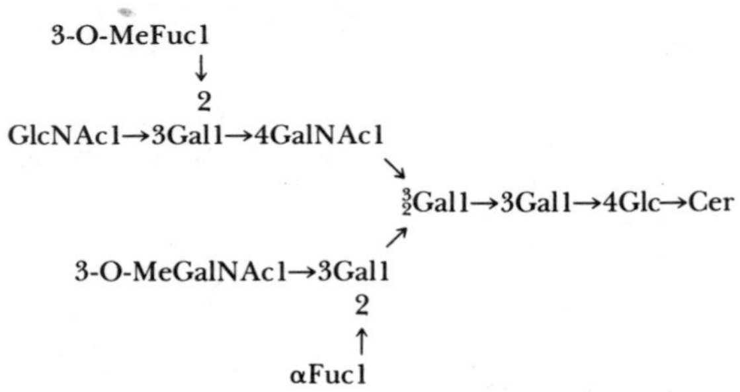

$$CH_2-O-P-CH_2-CH_2-NH-CH_3$$
$$O \quad OH$$

O—CH$_2$—CH—CH—CH—R'

NH   OH

CO

R

in various glycolipids of freshwater and seawater bivalves. (4) Unusual phosphonocarbon linkages were found in phosphoglycolipids of seawater and freshwater bivalve tissues.[522] (5) In addition, pentoses (arabinose and xylose) that are absent in higher-animal glycolipids are often present in those of water-living invertebrates. These unusual sugar compositions and structures may indicate that the carbohydrate structures of glycolipids may correlate to the stage of phylogenetic development.

Nagai and associates[444-446] described gangliosides of the sperm of sea urchins (*Pseudocentrotus depressus, Hermicentrotus pulcherrimus,* and *Anthocideris crassispina*) that have the structures sialosylglucosylceramide and disialosylglucosylceramide. Kochetkov *et al.*[447] described a similar ganglioside with different structure in the gonads of *Strongylocentrotus intermedius*. Surprisingly, they found the sialic acid as an internal residue of the carbohydrate chain.[447,508] A similar glycolipid with an internal sialosyl residue was also found in sea urchin eggs by Bergelson's group.[507] A tetrasaccharide, L-arabinosyl1→3(or 6)galactosyl1→3galactosyl1→4N-acetylneuraminic acid, and a lactosylceramide were isolated from a weak acid hydrolysate ganglioside isolated from hepatopancreas of a starfish (*Patiria pectinifera*). Thus, Kochetkov and associates[448,508] assigned the structure of this ganglioside as arabinosyldigalactosylsialic acid linked to a lactosylceramide. Independently, Sugita and Hori[510] isolated three gangliosides from whole tissue of a starfish (*Asterina pectinifera*). The ganglioside was degraded into lactosylceramide and tetra- or trisaccharide, arabinosyl1→6galactosyl1→4sialic acid, or arabinosyl1→ 6galactosyl1→4(galactosyl1→6)sialic acid on weak acid hydrolysis.[511,512] Thus, the presence of an unusual ganglioside having an internally located sialic acid was confirmed by three independent laboratories. A sulfated sialosylglucosylceramide was found in gonads of sea urchin (*Echinocardium cordatum*) by Kochetkov *et al.*[448] (see Table XXVI).

The chemical composition and structure of various sphingolipids of bivalves have been extensively investigated by two groups of investigators, Hori and his associates at Shiga University, and Hayashi and his associates at Kinki University, Japan. The former group studied freshwater bivalves and the latter group studied seawater bivalves. Glycolipids isolated from gonads and sperm of a bivalve (*Hyliopsis schlegelii*) or whole tissue of *Corbicula sandai* have been extensively investigated by Hori and his associates, and some of their structures have been characterized as Manβ1→4Glc, Manα1→4Manβ1→4Glc, or GlcNAcβ1→2Manβ1→3Manβ1→4Glc or their derivatives.[513,519] Mannose has never been found in higher-animal glycolipids.

Glycolipids of oysters have been investigated by Matsubara and Hayashi.[520] Methylation analysis of the one major oyster glycolipid showed a highly branched structure on galactose, since 4,6-di-*O*-methylgalactose was the major methylated sugar detectable.[520] Recently, these investigators found 3-*O*-methylgalactosamine and 3-*O*-methylfucose in the same oyster glycolipid,[523] and the entire structure of this highly branched glycolipid was determined as shown in Table XXVI (section C), although the structure is still tentative.[521]

Matsuura[522] described a new type of phosphonoglycolipid isolated from *Turbo cornutus* that was identified as *O*-[6'-(*N*-methylaminoethylphosphonyl)-galactopyranosyl]ceramide. Phosphonolipids have been found widely distributed in various bivalves and sea animals, almost replacing sphingomyelin; i.e., *N*-acyl-sphingosyl-1-*O*-(2'-aminoethyl)phosphonate(cereamide-2-aminoethyl-phosphonate) or *N*-acyl-sphingosyl-1-*O*-(*N*-methyl-2'-aminoethyl)phosphonate is the major sphingolipid in bivalve tissues.[524–526]

1.4.9. Plant Sphingolipids (Phytoglycosphingolipids) (see Table XXVII)

Through a series of classical studies by Carter *et al.*,[21,527–529] the presence of unique glycophosphosphingolipids in various plant seeds has been well established. This class of glycolipids is usually found in the phospholipid fraction and is characterized by the presence of an LCB similar to sphingosine that is now established as a phytosphingosine group (see Section 1.3.2). The phytosphingosine group was identical to the compound originally reported as "cerebrin base" that was isolated from fungus by Zellner[22] and from yeast by Reindel *et al.*[23] The glycolipid containing phytosphingosine was first isolated from a corn phospholipid fraction and named phytosphingolipid.[21] Subsequently, a similar compound with a similar chemical composition was isolated from soybean, peanuts, cotton seed, sunflower seeds,[527] and flax seeds.[528] The compound has been characterized by having glucosamine, glucuronic acid, mannose, and inositol phosphate and a ceramide with phytosphingosine as the common components. The compounds represent glycophosphoceramides containing phosphoinositol, and the glycolipids studied by Carter and associates were heterogeneous, but the major oligosaccharide was isolated in a large quantity from corn seeds after base and acid degradation, and was chemically characterized as tetrasaccharide with the following structure:

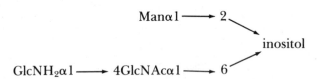

On the basis of this finding, the common core structure of phytosphingolipid was assumed by Carter *et al.*[529] to be that shown in Table XXVII. In addition to the core structure, it was suggested that other sugar components such as fucose, arabinose, and galactose could be attached to the nonreducing terminal sugar, either glucosamine or mannose (for a review, see Carter *et al.*[530]).

The chemistry of phytosphingolipids in tobacco leaves and fungi has been further developed by Lester and associates,[531-534] and the metabolism of yeast phytosphingolipids and inositides has been studied in association with inositol-less death of the fungus *Saccharomyces cervisiae* by the same investigators.[535] These workers first developed a preparation method of phytosphingolipids from tobacco leaves and obtained eight purified phosphoinositides through

Table XXVII. Structures of Phytoglycosphingolipids

Source	Structure	References
Corn and soybean seeds	Manα1 ↘ $\frac{2}{6}$Inositol—O—P—O→Cer ↗ GlcNH$_2$α1→4GlcUAα1	Carter *et al.*[529,530]
Peanuts	Gal ⌉ Arab ⌡ →GlcNH$_2$—GlcUA—Inositol—P—O→Cer Man	Wagner *et al.*[612]
Fungi *Saccharomyces cervisiae*	Man→Inositol—O—P—O→CerC$_{(26\,:\,0)}$ C$_{(26\,:\,0,\alpha—OH)}$	Smith and Lester,[536] Wagner and Zofcsik[611]
Neurospora crassa	Inositol—O—P—O→Inositol—O—P—O→Cer C$_{(26\,:\,0,\alpha—OH)}$ ↑ Man	Lester *et al.*[537]
	Gal→Gal→Gal→Glc→Cer C$_{(26\,:\,0,\alpha—OH)}$	Lester *et al.*[537]
Tobacco leaves (with mild procedure for isolation)	GlcNAcα1→4GlcUAα1→2Inositol 1→O—P—O→Cer (PSL-I) GlcNH$_2$α1→4GlcUAα1→2Inositol 1→O—P—O→Cer Arab ⌉ Gal ⌡ →GlcNAc (or GlcNH$_2$)α1→4GlcUAα1→2Inositol→1—O—P—O→Cer Man	Kaul and Lester[531,532] Hsieh *et al.*,[533] Lester *et al.*[534]
Oligosaccaride identified	Galβ1→4GlcNAcα1→4GlcUAα1→2Inositol Gal 1→6Galβ1→4GlcNAcα1→4GlcUAα1→2Inositol Arabl→6Galβ1→4GlcNAcα1→4GlcUAα1→2Inositol	Hsieh *et al.*,[613] Laine[614]

DEAE–cellulose chromatography followed by chromatography on "Porasil."[529,532] Chemical degradation studies with the major fractions (PSL-I and -II) indicated that: (1) the lipid-free fraction obtained was composed of phosphorus, inositol, glucosamine, and hexuronic acid; (2) all of the phosphorus was bound to inositol; and (3) the hexuronic acid was exclusively glucuronic acid. In collaborative work with Laine and his co-workers, the glucosamine was found to be linked to C-4 of the glucuronic acid.[533] The unusual part of this structure was that the glucuronic acid was shown to be attached to the 2-position of the inositol,[533] whereas Carter's findings with the corn "phytoglycolipid" indicated a mannose at this position and a uronic acid at the 6-position of inositol.[529] PSL-I differed from PSL-II only in having an *N*-acetyl function on the glucosamine. Of the glycophosphosphingolipids, 50% had a free amino group on this sugar.[533] Higher homologs of these structures have been studied further by collaboration between Laine and Lester. A major tetrasaccharide was shown to be Gal(β1→4)GlcNAc (α1→4)GlcUA(α1→2)Inositol,[584] and preliminary structures indicate that further substitutions on the galactose occur with arabinose in the pyranoside and furanoside forms, and galactose.[585]

A unique feature of these glycolipids is the attachment of the sugar chain to the ceramide via phosphate. Hence, these substances may be grouped under the descriptive term "glycophosphosphingolipids."[533,584] Plants have also been shown to contain mannosyl cerebrosides by Laine and Renkonen,[586] who isolated mono- to trihexaosyl ceramides containing mannose, but no phosphate or inositol. Thus, plants also contain true glycosphingolipids. Chloroplast membranes and other organelles of plants have been shown to contain large amounts of glycosyl diglycerides and other glycosides of hydrophobic aglycons such as steroids. Description of these is beyond the scope of this chapter.

Lester and associates[535–537] also studied sphingophosphoinositides of fungi and their metabolism. Three glycosphingolipids having mannosyl-inositol-phosphate have been isolated from *Saccharomyces cervisiae* (see Table XXVII). These lipids were characterized by the presence of a unique, very long fatty acid ($C_{26} = 0$ or $C_{26} = 0$ with a hydroxyl group,[535–537] and a novel tetrahexosylceramide, trigalactosyl glucosylceramide, and di(inositolphosphonyl)ceramide were isolated from *Neurospora crassa*.[537]

1.5. Chemical Synthesis and Modification of Sphingosines and Glycosphingolipids

To elucidate the diastereomeric structure of natural sphingosines, phytosphingosines, and their analogs, chemical synthesis of these compounds has been a subject of considerable interest among chemists (see Section 1.4.1). On the basis of chemical syntheses, the structure of natural sphingosine was determined as D-*erythro* (2*S*,3*R*) and that of phytosphingosine as D-*ribo* (2*S*,3*R*,4*R*) (see Tables XVIII and XIX). The synthesis of diastereomers of sphingosine and ceramide is also important for the study of metabolism and enzymatic conversion of sphingosine to ceramide and to various sphingolipids, the area extensively studied by Stoffel and associates and Kishimoto and associates (see Chapter 3). The major chemical syntheses of long-chain bases and sphingo-glycolipids are listed in Table XXVIII; the subject was reviewed by Shapiro[621] in 1969.

Table XXVIII. Chemical Synthesis of Sphingosine and Glycosphingolipids

A. Sphingosines

DL-sphingosine Isolation of *erythro*-isomer by fractional crystallization	Shapiro and Segal,[538] Shapiro *et al.*[66]	Reduction of DL-*erythro*-2-acetamido- 3-keto-4-*t*-octadecanoic acid
Synthesis and separation of D- and L-sphingosines	Shōyama *et al.*[553]	Resolution of DL-*erythro*-2-acetamido- 3-hydroxy-4-*t*-octadecanoate through L(+)- acetylmandeloylester
Differential synthesis of *trans*-, *erythro*-, and *threo*-; *cis*-, *erythro*-, and *threo*-	Grob and Gadient[541]	Condensation of 2-hexa-decanal and 2-nitro-ethanol, followed by reduction

B. Dihydrosphingosine

Dihydrosphingosine DL-*Threo, erythro*- mixture	Gregory and Malkin[542]	Oxamination of 3-keto- octadecanoate, followed by reduction with LiAlH₄
Dihydrosphingosine	Egerton *et al.*,[543] Grob *et al.*[544]	Condensation of palmital and β- nitro-ethanol
D-*Erythro*-dihydrosphingosine	Carter and Shapiro,[64] Carter *et al.*[65]	LiAlH₄ reaction of α-benzamido-β- hydroxyoctadecanoic acid methylester
D-*Erythro*- or D-*threo*-dihydro- sphingosine	Jenny and Grob[319,547]	Ammonolysis of *trans*- or *cis*-epoxide
D-*Erythro*-dihydrosphingosine	Shapiro *et al.*[66]	Reduction of β-*N*-acetyl-amino-β- keto-octanoate by LiAlH₄, followed by treatment with methyl-dichloroacetate and differential crystallization
¹⁴C-labeled DL-dihydro- sphingosine followed by racemic separation	Stoffel *et al.*[550]	Separation of DL-sphingosine of Shapiro *et al.*[66] into D- and L- diasteromers

C. Phytosphingosine

D-*Ribo*-D-amino-1,3,4-trihydroxy- sphinganine	Gigg *et al.*,[352] Gigg and Gigg[552]	Glucosamine derivative (methyl-α-2- benzamido-2-deoxygluco- furanoside) is converted to the oxazoline, the 2-allosamine derivative; compound is converted by oxidation to aldehyde, which is then coupled to olefin by Wittig reagent—triphenyl-tridecetyl phosphonium bromide; resulting product is reduced by catalytic hydrogenation

D. Cerebrosides

(cerasine, phrenosine, and glucocerebroside)	Shapiro and Flowers[68]	*N*-Fatty acyl-3-*O*-benzoyl- sphingosine condensed with acetobromo sugars by

(continued)

Table XXVIII. (Continued)

		Koenig–Knor type reaction; $Hg(CN)_2$ as a catalyst
E. Lactosylceramide (cytolipin H)	Shapiro *et al.*[67]	*N*-Fatty acyl-3-*O*-benzyol-sphingosine condensed with hepta-acetyllactosyl bromide; $Hg(CN)_2$ as a catalyst
F. Ganglio-triaosylceramide (Tay–Sachs' globoside)	Shapiro *et al.*[69]	Condensation of deca-acetylganglio-triosyl-chloride and *N*-stearoyl-3-*O*-benzoyl sphingosine
G. Sulfatide (3-*O*-sulfate-galactosylceramide)	Flowers[556]	Reaction of 2,4,6-tri-*O*-acetyl-α-D-galacto- pyranosyl bromide with *N*-octadecanoyl 3-*O*-benzoyl-DL-dihydro-sphingosine
H. Synthesis of glycolipid analogs the structure of which is absent in nature Galβ1→6Galβ1→1Ceramide Galβ1→3Galβ1→1Ceramide	Flowers[554] Flowers[555]	Reaction analygous for synthesis of lactosylceramide (see item E, above)

1.5.1. Synthesis of Long-Chain Bases

1.5.1.1. Synthesis of Sphingosine (Sphingenine)

The first total synthesis of DL-sphingosine (octadecasphingenine) and isolation of the *erythro*-isomer by fractional crystallization was performed by Shapiro and co-workers.[66,538] Myristic aldehyde, which was obtained by reduction of myristic acid chloride, was reacted with malonic acid by the Knoevenagel–Doebner condensation. The resulting *trans*-2-hexadecenoic acid was converted to its acyl chloride (Fig. 22, compound 1) and condensed with ethyl-sodio-aceto-acetic acid ($CH_3COCH_2COOC_2H_5$). The resulting α,α-diacyl ester (compound 2) was reacted with benzene diazonium chloride through the Japp–Klingermann reaction to give the phenylhydrazone compound. The phenylhydrazone compound (compound 3) was then converted to 2-*N*-acetyl-3-keto-octanoic acid ester (compound 4) by zinc and acetic acid. Compound 4 was reduced with sodium borohydride to give the *threo*- and *erythro*-isomers of 2-*N*-acetyl-3-hydroxy-octa-decenoic acid ethyl ester (compound 5), from which the pure *erythro*-isomer was isolated by fractional crystallization (compound 5′). Compound 5′ was hydrolyzed in ethanol HCl to give a pure 2-amino-3-hydroxy-4-*trans*-octadecenoic acid ethyl ester hydrochloride (compound 6). Compound 6 was finally reduced to DL-*erythro*-*trans*-sphingosine (DL-*erythro*-2-amino-3-hydroxy-4-*trans*-octadecene). The entire procedure is described in Fig. 22A.

Grob and Gadient[541] developed a quite different procedure for differential synthesis of *trans-erythro, trans-threo, cis-erythro,* and *cis-threo* isomers of sphingosine as shown in Fig. 23. The starting material was a 2-hexadecynal-1, an aldehyde containing a triple bond (compound 1) that was synthesized by the Grignard reaction of 1-penta-decyne. 2-Hexadecynal-1 was condensed with 2-nitro-ethanol and gave a mixture of *cis-* and *threo*-isomers containing

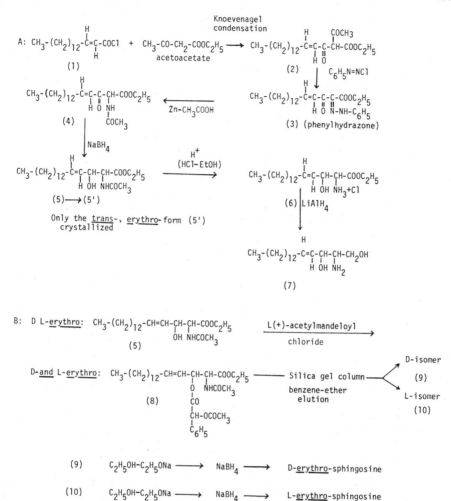

Figure 22. A, Synthesis of DL-sphingosine, followed by differential crystallization according to the methods of Shapiro and Segal.[538] B, Synthesis of DL-*erythro*-sphingosine and separation into D- and L- diastereomers through L(+)-acetylmandeloyl esters according to the method of Shōyama et al.[553]

the 2-nitro-1,3-diol structure (compounds 2 and 3). Compound 2 was reduced with aluminum-amalgam to give the *threo*-2-amino-1,3-diol compound (compound 4), which was converted to the *trans-threo* compound by reduction with Na–BuOH (compound 5). On the other hand, compound 4 was reduced with Pd–H$_2$ to a *cis-threo* compound (compound 6). Compound 3, the *erythro*-nitro-diol, was reduced by Zn–HCl to the *cis*-2-amino-1,3-diol compound (compound 7). The triple bond of compound 7 was reduced by LiAlH$_4$ to the *trans-erythro* compound (compound 8) and reduced with Pd–H$_2$ to the *cis-erythro* compound (compound 9). The procedure was based on differential reduction of triple bond to *cis-* and *trans-*olefin under different conditions. On the other hand, selection of *threo* and *erythro* was based on differential

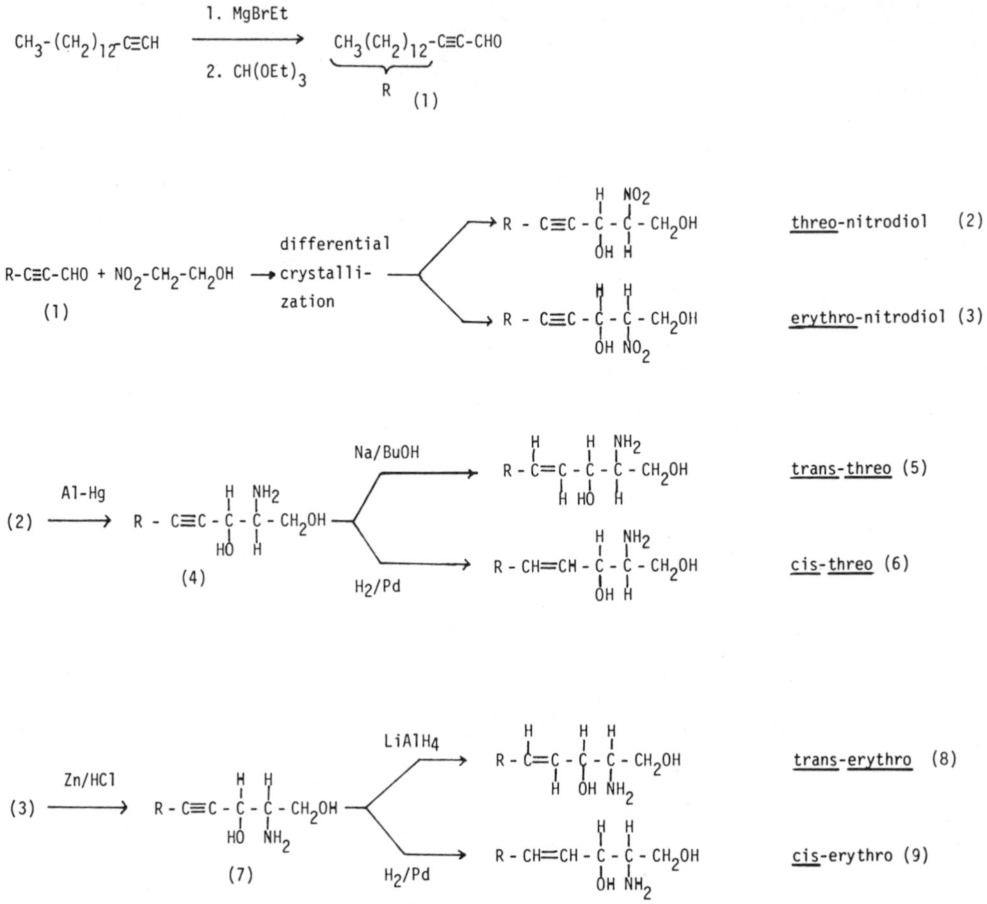

$$CH_3-(CH_2)_{12}-C\equiv CH \xrightarrow[\text{2. CH(OEt)}_3]{\text{1. MgBrEt}} \underbrace{CH_3(CH_2)_{12}-C\equiv C-CHO}_{R \quad (1)}$$

Reaction scheme (2) and (3):

$$R-C\equiv C-CHO + NO_2-CH_2-CH_2OH \xrightarrow[\text{zation}]{\substack{\text{differential}\\\text{crystalli-}}}$$

(1)

$$R - C\equiv C - \underset{\underset{OH}{|}}{\overset{\overset{H}{|}}{C}} - \underset{\underset{H}{|}}{\overset{\overset{NO_2}{|}}{C}} - CH_2OH \qquad \underline{threo}\text{-nitrodiol} \quad (2)$$

$$R - C\equiv C - \underset{\underset{OH}{|}}{\overset{\overset{H}{|}}{C}} - \underset{\underset{NO_2}{|}}{\overset{\overset{H}{|}}{C}} - CH_2OH \qquad \underline{erythro}\text{-nitrodiol} \quad (3)$$

$$(2) \xrightarrow{\text{Al-Hg}} R - C\equiv C - \underset{\underset{HO}{|}}{\overset{\overset{H}{|}}{C}} - \underset{\underset{H}{|}}{\overset{\overset{NH_2}{|}}{C}} - CH_2OH$$

(4)

$$\xrightarrow{\text{Na/BuOH}} R - \overset{\overset{H}{|}}{C} = \underset{\underset{H}{|}}{\overset{\overset{H}{|}}{C}} - \underset{\underset{HO}{|}}{\overset{\overset{H}{|}}{C}} - \underset{\underset{H}{|}}{\overset{\overset{NH_2}{|}}{C}} - CH_2OH \qquad \underline{trans\text{-}threo} \quad (5)$$

$$\xrightarrow{\text{H}_2/\text{Pd}} R - CH=CH - \underset{\underset{OH}{|}}{\overset{\overset{H}{|}}{C}} - \underset{\underset{H}{|}}{\overset{\overset{NH_2}{|}}{C}} - CH_2OH \qquad \underline{cis\text{-}threo} \quad (6)$$

$$(3) \xrightarrow{\text{Zn/HCl}} R - C\equiv C - \underset{\underset{HO}{|}}{\overset{\overset{H}{|}}{C}} - \underset{\underset{NH_2}{|}}{\overset{\overset{H}{|}}{C}} - CH_2OH$$

(7)

$$\xrightarrow{\text{LiAlH}_4} R - \overset{\overset{H}{|}}{C} = \underset{\underset{H}{|}}{\overset{\overset{H}{|}}{C}} - \underset{\underset{OH}{|}}{\overset{\overset{H}{|}}{C}} - \underset{\underset{NH_2}{|}}{\overset{\overset{H}{|}}{C}} - CH_2OH \qquad \underline{trans\text{-}erythro} \quad (8)$$

$$\xrightarrow{\text{H}_2/\text{Pd}} R - CH=CH - \underset{\underset{OH}{|}}{\overset{\overset{H}{|}}{C}} - \underset{\underset{NH_2}{|}}{\overset{\overset{H}{|}}{C}} - CH_2OH \qquad \underline{cis\text{-}erythro} \quad (9)$$

Figure 23. Differential synthesis of *trans-erythro, trans-threo, cis-erythro,* and *cis-threo* isomers of sphingosine according to the method of Grob and Gadient.[541]

crystallization. In the methods of both Shapiro and colleagues and Grob and Gadient, DL-isomers were not separated or well defined, although *threo, erythro,* and *cis* and *trans* configurations were well defined. A selective synthesis of D-*erythro-trans*-octadecasphingenine has not been performed until recently. An unequivocal synthesis of D- and L-sphingosine was performed by Shōyoma *et al.*[553] based on the separation of D-and L-diastereomers of *erythro*-2-acetamido-3-O-[L(+)-acetomandeloyl]-4-*trans*-octadecenoic acid ethylester through silicic acid chromatography. The isolated D-isomers of the L(+)-acetomandeloyl ester were readily converted to D-*erythro*-sphingosine after treatment with sodium ethoxide in ethanol and reduction with sodium borohydride (see Fig. 22B). Thus, D- and L-*erythro*- and *threo*-diasteromers of sphingosine and their N-fatty acyl derivatives (ceramides) have been totally synthesized.[553] When compound 9 in Fig. 22B is reduced with NaBH₄, it can be labeled with tritium when reduced with NaB[³H]₄.[553]

1.5.1.2. Synthesis of Dihydrosphingosine

Since the synthesis of well-defined diastereomers of sphingosines was very difficult and has not been performed until recently, the diastereomeric structure of natural sphingosine was identified through its dihydro derivative by comparison with synthetic dihydro derivatives. Dihydrosphingonine (octadecasphinganine) has been synthesized by a number of investigators through various procedures. Only two of them will be explained here. Carter *et al.*[64,65] obtained D-*erythro*-dihydrosphingosine (D-*erythro*-octadecasphinganine) through reduction of the sterically defined D-2-benzamido-3-keto-octadecanoic acid methylester (Fig. 24, compound 1) or D-2-benzamido-3-hydroxy-octadecanoic acid methylester (compound 2). The reduction product, D-2-benzamido-3-hydroxy-octadecanol (compound 3) was reduced by H_2–Pd to D-*erythro*-dihydrosphingosine. The sterically well-defined compound 1 or 2 was synthesized from palmitoyl chloride with oxazolin compound.

Jenny and Grob[547] developed another procedure for synthesis of D-

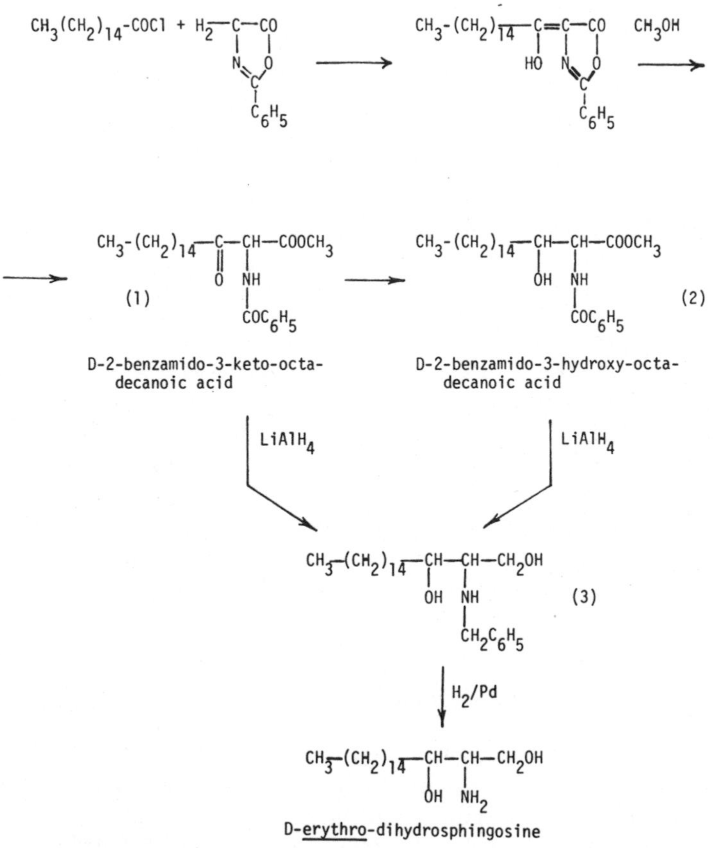

Figure 24. Synthesis of D-*erythro*-dihydrosphingosine according to the method of Carter and Shapiro,[64,546] and Carter *et al.*[65]

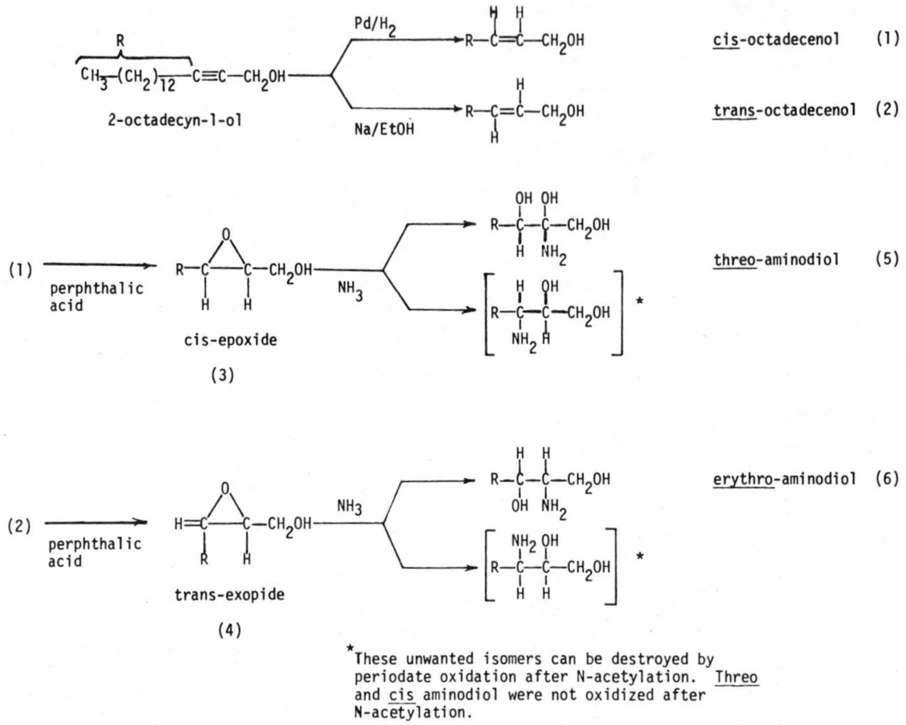

R

$CH_3-(CH_2)\overline{12}-C\equiv C-CH_2OH$

2-octadecyn-1-ol

Pd/H$_2$

$R-C=C-CH_2OH$ cis-octadecenol   (1)

$R-C=C-CH_2OH$ trans-octadecenol (2)

Na/EtOH

(1) $\xrightarrow{\text{perphthalic acid}}$ $R-C\overset{O}{—}C-CH_2OH$

cis-epoxide

(3)

NH$_3$

$R-\overset{OH}{\underset{H}{C}}-\overset{OH}{\underset{NH_2}{C}}-CH_2OH$ threo-aminodiol   (5)

$\left[R-\overset{H}{\underset{NH_2}{C}}-\overset{OH}{\underset{H}{C}}-CH_2OH\right]$ *

(2) $\xrightarrow{\text{perphthalic acid}}$ $H=C\overset{O}{—}C-CH_2OH$

trans-exopide

(4)

NH$_3$

$R-\overset{H}{\underset{OH}{C}}-\overset{H}{\underset{NH_2}{C}}-CH_2OH$ erythro-aminodiol   (6)

$\left[R-\overset{NH_2}{\underset{H}{C}}-\overset{OH}{\underset{H}{C}}-CH_2OH\right]$ *

*These unwanted isomers can be destroyed by periodate oxidation after N-acetylation. Threo and cis aminodiol were not oxidized after N-acetylation.

Figure 25. Synthesis of dihydrosphingosine according to the method of Jenny and Grob.[547]

erythro- and D-*threo*-octadecasphinganine based on ammonolysis of *cis* and *trans* epoxide as shown in Fig. 25. The starting material was 2-octadecyn-1-ol, which was converted to *cis*- and *trans*-octadecynol (compounds 1 and 2). Compound 1 was converted to *cis*-epoxide (compound 3) and compound 2 to *trans*-epoxide (compound 4) by perphthalic acid. The *cis*-epoxide was converted to *threo*-aminodiol [D-*threo*-aminodiol (D-*threo*-sphinganine) (compound 5)] and the *trans*-epoxide to *erythro*-aminodiol [D-*erythro*-sphinganine (compound 6)] by ammonolysis. During the ammonolysis of epoxides, 3-amino-1,2-diol compound was produced that was eliminated by periodate oxidation after N-acetylation.

1.5.1.3. Synthesis of Phytosphingosine

The structure of phytosphinganine was identified by Carter and Hendrickson[358] as D-*ribo*-1,3,4-trihydroxy octadecasphinganine. A stereospecific synthesis with a starting material of a known configuration based on D-glucosamine was performed by Gigg and colleagues.[352,552] The 5,6-*O*-isopropylidine derivative of phenyl-oxazoline-D-glucosamine was converted to phenyl oxazoline furanoside of glucosamine by removal of isopropylidene (Fig. 26, compound 1). Compound 1 was periodate oxidized to give an aldehyde compound of phenyl oxazoline derivative (compound 2), which was condensed with the Wittig reagent prepared from triphenyl-*N*-tridecylphosphonium bromide or *N*-pentadecyltriphenylphosphonium bromide to give the corresponding olefin derivative (compound 3). Compound 3 was reduced with H_2–Pd to the corresponding saturated derivative (compound 4). The acid hydrolysis of the oxazoline ring of compound 4 resulted in the salt of *O*-acyl amino alcohol (compound 5), and the conversion of the *O*-acyl amino alcohol into the *N*-acyl derivative (compound 6) under basic conditions is known to proceed without inversion of the configuration. Thus, compound 6 can be readily converted to D-*ribo*-1,3,5-trihydroxy-*N*-benzoyl octadecasphinganinn Fig. 26.

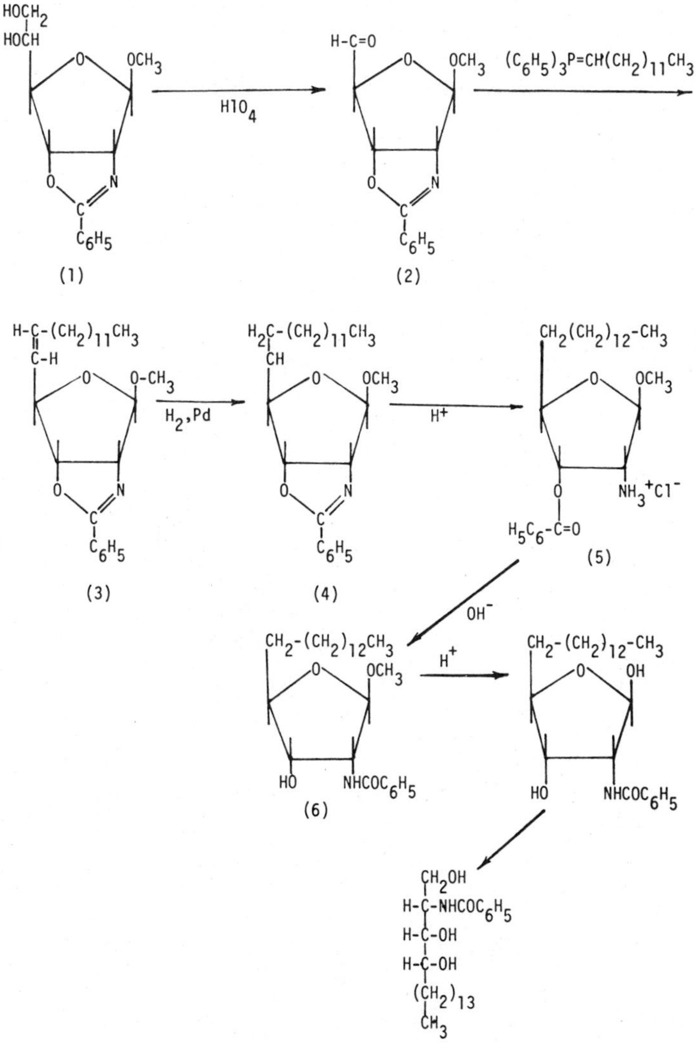

Figure 26. Synthesis of D-*ribo*-phytosphingosine according to the method of Gigg and Gigg.[552]

1.5.2. Synthesis of Glycosphingolipids

The first total synthesis of cerebroside was carried out by Shapiro and Flowers[61] through the Koenig–Knorr condensation of DL-*N*-acyl-3-*O*-benzoyl sphingosine base with 2,3,4,6-tetra-*O*-acetyl-D-galactopyranosyl bromide in the presence of mercuric cyanide [$Hg(CN)_2$]. The reaction product was deacylated to obtain a pure cerebroside with C_{24} and with C_{24} α-hydroxy fatty acids. It is interesting to note that the substituted oxazoline derivative, which was proved to be an excellent intermediate in the synthesis of sphingomyelin, was not a suitable aglycon in the synthesis of cerebrosides. The overall reaction is described in Fig. 27.

Employing a similar procedure, Shapiro *et al.*[67] successfully synthesized lactosylceramide and analog by the Koenig–Knorr condensation of 3-*O*-benzoyl-*N*-fatty acyl sphingosine and heptaacetyllactosyl bromide in the presence of $Hg(CN)_2$. The resulting compound was deacetylated to yield lactosylceramide with varying lengths of fatty acid. Lactosylsphingosine was also obtained

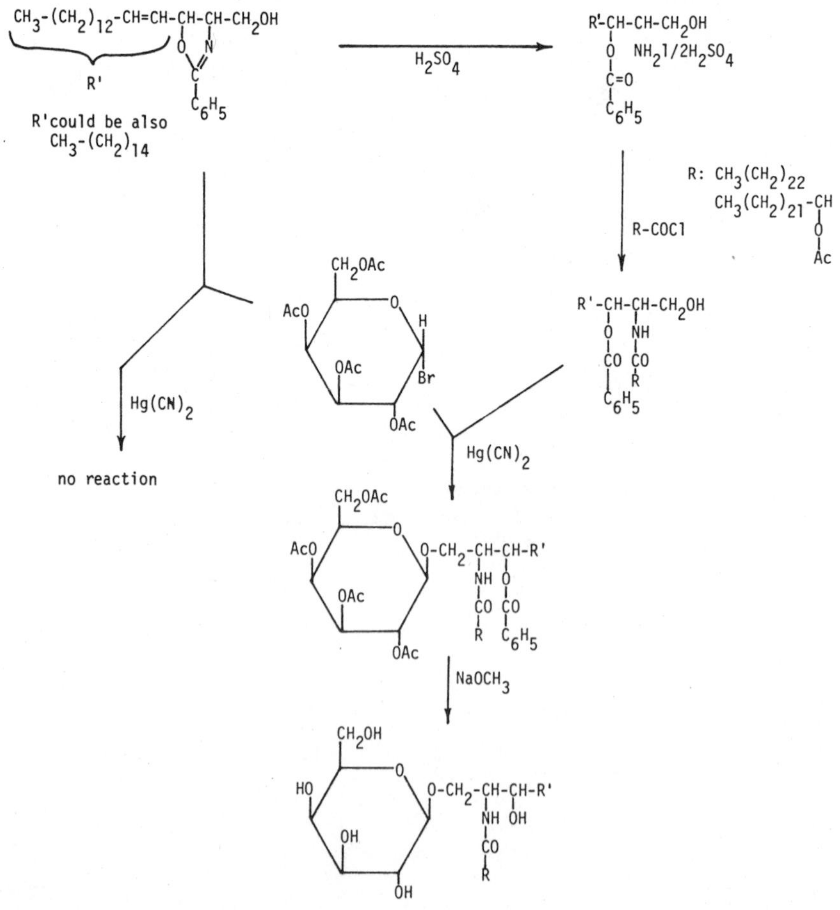

Figure 27. Synthesis of cerebroside according to the method of Shapiro and Flowers.[68]

through deacylation of a final compound. The product was immunologically identical to "cytolipin H" (see Chapter 5). Shapiro *et al.*[69] further extended their synthetic work in glycolipids with more complex carbohydrate structures. However, the major problem in obtaining a higher oligosaccharide and its condensation to ceramide is the great difficulty of making a β1→4Gal linkage, since Koenig's condensation to the axial hydroxy group is very sluggish. A successful synthesis of ganglio-*N*-triosylceramide (Tay–Sachs' globoside) was based on a successful synthesis of ganglio-*N*-triose (GalNAcβ1→ 4Galβ1→4Glc) that was carried out by condensation of *N*-dichloroacetyl-3,4,6-tribenzoyl-α-galactosaminopyranosyl bromide and 1,6-anhydrogalactopyranose. The 4-hydroxy group of 1,6-anhydrogalactose is equatorial, in contrast to the axial form of the 4-hydroxy group in other galactosides. Thus, the reactivity of the 4-hydroxy group was greatly enhanced. In contrast, the 3-hydroxy group of 1,6-anhydrogalactose decreased in reactivity because it is axial. The resulting disaccharide was converted to hepta-acetyl bromide and condensed with 1,6-anhydroglucose, and the resulting trisaccharide was converted to octa-acetyl bromide and, subsequently, condensed with 3-*O*-benzoyl-*N*-stearoyl sphingosine. The synthesis scheme is shown in Fig. 28. An analogous procedure has been used in the synthesis of various monosaccharides and synthetic disaccharides linked to ceramide by Flowers.[556] Thus, Galβ1→6Glcβ1→1Cer[556] and Galβ1→3Galβ1→1Cer[557] were synthesized through analogous reactions.

A sulfatide was synthesized by sulfation of 2,4,6-tri-*O*-acetylgalactopyranosyl ceramide followed by deacetylation. 2,4,6-Tri-*O*-acetyl-3-*O*-benzyl-D-galactopyranose was synthesized through benzylation of 4,6-ethylidene-1,2-*O*-isopropylidene-α-D-galactose followed by removal of ethylidene and isopropylidene groups and by acetylation. The resulting compound, 1,2,4,6-tetra-*O*-acetyl-3-*O*-benzylgalactose, was converted to 1,2,4,6-tetra-*O*-acetylgalactose by hydrozenolysis, and the product was converted into the unstable 2,4,6-tri-*O*-acetyl-α-D-galactopyranosyl bromide by boiling in chloroform and titanium bromide. This was in turn condensed with *N*-octadecaonyl-3-*O*-benzoyl-DL-dihydrosphingosine to give 1-*O*-(2,4,6-tri-*O*-acetyl-α-D-galactopyranosyl)-3-*O*-benzoyl-*N*-octadecaonyl-DL-dihydrosphingosine. The compound was sulfated with pyridine–sulfotrioxide, and the final compound was crystallized from methanol.[558] A cerebroside-6-sulfate was prepared previously by sulfation of cerebroside.[559]

1.6. Pioneers in Glycolipid Chemistry*

Johann Ludwig Wilhelm Thudichum

Johann Ludwig Wilhelm Thudichum was born August 27, 1829, at Büdingen, Ober-Hessen, Germany, the eldest son of a Lutheran minister and school principal, George Thudichum. In the summer of 1847, he matriculated in medicine at the University of Giessen, where he learned chemistry from Gustus von Liebig. In 1851, he obtained his medical degree at Giessen with

*Not including those in enzymology, immunology, and pathology.

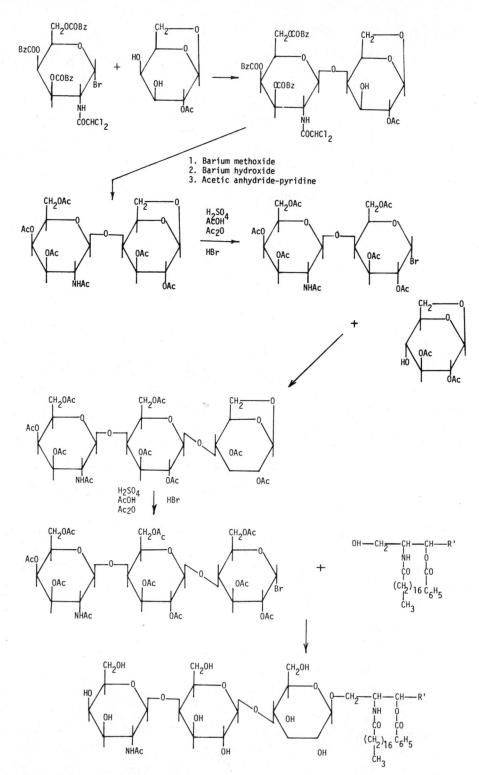

Figure 28. Synthesis of ganglio-*N*-triaosylceramide (asialo GM₂) according to the method of Shapiro *et al.*[(67)]

a thesis on "Fractures of the Upper End of the Humerus." Perhaps because of his unsuccessful application for the post at Giessen, he emigrated to London. He was appointed as Lecturer on Natural Philosophy, later as Professor of Chemistry in St. George and Grosvenor Place School of Medicine, which became defunct in 1863. In 1858, he published his first major book, *Pathology of the Urine,* and consequently, in 1860, was elected as a Fellow of the Royal College of Physicians. In 1864, he won the Hastings Gold Medal for his original work on urochrome, the normal pigment of urine. From 1865 to 1871, he was Lecturer and the first Director of the newly established Laboratory of Chemistry and Pathology of the St. Thomas Hospital School of Medicine. In 1866, he won the Silver Medal of the Society of Arts for his fine paper, "Diseases of meat as affecting the health of people." In 1869, he discovered and wrote a paper on "luteines," now known as carotenoids, the precursors of vitamin A. In the same year, he began his famous work on the chemistry of the brain. During the years 1870–1873, he discovered sphingosine, cerebroside, and sphingomyelin; in 1874, the first paper of his systematic studies on the chemistry of the brain was published. In 1872, he wrote a book entitled *Wines and Viticulture.* For his work on brain chemistry, he received violent and unfair criticism from Hoppe-Seyler, Maly, and Gamgee in 1877. In 1884, his famous book *Chemistry of the Brain* was published in complete form. On May 24, 1901, he published the enlarged and revised edition of his great classic, *Chemical Constitution of the Brain* (English edition). He died of a cerebral hemorrhage on September 7, 1901, only three and one-half months after his masterpiece appeared in print. (Abstracted from the biography by Drabkin.[4]

Herbert E. Carter

Herbert E. Carter was born in Morresville, Indiana, on September 25, 1910. He obtained his A.B. degree from DePauw University in 1930 and his M.S. (in 1931) and Ph.D. (1934) degrees in organic chemistry from the University of Illinois. He was appointed Professor of Biochemistry in 1945, and was Head of the Department of Chemistry and Chemical Engineering at the University of Illinois from 1954 to 1967. He was Vice Chancellor for Academic Affairs at the University of Arizona from 1967 to 1971, and Co-Coordinator of Interdisciplinary Programs from 1971 to 1977. He was the Chairman of the Department of Biochemistry at the University of Arizona from 1977 until his retirement in 1981. His early work concerned chemistry of hydroxyamino acids and azlactone. Beginning in 1940, he studied the isolation and structure of various antibiotics (e.g., streptomycin, streptotricin, chloromycetin, fumagillin, neomycin) and complex sphingolipids of plants and nerve tissue. The correct structures of sphingosines and cerebrosides were established in his early studies. Professor Carter also began work in the field of plant glycophosphosphingolipids, which he termed phyto-glycolipids. Only a few of these inositol-containing sphingolipids have been characterized from plants and fungi. A number of outstanding biochemists are his disciples, but relatively few of them have continued to work on glycolipids; among these are Charles C. Sweeley and Robert McCluer.

Gunnar Blix

Gunnar Blix was born on September 7, 1884, in Lund, Sweden, the son of a well-known Professor of Physiology at the University of Lund. He received his education in Lund and obtained a license to practice medicine in 1922. He defended his thesis for an M.D. degree in 1925. In 1926, he obtained a position as Associate Professor in Medicine and Physiological Chemistry at Uppsala University and became a full professor in 1930, retiring from his chair in 1961. Gunnar Blix was engaged in several fields of research. His most important contribution is the discovery of sialic acid. Other major papers concerned the isolation of pure sulfatide and determination of its chemical composition, and the discovery that blood lipids were concentrated in a specific plasma protein fraction called "lipoprotein."[618] The presence of hexosamine in a certain brain glycolipid fraction was also discovered by Blix, along with his colleague Svennerholm. This fraction was later described as a "ganglioside."

Gunnar Blix has combined his research with many administrative tasks and has taken part in the organization of two new medical schools in Sweden. A strong tradition of glycolipid research in Sweden, currently represented by Lars Svennerholm and Karl-Anders Karlsson among others, is attributed to, and directly or indirectly influenced by, this great pioneer. The photograph (p. 11), taken in 1959, and personal history were kindly sent by Professor Torvald C. Laurent.

Postscript: After this brief biography was written, the author received an obituary notice of Professor Blix, who died in December 1980.

Ernst Klenk

Ernst Klenk was born on October 14, 1896, at Pfalzgrafenweiler, district of Freudenstadt, Germany. He studied chemistry at the University of Tübingen, where he obtained his doctor's degree (Der. rer. nat.) in 1923 and worked as a collaborator of Hans Thierfelder and Franz Knoop at the Institute of Physiological Chemistry at the same university. In 1926, he received a docentship for physiological chemistry, and in 1931, he was promoted to Associated Professor. In 1936, he became a Professor of Physiological Chemistry at the University of Köln. From 1947 to 1948, he was the Dean of the Medical School and was subsequently elected as the Rector of the University in 1961. He has received a number of honors and honorary degrees, including the Heinrich-Wieland prize (1964), the American Oil Chemists' Society Award (1965), the Stouffer Award (1966), and the Otto Warburg Medal of the Gesellschaft für Biologische Chemie (1971). He was a member of the Deutsche Akademie der Natürforscher "Leopoldina," and an honorary member of the Gesellschaft für Biologische Chemie and of the Society of American Biological Chemists. He retired in 1965, and died on December 29, 1971.

The photograph (p. 12) and personal history were sent through the courtesy of Professor Hans-Dieter Klenk, at Giessen, the son of this great pioneer.

Tamio Yamakawa

Tamio Yamakawa was born on October 20, 1921, the son of Shōtaro Yamakawa, a celebrated physician and Professor of Medicine at the medical school in Sendai, Japan. He was matriculated at the Medical School of the University of Tokyo and graduated in September 1944. Since then, he has studied fatty acid metabolism and lipid composition of erythrocyte stroma at the Institute for Infectious Diseases at the University of Tokyo. Early in his career, he discovered hematoside. He earned the Dr. Med. Sci. degree in 1951 with a thesis on the metabolism of branched-chain fatty acids. A series of his works on glycolipids as the carrier of blood group haptens and on species differences of cellular glycolipid composition followed during the decade of 1950 to the early 1960s. He was promoted to Assistant Professor in 1957 and to full Professor in 1962 at the Institute of Infectious Diseases (now the Institute for Medical Sciences) of the University of Tokyo. In 1967, he became the Chairman and Professor of the Department of Biochemistry, Faculty of Medicine, University of Tokyo, a position of great prestige for a biochemist in Japan. In 1977, he was awarded the prize of the Japan Academy of Science and Art for his glycolipid studies. His work also includes investigations of glyceroglycolipids, of which the discovery of "seminolipid" was notable. The author feels that his major contribution was not only in his scientific work but also in the education of his colleagues. More than ten glycolipid researchers currently active as independent investigators at various institutions in Japan were former workers in his laboratory. They form a unique "Yamakawa school" that makes Japan one of the strong resources in glycolipid research.

Richard Kuhn

Richard Kuhn was born on December 3, 1900, in Vienna, Austria. He studied chemistry at the University of Vienna and subsequently at the University of Munich, Germany, and worked with Richard Wilstätter. He published a remarkable work on "specificity of enzymes" (amylase) for his thesis and earned his Dr. Phil. in 1922. In 1925, he had already been appointed as Professor of General and Analytical Chemistry at the Eidgenossischen Technischen Hochschüle in Zürich, Switzerland (in this appointment he succeeded, at the age of 25!, to the chair of the famous Hermann Staudinger). In 1929, he was appointed Director of the Institut für Chemie of the Kaiser-Wilhelm-Institut für Medizinische Forschung in Heidelberg, which is now the Max-Planck Institute. He was the Director of the institute until his death on July 31, 1967. It is not necessary to list his outstanding work on characterization of the vitamin B group (lactoflavin), carotenoids, vitamin A, and other substances. For these studies, he was awarded the Nobel prize for chemistry for the year 1938. His contribution to amino sugar chemistry, isolation and characterization of milk oligosaccharides, and the structural elucidation of brain gangliosides developed during his later career after World War II. The tradition of glycolipid research in Germany owes a great deal to his efforts and interest in ganglioside chemistry, which developed during his later career. His research has been succeeded by W. Wiegandt and H. Egge, among other workers.

The photograph (p. 14) and personal history were sent by Professor H. Egge.

Lars Svennerholm

Lars Svennerholm was born June 5, 1925, in Eskilstuna, Sweden, and studied medicine at the Faculty of Medicine and Science, University of Uppsala, where he obtained the M.D. degree in 1952. He was a Reader in Physical Chemistry at the University of Göteborg from 1952 to 1957 and received his Ph.D. in 1957. He has been Professor of Neurochemistry, Faculty of Medicine, Chemistry and Neurochemistry at St. Jorgen Hospital, Hisings Baca, since 1968. While he was working with Blix, he discovered the presence of galactosamine as a component of brain ganglioside and submaxillary mucin. A heterogeneity of brain ganglioside was first clearly demonstrated by chromatography on a cellulose column, and he correctly described the sugar sequence of the core structure of ganglioside independently of Kuhn and Wiegandt. More recently, he discovered a brain ganglioside with a pentasaccharide core structure, and membrane-bound sialidase directed to ganglioside in brain tissue. Among the "pioneers" described in this chapter, he is the youngest, and is still active in characterization of ganglioside and identification of ganglioside as a bacterial toxin receptor.

1.7. References

1. Liebrich, O. *Ann. Chem. Pharm.* **134,** 29 (1865).
2. Thudichum, J. L. W. *Reports of the Medical Officer of Privy Council and Local Government Board,* N. Ser., No. III, 113 (1874); N. Ser., No. VIII, 117 (1876).
3. Thudichum, J. L. W. *A Treatise on the Chemical Constitution of the Brain,* Bailliére, Tindall and Cox, London (1884); *Die chemische Konstitution des Gehirns des Menschen und der Tiere,* Franz Pieteker, Tübingen (1901).
4. Drabkin, D. L. *Thudichum; Chemist of the Brain,* University of Pennsylvania Press, Philadelphia (1958).
5. Rosenheim, O. *Biochem. J.* **7,** 604 (1913); **8,** 110 (1914).
6. Klenk, E. *Z. Physiol. Chem.* **185,** 169 (1929).
7. Thierfelder, H., and Klenk, E. *Die Chemie der Cerebroside und Phosphatide,* Julius Springer, Berlin (1930).
8. Kitagawa, F., and Thierfelder, H. *Z. Physiol. Chem.* **49,** 286 (1906).
9. Levene, P. A., and Jacobs, W. A. *J. Biol. Chem.* **12,** 389 (1912); **15,** 359 (1913).
10. Levene, P. A., and West, C. J. *J. Biol. Chem.* **16,** 549 (1913–1914).
11. Klenk, E., and Diebold, W. *Z. Physiol. Chem.* **198,** 25 (1931).
12. Carter, H. E., Glick, F. J., Norris, W. P., and Phillips, G. E. *J. Biol. Chem.* **142,** 449 (1942).
13. Carter, H. E., Glick, F. J., Norris, W. P., and Phillips, G. E. *J. Biol. Chem.* **170,** 285 (1947).
14. Carter, H. E., and Greenwood, F. L. *J. Biol. Chem.* **199,** 283–288 (1952).
15. Ohno, K. *Seikagaku (J. Jpn. Biochem. Soc.)* **19,** 1–4 (1946–1947).
16. Ohno, K. *Seikagaku (J. Jpn. Biochem. Soc.)* **20,** 38–40 (1948).
17. Nakayama, T. *J. Biochem. (Tokyo)* **37,** 309–319 (1950).
18. Prŏstenik, M. *Chem. Phys. Lipids* **5,** 1 (1970).

19. Seydel, P. V. Eidgenoessische Technische Hochschule, Zuerich für Kenntnis des Sphingosines, Dissertation, Druckeral A.-G., Gebr Üder Leeman, Zürich (1941).

20. Oda, T. *Yakugaku Zasshi (J. Pharm. Soc. Jpn.)* **72,** 142 (1952); *Chem. Abstr.* **46,** 6192 (1952).

21. Carter, H. E., Celmer, W. D., Lands, W. E. N., Müeller, K. L., and Tomizawa, H. H. *J. Biol. Chem.* **206,** 613 (1954).

22. Zellner, J. *Monatschrift* 32, 133 (1911); Rosenthal, R. *Monatschrift für Chemie* 43, 237 (1922); Reindel, F. *Annalen* **480,** 76 (1930); Wieland, H., and Coutelle, G. *Annalen der Chemie* **548,** 270 (1941).

23. Reindel, F., Weickmann, A., Picard, S., Luber, K., and Turula, P. *Annalen der Chemie* **544,** 116 (1940).

24. Bohonos, N., and Peterson, W. H. *J. Biol. Chem.* **149,** 295 (1943).

25. Proštenik, M., and Majhofer-Oreščanin, B. *Naturwissenschaften* **47,** 399 (1960).

26. Stanaćev, N. Ž., and Chargaff, E. *Biochim. Biophys. Acta* **98,** 168–181 (1965).

27. Lieb, H. *Z. Physiol. Chem.* **140,** 305 (1924).

28. Klenk, E., and Harle, R. *Z. Physiol. Chem.* **189,** 243 (1930).

29. Aghion, H. La maladie de Gaucher dans l'enfance, Thesis, Faculté de médicine de Paris (1934).

30. Halliday, N., Deuel, J. J., Jr., Tragelman, L. J., and Ward, W. E. *J. Biol. Chem.* **132,** 171–180 (1940).

31. Klenk, E. *Z. Physiol. Chem.* **235,** 24 (1935).

32. Sweeley, C. C., and Klionsky, B. *J. Biol. Chem.* **238,** 3148 (1963).

33. Blix, G. *Z. Physiol. Chem.* **219,** 82 (1933).

34. Yamakawa, T., Kiso, N., Handa, S., Makita, A., and Yokoyama, S. *J. Biochem. (Tokyo)* **52,** 226 (1962).

35. Stoffyn, P., and Stoffyn, A. *Biochim. Biophys. Acta* **70,** 218 (1963).

36. Klenk, E., and Rennkampf, F. *Z. Physiol. Chem.* **273,** 253 (1942).

37. Rapport, M. M., Graf, L., Skipski, V. P., and Alonzo, N. F. *Nature (London)* **181,** 1803 (1958).

38. Blix, G. *Z. Physiol. Chem.* **240,** 43 (1936).

39. Blix, G., Svennerholm, L., and Werner, L. *Acta Chem. Scand.* **6,** 358 (1952).

40. Walz, E. *Z. Physiol. Chem.* **166,** 210 (1927).

41. Levene, P. A., and Landsteiner, K. *J. Biol. Chem.* **75,** 607 (1927).

42. Blix, G. *Skand. Arch. Physiol.* **80,** 46 (1938).

43. Klenk, E. *Z. Physiol. Chem.* **268,** 50 (1941).

44. Klenk, E., and Langerbeins, H. *Z. Physiol. Chem.* **270,** 185 (1941).

45. Klenk, E. *Z. Physiol. Chem.* **273,** 76 (1942).

46. Klenk, E. *Z. Physiol. Chem.* **288,** 216 (1951).

47. Yamakawa, T., and Suzuki, S. *J. Biochem. (Tokyo)* **38,** 199 (1951).

48. Klenk, E., and Wolter, H. *Z. Physiol. Chem.* **291,** 259 (1951).

49. Klenk, E., and Lauenstein, K. *Z. Physiol. Chem.* **291,** 249 (1951).

50. Yamakawa, T., and Suzuki, S. *J. Biochem. (Tokyo)* **39,** 393 (1952).

51. Yamakawa, T. In:*Lipoid, 16 Colloquium der Gesellschaft Physiologische Chemie* (Schute, E., ed.), Mosbach Baden; Springer-Verlag, Berlin and New York (1966), pp. 87–111.

52. Folch-Pi, J., Arsove, S., and Meath, J. A. *J. Biol. Chem.* **191,** 819 (1951).

53. Rosenberg, A., and Chargaff, E. *J. Biol. Chem.* **232,** 1031 (1958).

54. Chatagnon, C., and Chatagnon, P. *Bull. Soc. Chim. Biol.* **35,** 1319 (1953).

55. Klenk, E., and Gielen, W. *Z. Physiol. Chem.* **319,** 283 (1960); **323,** 126 (1961).

56. Svennerholm, L. *Nature (London)* **177,** 524 (1956).

57. Svennerholm, L. *J. Lipid Res.* **5,** 145–155 (1964) in: *Methods in Carbohydrate Chemistry,* Vol. 4 (Whistler, R. L., and BeMiller, J. N., eds.) Academic Press, New York (1972), pp. 464–474.

58. Weicker, H., Dain, J., Schmidt, G., and Thannhauser, S. J. *Fed. Proc. Fed. Am. Soc. Exp. Biol.* **19,** 219 (1960).

59. Svennerholm, L. *Biochem. Biophys. Res. Commun.* **9,** 436 (1962).

60. Kuhn, R., and Wiegandt, H. *Chem. Ber.* **96,** 866 (1963).

61. Makita, A., and Yamakawa, T. *Jpn. J. Exp. Med.* **33,** 361 (1963).

62. Ledeen, R., and Salsman, K. *Biochemistry* **4,** 2225 (1965).

63. Carter, H. E., Greenwood, F., and Humiston, C. *Fed. Proc. Fed. Am. Soc. Exp. Biol.* **9,** 159 (1950).

64. Carter, H. E., and Shapiro, D. *J. Am. Chem. Soc.* **75,** 5131 (1953).

65. Carter, H. E., Shapiro, D., and Harrison, J. B. *J. Am. Chem. Soc.* **75,** 1007 (1953).
66. Shapiro, D., Segal, H., and Flowers, H. M. *J. Am. Chem. Soc.* **80,** 1194 (1958); **80,** 2170 (1958).
67. Shapiro, D., Rachaman, E. S., Rabinsohn, Y., and Diver-Haber, A. *Chem. Phys.* Lipids **1,** 54 (1966).
68. Shapiro, D., and Flowers, H. M. *J. Am. Chem. Soc.* **83,** 3327 (1961).
69. Shapiro, D., Acher, A. J., and Rabinsohn, Y. *Chem. Phys. Lipids.* **10,** 28 (1973).
70. Carter, H. E., and Hendrickson, H. S. *Biochemistry* **2,** 389 (1963).
71. Kuhn, R., and Wiegandt, H. *Z. Naturforsch.* **19b,** 256 (1964).
72. Yamakawa, T., Yokoyama, S., and Handa, N. *J. Biochem.* **53,** 28 (1963).
73. Yamakawa, T., Nishimura, S., and Kamimura, M. *Jpn. J. Exp. Med.* **35,** 201 (1965).
74. Hakomori, S., and Jeanloz, R. W. *J. Biol. Chem.* **239,** 3606 (1964).
75. Hakomori, S., and Strycharz, G. D. *Biochemistry* **7,** 1285–1286 (1968).
76. Karlsson, K.-A. *Lipids* **5,** 878 (1970).
77. Hakomori, S., Siddiqui, B., Li, Y.-T., Li, S.-C., and Hellerqvist, C. G. *J. Biol. Chem.* **246,** 2271 (1971).
78. Siddiqui, B., and Hakomori, S. *J. Biol. Chem.* **246,** 5766 (1971).
79. Hakomori, S., Stellner, K., and Watanabe, K. *Biochem. Biophys. Res. Commun.* **49,** 1061–1068 (1972).
80. Wiegandt, H. *Z. Physiol. Chem.* **354,** 1049–1056 (1973).
81. Naiki, M., and Marcus, D. M. *Biochem. Biophys. Res. Commun.* **60,** 1105 (1974).
82. Naiki, M., and Marcus, D. M. *Biochemistry* **14,** 4837–4841 (1975).
83. Marcus, D. M., Naiki, M., and Kundu, S. K. *Proc. Natl. Acad. Sci. U.S.A.* **73,** 3263 (1976).
84. Koscielak, J., Miller-Podraza, H., Krauze, R., and Piasek, A. *Eur. J. Biochem.* **71,** 9 (1976).
85. Gardas, A. *Eur. J. Biochem.* **68,** 177, 185 (1976).
86. Svennerholm, L., Månsson, J.-E., and Li, Y.-T. *J. Biol. Chem.* **248,** 740–742 (1973).
87. Wiegandt, H. *Eur. J. Biochem.* **45,** 367–369 (1974).
88. Folch-Pi, J., Lees, M., and Sloane-Stanley, G. H. *J. Biol. Chem.* **226,** 497 (1957).
89. Slomiany, B. L., and Slomiany, A. *Biochim. Biophys. Acta* **486,** 531–540 (1977).
90. Tettamanti, G., Bonali, F., Marchesini, S., and Zambotti, V. *Biochim. Biophys. Acta* **296,** 160 (1973).
91. Trams, E. G., and Lauter, C. J. *Biochim. Biophys. Acta* **60,** 350 (1962).
92. Djeter-Juszynski, M., Jarpaz, N., Flowers, H. M., and Saron, N. *Eur. J. Biochem.* **83,** 363–373 (1978).
93. Hamaguchi, H., and Cleve, H. *Biochim. Biophys. Acta* **278,** 271–280 (1972).
94. Wuthier, R. E. *J. Lipid Res.* **7,** 558 (1966).
95. Maxwell, M. A. N., and Williams, J. P. *J. Chromatogr.* **31,** 62 (1967).
96. Suzuki, K. *J. Neurochem.* **12,** 629 (1965).
97. Baumann, H., Nudelman, E., Watanabe, K., and Hakomori, S. *Cancer Res.* **39,** 2637 (1979).
98. Kanfer, J., and Spielvogel, C. *J. Neurochem.* **20,** 1483 (1973).
99. Svennerholm, L. *Acta Chem. Scand.* **17,** 239 (1963).
100. Quarles, R., and Folch-Pi, J. *J. Neurochem.* **12,** 543 (1965).
101. Radin, N. S., Brown, J. R., and Lavin, F. B. *J. Biol. Chem.* **219,** 977 (1956).
102. Dawson, R. M. C. *Biochem. J.* **75,** 45 (1960).
103. Vanier, M. T., Holm, M., Mansson, J. E., and Svennerholm, L. *J. Neurochem.* **21,** 1375 (1973).
104. Lees, M., Folch-Pi, J., Sloane-Stanley, G. H., and Carr, S. *J. Neurochem.* **4,** 9 (1959).
105. Svennerholm, L., and Thorin, H. *J. Lipid Res.* **3,** 483–485 (1962).
106. Saito, T., and Hakomori, S. *J. Lipid Res.* **12,** 257 (1971).
107. Krohn, K., Eberlein, K., and Gercken, G. *J. Chromatogr.* **153,** 550–552 (1978).
108. Rouser, G., Kritchevsky, G., Yamamoto, A., Simmon, G., Galli, C., and Bauman, A. *J. Methods in Enzymology,* Vol. 14 (Lowenstein, J. M., ed.), Academic Press, New York (1969), p. 272.
109. Yu, R. K., and Ledeen, R. W. *J. Lipid Res.* **13,** 680 (1972).
110. Kundu, S. K., and Roy, S. K. *J. Lipid Res.* **19,** 390–395 (1978).
111. Kawamura, N., and Taketomi, T. *J. Biochem. (Tokyo)* **81,** 1217–1225 (1977).

112. Penick, R. J., Meisler, M. H., and McCluer, R. H. *Biochim. Biophys. Acta* **116**, 279 (1966).
113. Svennerholm, L. In: *Methods in Carbohydrate Chemistry*, Vol. 6 (Whistler, R. L., and Bemiller, J. M., eds.), Academic Press, New York (1972), p. 464.
114. Momoi, T., Ando, S., and Nagai, Y. *Biochim. Biophys. Acta* **441**, 488 (1976).
115. Hakomori, S., and Watanabe, K. In: *Glycolipid Methodology* (Witting, L. A., ed.), American Oil Chemists' Society, Champaign, Illinois (1976), pp. 13–47.
116. Watanabe, K., Powell, M. E., and Hakomori, S. *J. Biol. Chem.* **253**, 8962–8967 (1978).
117. Ando, S., Isobe, M., and Nagai, Y. *Biochim. Biophys. Acta* **424**, 98 (1976).
118. Feizi, T., Childs, R. A., Hakomori, S., and Powell, M. E. *Biochem. J.* **173**, 245–254 (1978).
119. Stellner, K., Watanabe, K., and Hakomori, S. *Biochemistry* **12**, 656 (1973).
120. Watanabe, K., Laine, R. A., and Hakomori, S. *Biochemistry* **14**, 2725 (1975).
121. Hakomori, S., Wang, S.-H., and Young, W. W., Jr. *Proc. Natl. Acad. Sci. U.S.A.* 3023–3027 (1977).
122. Kato, M., Watanabe, K., and Naiki, M. *Proc. Jpn. Conf. Biochim. Lipids* **19**, 253 (1977).
123. Iwamori, M., and Nagai, Y. *Biochim. Biophys. Acta* **528**, 257–267 (1978).
124. Evans, J. E., and McCluer, R. H. *Biochim. Biophys. Acta* **270**, 565 (1972).
125. McCluer, R. H., and Jungalwala, F. B. In: *Advances in Experimental Medicine*, Vol. 68 (Volk, B. W., and Schneck, L., eds.), Plenum Press, New York (1976), pp. 533–544.
126. Tjaden, U. R., Krol, J. H., Van Hoeven, R. P., Oomen-Meulemans, E. P. M., and Emmelot, P. *J. Chromatogr.* **136**, 233 (1977).
127. Suzuki, A., Handa, S., and Yamakawa, T. *J. Biochem. (Tokyo)* **82**, 1185 (1977).
128. Kiehn, E. L. *J. Lipid Res.* **7**, 449 (1966).
129. Gahmberg, C. G., and Hakomori, S. *J. Biol. Chem.* **250**, 2438 (1975).
130. Hooghwinkel, G. J. M., Borri, P., and Riemersma, J. C. *Recl. Trav. Chim. Pays-Bas Belg.* **83**, 576 (1964).
131. Suomi, W. D., and Agranoff, B. W. *J. Lipid Res.* **6**, 211 (1965).
132. Taketomi, T., and Yamakawa, T. *Jpn. J. Exp. Med.* **37**, 11 (1967).
133. Ziolkowski, C. H. J., Fraser, B. A., and Mallette, M. F. *Immunochemistry* **12**, 297–302 (1975).
134. Hayashi, A., and Matsubara, T. *Tampakushitsu Kakusan Koso (Proteins Nucleic Acids and Enzymes)* **19**, 717 (1974).
135. Fredman, P., Nilsson, O., Tayot, J. L., and Svennerholm, L. *Biochim. Biophys. Acta* **618**, 42 (1980).
136. Taketomi, T., and Kawamura, N. *J. Biochem.* **72**, 189 (1972).
137. Gaver, R. C., and Sweeley, C. C. *J. Am. Oil Chem. Soc.* **42**, 294 (1965).
138. Carter, H. E., and Gaver, R. C. *Biochem. Biophys. Res. Commun.* **29**, 886 (1968).
139. Carter, H. E., and Gaver, R. C. *J. Lipid Res.* **8**, 391 (1967).
140. Weiss, B., and Stiller, R. L. *Lipids* **8**, 25 (1973).
141. Morrison, W. R., and Hay, J. D. *Biochim. Biophys. Acta* **302**, 460 (1970).
142. James, A. T. *Methods Biochem. Anal.* **8**, 1 (1960).
143. Carter, W. O. *Methods Biochem. Anal.* **17**, 135 (1969).
144. Ackman, R. G. *Prog. Chem. Fats Other Lipids*, **12**, 165 (1972).
145. Applewhite, T. H. *J. Am. Oil. Chem. Soc.* **42**, 321 (1965).
146. Morris, L. J., and Wharry, D. M. *J. Chromatogr.* **20**, 27 (1965).
147. Radin, N. S. *J. Am. Oil Chem. Soc.* **43**, 569 (1965).
148. Bouhours, J. F. *J. Chromatogr.* **169**, 462 (1979).
149. Kawanami, J. *Chem. Phys. Lipids* **7**, 159 (1971).
150. Tschöpe, G *Z. Physiol. Chem.* **354**, 1291 (1973).
151. Christie, W. W. *Lipid Analysis*, Pergamon Press, (1973), p. 121.
152. Laine, R. A., Young, D. N., Gerber, J. N., and Sweeley, C. C. *Biomed. Mass Spectrometry* **1**, 10 (1974).
153. Budzkiewitz, H., Kjerassi, C., and Williams, D. H. *Mass Spectrometry of Organic Compounds*, Holden-Day, San Francisco (1967).
154. Frigerio, A., and Castagnoli, N., (eds.) *Mass Spectrometry in Biochemistry and Medicine*, Raven Press, New York (1974).
155. Kawanami, J., and Otsuka, H. *Chem. Phys. Lipids* **3**, 135 (1969).
156. Niehaus, W. G., Jr., and Ryhage, R. *Anal. Chem.* **40**, 1840 (1968).

157. Polito, A. J., Makita, T., and Sweeley, C. C. *Biochemistry* **7**, 2609 (1968).
158. Sambasivarao, K., and McCluer, R. H. *J. Lipid Res.* **5**, 106–108 (1964).
159. Karlsson, K.-A., and Pascher, I. *J. Lipid Res.* **12**, 466 (1971).
160. Morrison, W. R. *Biochim. Biophys. Acta* **316**, 99 (1973).
161. Hayashi, A., and Matsubara, T. *Biochim. Biophys. Acta* **248**, 306 (1971).
162. Karlsson, K.-A. *Acta Chem. Scand.* **20**, 2884 (1966).
163. Hayashi, A., and Matsubura, T. *Chem. Phys. Lipids* **10**, 51 (1973).
164. Urdal, D. L., and Hakomori, S. *J. Biol. Chem.* **255**, 10,509 (1980).
165. White, D. C., Tucker, A. N., and Sweeley, C. C. *Biochim. Biophys. Acta* **187**, 527 (1969).
166. Polito, A. J., Naworal, J., and Sweeley, C. C. *Biochemistry* **8**, 1811 (1969).
167. Hara, S., and Matsushima, Y. *J. Biochem.* **71**, 907 (1972).
168. Mendershausen, P. B., and Sweeley, C. C. *Biochemistry* **8**, 2633 (1969).
169. Laine, R. A., and Renkonen, O. *Biochemistry* **12**, 1106 (1973).
170. Carter, H. E., Rothfus, J. A., and Gigg, R. *J. Lipid Res.* **2**, 228 (1961).
171. Klenk, E., and Huang, R. T. C. *Z. Physiol. Chem.* **250**, 1081 (1969).
172. Samuelsson, B., and Samuelsson, K. *Biochim. Biophys. Acta* **164**, 421 (1960).
173. Samuelsson, B., and Samuelsson, K. *J. Lipid Res.* **10**, 41–47 (1969).
174. Samuelsson, B., and Samuelsson, K. *Chem. Phys. Lipids* **5**, 44 (1970).
175. Hammarström, S., Samuelsson, B., and Samuelsson, K. *J. Lipid Res.* **11**, 150 (1970).
176. Hammarström, S., Samuelsson, B., and Samuelsson, K. *Eur. J. Biochem.* **15**, 581 (1970).
177. Hammarström, S. *J. Lipid Res.* **11**, 175 (1970).
178. Renkonen, O., and Hirvisalo, E. L. *J. Lipid Res.* **10**, 687 (1969).
179. Ando, S. *Yukagaku (Oil Chemistry)* **21**, 243 (1972).
180. Martin-Lomas, M., and Chapman, D. *Chem. Phys. Lipids* **10**, 152 (1973).
181. Karlsson, K.-A., Samuelsson, B. E., and Steen, G. O. *Biochim. Biophys. Acta* **316**, 336 (1973).
182. Chambers, R. E., and Clamp, J. R. *Biochem. J.* **126**, 1009 (1971).
183. Laine, R. A., Esselman, W. J., and Sweeley, C. C. *Methods Enzymol.* **28**, 159–167 (1972).
184. Yang, H.-J., and Hakomori, S. *J. Biol. Chem.* **246**, 1192 (1971).
185. Watanabe, K., Hakomori, S., Childs, R. A., and Feizi, T. *J. Biol. Chem.* **254**, 3221–3228 (1979).
186. Hellerqvist, C. G., and Ruden, U. *Eur. J. Biochem.* **25**, 96 (1972).
187. Sweeley, C. C., and Walker, B. *Anal. Chem.* **76**, 1461 (1964).
188. Ando, S., and Yamakawa, T. *J. Biochem.* **70**, 335 (1971).
189. Yu, R. K., and Ledeen, R. W. *J. Lipid Res.* **11**, 506 (1970).
190. Hakomori, S., and Saito, T. *Biochemistry* **8**, 5082 (1969).
191. Craven, D. A., and Gehrke, C. W. *J. Chromatogr.* **37**, 414 (1968).
192. Buscher, H., Casals-Stenzel, J., and Schauer, R. *Eur. J. Biochem.* **50**, 71 (1974).
193. Rauvala, H., and Kärkkäinen, J. *Carbohydr. Res.* **56**, 109 (1977).
194. Schauer, R. *Methods Enzymol.* **50**, 64–89 (1978).
195. Pfeil, R., and Schauer, R. In: *Glycoconjugates* (Schauer, R., Boer, P., Buddecke, E., Kramer, M. F., Vliegenthart, J. F. G., and Wiegandt, H., eds.), Georg Thieme, Stuttgart (1979), pp. 44–45.
196. Rauter, G., Pfeil, R., Kamerling, J. P., Schauer, R., and Vliegenthart, J. F. G. In: *Glycoconjugates* (Schauer, R., Boer, P., Buddecke, E., Kramer, M. F., Vliegenthart, J. F. G., and Wiegandt, H., eds.), Georg Thieme, Stuttgart (1979), pp. 89–90.
197. Ghidoni, R., Sonnino, S., Tettamanti, G., Baumann, N., Reuter, G., and Schauer, R. In: *Glycoconjugates* (Schauer, R., Boer, P., Buddecke, E., Kramer, M. F., Vliegenthart, J. F. G., and Wiegandt, H., eds.), Georg Thieme, Stuttgart (1979), pp. 51–52.
198. Wiegandt, H., and Baschang, G. *Z. Naturforsch. Teil B* **20**, 164 (1965).
199. Wiegandt, H., and Bücking, H. W. *Eur. J. Biochem.* **15**, 287 (1970).
200. Hakomori, S. *J. Lipid Res.* **7**, 789 (1966).
201. MacDonald, D., Patt, L., and Hakomori, S. *J. Lipid Res.* **21**, 642 (1980).
202. Ohashi, M., and Yamakawa, T. *J. Lipid Res.* **14**, 698 (1973).
203. Kishimoto, Y, and Mitry, M. T. *Arch. Biochem. Biophys.* **161**, 426 (1974).
204. Iwamori, M., and Nagai, Y. *Chem. Phys. Lipids* **20**, 193 (1977).

205. Nilsson, B., and Svensson, S. *Carbohydr. Res.* **62,** 377 (1978).
206. Nilsson, B., and Svensson, S. *Carbohydr. Res.* **69,** 292 (1979).
207. Lundblad, A., Svensson, S., Low, B., Messeter, L., and Cedergren, B. *Eur. J. Biochem.* **104,** 323 (1980).
208. Lindberg, B., Nilsson, B., Norberg, T., and Svensson, S. *Acta Chem. Scand.* **33,** 230 (1979).
209. Hakomori, S. *Chem. Phys. Lipids* **5,** 96 (1970).
210. Smith, L., and Unrau, A. M. *Chem. Ind. (London),* p. 881 (1959).
211. Iijima, Y., Muramatsu, T., and Egami, E. *Arch. Biochem. Biophys.* **145,** 150 (1971).
212. Robinson, D., and Thorpe, R. *FEBS Lett.* **45,** 191 (1974).
213. Li, Y.-T., and Li, S.-C. *J. Biol. Chem.* **243,** 3994 (1968).
214. Li, S.-C., and Li, Y.-T. *J. Biol. Chem.* **245,** 5153 (1970).
215. Li, Y.-T., and Li, S.-C. *Methods Enzymol.* **28,** 702 (1972).
216. Weissman, B., and Hinrichsen, D. F. *Biochemistry* **8,** 2034 (1969).
217. Gatt, S., and Rapport, M. M. *Biochim. Biophys. Acta* **113,** 567 (1966).
218. Li, S.-C., Mazzotta, M. Y., Chien, S.-F., and Li, Y.-T. *J. Biol. Chem.* **250,** 6786 (1975).
219. Uda, Y., Li, S.-C., Li, Y.-T., and McKibbin, J. M. *J. Biol. Chem.* **252,** 5194 (1977).
220. Kuhn, R., and Wiegandt, H. *Chem. Ber.* **89,** 2 (1963).
221. Wiegandt, H. *J. Neurochem.* **14,** 671 (1967).
222. Huang, R. T. C., and Klenk, E. *Hoppe-Seyler's Z. Physiol. Chem.* **353,** 679 (1972).
223. Tettamanti, G., Preti, A., Lombardo, A., Bonali, F., and Zambotti, V. *Biochim. Biophys. Acta* **306,** 466 (1973).
224. Öhman, R., Rosenberg, A., and Svennerholm, L. *Biochemistry* **9,** 3774 (1970).
225. Wenger, D. A., and Wardell, S. *J. Neurochem.* **20,** 607 (1973).
226. Rauvala, H. *FEBS Lett.* **65,** 229 (1976); *Eur. J. Biochem.* **97,** 555 (1979).
227. Uchida, Y., Tsukada, Y., and Sugimori, T. *J. Biochem. (Tokyo)* **86,** 1573 (1979).
228. Sugano, K., Saito, M., and Nagai, Y. *FEBS Lett.* **89,** 321 (1978).
229. Saito, M., Sugano, K., and Nagai, Y. *J. Biol. Chem.* **254,** 7845 (1979).
230. Kitamikado, M., and Ueno, R. *Bull. Jpn. Soc. Sci. Fish.* **36,** 1175 (1970).
231. Fukuda, M. N., and Matsumura, G. *J. Biol. Chem.* **251,** 6218 (1976).
232. Fukuda, M. N., Watanabe, K., and Hakomori, S. *J. Biol. Chem.* **253,** 6814 (1978).
233. Kuhn, R., and Egge, H. *Chem. Ber.* **96,** 3338 (1963).
234. Hakomori, S. *J. Biochem.* **55,** 205 (1964).
235. Björndal, H., Lindberg, B., and Svensson, S. *Carbohydr. Res.* **5,** 433 (1967).
236. Björndal, H., Hellerqvist, C. G., Lindberg, B., and Svensson, S. *Angew. Chem. Int. Ed. Engl.* **9,** 610 (1970).
237. Stellner, K., Saito, H., and Hakomori, S. *Arch. Biochem. Biophys.* **155,** 464 (1973).
238. Tai, T., Yamashita, K., and Kobata, A. *J. Biochem. (Tokyo)* **78,** 679 (1975).
239. Sanford, P. A., and Conrad, H. E. *Biochemistry* **5,** 1508 (1966).
240. Finne, J., Krusius, T., and Rauvala, H. *Biochem. Biophys. Res. Commun.* **74,** 405 (1977).
241. Caroff, M., and Szabo, L. *Biochem. Biophys. Res. Commun.* **89,** 410 (1979).
242. Hase, S., and Rietschel, E. T. *Eur. J. Biochem.* **63,** 93 (1976).
243. Finne, J., and Rauvala, H. *Carbohydr. Res.* **58,** 57 (1977).
244. Finne, J., Krusius, T., Rauvala, H., and Hemminki, K. *Eur. J. Biochem.* **77,** 319 (1977).
245. Rauvala, H. *Carbohydr. Res.* **72,** 257 (1979).
246. Angyal, S. J., and James, K. *Aust. J. Chem.* **23,** 1209 (1970).
247. Hoffman, J., Lindberg, B., and Svensson, S. *Acta Chem. Scand.* **26,** 661 (1972).
248. Laine, R. A., and Renkonen, O. *Biochemistry* **13,** 2837 (1974).
249. Laine, R. A., and Renkonen, O. *J. Lipid Res.* **16,** 102 (1975).
250. Oshima, M., and Ariga, T. *FEBS Lett.* **64,** 440 (1976).
251. Itasaka, O., Sugita, M., Yoshizaki, H., and Hori, T. *J. Biochem. (Tokyo)* **80,** 935 (1976).
252. Karlsson, K.-A., Samuelsson, B. E., and Steen, G. O. *Biochem. Biophys. Res. Commun.* **37,** 22 (1969).
253. Karlsson, K.-Z., Pascher, I., Samuelsson, B. E., and Steen, G. O. *Chem. Phys. Lipids* **9,** 230 (1972).
254. Sweeley, C. C., and Dawson, G. *Biochem. Biophys. Res. Commun.* **37,** 6 (1969).

255. Dawson, G., and Sweeley, C. C. *J. Lipid Res.* **12,** 56 (1971).
256. Karlsson, K.-A. In: *Glycolipid Methodology* (Witting, L. A., ed.), American Oil Chemists' Society, Champagne, Illinois (1976), pp. 97–122.
257. Karlsson, K.-A. *FEBS Lett.* **32,** 317 (1973).
258. Karlsson, K.-A. *Biochemistry* **13,** 3643 (1974).
259. Karlsson, K.-A., Pascher, I., Pimelott, W., and Samuelsson, B. E. *Biomed. Mass Spectrometry* **1,** 49 (1974).
260. Stoffyn, A., and Stoffyn, P. J. *Carbohydr. Res.* **23,** 251 (1972).
261. Karlsson, K.-A. In: *Ganglioside Function* (Porcellati, G., Ceccarelli, B., and Tettamanti, G., eds.), Plenum Press, New York (1976), pp. 15–25.
262. Holm, M., Pascher, I., and Samuelsson, B. E. *Biomed. Mass Spectrometry* **4,** 77 (1977).
263. Smith, E. L., McKibbin, J. M., Karlsson, K.-A., Pascher, I., Samuelsson, B. E., and Li, S.-C. *Biochemistry* **14,** 3370 (1975).
264. Smith, E. L., McKibbin, J. M., Breimer, M. E., Karlsson, K.-A., Pascher, I., and Samuelsson, B. E. *Biochim. Biophys. Acta* **398,** 84 (1975).
265. Ledeen, R. W., Kundu, S. K., Price, H. C., and Tong, J. W. *Chem. Phys. Lipids* **13,** 429 (1974).
266. Hanfland, P., *Chem. Phys. Lipids* **15,** 105 (1975).
267. Hanfland, P., and Egge, H. *Chem. Phys. Lipids* **16,** 201 (1976).
268. Ohashi, M., and Yamakawa, T. *J. Biochem. (Tokyo)* **81,** 1675 (1977).
269. Slomiany, B. L., Slomiany, A., and Marty, L. N. *Biochem. Biophys. Res. Commun.* **88,** 1092 (1979).
270. Nilsson, O., Månsson, J.-E., Tibblin, E., and Svennerholm, L. *FEBS Lett.* **133,** 197 (1981).
271. Carter, W. G., and Hakomori, S. *Biochemistry* **18,** 730 (1979).
272. Hounsell, E., Fukuda, M., Powell, M. E., and Hakomori, S. *Biochem. Biophys. Res. Commun.* **92,** 1143 (1980).
273. Oshima, M., Ariga, T., and Tamura, T. *Chem. Phys. Lipids* **19,** 289 (1977).
274. Ando, S., Kon, K., Nagai, Y., and Murata, T. *J. Biochem. (Tokyo)* **82,** 1623 (1977).
275. Breimer, M. E., Hansson, G. C., Karlsson, K.-A., Leffler, H., Pimlott, W., and Samuelsson, B. E. *Biomed. Mass Spectrometry* **6,** 231 (1979).
276. Breimer, M. E., Hansson, G. C., Karlsson, K.-A., Leffler, H., Pimlott, W., and Samuelsson, B. E. *FEBS Lett.* **89,** 42 (1978).
277. Hall, L. D. *Adv. Carbohydr. Chem.* **19,** 51 (1964).
278. Lemieux, R. U., and Stevens, J. D. *Can. J. Chem.* **43,** 2059 (1965).
279. Inch, T. D. *Annu. Rev. NMR Spectrometry* **2,** 35 (1969).
280. Bathbone, E. B., and Stephen, A. M. *Carbohydr. Res.* **21,** 73 (1972); **23,** 275 (1972).
281. Rowe, J. J. M., and Rowe, K. L. *Chem. Rev.* **70,** 1 (1970).
282. Kawanami, J. *J. Biochem.* **62,** 105 (1967).
283. Kawanami, J., and Tsuji, T. *Chem. Phys. Lipids* **7,** 49 (1971).
284. Falk, K.-E., Karlsson, K.-A., and Samuelsson, B.E. *Arch. Biochem. Biophys.* **192,** 164 (1979).
285. Falk, K.-E., Karlsson, K.-A., and Samuelsson, B. E. *Arch. Biochem. Biophys.* **192,** 177 (1979).
286. Falk, K.-E., Karlsson, K.-A., and Samuelsson, B. E. *Arch. Biochem. Biophys.* **192,** 191 (1979).
287. Urdal, D. L., and Hakomori, S. *J. Biol. Chem.* (1983), in press.
288. Dabrowski, J., Egge, H., Hanfland, P., and Kuhn, S. In: *Cell Surface Glycolipids* (Sweeley, C. C., ed.), American Chemical Society, Washington, D.C. (1980), pp. 55–64.
289. Clarke, J. T. R., Wolfe, L. S., and Perlin, A. S. *J. Biol. Chem.* **246,** 5563 (1971).
290. Breitmaier, E., Jang, G., and Voelter, W. *Angew. Chem. Int. Ed. Engl.* **10,** 673 (1971).
291. Rosenthal, S. N., and Fendler, J. H. *Adv. Phys. Org. Chem.* **13,** 279 (1976).
292. Perlin, A. S. In: *International Review of Science,* Vol. 7, *Carbohydrates* (Aspinall, G. D., ed.), Butterworths, London and Boston (1976), pp. 1–34.
293. Bundle, D. R., Jennings, H. J., and Smith, I. C. P. *Can. J. Chem.* **51,** 3812 (1973).
294. Dorman, D. E., and Roberts, J. D. *J. Am. Chem. Soc.* **92,** 1355 (1970).
295. Walker, T. E., London, R. E., Whatley, T. W., Baker, R., and Matwiyoff, N. A. *J. Am. Chem. Soc.* **98,** 5807 (1976).
296. Colson, P., Jennings, H. J., and Smith, I. C. P. *J. Am. Chem. Soc.* **96,** 8081 (1974).
297. Colson, P., and King, P. K. *Carbohydr. Res.* **47,** 1 (1976).

298. Dorman, D. E., and Roberts, J. D. *J. Am. Chem. Soc.* **93,** 4463 (1971).
299. Shapiro, Y. E., Viktorov, A. V., Volkora, V. I., Barsukov, L. I., Bystrov, V. F., and Bergelson, L. D. *Chem. Phys. Lipids* **14,** 227 (1975).
300. Jennings, H. J., and Smith, I. C. P. *J. Am. Chem. Soc.* **95,** 606 (1973).
301. Perlin, A. S., Cam, B., and Koch, H. J. *Can. J. Chem.* **48,** 2596 (1970).
302. Czarniecki, M. F., and Thornton, E. R. *J. Am. Chem. Soc.* **98,** 1023 (1976).
303. Sillerud, L. O., Prestegard, J. H., Yu, R. K., Schafer, D. E., and Koenigsberg, W. H. *Biochemistry* **17,** 2619 (1978).
304. Koerner, T. A. W., Cary, L. W., Li, S.-C., and Li, Y.-T. *J. Biol. Chem.* **254,** 2326 (1979).
305. Jardetsky, O., and Roberts, G. C. K., *NMR in Molecular Biology*, Academic Press, New York (1981).
306. Czarniecki, M. F., and Thornton, E. R. *J. Am. Chem. Soc.* **99,** 8273 (1977); **99,** 8279 (1977).
307. Sharom, F. J., and Grant, C. W. M. *Biochem. Biophys. Res. Commun.* **67,** 150 (1975).
308. Sharom, F. J., and Grant, C. W. M. *J. Supramol. Struct.* **6,** 249 (1977).
309. Pascher, I., and Sundell, S. *Chem. Phys. Lipids* **20,** 175 (1977).
310. Hakomori, S., Watanabe, K., and Laine, R. A. *Pure Appl. Chem.* **49,** 1215 (1977).
311. Taketomi, T. *Z. Allg. Mikrobiol.* **1,** 331 (1961).
312. Karlsson, K.-A. *Acta Chem. Scand.* **18,** 2397 (1964).
313. Michalec, C., and Kolman, Z. *Clin. Chim. Acta* **13,** 532 (1966).
314. Carter, H. E., Gaver, R. C., and Yu, R. K. *Biochem. Biophys. Res. Commun.* **22,** 316 (1966).
315. Carter, H. E., and Gaver, R. *Biochem. Biophys. Res. Commun.* **29,** 886 (1968).
316. Carter, H. E., and Hirschberg, C. B. *Biochemistry* **7,** 2296 (1968).
317. Mislow, K. *J. Am. Chem. Soc.* **74,** 5155 (1952).
318. Marinetti, G., and Stotz, E. *J. Am. Chem. Soc.* **76,** 1347 (1954).
319. Jenny, E. F., and Grob, C. A. *Helv. Chim. Acta* **36,** 1936 (1953).
320. Stoffel, W. *Chem. Phys. Lipids* **11,** 318 (1973).
321. Cahn, R. S., Ingold, C. K., and Prelog, V. *Angew. Chem. Int. Ed. Engl.* **5,** 385 (1966).
322. Kochetkov, N. K., Zhukova, I. G., and Glukhoded, I. S. *Biochim. Biophys. Acta* **60,** 431 (1962).
323. Kochetkov, N. K., Zhukova, I. G., and Glukhoded, I. S. *Biochim. Biophys. Acta* **70,** 716 (1963).
324. Klenk, E., and Löhr, J. P. *Z. Physiol. Chem.* **348,** 1712 (1967).
325. Kishimoto, Y., Wajda, M., and Radin, N. S. *J. Lipid Res.* **9,** 27 (1968).
326. Tamai, Y. *Jpn. J. Exp. Med.* **38,** 65 (1968).
327. Tamai, Y., Taketomi, T., and Yamakawa, T. *Jpn. J. Exp. Med.* **37,** 79 (1967).
328. Karlsson, K.-A., Samuelsson, B. E., and Steen, G. O. *J. Lipid Res.* **13,** 169 (1972).
329. Watanabe, K., Matsubara, T., and Hakomori, S. *J. Biol. Chem.* **251,** 2385 (1976).
330. Gray, G. M. *Biochim. Biophys. Acta* **239,** 494 (1971).
331. Costantino-Ceccarini, E., and Morell, P. *J. Biol. Chem.* **248,** 8240 (1973).
332. Handa, S., Ariga, T., Miyatake T., and Yamakawa, T. *J. Biochem. (Tokyo)* **69,** 625 (1971).
333. Li, Y.-T., Li, S.-C., and Dawson, G. *Biochim. Biophys. Acta* **260,** 88 (1972).
334. Slomiany, B. L., Slomiany, A., and Glass, G. B. J. *Eur. J. Biochem.* **78,** 33 (1977).
335. Martensson, E. *Biochim. Biophys. Acta* **116,** 521 (1966).
336. Stoffyn, A., Stoffyn, P., and Martensson, E. *Biochim. Biophys. Acta* **152,** 353 (1968).
337. Ishizuka, I., Suzuki, M., and Yamakawa, T. *Proc. Jpn. Conf. Biochem. Lipids* **14,** 61 (1972).
338. Ishizuka, I., Suzuki, M., and Yamakawa, T. *J. Biochem. (Tokyo)* **73,** 77 (1973).
339. Kornblatt, M. J., Schachter, H., and Murray, R. K. *Biochem. Biophys. Res. Commun.* **48,** 1489 (1972).
340. Knapp, A., Kornblatt, M. J., Schachter, H., and Murray, R. K. *Biochem. Biophys. Res. Commun.* **55,** 179 (1973).
341. Karlsson, K.-A. *Acta Chem. Scand.* **18,** 565 (1964).
342. Morrison, W. R. *Biochim. Biophys. Acta* **176,** 537 (1969).
343. Panganamala, R. V., Geer, J. C., and Cornwell, D. G. *J. Lipid Res.* **10,** 445 (1969).
344. O'Connor, J. D., Polito, A. J., Monroe, R. E., Sweeley, C. C., and Bieber, L. L. *Biochim. Biophys. Acta* **202,** 195 (1970).
345. Bieber, L. L., O'Connor, J. D., and Sweeley, C. C. *Biochim. Biophys. Acta* **187,** 157 (1969).
346. Karlander, S.-G., Karlsson, K.-A., Samuelsson, B. E., and Steen, G. O. *Acta Chem. Scand.* **23,** 3597 (1969).
347. Moscatelli, E. A., and Gilliland, K. M. *Lipids* **4,** 244 (1969).

348. Karlsson, K.-A. *Acta Chem. Scand.* **18,** 2395 (1964).
349. Popoviv, M. *Biochim. Biophys. Acta* **125,** 178, (1966).
350. Carter, H. E., and Humiston, C. *J. Biol. Chem.* **191,** 727 (1951).
351. Marinetti, G., and Stotz, E. *J. Am. Chem. Soc.* **76,** 1347 (1954).
352. Gigg, J., Gigg, R., and Warren, C. D. *J. Chem. Soc. (London) C* 1966: 1872.
353. Majhofer-Oreščanin, M., and Proštenik, M. *Croat. Chem. Acta* **33,** 219 (1961).
354. Gross, S. K., and McCluer, R. T. *Anal. Biochem.* **102,** 429 (1980).
355. Karlsson, K.-A. *Acta Chem. Scand.* **21,** 2577 (1967).
356. Karlsson, K.-A. *Chem. Phys. Lipids* **5,** 6–43 (1981).
357. Hayashi, A., and Matsubara, T. *Biochim. Biophys. Acta* **202,** 228 (1970).
358. Carter, H. E., and Hendrickson, H. E. *Biochemistry* **2,** 389 (1963).
359. Hirvisalo, E. L., and Renkonen, O. *J. Lipid Res.* **11,** 54 (1970).
360. Kornberg, R. D., and McConnell, H. M. *Biochemistry* **10,** 1111 (1971).
361. Klenk, E. *Z. Physiol. Chem.* **267,** 128 (1940).
362. Svennerholm, L. *Acta Chem. Scand.* **17,** 860 (1963).
363. Rapport, M. M., Graf, L., and Alonzo, N. F. *J. Lipid Res.* **1,** 301 (1960).
364. Yamakawa, T., Irie, R., and Iwanaga, M. *J. Biochem (Tokyo)* **48,** 490 (1960).
365. Makita, A., Suzuki, C., and Yosizawa, Z. *J. Biochem. (Tokyo)* **60,** 502 (1966).
366. Naoi, M., Lee, Y. C., and Roseman, S. *Anal. Biochem.* **58,** 571 (1974).
367. Vance, D., and Sweeley, C. C. *J. Lipid Res.* **8,** 621 (1967).
368. Li, S.-C., and Li, Y.-T. *J. Biol. Chem.* **246,** 3769 (1971).
369. Arita, H., and Kawanami, J. *J. Biochem. (Tokyo)* **76,** 1067 (1974); **81,** 1661 (1977).
370. Rapport, M. M., Schneider, H., and Graf, L. *Biochim. Biophys. Acta* **137,** 409 (1967).
371. Laine, R. A., Sweeley, C. C., Li, Y.-T., Kisic, A., and Rapport, M. M. *J. Lipid Res.* **13,** 519 (1972).
372. Siddiqui, B., Kawanami, J., Li, Y.-T., and Hakomori, S. *J. Lipid Res.* **13,** 657 (1972).
373. Kawanami, J. *J. Biochem. (Tokyo)* **64,** 625 (1968).
374. Brunius, F. E. Chemical studies on the true Forssman hapten, Doctoral thesis, Karolinska Institute, Stockholm (1936).
375. Papiermeister, B., and Mallette, M. F. *Arch. Biochim. Biophys.* **57,** 94 (1955).
376. Sung, S. J., Esselman, W. J., and Sweeley, C. C. *J. Biol. Chem.* **248,** 6528 (1973).
377. Taketomi, T., Hara, A., Kawamura, N., and Hayashi, M. *J. Biochem.(Tokyo)* **75,** 197 (1973).
378. Ziolkowski, C. H. J., Fraser, B. A., and Mallette, M. F. *Immunochemistry* **12,** 297 (1975).
379. Slomiany, B. L., Slomiany, A., and Horowitz, M. I. *Eur. J. Biochem.* **43,** 161 (1974).
380. Slomiany, B. L., and Slomiany, A. *Eur. J. Biochem.* **83,** 105 (1978).
381. Ando, S., Kon, K., Isobe, M., Nagai, Y., and Yamakawa, T. *J. Biochem. (Tokyo)* **79,** 625 (1976).
382. Joseph, K. C., and Gockerman, J. P. *Biochem. Biophys. Res. Commun.* **65,** 146 (1975).
383. Watanabe, K., and Hakomori, S. *J. Exp. Med.* **144,** 644 (1976).
384. Karlsson, K.-A. In: *Structure of Biological Membranes:* (Abrahamsson, S., and Pascher, I., eds.), Plenum Press, New York (1977), pp. 245–274.
385. Siddiqui, B., and Hakomori, S. *Biochim. Biophys. Acta* **330,** 147 (1973).
386. Ando, S., and Yamakawa, T. *J. Biochem. (Tokyo)* **73,** 387 (1973).
387. Sundsmo, J., and Hakomori, S. *Biochem. Biophys. Res. Commun.* **68,** 799 (1976).
388. Stellner, K., and Hakomori, S. *J. Biol. Chem.* **249,** 1022 (1974).
389. Eto, T., Ichikawa, T., Nishimura, K., Ando, S., and Yamakawa, T. *J. Biochem. (Tokyo)* **64,** 205 (1968).
390. Chien, J.-L., Li, S.-C., Laine, R. A., and Li, Y.-T. *J. Biol. Chem.* **253,** 4031 (1978).
391. Marcus, D. M. *Transfusion* **11,** 16 (1971).
392. Niemann, H., Watanabe, K., Hakomori, S., Childs, R. A., and Feizi, T. *Biochem. Biophys. Res. Commun.* **81,** 1286 (1978).
393. Koscielak, J., Piasek, A., Gorniak, H., Gardas, A., and Gregor, A. *Eur. J. Biochem.* **37,** 214 (1973).
394. Hanfland, P., Assmann, G., and Egge, H. *Z. Naturforsch.* **33c,** 73 (1978).
395. Hakomori, S., Koscielak, J., Bloch, K. J., and Jeanloz, R. W. *J. Immunol.* **98,** 31 (1967).
396. Marcus, D. M., and Cass, L. E. *Science* **164,** 553 (1969).

397. Hanfland, P. *Eur. J. Biochem.* **87**, 161 (1978).
398. Wherrett, J. R., and Hakomori, S. *J. Biol. Chem.* **248**, 3046 (1973).
399. Wherrett, J. R., and Brown, B. L. *Neurology* **19**, 489 (1969).
400. MacMillan, V. H., and Wherrett, J. R. *J. Neurochem.* **16**, 1621 (1969).
401. Wherrett, J. R. *Biochim. Biophys. Acta* **326**, 63 (1973).
402. Svennerholm, L., Bruce, A., Mansson, J., Rynmark, B., and Vanier, M.-T. *Biochim. Biophys. Acta* **280**, 626 (1972).
403. Li, Y.-T., Mansson, J. E., Vanier, M.-T., and Svennerholm, L. *J. Biol. Chem.* **248**, 2634 (1973).
404. Wintzer, G., and Uhlenbruck, G. *Z. Immunitaetsforsch. Allerg. Klin. Immunol.* **133**, 60 (1967).
405. Yamazaki, T., Suzuki, A., Handa, S., and Yamakawa, T. *J. Biochem. (Tokyo)* **86**, 803 (1979).
406. Rauvala, H., Krusius, T., and Finne, J. *Biochim. Biophys. Acta* **531**, 266 (1978).
407. Hakomori, S., Watanabe, K., and Laine, R. A. In: *Human Blood Groups*, (Mohn, J. F., ed.) (5th International Convocation on Immunology, Buffalo, New York 1977), S. Karger, Basel (1977), pp. 150–163.
408. Scandella, C. J., Devaux, P., and McConnell, H. M. Proc. Natl. Acad. Sci. **69**, 2056 (1972).
409. Karlsson, K.-A. In *Proceedings of the 27th IUPAC Meeting* (Helsinki, August 1979), (Varmavuori, A., ed.), Pergamon Press, Oxford (1980), pp. 171–183.
410. Hanfland, P., Egge, H., and Dabrowski, J. In: *Glycoconjugates* (Schauer, R., Boer, P., Buddecke, E., Kramer, M. F., Vliegenthart, J. F. G., and Wiegandt, H., eds.), Georg Thieme, Stuttgart (1979), p. 520.
411. Suzuki, C., Makita, A., and Yosizawa, Z. *Arch. Biochem. Biophys.* **127**, 140 (1968).
412. McKibbin, J. M. *Biochemistry* **8**, 679 (1969).
413. Watanabe, K., and Hakomori, S. *Biochemistry* **24**, 5502 (1979).
414. Hakomori, S., and Andrews, H. D. *Biochim. Biophys. Acta* **202**, 225 (1970).
415. Painter, T. J., Watkins, W. M., and Morgan, W. T. J. *Nature (London)* **199**, 282 (1963).
416. Watkins, W. M. *Science* **152**, 172 (1966).
417. Feizi, T., Childs, R. A., Watanabe, K., and Hakomori, S. *J. Exp. Med.* **149**, 975 (1979).
418. Slomiany, B. L., Slomiany, A., and Horowitz, M. I. *Eur. J. Biochem.* **56**, 353 (1975).
419. Slomiany, A., and Slomiany, B. L. *Biochim. Biophys. Acta* **388**, 135 (1975).
420. Smith, E. L., McKibbin, J. M., Karlsson, K.-A., Pascher, I., Samuelsson, B. E., and Li, S.-C. *Biochemistry* **14**, 3370 (1975).
421. McKibbin, J. M., Smith, E. L., Mansson, J.-E., and Li, Y.-T. *Biochemistry* **16**, 1223 (1977).
422. Smith, E. L., McKibbin, J. M., Karlsson, K.-A., Pascher, I., and Samuelsson, B. E. *Biochemistry* **14**, 2120 (1975).
423. Smith, E. L., McKibbin, J. M., Karlsson, K.-A., Pascher, I., Samuelsson, B. E., Li, Y.-T., and Li, S.-C. J. Biol. Chem. **250**, 6059 (1975).
424. Smith, E. L., McKibbin, J. M., Breimer, M. E., Karlsson, K.-A., Pascher, I., and Samuelsson, B. E. *Biochim. Biophys. Acta* **398**, 84 (1975).
425. Slomiany, B. L., and Horowitz, M. I. *J. Biol. Chem.* **248**, 6232 (1973).
426. Slomiany, A., Slomiany, B. L., and Horowitz, M. I. *J. Biol. Chem.* **249**, 1225 (1974).
427. Slomiany, B. L., Slomiany, A., and Horowitz, M. I. *Biochim. Biophys. Acta* **326**, 224 (1973).
428. Rauvala, H. *J. Biol. Chem.* **251**, 7517 (1976).
429. Slomiany, B. L., and Slomiany, A. *Biochim. Biophys. Acta* **486**, 531 (1977).
430. Slomiany, B. L., and Slomiany, A. *Chem. Phys. Lipids* **20**, 57 (1977).
431. Slomiany, B. L., and Meyer, K. *J. Biol. Chem.* **247**, 5062 (1972).
432. Ohashi, M., In: *Glycoconjugates* (Schauer, R., Boer, P., Buddecke, E., Kramer, M. F., Vliegenthart, J. F. G., and Wiegandt, H., eds.), Georg Thieme, Stuttgart (1979), p. 53.
433. Kornfeld, R. *Biochemistry* **17**, 1415 (1978).
434. Fukuda, M., Fukuda, M. N., and Hakomori, S. *J. Biol. Chem.* **254**, 5458 (1979).
435. Tadano, K., and Ishizuka, I. *Biochem. Biophys. Res. Commun.* **97**, 126 (1980).
436. Klenk, E., and Georgias, L. *Z. Phys. Chem.* **348**, 1261 (1967).
437. Gielen, W. *Z. Naturforsch.* **23b**, 1598 (1968).
438. Siddiqui, B., and McCluer, R. H. *J. Lipid Res.* **9**, 366 (1968).
439. Ledeen, R. W., Yu, R. K., and Eng, L. F. *J. Neurochem.* **21**, 829 (1973).
440. Yu, R. K., and Lee, S. H. *J. Biol. Chem.* **251**, 198 (1976).

441. Yu, R. K., and Yen, S. I. *J. Neurochem.* **25,** 229 (1975).
442. Yu, R. K., Ledeen, R. W., Gajdusek, D. C., and Gibbs, C. J. *Brain Res.* **70,** 103 (1974).
443. Yu, R. K., Ledeen, R. W., and Eng, L. F. *J. Neurochem.* **23,** 169 (1974).
444. Nagai, Y., and Hoshi, M. *Biochim. Biophys. Acta* **388,** 146 (1975).
445. Hoshi, M., and Nagai, Y. *Biochim. Biophys. Acta* **388,** 152 (1975).
446. Ohsawa, T., and Nagai, Y. *Biochim. Biophys. Acta* **389,** 69 (1975).
447. Kochetkov, N. K., Zhukova, I. G., Smirnova, G. P., and Glukhoded, I. S. *Biochim. Biophys. Acta* **326,** 74 (1973).
448. Kochetkov, N. K., Smirnova, G. P., and Chekareva, N. V. *Biochim. Biophys. Acta* **424,** 274 (1976).
449. Klenk, E., and Padberg, G. *Z. Physiol. Chem.* **327,** 249 (1962).
450. Yamakawa, T. *J. Biochem (Tokyo)* **43,** 867 (1956).
451. Lundblad, A., Svensson, S., Löw, B., Messeter, L., and Cedergren, B. *Eur. J. Biochem.* **104,** 323 (1980).
452. Klenk, E., and Heuer, K. *Z. Verdauungs-Stoffwechsel Krankheiten* **20,** 180 (1960).
453. Kuhn, R., Wiegandt, H., and Egge, H. *Angew. Chem.* **73,** 580 (1961).
454. Hakomori, S., and Saito, T. *Biochemistry* **8,** 5082 (1969).
455. Svennerholm, L. *J. Neurochem.* **10,** 613 (1963).
456. Handa, S., and Yamakawa, T. *Jpn. J. Exp. Med.* **34,** 293 (1964).
457. Veh, R. W., Sander, M., Haverkamp, J., and Schauer, R. In: *Glycoconjugate Research,* Vol. 1 (Gregory, J. D., and Jeanloz, R. W., eds.), Academic Press, New York and London (1979), pp. 557–579.
458. Feldman, G. L., Feldman, L. S., and Rouser, G. *Lipids* **1,** 21 (1966).
459. Robert, J., Freysz, L., Sensenbrenner, M., Mandel, P., and Rebel, G. *FEBS Lett.* **50,** 144 (1975).
460. Higashi, H., Naiki, M., Matuo, S., and Okouchi, K. *Biochem. Biophys. Res. Commun.* **79,** 388 (1977).
461. Handa, N., and Handa, S. *Jpn. J. Exp. Med.* **35,** 331 (1965).
462. Handa, S., and Burton, R. M. *Lipids,* **4,** 205 (1969).
463. Puro, K. *Biochim. Biophys. Acta* **189,** 401 (1969).
464. Holm, M., Månsson, J.-E., Vanier, M.-T., and Svennerholm, L. *Biochim. Biophys. Acta* **280** 356 (1972).
465. Holm, M., and Månsson, J.-E. *FEBS Lett.* **38,** 261 (1974).
466. Holm, M., and Månsson, J.-E. *FEBS Lett.* **46,** 200 (1974).
467. Holm, M., Pascher, I., and Samuelsson, B. E. *Biomed. Mass Spectrometry* **4,** 77 (1977).
468. Portoukalian, J, Zwingelstein, G., and Dore, J. F. *Biochimie* **58,** 1285 (1976).
469. Hakomori, S., Saito, T., and Vogt, P. K. *Virology* **44,** 609 (1971).
470. Hakomori, S. *Proc. Natl. Acad. Sci. U.S.A.* **67,** 1741 (1970).
471. Vance, W. R., Shook, C. P., and McKibbin, J. B. *Biochemistry* **5,** 435 (1966).
472. Avrova, N. F. *J. Neurochem.* **18,** 667 (1971).
473. Avrova, N. F., Li, Y.-T., and Obukhova, E. L. *J. Neurochem.* **32,** 1807 (1979).
474. Iwamori, M., and Nagai, Y. *J. Biol. Chem.* **253,** 8328 (1978).
475. Tettamanti, G., Bertona, L., Gualandi, V., and Zambotti, V. *Rend. Ist. del Lomb. Sci. Lett. Cl. Sci. Mat. Nat.* **99,** 173 (1965).
476. Svennerholm, L., and Raal, A. *Biochim. Biophys. Acta* **53,** 422 (1961).
477. Seyama, Y., and Yamakawa, T. *J. Biochem. (Tokyo)* **75,** 837 (1974).
478. Rosenfelder, G., Young, W. W., Jr., and Hakomori, S. *Cancer Res.* **37,** 1333 (1977).
479. Young, W. W., Jr., Hakomori, S., Durdik, J. M., and Henney, C. S. *J. Immunol.* **124,** 199 (1980).
480. Siddiqui, B., and Hakomori, S. *Cancer Res.* **30,** 2930 (1970).
481. Itaya, K., and Hakomori, S. *FEBS Lett.* **66,** 65 (1976).
482. Nigam, V. N., Lallier, R., and Brailovsky, C. *J. Cell Biol.* **58,** 307 (1973).
483. Gatt, S., and Berman, E. R. *J. Neurochem.* **10,** 43 (1963).
484. Sandhoff, K., Jatzkewitz, H., and Peters, G. *Naturwissenschaften* **56,** 356 (1969).
485. Sandhoff, K., Andreae, J., and Jatzkewitz, H. *Life Sci.* **7,** 283 (1968).
486. Kuhn, R., Egge, H., Brossmer, R., Gauhe, A., Klesse, P., Lochinger, W., Röhm, E., Trischmann, H., and Tschampel, D. *Angew. Chem.* **72,** 805 (1961).
487. Li, Y.-T., King, M. J., and Li, S.-C. *Adv. Exp. Med. Biol.* **125,** 93 (1980).

488. Ishizuka, I., Kloppenburg, M., and Wiegandt, H. *Biochim. Biophys. Acta* **210**, 299 (1970).
489. Ishizuka, I., and Wiegandt, M. *Biochim. Biophys. Acta* **260**, 279 (1972).
490. Drzeniek, R., *Histochem. J.* **5**, 271 (1973).
491. Stoffyn, A., Stoffyn, P., and Yip, M. C. M. *Biochim. Biophys. Acta* **409**, 97 (1975).
492. Yip, M. C. M. *Biochem. Biophys. Res. Commun.* **53**, 737 (1973).
493. Price, H., Kundu, S., and Ledeen, R. *Biochemistry* **14**, 1512 (1975).
494. Ando, S., and Yu, R. K. *J. Biol. Chem.* **252**, 6247 (1977).
495. Penick, R. J., and McCluer, R. H. *Biochim. Biophys. Acta* **106**, 435 (1965).
496. Ando, S., and Yu, R. K. *J. Biol. Chem.* **254**, 1224 (1979).
497. Klenk, E., and Naoi, M. *Z. Physiol. Chem.* **349**, 288 (1968).
498. Iwamori, M., and Nagai, Y. *J. Biochem. (Tokyo)* **84**, 1601 (1978).
499. Suzuki, A., Ishizuka, I., and Yamakawa, T. *J. Biochem. (Tokyo)* **78**, 947 (1975).
500. Ghidoni, R., Sonnino, S., Tettamanti, G., Wiegandt, H., and Zambotti, V. *J. Neurochem.* **27**, 511 (1976).
501. Ohashi, M., and Yamakawa, T. *J. Biochem. (Tokyo)* **81**, 1675 (1977).
502. Ando, S., and Yu, R. K. In: *Glycoconjugate Research*, Vol. I (Gregory, J. D., and Jeanloz, R. W., eds.), Academic Press, New York (1979), pp. 79–82.
503. Suzuki, K., and Chen, G. C. *J. Lipid Res.* **8**, 105 (1967).
504. Rapport, M. M., Graf, L., Hungund, B., Kisic, A., and Hung, Y. *Chem. Phys. Lipids* **17**, 233 (1976).
505. Gardas, A., and Koscielak, J. *Eur. J. Biochem.* **32**, 178 (1973).
506. Gardas, A., and Koscielak, J. *FEBS Lett.* **42**, 101 (1974).
507. Prokazova, N. V., Kosharov, S. L., Sadovskaya, V. L., Moshenski, J. V., Bergelson, L. D., and Zvezdina, N. E. *Bioorganichjeskaja (Bioorg. Chem.)* **5**, 458 (1979) (Russian with English abstract).
508. Kochetkov, N. K., and Smirnova, G. P. *Bioorganichjeskaja (Bioorg. Chem.)* **3**, 1048 (1977); **3**, 1280 (1977) (Russian with English abstract).
509. Macher, B. A., and Klock, J. C. *J. Biol. Chem.* **255**, 2092 (1980).
510. Sugita, M., and Hori, T. *J. Biochem. (Tokyo)* **80**, 637 (1976).
511. Sugita, M. *J. Biochem. (Tokyo)* **86**, 289 (1979).
512. Sugita, M. *J. Biochem. (Tokyo)* **86**, 765 (1979).
513. Hori, T., Itasaka, O., and Kamimura, M. *J. Biochem. (Tokyo)* **64**, 125 (1968).
514. Sugita, M., Shirai, S., Itasaka, O., and Hori, T. *J. Biochem. (Tokyo)* **77**, 125 (1975).
515. Karlsson, K.-A., and Larson, G. *J. Biol. Chem.* **254**, 9311 (1979).
516. Itasaka, O., and Hori, T. *J. Biochem. (Tokyo)* **85**, 1469 (1979).
517. Hori, T., Sugita, M., Kanbayashi, J., and Itasaka, O. *J. Biochem. (Tokyo)* **81**, 107 (1977).
518. Hori, T., Takeda, H., Sugita, M., and Itasaka, O. *J. Biochem. (Tokyo)* **82**, 1281 (1977).
519. Hori, T., Sugita, M., Ando, S., Kuwahara, M., Kumauchi, K., Sugie, E., and Itasaka, O. *J. Biol. Chem.* **256**, 10979 (1981).
520. Matsubara, T., and Hayashi, A. *J. Biochem. (Tokyo)* **74**, 853 (1973).
521. Matsubara, T., and Hayashi, A. *J. Biochem (Tokyo)* **83**, 1195 (1978).
522. Matsuura, F. *Chem. Phys. Lipids* **19**, 223 (1977).
524. Simon, G., and Rouser, G. *Lipids* **4**, 607 (1969).
525. Higashi, S., and Hori, T. *Biochim. Biophys. Acta* **152**, 568 (1968).
526. Matsubara, T., and Hayashi, A. *Biochim. Biophys. Acta* **296**, 171 (1973).
527. Carter, H. E., Celmer, W. D., Galanos, D. S., Gigg, R. H., Lands, W. E. M., Law, J. H., Mueller, K. L., Nakayama, T., Tomizawa, H. H., and Weber, E. *J. Am. Chem. Soc.* **35**, 335 (1958).
528. Carter, H. E., Galanos, D. S., Hendrickson, H. E., Jann, B., Nakayama, T., Nakazawa, Y., and Nichols, B. *J. Am. Oil Chem. Soc.* **39**, 107 (1962).
529. Carter, H. E., Strobach, D. R., and Hawthorne, J. N. *Biochemistry* **8**, 383 (1969).
530. Carter, H. E., Johnson, P., and Weber, E. J. *Annu. Rev. Biochem.* **34**, 109 (1965).
531. Kaul, K., and Lester, R. L. *Plant Physiol.* **55**, 120 (1975).
532. Kaul, K., and Lester, R. L. *Biochemistry* **17**, 3569 (1978).
533. Hsieh, T. C.-Y., Kaul, K., Laine, R. A., and Lester, R. L. *Biochemistry* **17**, 3575 (1978).
534. Lester, R. L., Becker, G. W., and Kaul, K. In: *Cyclitols and Phosphoinositides*, Academic Press, New York (1978), pp. 83–102.

535. Becker, G. W., and Lester, R. L. *J. Biol. Chem.* **252,** 8684 (1977).

536. Smith, S. W., and Lester, R. L. *J. Biol. Chem.* **249,** 3395 (1974).

537. Lester, R. L., Smith, S. W., Wells, G. B., Rees, D. C., and Angus, W. W. *J. Biol. Chem.* **249,** 3388 (1974).

538. Shapiro, D., and Segal, H. *J. Am. Chem. Soc.* **76,** 5894 (1954).

539. Matsumoto, M., and Taki, T. *Biochem. Biophys. Res. Commun.* **71,** 472 (1975).

540. Hirabayashi, Y., Taki, T., Matsumoto, M., and Kojima, K. *Biochim. Biophys. Acta* **529,** 96 (1978).

541. Grob, C. A., and Gadient, F. *Helv. Chim. Acta* **40,** 1145 (1957).

542. Gregory, G. I., and Malkin, T. *J. Chem. Soc. (London)* **1951,** 2453.

543. Egerton, M. J., Gregory, G. I., and Malkin, T. *J. Chem Soc. (London)* **1952,** 2272.

544. Grob, C. A., Jenny, E. F., and Utzinger, H. *Helv. Chim. Acta* **34,** 2249 (1951).

545. Koizumi, K., Ito, Y., Kojima, K., and Fuji, T. *J. Biochem. (Tokyo)* **79,** 739 (1976).

546. Vanier, M. T., Månsson, J.-E., and Svennerholm, L. *FEBS Lett.* **112,** 70 (1980).

547. Jenny, E. F., and Grob, C. A. *Helv. Chim. Acta* **36,** 1454 (1953).

548. Chien, J.-L., and Hogan, E. L. *Fed. Proc. Fed. Am. Soc. Exp. Biol.* **39** (6), 2183 (Abstr. 3040) (1980).

549. Hamanaka, S., Handa, S., and Yamakawa, T. *J. Biochem. (Tokyo)* **86,** 1623 (1979).

550. Stoffel, W., Sticht, G., and Heyn, G. Cited in Stoffel.[616]

551. Macher, B. A., Pacuszka, T., Mullin, B. R., Sweeley, C. C., Brady, R. O., and Fishman, P. H. *Biochim. Biophys. Acta* **588,** 35 (1979).

552. Gigg, J., and Gigg, R. *J. Chem. Soc. (London)* C **1966,** 1976.

553. Shoyama, Y., Okabe, H., Kishimoto, Y., and Costello, C. *J. Lipid. Res.* **19,** 250 (1978).

554. Flowers, H. M. *Carbohydr. Res.* **2,** 188 (1966).

555. Flowers, H. M. *Carbohydr. Res.* **5,** 126 (1967).

556. Flowers, H. M. *Carbohydr. Res.* **2,** 371 (1966).

557. Kasai, M., Iwamori, M., Nagai, Y., Okumura, K., and Tada, T. *Eur. J. Immunol.* **10,** 175 (1980).

558. Ishizuka, I., Inomata, M., Ueno, K., and Yamakawa, T. *J. Biol. Chem.* **253,** 898 (1980).

559. Holmgren, J., Svennerholm, L., Elving, H., Fredman, P., and Shannegard, Ö. *Proc. Natl. Acad. Sci. U.S.A.* **77,** 1947 (1980).

560. Yasue, S., Handa, S., Miyagawa, S., Inoue, J., Hasegawa, A., and Yamakawa, T. *J. Biochem. (Tokyo)* **83,** 1101 (1978).

561. Watanabe, K., and Arao, Y. *J. Lipid Res.* **22,** 1020–1024 (1981).

562. Watanabe, K., and Hakomori, S, in preparation.

563. Hamanaka, S., Handa, S., Inoue, J., Hasegawa, A., and Yamakawa, T. *J. Biochem. (Tokyo)* **86,** 696 (1979).

564. Merrick, J. M., Zadardik, K., and Milgrom, F. *Int. Arch. Allergy Appl. Immunol.* **57,** 477 (1978).

565. Nakahara, K., Ohashi, T., Oda, T., Hirano, T., Kasai, M., Okumura, K., and Tada, T. *N. Engl. J. Med.* **302,** 674 (1980).

566. Gatt, S., and Berman, E. R., *J. Neurochem.* **10,** 43 (1963).

567. McNeil, M., and Albersheim, P. *Carbohydr. Res.* **56,** 239 (1977).

568. Laine, R. A. In: *Proceedings of the 27th International Congress of Pure and Applied Chemistry* (Varmavuori, A., ed.), Pergamon Press, Oxford (1979), p. 194.

569. Breimer, M. E., Hansson, G. C., Karlsson, K.-A., and Leffler, H. In: *Cell Surface Glycolipids* (Sweeley, C. C., ed.), American Chemical Society Symposium Series, No. 128, American Chemical Society, Washington, D.C. (1980), pp. 79–104.

570. Slomiany, B. L., Slomiany, A., and Horowitz, M. I. *Biochim. Biophys. Acta* **348,** 388 (1974).

571. Slomiany, B. L., and Slomiany, A. *J. Biol. Chem.* **253,** 3517 (1978).

572. McKibbin, J. M. *Biochemistry* **8,** 679 (1969).

573. Levine, M., Bain, J., Narashimhan, R., Palmer, B., Yates, A. J., and Murray, R. K., *Biochim. Biophys. Acta* **441,** 134 (1976).

574. Kishimoto Y., and Radin, N. S. *J. Lipid Res.* **6,** 532 (1965).

575. Norton, W. T., and Podulso, S. E. *J. Neurochem.* **21,** 759 (1973).

576. Ishizuka, I., Inomata, M., Ueno, K., and Yamakawa, T. *J. Biol. Chem.* **253,** 898 (1978).

577. Abe, T., and Norton, W. T. *J. Neurochem.* **23,** 1025 (1974).

578. Karlsson, K-A. In *Structure and Function of Biological Membranes* (Abramsson, S., and Pascker, I., eds.), Plenum Press, New York (1977), p. 245.

579. Yamakawa, T., and Nagai, Y. *Trends Biochem. Sci.* **3,** 128 (1978).

580. Zalc, B. *FEBS Lett.* **92,** 92 (1978).

581. Li, Y.-T., Chien, J.-L., Wan, C. C., Laine, R. A., and Li, S.-C. *Trans. Am. Soc. Neurochem.* **8,** (1) (1977).

582. Sonnino, S., Ghidoni, R., Galli, G., and Tettamanti, G. *J. Neurochem.* **31,** 947 (1978).

583. Skipski, V. P. *Methods Enzymol.* **35,** 396 (1975).

584. Hsieh, T. C.-Y., Lester, T. L., and Laine, R. A. *J. Biol. Chem.* **256,** 7747–7755 (1981).

585. Laine, R. A., Hsieh, T. C.-Y., and Lester, R. L. In: *Cell Surface Glycolipids* (Sweeley, C. C., ed.), American Chemical Society Symposium Series, No. 128, American Chemical Society, Washington, D.C. (1980), pp. 65–78.

586. Laine, R. A., and Renkonen, O. *Biochemistry* **13,** 2837–2843 (1974).

587. Laine, R. A., *Anal. Biochem.* **116,** 383–388 (1981).

588. Nilsson, B., and Zopf, D. *Methods Enzymol.* **88,** 46–58 (1983).

589. Blomberg, J., Breimer, M. E., and Karlsson, K.-A. *Biochim. Biophys. Acta* **711,** 466 (1982).

590. Hanfland, P., Egge, H., Dabrowski, U., Kuhn, S., Roelcke, D., and Dabrowski, J. *Biochemistry* **20,** 5310–5319 (1981).

591. Ohashi, M. *J. Biochem. (Tokyo)* **88,** 583 (1980); in: *Advances in Experimental Medicine and Biology,* Vol. 152, *New Vistas in Glycolipid Research* (Makita, A., Handa, S., Taketomi, T., and Nagai, Y., eds.), Plenum Press, New York (1982), pp. 83–91.

592. Dabrowski, J., Hanfland, P., and Egge, H. *Biochemistry* **19,** 5652 (1980).

593. Young, W. W., Jr., MacDonald, E. M. S., Nowinski, R. C., and Hakomori, S. *J. Exp. Med.* **150,** 1008–1019 (1979).

594. Young, W. W., Jr., Portoukalian, J., and Hakomori, S. *J. Biol. Chem.* **256,** 10,967–10,972 (1981).

595. Magnani, J. F., Smith, D. F., and Ginsburg, V. *Anal. Biochem.* **109,** 399–402 (1980).

596. Fukuda, M. N., and Hakomori, S. *J. Biol. Chem.* **257,** 446–453 (1982).

597. Slomiany, A., and Slomiany, B. L. *Eur. J. Biochem.* **76,** 491–498 (1977).

598. Kon, K., Ando, S., Nagai, Y., and Yamakura, T. *Seikagaku (J. Jpn. Biochem. Soc.)* **48,** 707 (1976).

599. Ando, S., Kon, K., Nagai, Y., and Yamakura, T. In: *Advances in Experimental Medicine and Biology,* Vol. 152, *New Vistas in Glycolipid Research* (Makita, A., Handa, S., Taketomi, T., and Nagai, Y., eds.), Plenum Press, New York (1982), pp. 21–81.

600. Kundu, S. K., Marcus, D. M., Pascher, I., and Samuelsson, B. E. *Fed. Proc. Fed. Am. Soc. Exp. Biol.* **40,** 1545 (1981).

601. Iwamori, M., and Nagai, Y. *J. Biochem. (Tokyo)* **89,** 1253–1264 (1981).

602. Kannagi, R., Nudelman, E., Levery, S. B., and Hakomori, S. *J. Biol. Chem.* **257,** 1486 (1982).

603. Hakomori, S., Nudelman, E., Levery, S. B., Solter, D., and Knowles, B. B. *Biochem. Biophys. Res. Commun.* **100,** 1578–1586 (1981).

604. Hakomori, S., Nudelman, E., Kannagi, R., and Levery, S. B. *Biochem. Biophys. Res. Commun.* **109,** 36 (1982).

605. Kannagi, R., Levery, S. B., and Hakomori, S. *J. Biol. Chem.* (1983), submitted.

606. Watanabe, K., Powell, M. E., and Hakomori, S. *J. Biol. Chem.* **254,** 8223–8229 (1979).

607. Kannagi, R., Fukuda, M. N., and Hakomori, S. *J. Biol. Chem.* **257,** 4438–4442 (1982).

608. Kannagi, R., Levine, P., Watanabe, K., and Hakomori, S. *Cancer Res.* **42,** 5249 (1982).

609. Magnani, J. L., Brockhaus, M., Smith, D. F., Ginsburg, V., Blaszczyk, M., Mitchell, K. F., Steplewski, Z., and Koprowski, H. *Science* **212,** 55–56 (1981).

610. Magnani, J. L., Nilsson, B., Brockhaus, M., Zopf, D., Steplewski, Z., Koprowski, H., and Ginsburg, V. *J. Biol. Chem.* **257,** 14365 (1982).

611. Wagner, H., and Zofcsik, W. *Biochem Z.* **346,** 333 (1966).

612. Wagner, H., Pohl, P., and Munzing, A. *Z. Naturforsch.* **24,** 366–370 (1969); Wagner, H., Zofcsik, W., and Heng, I. *Z. Naturforsch.* **24,** 922–931 (1969).

613. Hsieh, T. C.-Y., Lester, R. L., and Laine, R. A. *J. Biol. Chem.* **256,** 7747–7753 (1981).

614. Laine, R. A. In: *Advances in Experimental Medicine and Biology,* Vol. 152, *New Vistas in Glycolipid Research* (Makita, A., Handa, S., Taketomi, T., and Nagai, Y., eds.), Plenum Press, New York (1982) pp. 115–120.

615. Ishizuka, I., and Tadano, K. In: *Advances in Experimental Medicine and Biology*, Vol. 152, *New Vistas in Glycolipid Research* (Makita, A., Handa, S., Taketomi, T., and Nagai, Y., eds.), Plenum Press, New York (1982), pp. 195–206.

616. Stoffel, W. *Chem. Phys. Lipids* **11**, 318 (1973).

617. Hakomori, S. (ed.). In: *International Review of Science*, Vol. 7, *Carbohydrates*, (Aspinall, G., ed.), Butterworths, London and Boston (1976), Chapt. 7, pp. 225–226.

618. Blix, G., Tiselius, A., and Svensson, H. *J. Biol. Chem.* **137**, 485 (1941).

619. Kotchetkov, N. K., Zhukova, I. G., and Glukhoded, I. S. *Biochim. Biophys. Acta* **60**, 431 (1962); **70**, 716 (1963).

620. Norton, W. T., and Brotz, M. *Biochem. Biophys. Res. Commun.* **12**, 198 (1963).

621. Shapiro, D. *Chemistry of Sphingolipids*, Vol. 7, *Chemistry of Natural Products*, Hermann, Paris, France (1969).

622. Kobata, A. *Anal. Biochem.* **100**, 1 (1979).

623. Kuhn, R. *Naturwissenschaften* **46**, 41 (1959).

624. Drzeniek, R. *Biochem. Biophys. Res. Commun.* **26**, 631 (1967).

625. Drzeniek, R., and Gauhe, A. *Biochem. Biophys. Res. Commun.* **38**, 651 (1970).

626. Taketomi, T., and Yamakawa, T. *J. Biochem.* **54**, 44 (1963).

627. Miyatake, T., Handa, S., and Yamakawa, T. *Jpn. J. Exp. Med.* **38**, 135 (1968).

628. Miyatake, T. *Jpn. J. Exp. Med.* **39**, 35 (1969).

629. Stothers, J. B., *Carbon 13 NMR Spectroscopy*, Academic Press, New York, 1972.

630. Barker, R., Nunez, H. A., Rosevear, P., and Serianni, A. S. *Methods Enzymol.* **83**, (1982) pp. 58–69.

631. Dabrowski, J., Hanfland, P., and Egge, H. *Methods Enzymol.* **83**, 69 (1982) pp. 69–86.

632. Macher, B. A., Klock, J. C., Fukuda, M. N., and Fukuda, M. *J. Biol. Chem.* **256**, 1968 (1981).

633. Yamada, A., Dabrowski, J., Hanfland, P., and Egge, H. *Biochim. Biophys. Acta* **618**, 473 (1980).

634. Dabrowski, J., Egge, H., Hanfland, P. *Chem. Phys. Lipids* **26**, 187 (1980).

635. Dabrowski, J., Hanfland, P., Egge, H., and Dabrowski, U. *Arch. Biochem. Biophys.* **210**, 405 (1981).

636. Lindberg, B., and Lönngren, J. *Methods Enzymol.* **50**, 3 (1978).

637. Jaques, L. W., Brown, E. B., Barrett, J. M., Brey, W. S., Jr., and Weltner, Jr. *J. Biol. Chem.* **252**, 4533 (1977).

638. Jaques, L. W., Riesco, B. F., and Weltner, W., Jr. *Carbohydr. Res.* **83**, 21 (1980).

639. Daman, M. E., and Kill, K. *Carbohydr. Res.* **102**, 47 (1982).

640. McNeil, M., Darvill, A. G., Åman, P., Franzén, L.-E., and Albersheim, P. *Methods Enzymol.* **83**, 3 (1982).

641. Rauvala, H., Finne, J., Krusius, T., Kärkkäinen, J., and Järnefelt, J. *Adv. Carbohy. Chem. Biochem.* **38**, 389 (1981).

642. Young, W. W., Jr., and Hakomori, S. *Science* **211**, 487 (1981).

643. Abraham, R. J., and Loftus, P. *Proton and Carbon 13 NMR Spectroscopy: An Integrated Approach*, Heyder, London (1978).

644. Breimer, M. E., Hansson, G. C., Karlsson, K.-A., and Leffler, H. In: *Cell Surface Glycolipids* (Sueeley, C. C., ed.) American Chemical Society #128 (1980), pp. 79–104.

Suggested Readings

The following review articles and monographs on the structural chemistry of glycolipids have been published during the past few years:

Brunngraber, E. G. In: *Neurochemistry of Aminosugars*, Charles C. Thomas, Springfield, Illinois (1979).

Hakomori, S., and Ishizuka, I. In: *Handbook of Biochemistry and Molecular Biology, The Table of Glycolipids* (Fasman, G. D., ed.) CRC Press, Cleveland, Ohio (1975).

Hakomori, S. In: *International Review of Science*, Vol. 7, *Glycolipids of Animal Cell Membranes* (Aspinall, G. O., ed.) Butterworths, London and Boston (1976), pp. 223–249.

Ledeen, R. W. In: *Research Methods in Neurochemistry*, Vol. 4, *Methods for Isolation and Analysis of Gangliosides* (Marks, N., and Rodnight, R., eds.) Plenum Press, New York (1978), pp. 371–410.

Ledeen, R. W., and Yu, R. K. In: *Methods in Enzymology*, Vol. 83, *Gangliosides: Structure, Isolation, and Analysis*, Academic Press, New York (1982), pp. 139–192.

Macher, B. A., and Sweeley, C. C. In: *Methods in Enzymology*, Vol. 50, *Glycosphingolipids: Structure, Biological Source, and Properties*, Academic Press, New York (1978), pp. 236–251.

Makita, A., Handa, S., Taketomi, T., and Nagai, Y. (eds.) In: *Advances in Experimental Medicine and Biology*, Vol. 152, *New Vistas in Glycolipid Research*, Plenum Press, New York (1982).

Svennerhom, L., Mande, P., Dreyfus, H. H., and Urban, P-F. (eds.) In: *Advances Experimental Medicine and Biology*, Vol. 125, *Structure and Function of Gangliosides*, Plenum Press, New York (1980).

Sweeley, C. C. *Cell Surface Glycolipids*, American Chemical Society publication, #128 (1980).

Sweeley, C. C. and Siddigui, B. In: *Biochemistry of Mammalian Glycoproteins and Glycolipids*, (Pigman, W. and Horowitz, M. I., eds.) Academic Press, New York (1977), pp. 459–475.

Weigandt, H. In: *Advances in Neurochemistry*, Vol. 4, *The Gangliosides* (Agranoff, B. W., and Aprison, M. H., eds.) Plenum Press, New York (1982), Chapter 3, pp. 149–223.

Witting, Lloyd A. (ed.) In: *Glycolipid Methodology*, American Oil Chemists Society, Champagne, Illinois (1976).

Chapter 2

Sphingolipid Metabolism

Julian N. Kanfer

2.1. Sphingosine Bases

The principal base encountered in mammalian tissue is sphingosine (D-*erythro*-1,3-dihydroxy-2-amino-*trans*-4-octadecene). A novel feature of this molecule is the presence of both a polar and a nonpolar portion. There are no authenticated reports on the occurrence of this material in its unsubstituted state in nature. The preponderant species found in lipids has 18 carbon atoms, 2 hydroxy groups, 1 amino group, and an unsaturated bond. The pathway for the *in vitro* biosynthesis of this material has been established through intensive studies by several laboratories in the past few years. Information available from *in vivo* experiments carried out nearly 30 years ago provided the basis for the general pathway elucidated with cell-free preparations. With the availability of radioactive precursors, it became possible to trace the incorporation of various substrates into sphingolipids[1-3] with intact animals. In most instances, a mixed sphingolipid fraction was isolated from the experimental animals. It was subjected to hydrolysis and the sphingosine base fraction was studied in detail either as such, or after catalytic hydrogenation.

2.1.1. In Vivo Studies

[1-^{14}C]Acetate was found to label principally carbon atoms 3–18 with little appearance of the tracer in C-1 or C-2. The radioactivity from [^{14}C]formate was found mainly in the primary hydroxy group. These general observations led investigators to reason that carbons 3–18 were probably derived from palmitic acid while carbons 1 and 2 originated from a small molecule (see Table I).

$$CH_3-(CH_2)_{12}-CH{=}CH-\underset{\underset{OH}{|}}{\overset{\overset{H}{|}}{C}}-\underset{\underset{NH_2}{|}}{\overset{\overset{H}{|}}{C}}-CH_2OH$$

18   3   2   1

Table I. Distribution of Isotope in Rat Brain Sphingosine[a]

| | | | CH₂—CH—CH—CH₂—CH₂—CH₂—CH₂ | | | | |
| | | $\begin{array}{ccc} & \vert & \vert & \vert \\ & OH & NH_2 & OH \end{array}$ | | | | |
Isotope donor (* designates isotope)	1 %	2 %	3 %	4 %	5 %	18 %
	C-1	C-2	C-3	C-4	C-5	C-3–18
CH₃————C*OOH	0	0	—	—	—	99
C*H₃————COOH	3	2	2	10	3	—
H————C*OOH	64	10	—	—	—	16
C*H₂—C*H₂ / OH NH₂	7	12	—	—	—	57
C*H₂—CH₂ / OH NH₂	2½	2½	—	—	—	95
CH₂————C*OOH / NH₂	7	8	—	—	—	92
C*H₂————COOH / NH₂	27	50	—	—	—	23
C*H₂—CH————COOH / OH NH₂	65	2	—	—	—	33

[a] Modified from Ref. 4.

The simplest compound and most logical low-molecular-weight precursor would be ethanolamine itself; however, when [^{14}C]ethanolamine was administered, little radioactivity was recovered in carbons 1 and 2 of sphingosine. The amino acid glycine, labeled in the carboxyl group, was tested and 90% was found in carbons 3–18; when [2-^{14}C]glycine was used, 50% of the radioactivity was present in carbon 2 and 27% in carbon 1. These results were interpreted as indicating that serine was the most probable precursor of carbons 1 and 2. The finding of 65% of the label in carbon 1 after the administration of [3-^{14}C]serine confirmed this hypothesis. This suggested that sphingosine was assembled from serine and palmitate.

2.1.2. In Vitro Studies

These in vivo data were rapidly exploited in attempts to demonstrate that a condensation occurs between serine and palmitate with cell-free rat brain preparations. The product formed in one of these early studies appeared to be ceramide,[5] while Brady and Koval[6] isolated a mixed sphingolipid fraction. In both cases, hydrolysis of a mixed lipid fraction was employed and the radioactivity in the isolated sphingosine base was determined. Typical incubation mixtures included either Tris or phosphate buffer, Mg^{2+}, pyridoxal phosphate, ATP, TPN, nicotinamide, UTP, DPN, CDP-choline, glucose-6-phosphate and glucose-6-phosphate dehydrogenase, CoASH, palmitate or palmitoy CoA, and DL-[3-^{14}C]serine. Employing a rat brain particulate fraction (microsomes) suitably fortified, nearly 90% of the radioactivity was reported present in the first two carbon atoms.[6] A requirement for pyridoxal phosphate, CoASH, ATP, and glucose-6-phosphate as well as a marginal stimulation by Mg^{2+} or Mn^{2+} was reported (Table II). Cell-free extracts dialyzed against cystine were inactive, implying an essential role for pyridoxal phosphate, and the addition of this vitamin resulted in some reactivation of the system. There was no exogenous source of palmitic acid reported in these experiments, suggesting that the particles had sufficient quantities. The requirements for CoASH and ATP could be replaced by the addition of palmitoy CoA to the incubation tubes. Under these conditions, a requirement for TPNH was still observed. Brady and Koval postulated that a reduced pyridoxine nucleotide was responsible for the reduction of the palmitoy group of palmitoy CoA to the level of thiosemialdehyde or free fatty aldehyde. The presence in the particles of a palmitoy CoA-dependent oxidation of TPNH was demonstrated. The formation of a thiosemialdehyde or palmitaldehyde under these conditions in support of this hypothesis was not reported. Recent evidence has indicated that TPNH is required for the reduction of 3-ketodihydrosphingosine to dihydrosphingosine. Attempts to replace palmitoy CoA with either palmitaldehyde or *trans*-2-hexadecanol were unsuccessful.

Table II. Cofactors Required for the Conversion of Serine to Sphingosine[a,b]

Reactant omitted	Radioactivity of recovered sphingosine	
	Total (cpm/μmole)	Carbons 1 and 2 (%)[c]
None	270	88
CoASH	10	
ATP	49	78
PyrPO$_4$	65	34
Glucose-6-phosphate	97	21
MgCl$_2$	123	81
MnCl$_2$	130	82
TPN	162	85
CDP-choline	188	86
Glucose-6-phosphate dehydrogenase	192	84
DPN	206	90
Nicotinamide	210	85

[a] Modified from Ref. 6.

[b] The reaction mixtures contained 150 μmoles of potassium phosphate buffer (pH 7.8), 4 μmoles of DL-[3-^{14}C]serine (4.0 μCi), 0.6 μmole of CoASH, 4 μmoles of ATP, 1 μmole of PyrPO$_4$, 20 μmoles of glucose-6-phosphate, 5 μmoles of MgCl$_2$, 0.5 μmole of MnCl$_2$, 0.3 μmole of TPN, 2 μmoles of CDP-choline, 2 mg of glucose-6-phosphate dehydrogenase (2.0 U/mg protein), 0.6 μmole of DPN, 20 μmoles of nicotinamide, and 24 mg of enzyme protein. The total volume was 2.0 ml and the incubation time was 3 hr at 30°C in air. After extracting the sphingolipids with butanol, 18 μmoles of sphingosine was added to each sample as carrier.

[c] Determined by difference after degrading the sphingosine with sodium periodate.

Kinetic studies suggested that dihydrosphingosine may have been the initial product of the condensation, which was then converted to sphingosine (Fig. 1). The inability to demonstrate a palmitaldehyde dependence of [^{14}C]serine incorporation into sphingosine was ascribed to the relative water-insolubility of the lipid aldehyde. The addition of small amounts of Tween 20 (0.6 mg/ml) facilitated the conversion of palmitaldehyde to sphingosine.[7] However, this activity was inhibited both by CoA and by pyridine nucleotide, and palmitoy CoA was still more effective than hexadecanal. The authors presented evidence indicating that the enzyme system had the capacity to catalyze the palmitaldehyde-dependent reduction of DPN. More recent investigations have supported the role of palmitoy CoA rather than palmitaldehyde in this condensation reaction.[8] The rat brain particles catalyzed an active, pyridoxal phosphate-stimulated release of $^{14}CO_2$ from [U-^{14}C]serine, which might suggest the involvement of ethanolamine as the active precursor of carbon atoms 1 and 2. It should be noted that there is a spontaneous nonenzymatic decarboxylation of L-serine catalyzed by Mn^{2+} and pyridoxal phosphate. However, a comparative study with serine and ethanolamine indicated the latter was ineffective.

As a result of these studies published in 1958, the following reaction was believed responsible for sphingosine biosynthesis:

$$\text{palmitaldehyde} + \text{L-serine} \xrightarrow[\text{pyridoxal phosphate}]{Mn^{2+}} \text{dihydrosphingosine} + CO_2$$

The carboxyl group of serine is lost during this condensation so that the β-carbon becomes C-1 and the α-carbon becomes C-2 of the sphingosine base. Approximately four years later, a report appeared addressing itself to the configuration of the sphingosine base formed *in vitro*.[9] The investigators

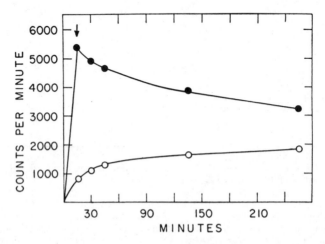

Figure 1. Kinetic study of the conversion of L-[U-^{14}C]serine to dihydrosphingosine (●) and sphingosine (○). (From Ref. 7.)

were able to exploit the relatively new technique of thin-layer chromatography for separation of various sphingosine bases. These studies employed DL-[3^{14}C]serine as the precursor and a fortified rat brain homogenate, and it was found that one-third to one-fifth of the radioactivity recovered in the total lipids was as sphingosine. Hydrolysis of a crude lipid mixture in order to liberate lipid bases was not employed in these experiments. The remainder probably was in the phospholipid fraction due to the widespread base-exchange reaction, which employs serine as the substrate and results in the formation of phosphatidylserine. Carrier *erythro*- or *threo*-sphingosines were added to aliquots of the metabolic products and the mixtures were converted to their N-benzoylsphingosines. After catalytic hydrogenation, the saturated N-benzoyldihydrosphingosines were converted to their tribenzoyl derivatives. This derivative is extremely useful for such differentiations since the *erythro* compound is a crystalline product while the *threo* compound is an oil. These investigations provided evidence for the *erythro* configuration of the metabolic product formed through the condensation of L-serine and palmitoy CoA.

In 1963, Weiss[10] reported the distribution of label in sphingosine bases obtained upon hydrolysis of mixed sphingolipids after the subcutaneous administration of DL-[2,3-^{3}H]serine. The discussion of these results considered as one possible mechanism for the condensation an intermediate formation of a 3-keto derivative. Publications on the chemical synthesis of 3-keto derivatives of N-acetylsphingosine and N-acetyldihydrosphingosine[11,12] provided the basis for subsequent identification of this metabolic product. The biosynthesis of sphingosine was not reinvestigated until a report appeared describing experiments with an *in vitro* system using the yeast *Hansenula ciferri*. This organism was employed since it excretes large quantities of acetylated phytosphingosine and dihydrosphingosine into the growth medium.[13] Labeling experiments had indicated that both palmitic acid and serine were incorporated into sphingosine bases in a manner analogous to the early studies with rats.[14] A 100,000g particulate fraction from *H. ciferri* was found to catalyze a palmitoy CoA/TPNH-stimulated incorporation of [^{14}C]serine into an alkali-stable material.[15] Under these conditions, palmitaldehyde was ineffective (Table III). Particles treated with cysteine were inactive, presumably because of complex formation between this amino acid and PLP, which has been well established. The addition of PLP to these preparations resulted in full restoration of catalytic activity. Braun and Snell postulated as a working hypothesis the formation of a 3-keto derivative, which is then reduced by TPNH to yield dihydrosphingosine. The *in vitro* yeast system yielded certain unanticipated results in that *erythro*-dihydrosphingosine and sphingosine were the sole bases formed in the presence of a TPNH-generating system. Sphingosine itself does not appear to be a usual component of *H. ciferri*; phytosphingosine was not formed *in vitro* although it is the principal sphingolipid base present in this organism. The products of the biosynthetic reaction were characterized by a combination of both thin-layer chromatographic examination of either the free base or its dinitrophenyl derivative and carrier dilution with subsequent recrystallization of the triacetyl derivative. In the absence of TPNH, however, these products were not obtained, but rather a

Table III. Effect of Palmitoy CoA, Sodium Palmitate, Palmitaldehyde, and Various Cofactors on the Conversion of [^{14}C]Serine to Sphingolipids^{a,b}

Experiment	Additions to reaction mixtures	^{14}C incorporation into sphingolipids (dpm/mg protein)
1	Palmitoy CoA (0.35 mM) + TPNH-GS^c	8310
	Sodium palmitate (0.5 mM) + TPNH-GS	135
	+ ATP (1 mM)	252
	+ CoA (0.25 mM)	440
	+ ATP (1 mM) + CoA (0.25 mM)	6125
2	Palmitaldehyde (0.25 mM)	52
	+ TPNH (1 mM)	56
	+ DPNH (1 mM)	40
	+ DPN$^+$ (0.5 mM)	28
	+ TPN$^+$ (0.5 mM)	52
	+ TPNH (1 mM) + MnCl$_2$ (0.5 mM)	74
	+ TPNH-GS + ATP (1 mM) + CoA (0.25 mM)	6218
3	Palmitaldehyde (0.25 mM) + TPNH-GS	—
	+ ATP (1 mM)	560
	+ CoA (0.25 mM)	980
	+ ATP (1 mM) + CoA (0.25 mM)	4406
4	Palmitaldehyde (0.25 mM) + ATP (1 mM) + CoA (0.25 mM)	1382
	+ TPN$^+$ (0.5 mM)	3742
	+ TPNH-GS	4406
5	Palmitoy CoA (0.25 mM)	130
	+ TPNH-GS	1607
	+ TPN$^+$ (0.5 mM)	337
	+ DPN$^+$ (0.5 mM)	236
	+ DPNH (1 mM)	140

^{a} Modified from Ref. 15.

^{b} The incubation mixtures (2.0 ml) contained substrates and cofactors as indicated, together with potassium phosphate buffer (0.1 M), dithiothreitol (0.5 mM), MgCl$_2$ (0.4 mM), DL-[3-^{14}C]serine (6 mM, 6 μCi), and 4–5 mg protein (100,000g pellet). When palmitaldehyde was present, the reaction mixtures also contained 0.5 mg Cutscum. The pH was 7.5. All tubes were incubated for 1 hr at 30°C.

^{c} The TPNH-generating system (TPNH-GS) was composed of TPN$^+$ (0.5 mM), glucose-6-phosphate (0.5 mM), and glucose-6-phosphate dehydrogenase (6 U).

more polar radioactive material that migrated farther than dihydrosphingosine on thin-layer chromatograms.[16] The original assay conditions were modified in order to avoid losses of the 3-keto intermediate due to the chemical instability when exposed to strong alkali. Indeed, the ketonic intermediates themselves were found too unstable for the rigors of unequivocal identification. However, they could readily be converted to *N*-acetyl derivatives, which are amenable to the manipulations required for careful characterization (Fig. 2). This was largely based upon thin-layer chromatography, gas–liquid chromatography of the trimethylsilyl and 2,4-dinitrophenyl derivatives, as well as the products formed as a result of sodium borohydride reduction.[17,18] The rate of formation of the 3-ketodihydrosphingosine bases as a function of acyl CoA chain length revealed that palmitoy CoA was approximately twice as effective as stearoy CoA.[19] Acyl CoA derivatives are regarded as being non-

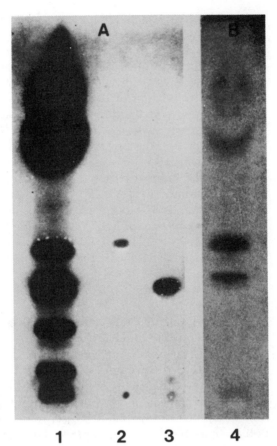

Figure 2. Chromatogram (A) and the corresponding autoradiogram (B) of a sphingolipid extract containing the "unstable" intermediate after acetylation. A2 and A3 are authentic standards of N-acetyl-3-ketodihydrosphingosine and N-acetyldihydrosphingosine, respectively. The chromatogram was on silica gel G developed with chloroform–methanol (90 : 10). Spots were visualized with iodine vapor (A) or by autoradiography (B). (From Ref. 16.)

specific inhibitors of a variety of enzymes, and serum albumin has commonly been added to incubation mixtures to overcome this inhibition and allow saturating levels of palmitoy CoA to be used. These assays were conveniently monitored by measuring the acyl CoA-dependent release of $^{14}CO_2$ from L-[1-^{14}C]serine.

The formation of 3-ketodihydrosphingosine as a product of the condensation of palmitoy CoA and L-serine in mammalian tissues was first documented by Braun *et al.*[20] Microsomes that had been dialyzed overnight against sucrose were isolated from the brains of 16-day-old mice and employed as the enzyme source. ^{14}C-labeled acyl CoA's were used as the substrate in order to assay the activity and the C_{16} derivative was found to be more effective than the C_{18}. Thin-layer chromatography and autoradiography of the lipids in such reaction mixtures revealed the presence of approximately 10 products, one of which cochromatographed with authentic samples of 3-ketodihydrosphingosine. Essentially similar results were obtained using L-[^{14}C]serine and a particulate fraction from young rat brain.[21] The product was identified via characterization of the $NaBH_4$ reduction products, as well as by N-acetylation and dinitrophenylation. The product formed appeared to be exclusively the saturated base with virtually no evidence for the formation of unsaturated base in this system. In contrast to the many additions present in reaction tubes of earlier studies on this reaction in rat brain, the more recent incubations contained merely buffer, L-serine, palmitoyl CoA, and an enzyme source. Substantial stimulation of dependence upon pyridoxal phosphate or metal ions was not observed, and partial purification of the enzyme from rat brain particles has been reported. Employing L-[^{14}C]serine as substrate, two products are formed in the presence of palmitoyl CoA. One of these is phosphatidylserine, which is produced as a consequence of the base-exchange reaction present in the brain particles. This product could be completely eliminated by the addition of EDTA at 50 µmoles/ml to the incubation mixtures. The second product is 3-ketodihydrosphingosine, the synthesis of which is unaffected by the presence of EDTA, suggesting that a metal, if required, is tightly bound.

Current evidence favors the following reaction:

$$\text{L-serine} + \text{palmitoy CoA} \longrightarrow \text{3-ketodihydrosphingosine} + CO_2$$

Pyridoxal phosphate is undoubtedly involved; however, evidence for this requirement is indirect and has been derived largely through studies employing semispecific inhibitors. More recently, this reaction was studied in brain particles prepared from vitamin B_6-deficient rats. The activity was significantly reduced in the experimental animals and was restored *in vitro* by the addition of pyridoxal phosphate.[22] The mechanism of this condensation has been studied by incubating [2,3,3-^{2}H$_3$]serine and palmitic acid with particulate preparations of rat liver or *H. ciferri*. The primary hydroxyl group contained two tritium atoms. When these preparations were carried out in ^{2}H$_2$O, the deuterium was present at C-2. It was suggested that the α-hydrogen group and the carboxyl group of serine are replaced by a proton from the medium and a palmitoyl group during 3-ketodihydrosphingosine formation.[22a]

2.1.2.1. Reduction of 3-Ketodihydrosphingosine

As mentioned above, in the absence of a TPNH-generating system, 3-ketodihydrosphingosine is the major product obtained as a result of the condensation of L-serine and palmitoyl CoA, while in the presence of reduced pyridine nucleotide, dihydrosphingosine is recovered. Employing chemically synthesized radioactive 3-ketodihydrosphingosine, Stoffel *et al.*[17] demonstrated the TPNH-dependent reduction of this substrate to *erythro*-dihydrosphingosine. DiMari *et al.*,[19] also employing a radioactive synthetic substrate, reported a 40–80% conversion to dihydrosphingosine by *H. ciferri* extracts. This reaction has been further characterized with rat liver microsomes,[23] in which either a radioactive assay based on 3-keto[^{14}C]dihydrosphingosine conversion or a spectrophotometric assay based on TPNH oxidation measured at 340 nm has been employed. In rat tissue, the order of this reductive activity is liver > muscle > brain > lung > spleen > kidney, while at the subcellular level, the microsomal fraction has the highest specific activity. Stoffel[24] has reported an 83-fold purification of the 3-ketodihydrosphingosine reductase from either rat or beef liver. The enzyme was solubilized with Triton X-100 and precipitated between 25 and 40% $(NH_4)_2SO_4$ concentration. Two negative absorption treatments with calcium apatite gel followed by a positive absorption and elution from the gel resulted in a 25-fold purification. The final step involved banding on a continuous sucrose gradient. Some of the characteristics of the enzyme were: a pH optimum of 7.0; a K_m for C_{18}-3-ketodihydrosphingosine of 1.5×10^{-5} M; and a K_m for C_{20}-3-ketodihydrosphingosine of 3×10^{-5} M. The synthetic substrate utilized in such studies was DL or a racemic mixture at carbon 2. There is a stereospecific reduction in that the hydrogen from the B side of TPNH is utilized, rather than the hydrogen from the A side. The reduction of the keto group introduces another asymmetric carbon atom when the hydroxy group at C-3 is introduced into the molecule. Carrier DL-*erythro*-dihydrosphingosine was added to the reaction mixture and advantage was taken of the ability to separate the diastereoisomeric glutamates.[25] Thus, the addition of D-glutamate results in the crystallization only of the D-*erythro*-dihydrosphingosine, which upon recrystallization contained a large amount of radioactivity. The L isomer, which crystallized from the supernatant after L-glutamate addition, had only one-fifth the activity of the D isomer. These data indicate a stereospecific formation of D-*erythro*-dihydrosphingosine:

$$\text{3-ketodihydrosphingosine} + \text{TPNH} \longrightarrow \text{D-}erythro\text{-dihydrosphingosine}$$

Recent *in vivo* studies suggest that the reduction of the keto group of N-acyl-3-ketosphingosine (3-ketoceramide) may be an alternative. Doubly labeled material, $[1-^{14}C]$ lignoceroyl 3-keto$[1-^3H]$sphingosine, was prepared and injected into the circulatory system of rats. Approximately 3% of the administered dose was recovered as ceramide from liver tissue after 1 hr.[25a]

2.1.2.2. Dihydrosphingosine-1-phosphate

Publications from several laboratories reported the loss of dihydrosphingosine from incubation mixtures containing cell-free extracts from rat tissues and this loss was claimed to be stimulated by the addition of ATP and Mg^{2+}.[26–28] These conclusions, however, were not unequivocally supported by the experimental evidence provided in these publications. These experimental observations were largely based upon the appearance of free fatty acids from various dihydrosphingosines employed as substrates that were specifically labeled at some position from C-3 to C-18 in the chain.

Evidence for the formation of dihydrosphingosine-1-phosphate was obtained with an eightfold purified system from bovine kidney.[29] The product formed was found only sparingly soluble in organic and aqueous solutions. After incubation with $[^{32}P]$-ATP, the precipitate obtained after the addition of trichloroacetic acid to the reaction mixture was saponified in order to remove the bulk of phospholipids. Reproducible analytical results were then obtained with a chloroform extract after acidification of the residue. In experiments with radioactive dihydrosphingosine as the substrate, the trichloroacetic acid precipitate was first extracted with acetone and then ether prior to saponification in order to remove any unreacted substrate. Employing these protocols, a dihydrosphingosine-dependent appearance of radioactivity from [^{32}P]-ATP was demonstrated and vice versa. The pH optimum appeared to be 7.0, and Mg^{2+} and NaF enhanced the incorporation. The ratio of ^{32}P and ^{3}H was approximately 1 : 1 in the reaction products from experiments in which both labeled nucleotide and base were incubated. Periodate oxidation indicated that the location of the phosphate was on the primary hydroxy group and a compound similar to glycolaldehyde was formed:

$$CH_3-(CH_2)_{14}-\overset{\overset{\displaystyle H}{|}}{\underset{\underset{\displaystyle HO}{|}}{C}}-\overset{\overset{\displaystyle H}{|}}{\underset{\underset{\displaystyle NH_2}{|}}{C}}-\overset{\overset{\displaystyle H}{|}}{\underset{\underset{\displaystyle H}{|}}{C}}-O-PO_3H \xrightarrow{\text{HIO}_4} CH_3-(CH_2)_{14}-CHO+CHO-CH_2OPO_3H_2$$

These observations were corroborated and extended with a partially purified rat liver enzyme preparation.[30] The identification of the phosphate ester was accomplished by demonstration of ethylene glycol phosphate after an $NaBH_4$ reduction of periodate oxidation mixtures. Palmitaldehyde produced during this oxidation procedure was identified by gas–liquid chromatography. In addition, the product was purified by both silicic acid and cellulose powder column chromatography. A modest purification of the mammalian enzyme has been reported; however, it has not been extensively purified or characterized. More recently, certain properties of a cell-free preparation of *T. pyriformis* have been documented.[31] Human and rabbit red blood cells[32] as well as platelets[32a] have been shown to catalyze the phosphorylation of dihydrosphingosine, sphingosine, and phytosphingosine. A 200-fold purification of the bovine brain kinase has been achieved. It exists in multiple forms and the most highly purified fraction has an apparent molecular weight of 190,000.[32b]

2.1.2.3. Dihydrosphingosine-1-phosphate Aldolase (Lyase)

Early studies had indicated the loss of dihydrosphingosine from a variety of *in vitro* systems as measured by palmitic acid appearance. In addition, a role for pyridoxal phosphate was suggested since inactive dialyzed preparations regained the capacity for catalyzing this reaction by the addition of this vitamin.[28] In order to understand the reaction, Stoffel *et al.*[26] employed a variety of chemically synthesized dihydrosphingosine-1-phosphates with label at specific positions. With [1-^{14}C]dihydrosphingosine-1-phosphate as substrate and rat liver mitochondria, ethanolamine-1-phosphate was reported as the major cleavage product, as judged by electrophoresis (Fig. 3). Experiments employing [3-^{14}C]dihydrosphingosine-1-phosphate as substrate yielded palmitaldehyde as the major product, which was characterized directly and as its dimethylacetal derivative by thin-layer chromatography. Further evidence

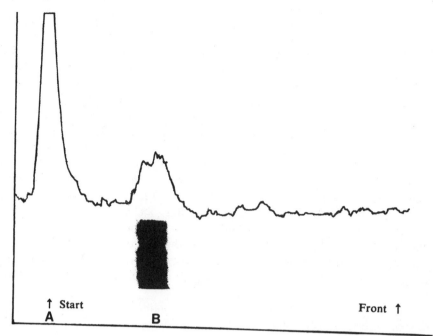

Figure 3. Electrophoretic separation and distribution of radioactivity after incubation of [1-^{14}C]-dihydrosphingosine-1-phosphate with rat liver mitochondria, (A) dihydrosphingosine; (B) ethanolamine phosphate. The incubation mixture contained, in a volume of 2.0 ml, 1.2 μmoles of [1-^{14}C]dihydrosphingosine-1-phosphate buffer, pH 7.2; 2 hr at 37°C. (From Ref. 26.)

was obtained for this formation when [3-^{14}C, 3-^{3}H]dihydrosphingosine-1-phosphate (^{3}H/^{14}C ratio of 12 : 1) was incubated and the isolated palmitaldehyde dimethylacetal was found to retain the starting ratio. This provided evidence for the formation of the aldehyde rather than the acid, which might have been subsequently reduced.

$$CH_3-(CH_2)_{14}-\underset{\underset{OH}{|}}{CH}-\overset{\overset{H}{|}}{\underset{\underset{NH_2}{|}}{C}}-CH_2-O-\overset{\overset{O}{\|}}{\underset{\underset{O}{|}}{P}}-OH \longrightarrow$$

$$\underset{\underset{NH_2}{|}}{H_2C}-CH_2-OPO_3H_3 \ + \ CH_3-(CH_2)_{14}-COOH$$

Mammalian tissues have also been reported to contain a phosphatase that cleaves the ethanolamine phosphate ester. However, it is not evident whether this is a substrate-specific or general phosphatase.

Although the aldolase has not been purified, certain characteristics of the systems have been described.[33] Mitochondria and microsomes isolated from rat liver are reported to possess lyase activity, while the cytosol is inactive. Both the inner and the outer mitochondrial membranes catalyze the hydrolysis, the latter having the highest specific activity. Dialyzed particulate preparations were unable to catalyze this cleavage; the addition of pyridoxal phosphate was reported to restore their activity. Deoxypyridoxine phosphate, in the presence of equimolar concentration of pyridoxal phosphate, resulted in a 50% inhibition of the enzyme. NaF inhibits the phosphatase and therefore appears to be stimulatory for the lyase. A variety of sulfhydryl-binding reagents such as *p*-chloromercuribenzoate and *N*-ethylmaleimide are inhibitory. Both the *erythro* and the *threo* isomers of dihydrosphingosine-1-phosphate are cleaved. The presence of either nonradioactive palmitaldehyde or ethanolamine-1-phosphate did not inhibit the lyase. Sphingosine-1-phosphate and phytosphingosine-1-phosphate appear to undergo a similar cleavage producing ethanolamine-1-phosphate and hexadecen-1-al and hydroxyhexadecenal, respectively.[34,35] It has been demonstrated that during the hydrolysis there is a stereospecific incorporation of one hydrogen atom from the solvent into ethanolamine-1-phosphate[35a] and this appears to be into the *R* configuration.[35b]

The following mechanism has been postulated for the cleavage of dihydrosphingosine-1-phosphate:

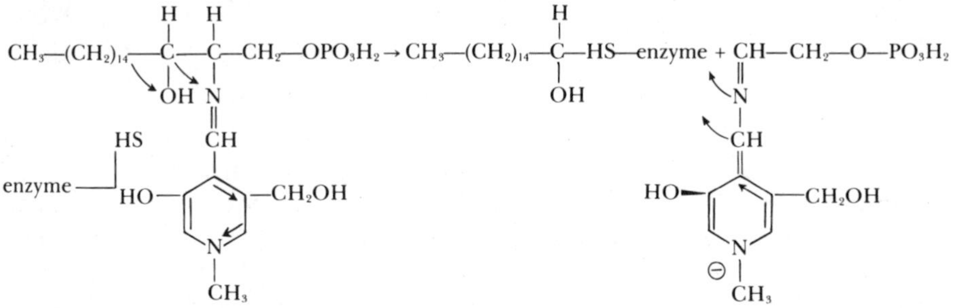

The enzyme-bound alkane derivative can then dissociate to yield palmitaldehyde and the pyridoxal complex can dissociate to yield ethanolamine-1-phosphate.

2.1.2.4. Introduction of the Double Bond

The information thus far available derived from *in vitro* studies with mammalian systems indicated the formation almost exclusively of the saturated sphingosine bases. As indicated earlier, Brady and Koval[6] suggested that dihydrosphingosine was the first product and was subsequently converted to sphingosine. This sequence was based on measurements of the kinetics of L-serine incorporation into these two bases. Further evidence supporting this was the ability of the enzyme system to reduce the dye safranin T in the presence of dihydrosphingosine but not in the presence of sphingosine. Because of the electrochemical potentials, it was then postulated that flavin nucleotides were involved.

$$CH_3-(CH_2)_{12}-CH_2-CH_2-\underset{\underset{OH}{|}}{\overset{\overset{H}{|}}{C}}-\underset{\underset{NH_2}{|}}{\overset{\overset{H}{|}}{C}}-CH_2OH \xrightarrow{\text{flavin}}$$

$$CH_3-(CH_2)_{12}-CH=CH-\underset{\underset{OH}{|}}{\overset{\overset{H}{|}}{C}}-\underset{\underset{NH_2}{|}}{\overset{\overset{H}{|}}{C}}-CH_2OH \ + \ \text{flavin-H}_2$$

More recently, Fujino and Nakano[36,37] have reported that upon incubating DL-[3-^{14}C]serine and palmitoyl CoA with a variety of cofactors and rat liver microsomes, the major products are 3-ketosphingosine and sphingosine. The total absence of radioactivity in phosphatidylserine and phosphatidylethanolamine is remarkable in view of the known capacity of such particles to catalyze the base-exchange reaction[38,39] Curiously, no 3-ketodihydrosphingosine was reported to be formed in these reaction mixtures. These same workers also have a preliminary report on the flavin (FAD or FMN)-catalyzed conversion of 3-ketodihydrosphingosine to 3-ketosphingosine by such particle preparations. In these studies, approximately 5000 cpm (ca. 5 nmoles) of biosynthetically produced 3-ketodihydrosphingosine was incubated with rat liver microsomes. A mixture of carrier sphingosine, ceramide, and sphingomyelin was added to the lipid extract after incubation, the total lipid mixture N-acetylated, and the N-acetyl keto bases purified by column chromatography. These were then separated into the saturated and unsaturated derivatives, and, in the presence of FAD, 3-ketosphingosine was found on autoradiography of thin-layer chromatograms. Iodine vapor revealed the presence in all the samples of both N-acetyl-3-ketosphingosine and N-acetyl-3-ketodihydrosphingosine. These could be regarded as preliminary unconfirmed observations that are recorded in the scientific literature. The authors suggest the following:

$$CH_3-(CH_2)_{12}-CH_2CH_2-\underset{\underset{O}{\|}}{C}-\underset{\underset{NH_2}{|}}{CH}-\underset{\underset{OH}{|}}{CH_2} \xrightarrow[\text{FADH}_2]{\text{FAD}}$$

$$CH_3-(CH_2)_{12}-CH=CH-\underset{\underset{O}{\|}}{C}-\underset{\underset{NH_2}{|}}{CH}-\underset{\underset{OH}{|}}{CH_2}$$

Another hypothesis for the formation of the unsaturated sphingosine base postulates that this occurs at the level of the fatty acyl CoA derivative. Thus, it is conceivable that 2-hexadecenoyl CoA rather than palmitoyl CoA condenses with serine to give rise to 3-ketosphingosine rather than 3-keto-dihydrosphingosine.

$$CH_3\text{—}(CH_2)_{12}\text{—}CH\text{=}CH\text{—}\underset{\displaystyle O}{\overset{\displaystyle O}{C}}\text{—}S\text{—}CoA \;+\; \underset{\underset{NH_2}{|}}{\overset{\overset{H}{|}}{CH_2\text{—}C\text{—}COOH}} \longrightarrow$$

with the CH_2 bearing an OH group:

$$CH_3\text{—}(CH_2)_{12}\text{—}CH\text{=}CH\text{—}\underset{O}{\overset{}{C}}\text{—}\underset{NH_2}{\overset{H}{C}}\text{—}CH_2OH$$

$$CH_3\text{—}(CH_2)_{12}\text{—}CH_2\text{—}CH_2\text{—}\overset{\displaystyle O}{C}\text{—}S\text{—}CoA \;+\; \underset{\underset{NH_2}{|}}{\overset{\overset{H}{|}}{CH_2\text{—}C\text{—}COOH}} \longrightarrow$$

$$CH_3\text{—}(CH_2)_{12}\text{—}CH_2\text{—}CH_2\text{—}\underset{O}{\overset{}{C}}\text{—}\underset{NH_2}{\overset{H}{C}}\text{—}CH_2OH$$

Attempts at obtaining direct evidence for this possibility have been unsuccessful in *H. ciferri*[19] since cell-free preparations from these organisms unfortunately have the capacity to interconvert palmitic and 2-hexadecenoic acids. In addition, when 2-hexadecenoyl CoA was incubated with L-serine, the ratio of 3-ketodihydrosphingosine and 3-ketosphingosine was identical to that obtained in incubations with palmitoyl CoA. The *in vivo* conversion of dihydroceramides to ceramide has been observed in two separate laboratories. Dual-labeled dihydroceramides containing either N-[^{14}C]palmitoyl or stearoyl[^{3}H]dihydrosphingosine were administered to rats. The corresponding sphingosine-containing ceramides with isotope ratios similar to the precursor were subsequently isolated. This strongly suggests that the introduction of the characteristic double bond present in sphingosine may occur at the ceramide level rather than the free base.[39a,b)]

$$CH_3\text{—}(CH_2)_{14}\text{—}\underset{OH}{\overset{H}{C}}\text{—}\underset{NH}{\overset{H}{C}}\text{—}CH_2OH \rightarrow CH_3\text{—}(CH_2)_{12}\text{—}CH\text{=}CH\text{—}\underset{OH}{\overset{H}{C}}\text{—}\underset{NH}{\overset{H}{C}}\text{—}CH_2OH$$

with each NH bearing:
$$\underset{R}{\overset{}{C\text{=}O}}$$

2.1.3. In Vivo Studies on Sphingosine Base Utilization

The results of intracerebral administration of chemically synthesized[40] DL-*erythro*- and -*threo*-[4,5-³H]sphingosines into young rats were reported. The incorporation of these bases *in vivo* only into brain ceramide, sphingomyelin,[41] and gangliosides[42] was documented (Tables IV, V). The sphingolipid fractions were purified chromatographically, the relevant carrier nonradioactive sphingolipid was added, and the mixtures were recrystallized to constant specific activity. This observation was subsequently confirmed by other workers[43,44] employing several specifically labeled sphingosine bases. It was noted that a significant amount of the hydrophobic portion of the base was also converted to fatty acid.[43] The chemical synthesis of a variety of specifically labeled sphingosine bases, which have been used in a variety of *in vivo* studies, has been described.[45] Intravenous administration of DL-*erythro*-[3-¹⁴C]dihydrosphingosine or DL-*erythro*-[7-³H]sphingosine,[46] or [3-¹⁴C]-3ketodihydrosphingosine led to the labeling of palmitic acid, which was recovered from a variety of lipids. However, when [1-³H]-[46] or [1-¹⁴C]dihydrosphingosine[47] was utilized, appreciable radioactivity was found in the base portion of phosphatidylethanolamine. The phospholipid fraction was isolated from liver lipids and was subjected to phospholipase C *(B. cereus)* hydrolysis.

$$
\begin{array}{l}
\quad\quad\quad\quad\quad\overset{\textstyle O}{\underset{\textstyle \|}{}} \\
\overset{O}{\underset{\|}{}}\quad H_2C\!-\!O\!-\!C\!-\!R \\
RC\!-\!O\!-\!CH \quad\quad \overset{O}{\underset{\|}{}} \\
\quad\; H_2C\!-\!O\!-\!\overset{\textstyle}{\underset{\textstyle O^{\ominus}}{P}}\!-\!O\!-\!C^*H_2\!-\!CH_2NH_2 \rightarrow H_2OP\!-\!O\!-\!C^*H_2\!-\!CH_2NH_2
\end{array}
$$

The radioactive product recovered after dialysis was found to cochromatograph with appropriate standards of ethanolamine phosphate upon electrophoretic separations. Intraperitoneal injection of DL-[1-³H]sphingosine led to the labeling of brain sphingolipid without any free sphingosine being recovered, whereas in liver, some of the administered material was recovered

Table IV. Ceramide Carrier Dilution Data[a]

	Derived from animals receiving	
	DL-*Erythro*-[³H]sphingosine (cpm/mg)	DL-*Threo*-[³H]sphingosine (cpm/mg)
First crystallization	286	366
First recrystallization	176	246
Second recrystallization	159	249
Third recrystallization	163	—

[a] Modified from Ref. 41.

Table V. Sphingomyelin Carrier Dilution Data[a]

	Derived from animals receiving	
	DL-*Erythro*-[³H]sphingosine (cpm/mg)	DL-*Threo*-[³H]sphingosine (cpm/mg)
First crystallization	98	793
First recrystallization	91	670
Second recrystallization	101	665

[a]Modified from Ref. 42.

unchanged.[48] Synthetic [1-¹⁴C]dihydrosphingosine, when administered to rats, gives rise to phosphorylethanolamine as the principal water-soluble radioactive material recoverable from the liver.[49] It has been demonstrated that carbon atoms 3 through 18 of [3-³H, 3-¹⁴C]dihydrosphingosine are a precursor of the alk-1-enyl chain of brain plasmalogens.[50] The ratio of the label in the recovered palmitaldehyde was 50% that of the starting material, suggesting a reduction to hexadecenal prior to incorporation in the phospholipids.[50a]

Thus, the results of both *in vivo* and *in vitro* studies on sphingosine metabolism are in agreement and are compatible with the following pathway:

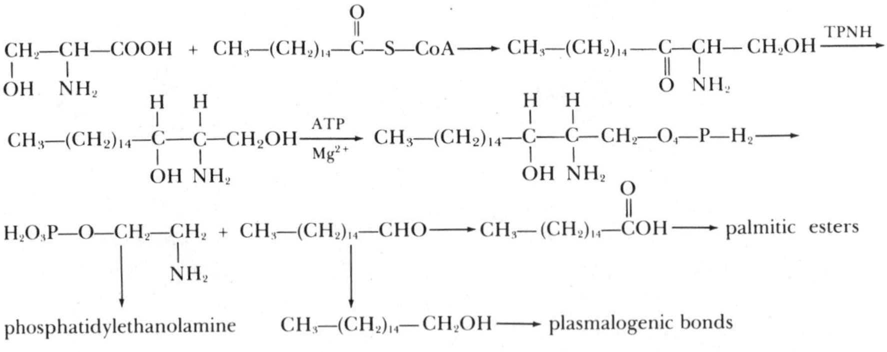

In an elegant attempt to study the specificities of sphingosine base utilization *in vivo*, Stoffel's laboratory chemically synthesized four dihydrosphingosines with different chiral configurations. These were administered to rats and the distribution in hepatic sphingolipids determined. The results are shown in Table VI.

It appears from the data of Table VI that little specificity exists for ceramide biosynthesis, since all four bases were utilized. A 2S configuration for the amino group was required for sphingomyelin and presumably for glucosylceramide formation.

Table VI. *Distribution of Four Synthetic Dihydrosphingosines in Rat Heptic Sphingolipids*

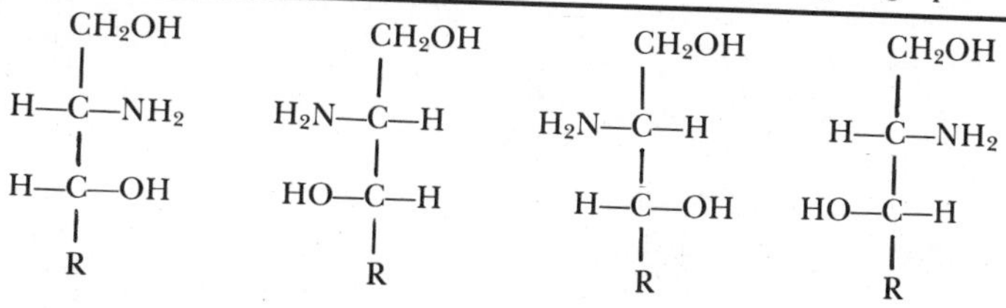

	CH$_2$OH H—C—NH$_2$ H—C—OH R D(+)-*Erythro* (2S,3R)	CH$_2$OH H$_2$N—C—H HO—C—H R L(−)-*Erythro* (2R,3S)	CH$_2$OH H$_2$N—C—H H—C—OH R D(+)-*Threo* (2R,3R)	CH$_2$OH H—C—NH$_2$ HO—C—H R L(−)-*Threo* (2S,3S)
Ceramides	12	53	44	90
Sphingomyelin	38	—	—	10
Cerebroside	4	—	—	—
Unchanged starting material	—	25	12	—
Starting material-1-phosphate	—	22	44	—
4-t-Sphingosine base	65.80	—	—	32

2.2. The Psychosines

Galactosylsphingosine, commonly known as psychosine, has never been unequivocally demonstrated to occur in normal biological material. It is a product that can be derived from galactosylceramide after treatment with fairly strong alkaline hydrolysis under very specific conditions:

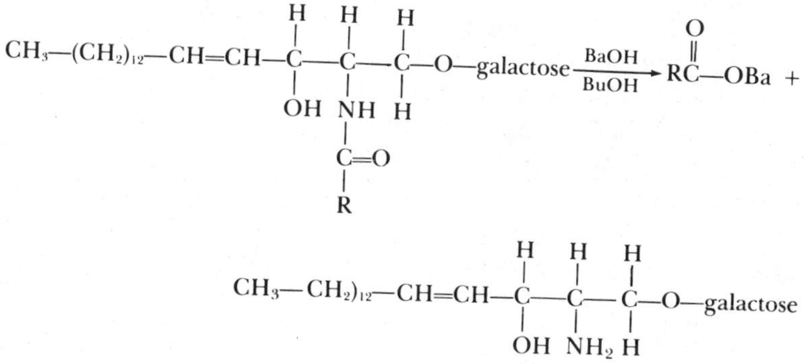

A similar product in which glucose is substituted for galactose can be produced from glycosylceramide. Classically, when the term "psychosine" is used, and the carbohydrate moiety is unspecified, it is assumed that this refers to galactosylsphingosine. The chemical and physical properties of psychosine have not been extensively documented. The general impression gained is that the removal of the fatty acid residue markedly increases the water solubility. Cerebrosides are present only in low concentration in the aqueous methanol upper phase of a Folch partition system; rather, they are retained in the lower chloroform phase. Psychosine, however, is found in this aqueous layer at either neutral or especially acidic pH's. However, under alkaline conditions, psychosine as well as sphingosine will be retained in the organic layer. As a result of this solubility property, the occurrence of such a compound in a conventional lipid extract would depend on the methodology employed.

2.2.1. Galactosylsphingosine Formation

The principal interest in galactosylsphingosine is its potential role as an intermediate in galactosylceramide biosynthesis. Early *in vitro* experiments with cell-free preparations of brain indicated that [14C]galactose from UDP-[14C]galactose was incorporated into a product that appeared to be galactosylceramide.[51] These authors did not attempt to add an exogenous source of lipid acceptor and merely relied upon an unidentified endogenous material. Dr. Radin indicated, in a discussion section of a symposium on myelin that appeared in 1959, that his unpublished results "some years ago" indicated that ceramide but not sphingosine stimulated [14C]galactose incorporation into cerebroside. The ceramide was isolated from the droppings of chickens that had been fed large amounts of cerebroside, and this potential substrate was coated on Celite[52] prior to the *in vitro* incubations.

The biosynthesis of galactosylsphingosine *in vitro* was documented employing a cell-free system from guinea pig brain by Cleland and Kennedy.[53] An alkaline solution was employed in the assay procedure in order to keep the product in an organic phase. D-*Erythro-trans*-sphingosine and dihydrosphingosine were equally active, while the *threo* isomer was only 30% as effective. The particulate fraction sedimenting at 10,000 to 28,000g for 90 min possessed the greatest activity. The psychosine formation was stimulated by Mg^{2+} at 2–6 μmoles/ml, Tween 20 at 2–4 mg/ml, and a very broad pH curve from 7 to 9 was observed. Sphingosine base at 4 μmoles/ml appeared to be the optimal concentration for this acceptor and the K_m observed for UDP-galactose was 2.5×10^{-4} M. The product was isolated and identified chromatographically both as a free amino compound and as the dinitrophenyl derivative. It was suggested that because of its solubility properties, psychosine may be a suitable intermediate in the pathway involved in cerebroside synthesis. The discussion indicated that preliminary experiments in which acyl CoA derivatives were added to the psychosine-synthesizing system resulted in the formation of a neutral lipid, possibly cerebroside. Crude tissue homogenates possessed some ability to use *N*-acetylsphingosine as an acceptor for galactose.

Evidence for the specificity of galactosylsphingosine synthesis was provided as a result of experiments with rat brain particles.[54] UDP-[¹⁴C]galactose incorporation into psychosine was dependent only upon the presence of *erythro* but not *threo*-sphingosine. Under these conditions, neither UDP-glucose nor glucosylsphingosine additions to the incubation mixtures inhibited psychosine formation; however, the product, galactosylsphingosine, was an effective inhibitor. In the absence of an exogenous acceptor, galactosylceramide appeared to be the product; however, ceramide addition did not increase this formation. The product, galactosylsphingosine, was identified by its conversion to galactosyl-*N*-palmitoylsphingosine, which was isolated by column chromatography and was recrystallized to constant specific activity. Under identical conditions, the presence of *erythro*-sphingosine stimulated the incorporation of [¹⁴C]glucose from UDP-[¹⁴C]glucose into glycosylceramide. That a glucosylceramide was not involved as an intermediate was suggested by (1) no inhibition of glucose incorporation by UDP-galactose, glucosylsphingosine, or galactosylsphingosine and (2) no conversion in this system of [¹⁴C]-sphingosylglucoside into cerebroside. The presence of palmitoyl CoA did not have any effect on any of these reactions. The developmental appearance of this enzyme reaction may be correlated chronologically with myelination in the rat[55] and may be involved in cerebroside biosynthesis.

The hydrolysis of galactosylsphingosine has been reported with a crude mitochondrial–lysosomal pellet that had been sonicated and freeze–thawed in order to liberate the hydrolytic activity in soluble form.[56] The substrate employed was [6-³H]galactosylsphingosine obtained by the alkaline hydrolysis of [6-³H]galactosylceramide, which had been prepared by a galactose oxidase oxidation and subsequent sodium borotritide reduction. The enzyme assay was based upon the appearance of radioactivity in the aqueous phase of an alkaline Folch partitioning since, as indicated above, the substrate, galactosylceramide, would be expected to remain in the lower phase. The pH optimum was 4.2–4.5, the reaction was linear for 5 hr, it was stimulated 100% by oleic acid and 400% by sodium taurocholate, and it had a K_m of 1.1×10^{-5} M. Galactosylceramide was reported to be a competitive inhibitor with a K_i of 2×10^{-5} M, suggesting that a common protein may be responsible for the hydrolysis of the β-galactosidic linkage present in these two glycosphingolipids. Subsequent observations from these investigators reinforce this probability.[56a] However, the ability of other β-galactosides to inhibit the hydrolysis was not reported.

$$\text{sphingosine} + \text{UDP-galactose} \longrightarrow \text{galactosylsphingosine} + \text{UDP}$$
$$\text{galactosylsphingosine} \longrightarrow \text{sphingosine} + \text{galactose}$$

Preliminary *in vivo* evidence for the psychosine formation has appeared. Young rats received L-[¹⁴C]serine intracerebrally and a radioactive product that cochromatographed with carrier galactosylsphingosine was observed. The amount of radioactivity present in this material decreased from 15 min to 3 hr postinjection.[56b]

2.2.2. Glucosylsphingosine Formation

Microsomal preparations from young rat brains were shown to produce glucosylsphingosine from UDP-[^{14}C]glucose in the presence of added sphingosine. The properties of the reaction were not investigated and product identification relied upon chromatographic migration similar to known standards.[56c] The hydrolysis of glucosylsphingosine appears to be catalyzed by the same enzyme that hydrolyzes glucosylceramide.[56d]

Glucosylsphingosine has been isolated from Gaucher's spleen samples. Reasonably unequivocal identification was based upon (1) chromatographic properties of the material and derivatives, (2) gas–liquid quantitation of products of hydrolysis as glucose and sphingosine, (3) hydrolysis by a purified glucosidase, and (4) combined gas–liquid chromatography–mass spectra. This was the first report of the natural occurrence of a psychosine in biological material.[56e]

2.2.3. Galactosylsphingosine Acylation

The acylation of psychosine has been a subject of some controversy since it was reported that a 6000–100,000g particulate fraction from 14-day-old rats was useful as the enzyme source for this reaction.[57] The assay procedure for the activity was based upon the conversion of [1-^{14}C]stearoyl CoA to cerebroside. Carrier cerebroside (galactosylceramide) was added to the incubation tubes, at the time of terminating the reaction, the mixtures were subjected to chromatography on Florisil, on mixed-bed ion-exchange, and on silicic acid columns followed by recrystallization of the purified material to constant specific activity. Under the experimental conditions, there was a psychosine (galactosylsphingosine)-dependent, ATP-stimulated incorporation of the radioactive precursor into cerebrosides. Sphingosine sulfate and UDP-glucose were reported to be ineffective.

Several attempts to reproduce these results had been unsuccessful.[54,58,59] Interest in this reaction has been rekindled by the report that chemically synthesized deuterium-labeled DL-galactosylsphingosine in the presence of suitably fortified brain particles and [14C]stearoyl CoA gave rise to galactosylceramide. The identification was based upon the GLC–mass spectrum of the isolated product.[60] Subsequently, this same author has described the nonenzymatic potassium phosphate-dependent formation of cerebroside from psychosine and [14C]stearoyl CoA.[61] Six times as much incorporation occurred in the absence of microsomes as in the presence of boiled or unboiled particles. This was ascribed to nonspecific binding of acyl CoA to protein; however, control studies with serum albumin or other noncatalytic proteins were not reported. Similarly, both sphingosine and 1-N-dodecylamine were nonenzymatically acylated to yield N-stearoylsphingosine and N-dodecyloctadecamide.

The role of galactosylsphingosine, if any, in galactosylceramide biosynthesis remains somewhat in question. The biosynthesis of this material has been firmly demonstrated *in vitro*, while the formation of its glucosyl analog is not as well documented. It has not been shown to occur naturally and its acylation to cerebroside is of questionable biochemical significance according to current accepted pathways of galactosylceramide biosynthesis.

2.2.4. Glucosylsphingosine Acylation

A microsomal fraction prepared from embryonic chick brain has been shown to acylate glucosylsphingosine. The assay was based upon the conversion of [1-14C]stearoyl CoA to a material that chromatographed with glucosylceramide in the presence of added glucosylsphingosine. A considerable level of conversion occurred in the absence of protein. However, this was considerably reduced by the inclusion of boiled particles.[61a]

2.3. Ceramide (N-Acylsphingosine)

The *in vivo*[62] and *in vitro*[5] incorporation of radioactive precursors into ceramide have been well documented. The problems of characterizing the enzyme reaction mechanisms responsible for the formation of N-acylsphingosine were biased due to early studies on the enzyme ceramidase.[63] Activity for the hydrolysis of free fatty acids from ceramide was released from a rat brain particulate fraction that sedimented between 1500 and 15,000g. Sonication in the absence and presence of detergent "solubilized" the enzyme, which was subsequently purified by ammonium sulfate fractionation with a purification of 110-fold over the starting homogenate. The soluble preparation required the presence of a bile acid and had a pH optimum of 5 and had no metal requirement. Several substrates were employed and the relative activities found were N-oleoyl- > N-palmitoyl- > N-palmitoyldihydro- > N-stearoylsphingosine, while N-acetyl- and N-lignoceroyldihydrosphingosine were inactive. The free fatty acid released during the reaction was found to be inhibitory.

These preparations, in addition to hydrolyzing the amide bond, were reported to catalyze the synthesis of ceramides from sphingosine and free fatty acids (Fig. 4). Either [^{14}C]palmitic acid and nonradioactive sphingosine or nonradioactive palmitic acid and [^{3}H]sphingosine were converted to radioactive N-palmitoylsphingosine. The pH optimum also was 5.0 and cholate was required. This condensation occurred in the presence of ceramide and was independent of CoA, ATP, Mg^{2+}, and a fatty acid activating system.

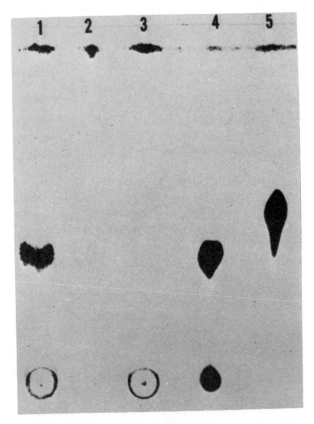

Figure 4. Enzymatic synthesis of ceramide from sphingosine and fatty acid. The reaction mixture contained, in a volume of 1 ml, 0.1 ml of a 1 M solution of a universal buffer (acetate–Tris–ethanolamine), pH 4.8; 0.8 mg of Triton X-100; 1.4 mg of sodium cholate; 0.3 μmole of potassium palmitate; 1.2 μmoles of DL-*erythro*-sphingosine; and 0.70 mg of enzyme. After incubation for 2 hr at 37°C, the reaction was terminated with 5 ml of Dole's reagent; 2 ml of water and 3 ml of heptane were then added, the phases were separated, and the upper phase was shaken three times with 5 ml of 0.1 N NaHCO$_3$ in ethanol–water (1 : 1). The heptane was evaporated, the residue was taken up in chloroform–methanol (2 : 1) and applied to thin-layer silica gel, and the plate was chromatographed in a mixture of chloroform–acetic acid–methanol (94 : 5 : 1). It was subsequently dried, sprayed with 2 N H$_2$SO$_4$, and charred at 180°C. Spot 1, complete reaction mixture; spot 2, sphingosine omitted; spot 3, enzyme omitted; spot 4, N-palmitoylsphingosine; spot 5, N-palmitoyldihydrosphingosine. (From Ref. 63.)

Palmitoyl CoA was less effective than palmitic acid itself and an exchange between [^{3}H]sphingosine and ceramide in the absence of fatty acid was also observed.

$$
\begin{array}{c}
\quad\quad\quad\quad\quad \overset{\displaystyle H}{|}\;\;\overset{\displaystyle H}{|}\;\;\overset{\displaystyle H}{|} \\
CH_3-(CH_2)_{12}-CH{=}CH-C-C-C-OH \rightleftharpoons \\
\quad\quad\quad\quad\quad \overset{|}{OH}\;\;\overset{|}{NH_2}\;\overset{|}{H} \\
\quad\quad\quad\quad\quad\quad\quad\quad \overset{|}{C}{=}O \\
\quad\quad\quad\quad\quad\quad\quad\quad \overset{|}{R}
\end{array}
$$

$$
\begin{array}{c}
\quad\quad\quad\quad\quad \overset{\displaystyle H}{|}\;\;\overset{\displaystyle H}{|}\;\;\overset{\displaystyle H}{|}\quad\quad\quad\quad\quad \overset{\displaystyle O}{\|} \\
CH_3-(CH_2)_{12}-CH{=}CH-C-C-C-OH + R-C-OH \\
\quad\quad\quad\quad\quad \overset{|}{OH}\;\;\overset{|}{NH_2}\;\overset{|}{H}
\end{array}
$$

The most unusual aspect of this condensation was the ability to carry out the non-energy-dependent biosynthesis of ceramide.[63] The ability of ceramidase to produce ceramide has recently been challenged. These authors claim that rather than N-acylsphingosine production they find N-acylethanolamine formation.[63a] Ethanolamine is a constituent of the buffer system originally employed by Gatt.

However, this does not invalidate the palmitate-dependent conversion of [^{3}H]sphingosine to [^{3}H]ceramide reported by Gatt.[63]

Sribney,[64] employing a washed 10,000–100,000*g* particulate fraction, provided evidence for another mechanism of ceramide formation, i.e., the incorporation of [1-^{14}C]palmitic acid from its CoA thioester into ceramide in the presence of sphingosine base. The pH optimum appeared to be 7.5, employing Tris buffers, and a 100% stimulation occurred upon the addition of Mg^{2+}. Both *erythro*- and *threo*-sphingosine were acceptors for either palmitoyl or nervonyl CoA.

$$
\begin{array}{c}
\quad\quad\quad\quad\quad \overset{\displaystyle H}{|}\;\;\overset{\displaystyle H}{|}\;\;\overset{\displaystyle H}{|}\quad\quad\quad\quad \overset{\displaystyle O}{\|} \\
CH_3-(CH_2)_{12}-CH{=}CH-C-C-C-OH \;\; + \;\; R-C-S-CoA \xrightarrow{Mg^{2+}} \\
\quad\quad\quad\quad\quad \overset{|}{OH}\;\;\overset{|}{NH_2}\;\overset{|}{H}
\end{array}
$$

$$
\begin{array}{c}
\quad\quad\quad\quad\quad \overset{\displaystyle H}{|}\;\;\overset{\displaystyle H}{|}\;\;\overset{\displaystyle H}{|} \\
CH_3-(CH_2)_{12}-CH{=}CH-C-C-C-OH + CoASH \\
\quad\quad\quad\quad\quad \overset{|}{OH}\;\;\overset{|}{NH}\;\overset{|}{H} \\
\quad\quad\quad\quad\quad\quad\quad\quad \overset{|}{C}{=}O \\
\quad\quad\quad\quad\quad\quad\quad\quad \overset{|}{R}
\end{array}
$$

   The properties of ceramidase were further investigated using either *N*-oleoyl[^{3}H]dihydrosphingosine or *N*-[1-^{14}C]palmitoyl sphingosine as substrate.[65] The basic assay depended upon the preferential distribution of the fatty acid released between an organic (heptane) and an aqueous phase as a function of pH.[66] The brain particulates, which sedimented between 800 and 15,000*g*, had the highest specific activity and were employed for further enzyme purification to yield a 100-fold overall enrichment. Triton X-100, Tween 20, and sodium cholate were either required or stimulated ceramide hydrolysis. The K_m for ceramide was 3×10^{-4} M, the V_{max} 170 nmoles/mg protein/hr, and sphingosine, laurate, and palmitate inhibited at pH 4.8. However, at pH 8, palmitate and laurate were no longer inhibitory. The pH optimum curve indicated that at pH 8.0, ceramidase was 60% as active as at pH 4.8. Neither cerebroside nor sphingomyelin was a substrate, suggesting the requirement for an unsubstituted primary hydroxyl group. With either a brain homogenate or a soluble enzyme, and using [^{3}H]dihydrosphingosine as acceptor, palmitic acid appeared to result in ceramide formation; the addition of ATP, CoA, or palmitoyl CoA was not stimulatory. A bell-shaped pH curve was obtained with a crest extending from pH 4 to pH 7 with sphingosine and [1-^{14}C]palmitic acid. The enzyme was activated by cholate and was not inhibited by ceramide or sphingosine; the K_m for sphingosine was 2×10^{-4} M and for dihydrosphingosine, 3×10^{-3} M. Attempts to establish the kinetics for palmitic acid were complex at pH 8.0 but a K_m around 4×10^{-5} M was calculated, while at pH 5.0, inhibition occurred. This phenomenon may be due to the precipitation of the fatty acid at pH 5.0, while a salt which would be expected to be soluble at 8.0 was an effective substrate. The effect of various activators at pH 8.0 on the incorporation of the CoA thioester derivatives was not reported. An investigation relating fatty acid chain length to the formation of ceramide at pH 8.0 and 5.0 was published. In general, the incorporation was much greater at the higher pH than at the lower pH, and the order reported was $C_{12} \gg C_{16} > C_{18}$ to C_{24}. The ceramide formed in the reaction appeared to be well characterized.

   Some of the properties of a more highly purified rat brain ceramidase appeared.[67] Although earlier studies indicated the pH optimum for the hydrolysis as being 5.0, these studies were carried out at 7.4 in order to "facilitate homogeneous dispersions of substrates and detergents." The enzyme was purified to a greater degree than previously through the introduction of a trypsin and chymotrypsin digestion step with a final enrichment of greater than 200-fold. The stability to heating at 58°C decreased substantially with purification: a 12-fold purified preparation lost 20–25% for 2 hr, while a 112-fold purified enzyme lost 75–80% over the same time period. A 30% decrease in molecular weight occurred after proteolytic hydrolysis as estimated by the elution position on Bio-Gel P-150 columns. It was claimed that the enzyme preparation that had been previously employed[65] possessed an enzyme that actively cleaved the acyl CoA substrates to free fatty acids and free CoA at a rate 5-fold greater than the rate of ceramide synthetase. Although conclusive evidence is not provided, it is suggested that the acyl CoA thiolase is destroyed by proteolytic enzyme treatment employed in this purification. A comparison of partially purified fractions from both trypsin-treated and nontreated preparations indicated that oleoyl CoA is as effective as oleic acid for ceramide formation only in the untreated sample. This is interpreted as indicating that acyl CoA is cleaved to free fatty acid prior to incorporation into ceramide. The ratios of synthesis and hydrolysis of such preparations were claimed to be similar; however, a table of purification listing the ratios at each step was not provided. Therefore, it was concluded that both ceramidase and ceramide synthetase activities are catalyzed by the same protein. Equilibrium constants obtained for the synthetic capacity with sphingosine and oleic acid were reported to be 2×10^{-4} M, while the hydrolysis with oleoylsphingosine as substrate was $4–8 \times 10^{-6}$ M. These are "apparent" values since it is not possible to estimate the actual quantity of active species presented to the enzyme due to "micelle"[68] formation and the presence of detergents. The results of these studies on ceramidase suggest that the enzyme is capable of catalyzing both the hydrolysis and the synthesis of the N-acylsphingosine bond.

Homogenates of human cerebellum appear to possess ceramidase activity with two distinct pH optima, one at 4.0 and another at 9.0. Both the synthesis from sphingosine and free fatty acids and the hydrolysis of ceramide occurred at these pH's. Kidney homogenates possessed only the pH 4.0 activity. Acyl CoA-dependent activity had a neutral pH optimum in both tissues.[68a]

Studies employing particulate preparations from mouse brain, although not contradicting these studies on ceramidase, have led to the conclusion that the "synthetic capacity of ceramide hydrolase has little or no physiological significance.[69] Detergents were not employed in the biosynthetic experiments and the lipid substrates were coated on Celite; for ceramide hydrolysis, emulsions of N-[1-^{14}C]stearoyl sphingosine were used as substrate. The assay procedures were based upon thin-layer chromatographic separation of ceramide and fatty acids. It is difficult to estimate the percent conversion of any one precursor to ceramides employing a variety of radioactive acyl CoA substrates since a large quantity of the radioactivity originally added is recovered in an appreciable number of bands including an "X" or unidentified product both with C_{16}- and C_{18}-CoA derivatives. The ability to distinguish various ceramides differing in both acyl chain length and base composition by thin-layer chromatography is documented. Although a variety of details are provided, there is no indication that either zero-time, boiled enzyme, or buffer controls were examined in these studies. With dihydrosphingosine as the acceptor and [1-^{14}C]stearoyl CoA as the donor, no stimulation by Mg^{2+} or ATP was found and EDTA was noninhibitory. Tris buffer, presumably due to structural similarity to dihydrosphingosine, was less effective than phosphate buffer; most detergents were not required and bile acids inhibited. With dihydrosphingosine as acceptor, $C_{18:0}$ is a more effective donor for ceramide formation than either $C_{24:0}$, $C_{16:0}$, or $C_{18:1}$ fatty acids (Fig. 5). The rates of radioactivity incorporation of a given acyl CoA into free fatty acid, phospholipids, and ceramide were comparable. An inverse relationship between total fatty

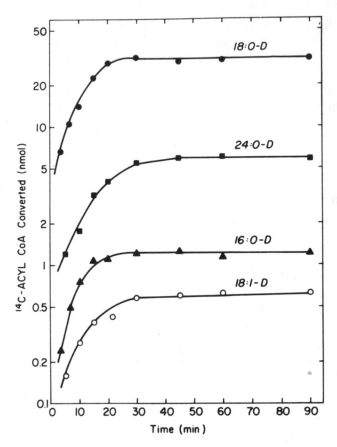

Figure 5. Time course of conversion of radioactive acyl CoA to ceramide. The acceptor was dihydrosphingosine, and specific activity of each radioactive acyl CoA was 1250 cpm/nmole. Total lipids were isolated and separated by thin-layer chromatography. After autoradiography, regions containing radioactive ceramide were scraped and counted. (From Ref. 69.)

acid formed and ceramide synthesis was apparent, suggesting that C_{18}-CoA may have appeared the best ceramide precursor since it is the poorest substrate for the thiolase. In a comparison of base acceptor specificity, 3-ketodihydrosphingosine was reported to be active with [^{14}C]stearoyl CoA; however, in general, there was a low specificity. Data were presented comparing the ability of various brain subcellular fractions to catalyze the synthesis of either stearoyl- or lignoceroyldihydrosphingosine and the hydrolysis of N-[^{14}C]stearoyl sphingosine. Two distinct distributions were observed, the synthesis being localized in a "microsomal" fraction possessing little hydrolase activity and a "mitochondrial–lysosomal" fraction with appreciable hydrolase activity and little synthetic capacity. These studies provide evidence for the presence of an enzyme system in brain that catalyzes the acyl CoA-dependent synthesis of ceramide from dihydrosphingosine. These preparations were not assayed for the synthetic capacity of ceramidase but provide evidence for the existence of an additional pathway in brain tissue for ceramide formation. The authors advise investigators on the problem of carrying out kinetic analysis with acyl CoA's and bases that are known to bind nonspecifically to microsomal protein.

A suggestion has been put forth that the specific fatty acid distribution found in the ceramide portion of sphingolipids may be the result of acyl transferase specificity.[70] The studies on the conversion of normal fatty acid to ceramides were expanded to the CoA derivatives of hydroxy fatty acids[71] and a detailed description of the synthesis of these thioesters is provided. D-*Erythro*-sphingosine was employed as the base acceptor in these experiments and problems of adequate thin-layer chromatographic resolution of the products formed were overcome through the use of borate-impregnated plates. Curiously enough, using either $C_{18:h0}$ or $C_{24:h0}$ acyl CoA's, a nonenzymatic synthesis of ceramides occurred that was 30% as effective with the C_{18} derivative and better than with the C_{24} derivative as with heat-inactivated microsomes. This efficient nonenzymatic synthesis appeared to require phosphate buffer and the activity decreased with increasing amounts of protein, presumably due to nonspecific binding of acyl CoA thioesters to proteins. A developmental comparison of ceramide synthesis with mouse brain microsomes indicated that $C_{18:0}$ was the best acyl donor and questionable differences in the chronological profile were seen. In general, neuronal cells appeared to be from 3 to 25 times more active than glial cells in ceramide synthesis. The validity of such comparison is open to question since the neurons were homogenized and a 100,000g pellet employed as the enzyme system, while only intact glia were used. Therefore, these differences noted could be due to an intact acyl CoA thioester having to traverse the glial plasma membranes. Experiments were not reported comparing intact cells exclusively. Since lignoceroyl CoA and stearoyl CoA did not inhibit one another, the presence of two distinct acyl transferases was postulated. However, h18:0 CoA was a noncompetitive inhibitor of both these substrates with an observed K_i of 1.2–2 × 10^{-4} M. Similarly, both nonhydroxy acyl CoA thioesters were noncompetitive inhibitors of the h18:0 CoA transferase. Therefore, these observations were interpreted as demonstrating separate acyl transferases existing for hydroxy or nonhydroxy acyl CoA thioesters, solely based upon the noncompetitive nature of the inhibition.

The data currently available suggest that separate acyl transferases specific for the fatty acid residue may exist for ceramide synthesis. However, since these studies have been carried out with whole microsomal suspensions, they cannot be regarded as conclusive. The corroboration of this hypothesis will have to await successful purification of the individual enzymes involved.

Thus, at least two mechanisms for ceramide formation have been documented; one of these is the reversibility of ceramidase, and the other the acyl CoA-dependent synthesis. That ceramidase can catalyze both the formation and the hydrolysis of N-acylsphingosine bonds is evident from studies on Farber's tissue.

The conversion by rat brain particles of sphingosine and lignoceric acid to lignoceroylsphingosine requires reduced pyridine nucleotide in addition to two soluble components. This activity is abolished by common inhibitors of oxidative phosphorylation, suggesting that the synthesis is coupled in some manner to the electron transport chain,[71a] and certain of the properties of this complicated system have been examined.[71b] The subject of brain ceramide metabolism has been reviewed.[71c]

2.4. Galactosylceramide

2.4.1. In Vivo Studies

Studies on the *in vivo* incorporation of radioactive precursors into cerebral glycosphingolipids have been continually reported since 1957. Radin *et al.* [72] were the earliest investigators to document the conversion of radioactive [1-^{14}C]galactose into cerebrosides. The isotope was administered intraperitoneally into young female rats (35–40 days) and at various intervals the animals were sacrificed. The cerebral lipids were grossly fractionated into a "strandin" or ganglioside fraction, a partially purified galactosylceramide fraction, and a sulfatide-containing fraction (Fig. 6). These were submitted to hydrochloric acid hydrolysis, carrier galactose was added, and the mixture was then converted to mucic acid, which was counted. The results indicated that [1-^{14}C]galactose was incorporated into the galactose moiety of cerebroside, but some radioactivity presumably was also present in the lipidic portion of the sphingosine as well. A turnover rate of 13 days for 50% replacement of the galactose was observed with a maximal incorporation occurring between 8 and 24 hr.

Studies employing younger animals documented the conversion of both [1-^{14}C]galactose and [U-^{14}C]glucose to "neutral glycolipids" *in vivo*.[51] It was found that maximal incorporation of tracer by brain tissue occurred in animals 10–20 days old. Most of the radioactivity was found in the "microsomal" fraction after administration of the [1-^{14}C]galactose. These observations were corroborated and further extended[73] in studies where mice were given either radioactive glucose or galactose intraperitoneally and sacrificed 1 hr later. Under these conditions, [1-^{14}C]glucose was twice as effective as [1-^{14}C]galactose for labeling the galactose moiety of galactosylceramide. In addition, it was reported that [6-^{14}C]glucose was incorporated without appreciable randomization into this galactose. Subsequent papers have essentially reinforced these early observations on the ability of carbohydrates to label cerebrosides.[74–77]

The ability of [^{14}C]acetate and a variety of ^{14}C-labeled fatty acids to label the acyl moiety of galactosylceramide has been well documented,[62,78–81] and these studies also indicated certain metabolic interrelationships among the cerebral fatty acids. These studies usually allowed several hours or days to elapse after the administration of isotope prior to the sacrifice of the animals.

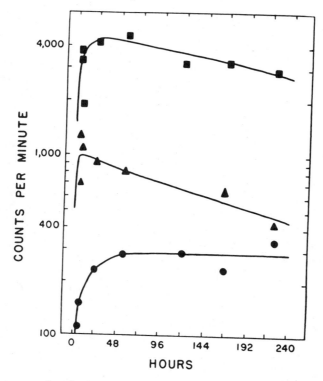

Figure 6. ^{14}C content of rat brain lipid galactose as a function of time. (■) Cerebrosides; (▲) strandin; (●) sulfatides. (From Ref. 72.)

A study has appeared in which [^{14}C]acetate was given intraperitoneally to 3-week-old mice, brain sample was removed from ½ to 15 min, and the labeling of cerebroside hydroxy fatty acids, nonhydroxy fatty acids, sphingosine bases, and galactose was measured.[82] Radioactivity was absent from the long-chain fatty acids (C_{24}), although much appeared in the C_{16}–C_{18} group. Presumably at longer times the former would have been labeled. Labeling of cerebrosides and ceramides by [^{14}C]acetate 15 sec after a carotid injection has been reported.[82a]

These *in vivo* tracer studies indicated that both fatty acids and carbohydrates could be incorporated intact into brain galactosylceramide. However, when N-[1-^{14}C]stearoyl sphingosine was administered intracerebrally to experimental animals, no incorporation into galactosylceramide was observed.[83] This tended to support the "psychosine" pathway as being responsible for cerebroside biosynthesis rather than a direct galactosylation of ceramide.

2.4.2. Biosynthesis in Vitro

The ability of a crude microsomal preparation from young mice to catalyze the galactosylation of hydroxy fatty acid-containing ceramide was documented.[84] Ceramide acceptors were prepared from naturally occurring galactosylceramide by a periodate oxidation procedure,[85] and were separated on silicic acid columns into ceramides containing hydroxy fatty acids and those containing nonhydroxy fatty acids. The ceramides were coated on Celite in order to disperse them sufficiently to be suitable substrates for the microsomal suspension. [14C]-Galactose-1-phosphate was employed as the radioactive tracer, in the presence of ATP and UDP-glucose. The assumption was that the crude particulate fraction possessed the necessary enzyme complement to form UDP-[14C]galactose, the presumed active carbohydrate donor. Nonhydroxy fatty acid ceramide did not stimulate under these conditions, while hydroxy fatty acid ceramide was active in promoting cerebroside formation (Table VII). The product appeared to be the corresponding galactosylceramide as judged by thin-layer chromatography. In the presence of long-chain base, the synthesis of psychosine occurred. A study employing both "mitocondria" and "microsomes" from young rat brain with UDP-[14C]galactose as donor compared a variety of N-acylsphingosines as acceptor for cerebroside formation.[86] It was reported that of nine compounds employed, N-acetyl-*erythro*-sphingosine was the best substrate. Employing a particulate fraction of 19-day-old chick embryo brain isolated from a discontinuous sucrose gradient, certain characteristics of this reaction were examined.[87] The active hydroxy fatty acid ceramide employed as acceptor in these studies was obtained from commercial sources. A detergent combination of Cutscum and Triton X-100 was required for maximul activity since the sphingolipid was not coated on Celite; also required was a metal ion such as Mg^{2+} or Ca^{2+}. The pH optimum appeared to be 7.8 and a K_m for ceramide of 1.1×10^{-4} M and a K_m for UDP-[14C]galactose of 0.4×10^{-4} M were reported. Neither [14C]galactose, [14C]galactose-1-phosphate, ADP-[14C]galactose, nor UDP-[14C]glucose was found to substitute for UDP-[14C]galactose in these experiments. Hydroxy

Table VII. Acceptors for Incorporation of [14C]Galactose into Cerebroside[a,b]

Acceptor	[14C]Galactose incorporated (nmoles)			
	I	II	III	IV
HFA ceramide	1.9	3.1	2.9	2.0
HFA ceramide (purified by thin-layer chromatography)	1.6	2.9	—	—
NFA ceramide	0.4	—	0.7	—
NFA ceramide (purified by thin-layer chromatography)	0.4	0.5	—	—
Stearoyl LCB	0.5	0.5	—	0.4
LCB	0.6[c]	—	—	0.6[c]
NFA ceramide + HFA ceramide	2.2	—	3.2	—
None	0.3	0.4	0.5	0.3

[a] Modified from Ref. 85.
[b] Incubations contained 0.5 mg of each of the indicated lipids coated on Celite.
[c] When LCB was present as the acceptor, approximately 1.2 nmoles of [14C]galactose was incorporated into psychosine. No psychosine was formed in any of the other incubations.

fatty acid ceramide rather than nonhydroxy fatty acid ceramide was found to be the most effective acceptor. Sphingosine was reported to be an inhibitor, although the formation of psychosine under these conditions was not documented. This reaction has subsequently been corroborated using mass spectrometric techniques,[88] white matter having greater enzyme activity than gray matter.[89] Using other incubation and assay conditions, the ability of nonhydroxy fatty acid ceramides to accept [^{14}C]galactose from UDP-[^{14}C]galactose was reported.[58,90] Dolichol monophosphates, presumably involved in glycoprotein biosynthesis, are not involved as intermediates in cerebroside formation.[91] The rate of galactosylceramide synthesis in kidney homogenates from male C57BL mice is higher than that of female C57BL mice,[92] and testosterone administration *in vivo* results in increased *in vivo* synthesis in both sexes.[93] These studies all support the galactosylation of both hydroxy and nonhydroxy fatty acid-containing ceramide as being responsible for the formation of the corresponding galactosylceramide.

$$\text{ceramide} + \text{UDP-galactose} \xrightarrow{\text{Mg}^{2+}} \text{galactosylceramide} + \text{UDP}$$

A UDP-galactose : ceramide galactosyltransferase has been solubilized from rat brain microsomes by detergent extraction and purified 105-fold and the general properties of the purified preparations documented.[93a] Phospholipids stimulate the transferase activity in both its soluble[93a] and particulate[93b] forms. This galactosyltransferase has been found associated with "highly purified" myelin,[93c] one of its subfractions,[93d] and CNS axolemma.[93e] UDP-galactose : ceramide galactosyltransferase activity levels are higher in isolated neuronal cells than in glial cells. This activity is virtually undetected in these culture cells.[93f]

The brain microsomal transferase has been purified some 100-fold. This has been accomplished by detergent extraction, batchwise negative-ion-exchange cellulose treatment, and Pronase digestion. The stability of the purified preparation is enhanced by glycerol and contains about 10 mg phospholipid/mg protein. Phospholipase treatment either markedly decreases or destroys the transferase activity, suggesting that the preparation is composed of lipoprotein or membrane fragments.[93g]

The introduction of the hydroxy group in α-hydroxy acids characteristically found in galactosylceramide is not well understood although a mechanism and stereochemistry have been reported.[93h] It has been suggested that hydroxy ceramide formed through the α-hydroxylation of lignoceroyl CoA and condensation with sphingosine is the precursor of hydroxy cerebrosides.[93i]

In vivo studies have not provided supportive evidence for the role of hydroxy fatty acid-containing ceramides in galactosylceramide formation. The specific activity of nonhydroxy fatty acid and hydroxy fatty acid galactosylceramides increased between 5 and 10 hr after the intracerebral injection of L-[14C]serine in young rats. Analysis of the ceramides from 10 min to 10 hr indicated that only the nonhydroxy fatty acid species were labeled, suggesting that hydroxy fatty acid ceramides may not be cerebroside precursors.[93j]

Structurally related inhibitors of galactosylceramide biosynthesis have been described.[94] The most active derivative was *N*-bromoacetyl-DL-*erythro*-3-phenyl-2-amino-1,3-propanediol:

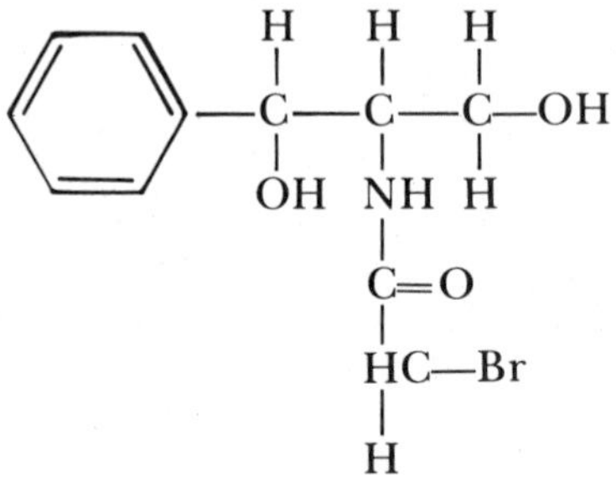

Presumably, the noncompetitive inhibition, K_i 2 × 10^{-4} M, observed with this material is due to the structural similarity to ceramide.

The hydrolysis of UDP-[14C]galactose and UDP-[14C]glucose to the corresponding hexose-1-phosphate as well as the chemical decomposition to their corresponding hexose 1,2 cyclic phosphate derivative occurs in addition to their transfer to glycoconjugate under standard incubation conditions.[94a]

2.4.3. Hydrolysis in Vitro

Early studies on the incorporation of [14C]galactose into brain galactosylceramide indicated a turnover rate of 13 days,[72] suggesting the presence of a hydrolytic enzyme activity for this lipid in cerebral tissue. N-[^{14}C]Stearoyl galactosylsphingosine was administered intracerebrally to rats and radioactive ceramide was subsequently isolated from the brain tissue,[83] also suggestive of degradative activity.

An enzyme was purified approximately 1000-fold from rat intestinal tissues employing [1-^{14}C]galactosyl ceramide as substrate.[95] The preparation was demonstrated to catalyze the release of radioactive galactose (identified by paper chromatography and by carrier dilution studies) and ceramide (identified by thin-layer chromatography and quantitative colorimetry) in stoichiometric amounts. The enzyme is apparently lysosomal[96] with a pH optimum of 5–6 and is stimulated by cholate. Copurification of an enzyme activity catalyzing the hydrolysis of glucosylceramide was reported and the K_m values for both substrates were approximately identical. In addition, p-nitrophenyl-β-D-glucoside, β-D-galactoside, and β-D-glucosaminide were hydrolyzed by the purified preparation. The hydrolysis of either labeled sphingolipid was inhibited by several nonradioactive sphingolipids. Although the data suggest that the hydrolysis of both cerebrosides is catalyzed by the same protein, specific enzyme deficiencies in distinct genetic diseases would indicate there are separate specific enzymes. The substrates employed in these particular studies were produced by a fairly sophisticated chemical synthesis. It has been suggested that this glycosylceramidase may be identical to the phlorizin hydrolase of intestinal tissues.[96a]

The presence of galactosylceramide : β-galactosidase was also detected in brain tissue, using [6-^{3}H]galactosylceramide as substrate.[97] This material was synthesized by utilizing the ability of galactose oxidase to catalyze the oxidation of the hydroxy group on carbon 6 of galactosylceramide.[98,99] The enzymatically produced aldehyde could then be specifically reduced with NaB^{3}H$_4$ with the introduction of tritium specifically at C-6 of the galactose. The hydrolytic activity was liberated from brain tissue through the use of pancreatic lipase[100] and a 300-fold overall purification was achieved. Taurocholate was required, and activity toward glucosylceramide (40%), lactosylceramide (10%), and O-nitrophenylgalactoside (1100%) was also present in the purified preparation. Papers reporting improved assay conditions have appeared,[101,102] and developmental studies suggest that the initial appearance of the enzyme occurs prior to active cerebroside deposition in the brain.[101]

The problem of specificity of lysosomal enzymes is illustrated by the galactosidase involved in galactosylceramide hydrolysis. It appears that this also hydrolyzes galactosylsphingosine.[56a] This protein may also possess β-galactosidic activity toward lactosylceramide. However, the ability to detect this hydrolysis is solely dependent upon the assay mixtures.[101a] This activity has been termed lactosylceramidase I.

In common with many hydrolases, the enzyme appears to catalyze a transglycosylation reaction and has the ability for producing galactosylceramide. Partially purified brain preparations can utilize ceramide as acceptor and several β-galactosides as donors.[101b] The synthesis of an effective inhibitor,[103] N-decanoyl-DL-*erythro*-3-phenyl-2-amino-propanediol:

$$\text{C}_6\text{H}_5-\overset{\overset{\displaystyle H}{|}}{\underset{\underset{\displaystyle OH}{|}}{C}}-\overset{\overset{\displaystyle H}{|}}{\underset{\underset{\displaystyle NH}{|}}{C}}-\overset{\overset{\displaystyle H}{|}}{\underset{\underset{\displaystyle H}{|}}{C}}-\text{OH} \quad (K_i = 4 \times 10^{-4}\ \text{M})$$

$$\begin{array}{c} \text{C}=\text{O} \\ | \\ (\text{CH}_2)_8 \\ | \\ \text{CH}_3 \end{array}$$

and a stimulator,[104,104a] *N*-decanoyl-2-amino-methylpropanol:

$$\begin{array}{c} \text{CH}_3 \\ | \\ \text{CH}_3-\text{C}-\text{CH}_2\text{OH} \\ | \\ \text{NH} \\ | \\ \text{C}=\text{O} \\ | \\ (\text{CH}_2)_8 \\ | \\ \text{CH}_3 \end{array}$$

of galactosylceramide : β-galactosidase has been undertaken. This enzyme has been reported absent in Krabbe's disease.

The ability of both neurons and glia to catalyze this reaction has been documented.[105]

2.5. Ceramide-galactoside-3-sulfate (Sulfatide)

2.5.1. In Vivo Studies

The specificity of $^{35}\text{SO}_4$ labeling of cerebral sulfatide makes it technically simple to undertake *in vivo* studies. In 1957, Radin *et al.*[72] demonstrated the incorporation into brain sulfatide of intraperitoneally administered $^{35}\text{SO}_4$, and indicated that during the experimental period, there was no detectable turnover of sulfatide, in contrast to cerebrosides. Cerebral sulfatide has a very slow turnover, while visceral organs such as kidney show a very rapid turnover.[106] An interesting *in vivo* comparison of $^{35}\text{SO}_4$ and L-[^{35}S]methionine labeling of sulfatide after intraperitoneal injection indicated that the labeled inorganic precursor was less effective than the labeled amino acid precursor.[107] The reverse was observed with intracranial injection into 50-day-old rats, and a negligible turnover of sulfatides in brain was seen. The maximum *in vivo* incorporation of intraperitoneally administered $^{35}\text{SO}_4$ into brain sulfatides occurs in 14- to 15-day-old rats.[108] Adult rats were characterized by slow turnover and low synthetic rates. When timed studies on the labeling patterns of individual cerebral subcellular particles were carried out, an appreciable turnover of the sulfolipids was seen with microsomes and mitochondria. However, the amount of labeling in myelin continued a slow steady increase.[109] The bulk of the sulfatide present in brain is a structural component of the myelin sheath. A serious problem for these early workers was a reliable methodology for the chemical quantitation of sulfatide mass. The general procedures available were based upon an acid hydrolysis of the lipid mixtures and the sulfate liberated was quantitated turbidimetrically as an insoluble sulfate salt. This technique was tedious and difficult to use with any degree of confidence on the small quantities of material available from these biochemical studies. The labeling by $^{35}SO_4$ administered to animals at this point merely demonstrated that sulfatides were synthesized and indicated a slow rate of turnover in brain. In an attempt to deduce the metabolic precursor of this sulfolipid, Hauser administered [6-^{14}C]galactose to 20-day-old rats and determined the specific activity of both galactosylceramide and sulfatide. The greater value found for the unsulfated lipids was "in accord with the postulate that cerebrosides are the direct precursors of sulfatide."[110] Regional selectivity of sulfatide labeling by $^{35}SO_4$ appears to exist, since the brainstem in experimental animals up to 10 days old contains around 50% of the total radioactivity, the cerebral hemispheres have 22%, and both the ventral area and the cerebellum each have about 13%.[111] The subcellular distribution in cerebral tissue indicated that 70% of the sulfatide was present in "myelin." It should be noted that the methodology employed for the subcellular fractionation was somewhat unsophisticated and that confirmatory evidence for its identity by marker enzyme activities of the organelles was not provided. Studies reporting the 10- to 30-fold greater effectiveness of $^{35}SO_4$ rather than galactose-6-$^{35}SO_4$ for labeling brain sulfatide were interpreted as rejecting a role for preformed hexose sulfate in sulfatide formation.[112] Although this approach was novel and supported the concept of a direct conversion of cerebroside to sulfatide, the accepted structure of the sulfolipid was in error at the time of these experiments. The sulfate residue was believed esterified at carbon 6 of the galactose ester, but it was subsequently unequivocally proven to be the 3-sulfate ester.

An interesting observation implicating a lipoprotein in sulfatide metabolism has been suggested from certain *in vivo* studies. There appears to be present in the cytosol of rat brain a soluble lipoprotein that contains sulfatide. Young (17-day-old) rats were given $^{35}SO_4$ intraperitoneally followed 2 hr later by a large amount of nonradioactive inorganic sulfate, in an attempt to carry out a classical "chase" experiment.[113] The specific activity of sulfatide in the microsomes, the soluble fraction, and the myelin fraction was determined as a function of time (Fig. 7). A very active uptake into and loss from microsomes was observed, followed by the appearance of labeling in the supernatant and finally in the myelin. The kinetics obtained are in accord with a precursor–product relationship,[114] suggesting that sulfatide originally synthesized

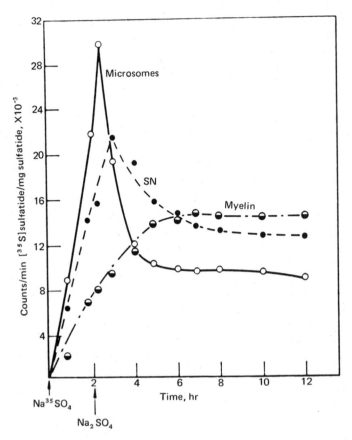

Figure 7. Turnover of [^{35}S]sulfatide in subcellular fractions and SN. Animals received an intraperitoneal injection of 2 μCi of NA$_2$^{35}SO$_4$/g body wt and after 2 hr a second intraperitoneal injection of 0.5 ml of 7% Na$_2$SO$_4$. Animals were killed by decapitation at various times after the first injection. Homogenates from two brains were pooled, subcellular fractions were prepared, and analyses were made on each fraction in duplicate. For each time point, the entire procedure was repeated. A 2-hr control experiment was included for each series of animals and the results were normalized to a given level of [^{35}S]sulfatide in homogenate at 2 hr. (From Ref. 113.)

in the microsomes is transported by a specific lipoprotein of the cytosol to this sulfolipid's final resting place, myelin. Puromycin, an inhibitor of protein synthesis, was employed to examine the possibility that both protein and sulfatide biosynthesis had to occur simultaneously. It was found that under conditions of protein inhibition, although protein synthesis was effectively inhibited, there was negligible reduction in sulfatide formation. These workers also succeeded in demonstrating that different populations of cerebral microsomes, subfractionated on CsCl$_2$ gradients, catalyzed the synthesis of either proteins or sulfatide but not of both.[115] The interpretation from these data that sulfatide and protein are synthesized independently is subject to criticism on two grounds: (1) there is no *a priori* reason to assume that the same microsomal subfraction must synthesize both proteins and sulfatide, and (2) with glycoproteins, puromycin does not inhibit the incorporation of carbohydrates into the polypeptide chains still attached to ribosomes.

[^{3}H]Cerebroside [^{35}S]sulfate was chemically synthesized and injected into the brains of young rats. Eight hours later, the tritium was recovered in the starting material and hydroxy and nonhydroxy fatty acid cerebrosides and ceramides. The radioactive sulfate was present in the starting material, seminolipid, and cholesterol sulfate.[115a]

2.5.2. Biosynthesis in Vitro

The role of galactosylceramide in the biosynthesis of sulfatide became evident from *in vitro* investigations. Earlier studies with slices had indicated that ^{35}SO$_4$ could be incorporated into sulfatides *in vitro*. The active sulfate donor has been demonstrated to be 3'-phosphoadenosine-5'-phosphosulfate (PAPS), which is synthesized at the expense of 2 moles of ATP and 1 mole of an inorganic sulfate according to the following equations[116]:

$$\text{ATP} + \text{SO}_4^{2-} \longrightarrow \text{APS} + \text{pyrophosphate}$$

$$\text{APS} + \text{ATP} \longrightarrow \text{PAPS} + \text{ADP}$$

In vitro studies commonly have employed "a PAPS-generating system," which generally consists of a 100,000g soluble tissue preparation more frequently than PAPS itself. This has been necessary since PAPS is only sporadically available commercially as a radioactive product and is unavailable in a non-radioactive form. Its preparation and analysis are laborious and its stability under a variety of conditions is questionable.

Almost simultaneously, two publications appeared reporting the *in vitro* formation of sulfatide. McKhann *et al.*[117] employed a deoxycholate-solubilized system from brain microsomes of 16- to 18-day-old rats as a source of the cerebroside-sulfating enzyme and a 100,000g supernatant as the PAPS-generating system. An emulsifying agent, Brij 96, was employed to solubilize the endogenous galactosylceramide added as acceptor to the incubation tubes. The sulfated galactolipids were isolated from the reaction mixtures via column and thin-layer chromatography and presumably represented ceramide galactoside-3-SO$_4$. With intact microsomes, a 3-fold increase in ^{35}SO$_4$ incorporation into sulfatides was reported when galactosylceramide was added, while with the solubilized 27-fold purified preparation an 8- to 20-fold increase occurred. The addition of N-palmitoylglucosylsphingosine, sulfatide, cholesterol, or sphingomyelin did not result in any stimulation of sulfatide formation. PAPS itself was not employed in these studies.

The white matter of sheep brain was also employed as an enzyme source, since it appeared to be more active than gray matter. It should be recalled that sulfatide has been termed a "myelin component" and is found in highest concentration in white matter. An extract was subjected to ammonium sulfate fractionation and employed as the enzyme source after dialysis. However, no information was provided regarding the purification achieved. The substrate used was [^{35}S]-PAPS and the assay was dependent upon the formation of lipid-soluble radioactivity. It was found that both EDTA and a sulfhydryl compound such as cysteine stimulated such formation.[118] These investigators did not employ an endogenous source of lipid acceptor since adequate amounts were presumably present in the enzyme preparation. Indirect evidence was provided supporting this presumption. Galactose oxidase has been known to catalyze the oxidation of the C-6 hydroxy group of the galactose moiety of cerebrosides. When the sulfotransferase enzyme was treated, it no longer carried out the synthesis of sulfatide. The addition of cerebroside to the treated enzyme did not restore this enzyme activity. Therefore, it was assumed that the enzyme activity and acceptor were firmly associated with one another. All attempts to separate the endogenous cerebrosides from the preparation were unsuccessful. More recently, separation of the transferase and acceptor was reported utilizing Sephadex G-200 gel filtration. Using recombination experiments, it was suggested that the decrease in detectable galactosylcer- amide consumed as substrate was approximately equivalent to the amount of sulfatide synthesized.[118a] The enzyme from rat brain particles has been sol- ubilized and enriched sevenfold. Many of the properties examined are similar to those reported for the bound forms.[118b] Developmental studies both *in vivo*, measuring $^{35}SO_4$ incorporation, and *in vitro*, measuring the galactosyl- ceramide-stimulated[^{35}S]-PAPS incorporation into sulfatides, have revealed a very rapid increase in rate of seven- to eightfold between 8 and 15 days in rats.[119] In contrast, the kidney enzyme activity was very low at 20 days and rapidly increased. These studies suggest the concept of an age-dependent myelination process. The sulfotransferase activity is enriched in neuronal cells as compared to glial cells and is present in cultured cells of both types.[93e]

A fairly detailed study compared several potential galactosides as accep- tors for [^{35}S]-PAPS.[120] All subcellular particles from young rat brain (16–20 days old) were found to catalyze the sulfation of galactosylceramide, lacto- sylceramide, and lactose with a specific activity comparable to that of the whole homogenates. For convenience, the microsomal fraction was further studied since a portion of the total activity could be rendered soluble with deoxycho- late. The soluble preparation was employed for lipid acceptors and the in- soluble preparations for water-soluble acceptors. It was found that $Na_2{}^{35}SO_4$ could not substitute for [^{35}S]-PAPS, detergents were required for the lipid but not the water-soluble acceptors, and Mg^{2+} stimulated 100%. The following materials appear to be potential acceptors: *p*-nitrophenyl-β-D-galactoside > lactose > galactose > neuraminlactose for the water-soluble; galactosylsphin- gosine > lactosylceramide > galactosylceramide > tetrahexosylceramide > trihexosylceramide for the lipid acceptors. All substrates appeared to have pH optima around 6.8 with phosphate buffers. The K_m of [^{35}S]-PAPS for lactosylceramide was 2.4×10^5 cpm/ml, for lactose 1×10^6 cpm/ml, indicating a fourfold greater affinity for the lipid acceptor than for the water-soluble acceptor. The unusual manner of expressing these values in terms of cpm/ml is due to the uncertainty of the molar concentration of the PAPS. The apparent K_m's for the acceptors were lactosylceramide 5.6×10^{-5} M, galactosylsphingosine 4.5×10^{-5} M, galactosylceramide $3.3–8.5 \times 10^{-5}$ M, lactose 1.2×10^{-2} M, and galactose 1.12×10^{-1} M. It appeared that the lipid acceptors inhibited the [^{35}S]-PAPS incorporation into one another. Similarly, the water-soluble acceptors inhibited the synthesis of each other. Although the lipid-soluble substrates inhibited the incorporation into the water-soluble substrates, the reverse inhibitions did not occur. Information pertinent to K_i's and types of inhibition, competitive vs. noncompetitive, was not reported; however, these workers concluded that two different sulfotransferases were probably present, one for the water-soluble acceptors and the other for the lipid-soluble acceptors.

The product formed *in vitro* in these studies of condensation between galactosylceramide and PAPS was assumed to be sulfatide principally based on chromatographic data. Unequivocal evidence for the identity with naturally occurring sulfatide was provided by Stoffyn *et al.*[121] These workers synthesized [1-^{14}C]galactosylceramide in a sequence of reactions:

Galactosylceramide (naturally occurring) ⟶ ceramide ⟶ 1-*O*-tritylceramide ⟶

3-*O*-benzoyl-1-*O*-tritylceramide ⟶ 3-*O*-benzylceramide +

tetra-*O*-acetyl [1-^{14}C]galactosyl bromine ⟶ [1-^{14}C]galactosylceramide

This radioactive product was then incubated according to the system already described[119] and the radioactive sulfatides formed during the reaction isolated. After addition of carrier galactosylceramide sulfates, the mixture was permethylated and the methylated galactose obtained after hydrolysis. A single radioactive spot was found in thin-layer chromatography corresponding to 2,4,6-trimethylgalactose, proving that the sulfate had been incorporated into position 3 of the galactosylceramide.

Both α-hydroxy fatty acid- and nonhydroxy fatty acid-containing galactosylceramide are capable of being sulfate acceptors with Triton X-100-solubilized rat brain microsomal preparations.[122] Sulfatide labeled with both ^{3}H and ^{35}S was produced when [6-^{3}H]galactosylceramide and [^{35}S]-PAPS were incubated together. Rat brain microsomes may contain two distinct sphingolipid sulfotransferases, one for the "psychosine"-type sphingolipids and the other for the *N*-acylated sphingolipids. Although the V_{max}'s were identical for hydroxy fatty acid ceramides and nonhydroxy fatty acid ceramides, the K_m's were 5 and 30, respectively.[122a] Further characterization of the sulfation of galactosylsphingosine to form psychosine sulfate has been reported.[123] This sulfotransferase is associated with the Golgi apparatus of rat kidney[123a] but not rat liver.[123b]

$$CH_3—(CH_2)_{12}—CH{=}CH—\underset{\underset{OH}{|}}{\overset{\overset{H}{|}}{C}}—\underset{\underset{NH_2}{|}}{\overset{\overset{H}{|}}{C}}—\underset{\underset{H}{|}}{\overset{\overset{H}{|}}{C}}—O—galactose + PAPS \longrightarrow$$

$$\underset{\underset{\underset{R}{|}}{\overset{|}{C{=}O}}}{}$$

$$CH_3—(CH_2)_{12}—CH{=}CH—\underset{\underset{OH}{|}}{\overset{\overset{H}{|}}{C}}—\underset{\underset{NH_2}{|}}{\overset{\overset{H}{|}}{C}}—\underset{\underset{H}{|}}{\overset{\overset{H}{|}}{C}}—O—galactose\text{-}3\text{-}SO_4$$

$$\underset{\underset{\underset{R}{|}}{\overset{|}{C{=}O}}}{}$$

2.5.3. Hydrolysis in Vitro

Ceramide galactosyl-3-sulfate : sulfohydrolase activity has only rarely been assayed directly with sulfatide itself as substrate. Usually a synthetic sulfate ester such as nitrocatechol sulfate has been employed under assay conditions for measuring arylsulfatase A.

A 6000-fold purification of an enzyme was achieved from pig kidney[124] employing ^{35}S-labeled sulfatide as substrate isolated from rabbit brain after ^{35}SO$_4$ administration. The enzyme activity found in tissues presumably is associated with the lysosomes. Final purification was obtained with carrier-free electrophoresis, which resulted in the separation of arylsulfatase A and sulfatidase from arylsulfatase B. When material present in column chromatographic fractions that had no detectable enzyme activity was added to the arylsulfatase A fractions, there was an appreciable stimulation of sulfatide hydrolysis without affecting the hydrolysis of nitrocatechol sulfate (Table VIII).[125] This heat-stable "complementary" fraction has been partially purified but has not been further characterized. The activator obtained from human liver has the capacity to stimulate arylsulfatase purified from six invertebrates. The intriguing problem of modifying the substrate specificity of an enzyme by a noncatalytic "specific" protein has not been extensively studied. Support for the relationship between sulfatidase and arylsulfatase A comes from studies on metachromatic leukodystrophy, which have indicated a severe reduction of both enzyme activities in affected individuals.[126,127] This conclusion is supported by studies on substrate specificity of purified enzymes from human liver and kidney.[127a] An interesting approach for the purification of arylsulfatase A and presumably sulfatidase, employing affinity chromatography, has been reported. The basic principle exploited is that a substrate or substrate-like analog is coupled to an inert support, which specifically absorbs the enzyme.[128] Breslow and Sloan[129] indicated that psychosine sul-

Table VIII. *Effect of the Complementary Fraction on Release of Sulfate from Cerebroside-3-Sulfates by Arylsulfatases A and B from Pig Kidney*[a,b]

Preparation (units)	Addition of complementary fraction (μg/ml)	Sulfate released from	
		Cerebroside sulfate (cpm)	*p*-Nitrocatechol sulfate (μmoles)
Arylsulfatase A	—	1,460	10.2
(0.68)	80	10,700	11.4
Arylsulfatase B	—	<10	23.0
(1.53)	80	420	
	80	90[c]	
Complementary fraction	80	<10	0.0

[a] Modified from Ref. 125.
[b] Temperature 37°C and pH 4.5.
[c] Incubation at pH 6.1 (pH optimum for arysulfatase B).

fate purchased from Pierce Chemical Company was coupled to agarose for this purpose. Unfortunately, the exact structure of the commercial product was not provided. They assumed it was sphingosine-galactose-3-sulfate presumably derived from sulfatide. Quite likely this was galactosylsphingosine. Indeed, only a very modest purification at this step was achieved, although the final preparation was homogeneous on electrophoresis.

Arylsulfatase A has been highly purified from rabbit liver,[128a] human liver,[128b] ox liver,[128c] and human urine.[128d] General properties include anomalous kinetics with natural and artificial substrates, acidic pH optimum, and monomer molecular weights of around 100,000. A histidine residue appears essential,[128e,f] and the enzyme is probably a glycoprotein.[128g] Arylsulfatase is believed to hydrolyze ascorbic acid 2-sulfate[128h,i] and seminolipid [1-*O*-alkyl-2-*O*-acyl-3-*O*-(β-D-galactoside-3′-sulfate)-glycerol].[128j,k]

A role for vitamin A in sulfatide production has been postulated, based upon differences in the sulfatide content of animals from dams raised on a deficient diet.[130] Subsequent reports have indicated that in carefully controlled studies, where care and attention to pair feeding was assured, no differences in sulfatide levels were observed.[131]

2.6. Glucosylceramide

Although galactosylceramide represents the principal monohexosyl derivative of ceramide in the central nervous system, it is present only in trace quantities in peripheral organs. The major cerebroside of extraneural tissue is glucosylceramide.[132] A large number of publications have appeared concerning the incorporation of radioactive tracers into galactosylceramide *in vivo;* however, little *in vivo* information is available concerning glucosylceramide formation. The efficient conversion of [14C]glucose to the galactose portion of brain galactosylceramide was described.[73] L-[14C]serine has been shown to be an effective precursor of glucosylceramide after intracerebral administration.[93j] The bulk of glucose bound to ceramide is present in the complex group of sphingoglycolipids, the gangliosides. Indeed, this is the major repository of β-glucosidic bonds in mammalian tissues. The metabolism of gangliosides has been well studied, but glucosylceramide itself has not been the subject of intensive investigations.

Slices from young guinea pig and rat cerebral cortex were employed in one of the few reports of glucose labeling of a glucosylceramide. Both [^{14}C]glucose and [^{14}C]galactose were found to label the cerebrosides, which had been fractionated on silicic acid columns into both the "kerasine"-saturated nonhydroxy fatty acid and "phrenosine" hydroxy fatty acid-containing cerebrosides. The methyl glycosidetrimethylsilyl ethers were subjected to gas–liquid chromatography and a significant amount of radioactivity from both [^{14}C]glucose and [1-^{14}C]galactose was found in the glucose derivative obtained only from the kerasine type.[133] Similar studies were carried out with rat brain slices and [U-^{14}C]glucose. Separation of the cerebrosides on borate-impregnated thin-layer chromatograms indicated that more radioactivity was present in the glucocerebrosides than in the galactocerebrosides. The purified glucosylceramides did not contain any hydroxy fatty acids but only shorter chain nonhydroxy fatty acids.[133a] The transfer of newly biosynthesized glycosylceramides to other cellular membranes seems to be facilitated by "translocation" proteins present in spleen[133b] and brain[133c] tissues.

2.6.1. Biosynthesis in Vitro

The *in vitro* biosynthesis of glucosylceramide was documented with a particulate fraction purified sixfold over the original homogenate, from 13-day-old chick embryo brains.[134,134a] In the presence of a mixture of detergents, Cutscum and Triton, with ceramide as an acceptor, approximately 0.3 nmole of UDP-[^{14}C]glucose/mg protein per hr was incorporated. There did not appear to be any metal requirements, since 0.025 M EDTA had no inhibitory effect. The pH optimum was reported to be 7.8, the K_m for UDP-glucose was 1.2×10^{-4} M, the K_m for ceramide 0.8×10^{-4} M, and the predicted product was found to cochromatograph with known standards. Sphingosine and dihydrosphingosine appeared to stimulate glucosylsphingosine formation, but care was not taken to ensure that there were no losses of this material in the assay procedure. The ability of rodent brain microsomes to catalyze the ceramide-stimulated incorporation of glucose from UDP-[^{14}C]glucose into glucosylceramide has been reported.[58,89,135,136] Both hydroxy and nonhydroxy fatty acid ceramides were active as acceptors. Ceramides with varying chain lengths as well as structural homologs of decasphingosine were tested as substrates for glucosylceramide formation. N-Octanoylsphingosine and N-lauroyldecasphingosine were the most effective of this series of compounds.[136a] *In vivo* studies indicated the ability of puromycin to inhibit incorporation of several radioactive precursors into gangliosides.[137,138] Subsequently, it was reported that puromycin reduced the level of glucosylceramide-synthesizing capacity of microsomes from drug-treated animals.[139]

$$CH_3-(CH_2)_{12}-CH=CH-\underset{\underset{OH}{|}}{\overset{\overset{H}{|}}{C}}-\underset{\underset{NH}{|}}{\overset{\overset{H}{|}}{C}}-CH_2OH + \text{UDP-glucose} \longrightarrow$$

$$\underset{\underset{\overset{\displaystyle C=O}{|}}{\displaystyle R}}{}$$

$$CH_3-(CH_2)_{12}-CH=CH-\underset{\underset{OH}{|}}{\overset{\overset{H}{|}}{C}}-\underset{\underset{NH}{|}}{\overset{\overset{H}{|}}{C}}-\underset{\underset{H}{|}}{\overset{\overset{H}{|}}{C}}-O-\text{glucose}$$

$$\underset{\overset{\displaystyle C=O}{|}}{\underset{\displaystyle R}{}}$$

2.6.2. *Hydrolysis in Vitro*

N-Stearoyl[1-^{14}C]glucosylsphingosine was chemically synthesized in order to study the *in vivo* metabolism of glucosylceramide.[140] When this material was administered intracranially to young rats, the sole radioactive sphingolipid recovered was the starting material, indicating that intact glucosylceramide was not converted to a polyhexosyl derivative *in vivo*. In contrast, when N-stearoyl[1-^{14}C] glucosylsphingosine was employed, in addition to starting material, radioactive N-stearoyl[1-$^{1-4}$C] sphingosine was also obtained. This suggested the presence of a hydrolytic enzyme that catalyzed the removal of the β-glucosidic moiety. β-Glucosidase was purified 82-fold from both rat and human spleen tissue by conventional techniques.[141] Fortunately, the tissue was originally homogenized in 0.1 M phosphate buffer, which rendered the activity soluble, since later studies indicated that this is a lysosome-bound enzyme.[96] The availability of a substrate specifically labeled in the carbohydrate portion allowed for a convenient assay system to be developed, which merely measured the appearance of water-soluble radioactivity from a water-insoluble compound. The final enzyme preparation was free of galactosidase activity but was enriched in hexosaminidase activity. The hexosaminidase and glucosidase activities were thought to be distinct, since studies with mixed substrates indicated that no inhibition had occurred. The optimum pH for the glucosylceramide : β-glucosidase was around 6.0, it was inhibited by Tris at pH 8.0, and it had a K_m of 4.2 $\times$ 10^{-5} M. The products of the reaction, ceramide and glucose, were characterized both chromatographically and after carrier dilution. A highly purified glucosylceramide : β-glucosidase preparation appears to be an intestinal enzyme preparation that was purified 2000-fold and also had the ability to cleave galactosylceramide.[95] This enzyme has been reinvestigated and shown to possess β-galactosidase, β-glucosidase, β-xylosidase, β-fucosidase, and α-arabinosidase activity. It also hydrolyzed β-glucosyl or β-galactosyl linkages to either sphingosine or ceramide, lactose, and phlorizin.[95a]

Glucocerebrosidase has been extensively purified from human placenta and hydrolyzes 1 mmole glucosylceramide/mg protein per hr. It may be a tetramer composed of monomers of 60,000 daltons and is claimed to be absolutely specific for glucosylceramide.[141a] Bovine spleen β-glucosidase was purified on an affinity column containing D-gluconolactone as the ligand.[141b] The ability to catalyze the formation of glucosylceramide from ceramide as acceptor and various glucosidases as donor has been demonstrated. This trans-glucolytic activity has been shown to be identical to the hydrolytic activity of the enzyme. The β-glucosidase appears to hydrolyze *p*-nitrophenyl-β-D-glucoside, 4-methylumbelliferyl-D-glucoside, glucosylsphingosine, glucosylceramide, and steroid β-D-glucosides.[141c] An enzyme purified from brain tissue has similar properties.[141d] Thus, it appears that there may not be a specific glucosylceramide : glucosidase but rather a relatively nonspecific β-glucosidase.[141e] The ability to catalyze transglycosylation reactions is a useful technique for evaluating the substrate specificity of a hydrolytic enzyme.

A glycoprotein (or glycopeptide) fraction prepared from several mammalian sources stimulates β-glucosidase,[141c] which in association with phospholipids is claimed to give active enzyme.[141f] This glycoprotein has been employed to prepare an affinity column for β-glucosidase purification.[141g]

A compound, *N*-hexyl-*O*-glucosylceramide, has been synthesized and claimed to be a somewhat specific, probably competitive, inhibitor of glucosylceramide : glucosidase.[141h]

Several other analogs containing an *N*-alkyl group of up to 18 carbon atoms were also inhibitory. Another class of inhibitors based upon an *N*-decyl-DL-*erythro*-3-phenyl-2-amino-1,3-propanediol structure have been synthesized and examined as β-glucosidase inhibitors.[141i]

A β-glucosidase was obtained by cholate solubilization from an 800–13,000*g* particulate fraction of rat brain.[141j] When the purification procedure was applied to calf brain, only a 6- to 10-fold purification was achieved. The substrate employed for these studies was *p*-nitrophenyl-β-D-glucoside. This type of an "artificial" substrate has been extensively employed for the purification of lysosomal acid hydrolases.

$$O_2N-\!\!\left\langle\!\!\!\bigcirc\!\!\!\right\rangle\!\!-O-glucose \longrightarrow O_2N-\!\!\left\langle\!\!\!\bigcirc\!\!\!\right\rangle\!\!-O-Na \; + \; glucose$$

(yellow)

The brain β-glucosidase activity was stimulated by detergents such as Triton X-100 and bile salts. This requirement, it was speculated, is due to the dispersion of the enzyme preparation, which was insoluble at the pH optimum of 4.5–5.0. A K_m of 7×10^{-4} M and a K_i for γ-gluconolactone of 3×10^{-5} M were obtained; phenyl- and methylglucosides, cellibiose, gentiobiose, and salicin were not hydrolyzed. The intriguing possibility that a reverse or "transferase" reaction might be catalyzed was also considered. The β-glucosidase was incubated with *p*-nitrophenyl-β-D-glucoside and either glucose or galactose. Paper chromatography was used in an attempt to reveal the formation of disaccharides. It was questionable whether these experiments could have been successful since the potential carbohydrate acceptors utilized were found to be contaminated with disaccharides. The enzyme preparation was found to catalyze the release of glucose, as measured with glucose oxidase, from glucosylceramide that had been isolated from Gaucher's spleen tissues.[142,143] Stimulation of hydrolysis occurred with either Triton X-100 or bile acids, the pH optimum was around 5.0, the K_m was 1.8×10^{-4} M, and the V_{max} was 415 nmoles/mg protein per hr. No exchange with [14C]glucose to form glucosylceramide occurred, and gluconolactone was a competitive inhibitor with a K_i of 1.5×10^{-4} M, as was sphingosine.

The diminution of this enzyme activity is believed responsible for the accumulation of glucosylceramide in tissues of individuals with Gaucher's disease. It presumably functions in removing the final carbohydrate molecule from parent complex sphingoglycolipids such as globosides and gangliosides.[144]

2.7. Ganglioside Metabolism

2.7.1. In Vivo Studies

In vivo experiments indicated an effective labeling of gangliosides after administration of [14C]glucose or [14C]galactose[72,73] (Fig. 6). The degree of tracer incorporation into the individual sugar components of these complex lipids was not examined; however, a turnover was noted. The complex nature of gangliosides, which are composed of glucose, galactose, N-acetylgalactosamine, N-acetylneuraminic acid, sphingosine bases, and fatty acids, makes it technically difficult to distinguish whether a specific component of the intact molecule is undergoing turnover. This approach has not provided a great deal of information regarding precursor–product relationships. These investigations demonstrated an active period of precursor incorporation that coincided chronologically with active myelination. A comparative study with several potential radioactive precursors attempted to examine the incorporation into the specific components of ganglioside. L-[14C]serine when administered was found to label NeuAc, sphingosine, and all the carbohydrate moieties, [14C]glucose and [14C]galactose labeled all the carbohydrates, and [14C]glucosamine labeled principally NeuAc and GalNAc.[74] Unfortunately, the identification and designation of the individual ganglioside species were undefined in these studies. Paper chromatography was employed to separate the gangliosides into two spots designated α (GD_{1b}?) and β (GD_{1a}?), both of which were converted to γ (GM_1) with neuraminidase. This neuraminidase-lability would suggest that α and β were disialo- or trisialogangliosides. Peculiarly, these investigators found only α- and β-gangliosides to be present in brain of the rats employed in these studies, which ranged from 5 to 30 days of age. Developmental studies have indicated that at 20–30 days of age, GM_1

represents an appreciable proportion of the total galglioside NeuAc.[145–147] Their inability to detect any monosialoganglioside in these preparations is difficult to explain, and could affect some of these conclusions from this study. A turnover of 10 days based upon glucose and 24 days based upon glucosamine labeling was calculated.[148]

[1-^{14}C]Acetate was administered to rats at 7, 13, and 22 days of age, and the ganglioside fatty acids were isolated from each of these groups over a 30-day period. A very rapid labeling and turnover of palmitate with a subsequent increase in stearate labeling was observed, suggesting chain elongation of C_{16} to higher homologs.[149,150] Different rates of loss of label from ganglioside stearate occurred during the 1-month experimental period depending upon the age of the animals when initially injected, 50% with the youngest, 10–20% with the 13-day-old group, and nearly none with the oldest group.

The *in vivo* labeling by [U-^{14}C]glucose of all the carbohydrate components of the four major brain ganglioside types was undertaken with methodological improvement for the purification and isolation of these sphingolipids.[151,152] Rats 7 days of age were injected intraperitoneally with [^{14}C]glucose and then sacrificed at 1, 3, and 5 hr. The ganglioside fraction was separated by thin-layer chromatography into GM_1, GD_{1a}, GD_{1b}, and GT components. Neuraminidase was employed to liberate the labile NeuAc, and the monosialogangliosides produced were isolated and subjected to acid hydrolysis. The specific activities of glucose, galactose, and the neuraminidase-resistant NeuAc were roughly comparable in each ganglioside type at each time point investigated. The galactosamine and the neuraminidase-labile NeuAc specific activities were similar but had a specific activity only 50% of the other carbohydrate moieties. These studies did not provide any information concerning the metabolic pathway for ganglioside biosynthesis. The authors also reported that [U-^{14}C]glucose was a very effective precursor of a "protein residue" since this fraction was at least 50 times more highly labeled than gangliosides. It was suggested that glucose had been converted to a variety of amino acids. However, carbohydrate labeling of glycoproteins, glycopeptides, or mucopolysaccharides is an additional explanation.

The metabolism of individual gangliosides was investigated employing more sophisticated[153] isolation techniques, including enzyme degradation and removal of contaminating nucleotide sugars.[154] [U-^{14}C]Glucose or [1-^{14}C]-glucosamine was administered intraperitoneally to 2-day-old rats and the radioactivity present in the individual gangliosides was determined 1, 8, 16, 30, and 60 days later.[155] Expressing the data in terms of dpm/μg NeuAc, a very large loss of label, from 95 to 99%, from day 1 to 61 is observed. However, this is an operational artifact due merely to the dilution that occurs during the accretion of gangliosides in the animals during this time period. The data, when expressed as dpm/brain in the different gangliosides, indicated a great similarity in turnover rate ($t_{1/2}$) of about 20 days, although certain glaring individual inconsistencies are also seen. These studies do not provide any information concerning precursor–product relationships and suggest independent metabolism for each ganglioside species.

The distribution of [U-^{14}C]glucosamine in mixed gangliosides of various cerebral subcellular particles was studied.[156] Aside from the interesting observation that "synaptosomes" were as active as "microsomes," no information of ganglioside metabolism was obtained. With [U-^{14}C]glucose in short-term experiments, there was a suggestion that GM_1 might be a precursor of the higher homologs.[157] Cultured oligodendrocytes, isolated from lamb brain, incorporated [^{3}H]galactose principally into GM_3, GM/GD_3, and GD_{1a} without any appreciable label appearing in GM_2.[157a]

These studies had employed a relatively nonspecific labeling agent for gangliosides. Therefore, it would be technically and theoretically quite difficult to obtain more than gross impressions concerning the metabolism of these complex sphingoglycolipids. When *N*-acetyl[^{3}H]mannosamine, a more specific labeling agent, is injected intracerebrally into young rats, only the sialic acid moieties become radioactive.[158] This precursor was injected into 15-day-old rats intracerebrally and the animals were sacrificed at 30 min, 1, 2, 4, 8, and 24 hr later. The gangliosides were isolated and separated into their individual types by thin-layer chromatography. Bands corresponding to standards were scraped and separate portions were used for radioactivity and sialic acid quantitation. Surprisingly, only GM_2 had appreciable radioactivity at the first three time points; the other gangliosides showed a 2-hr lag period prior to significant labeling. It was suggested that GM_2 is a precursor of GM_1, which in turn is converted to both GD_{1a} and GD_{1b}. GM_3 could not be a precursor of GM_2 as judged from the data obtained (Fig. 8).[159]

[^{14}C]-NeuAc, when administered intracerebrally, is less effective than either hexoses or hexosamines for labeling cerebral gangliosides.[160] [6-^{3}H]-Glucosamine was administered intracisternally to either 8- or 18-day-old rats and the radioactivity present in the individual gangliosides at various time periods was estimated.[161] The specific activities of both the neuraminidase-labile and -resistant sialic acids were found to be identical, in contrast to those reported previously.[151] These studies also indicated that the specific activity of GM_3 was appreciably lower than the specific activities of other gangliosides. This observation would tend to rule out this material as a general precursor. However, in an attempt to reconcile *in vivo* observations with *in vitro* studies on ganglioside formation, the investigators argue that due to relative pool sizes, this may not be true. A slow labeling and turnover of brain gangliosides, in contrast to brain glycopeptides, was reported when 15-day-old rats were given [^{14}C]glucosamine intraperitoneally.[162] A fairly ambitious examination of *in vivo* ganglioside metabolism has been undertaken. [^{3}H]Acetate has been used to study the turnover of the sphingosine and stearic acid moieties of the individual gangliosides, and [^{14}C]glucosamine employed to monitor both the neuraminidase-stable and -labile NeuAc and hexosamines.[163] Short-term experiments lasting from $1\frac{1}{2}$ to 12 hr, after intracerebral injection, have not provided data on precursor–product relationships. Long-term experiments revealed different turnover rates that were dependent upon the age of the animals. The neuraminidase-labile and -resistant NeuAc showed some differences in specific activity that also appeared to be age-dependent.

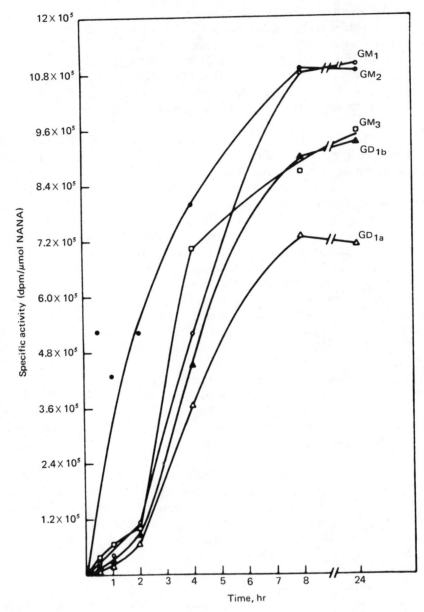

Figure 8. Specific activity of the individual gangliosides as a function of time after the intracerebral administration of N-acetyl-[^{3}H]mannosamine to 15-day-old rats. The curve labeled GD_{1b} represents the mixed GD_{1b} and GT_1. (From Ref. 159.)

The general conclusions from *in vivo* studies concerning ganglioside metabolism are (1) a variety of radioactive precursors can label either specifically or nonspecifically; (2) it is difficult to compare different studies due to variables of age of animals, radioactive precursors employed, mode of administration, and duration of experiment; (3) little information elucidating precursor–product relationships has been provided; (4) at least two reports tend to rule out GM_3 as precursor, while GM_1 itself could be a precursor of GD_{1a}, GD_{1b}, and GT; and (5) the half-lives appear to be age-dependent, approximately 20 days for young animals and perhaps 50–70 days for adults.

GM_1 labeled in the terminal Gal residue with tritium, was administered to mice and the recovery of volatile and nonvolatile radioactive materials in several organs established. There was appreciable formation of ^3H_2O with label also present in unaltered GM_1 as well as in phospholipids and triglycerides.[163a]

2.7.2. Biosynthesis in Vitro

2.7.2.1. Lactosylceramide

The *in vivo* labeling of this lipid has not been reported. However, the ability of intact human white blood cells to carry out the incorporation of labeled hexose into this sphingolipid has been documented.[164] Incubating either [^{14}C]glucose or [^{14}C]galactose with leukocytes resulted in an active incorporation of radioactivity into the glucose of glucosylceramide and the galactose of lactosylceramide. Because of the observed differences in specific activities, it was speculated that the intact monohexosylceramide was converted to the dihexosylceramide through the addition of galactose. The emphasis in this study was identification of the products and the nature of endogenous acceptors. The role of metals or sugar donors was not investigated.

The presence of such a galactosyltransferase enzyme was initially demonstrated with spleen homogenates.[165] UDP-[1-^{3}H]galactose was found to label lactosylceramide in the absence of added acceptor, and this activity was increased further by the addition of acceptor glucosylceramide. The product of the reaction was found to cochromatograph with lactosylceramide standard, and only radioactive galactose was liberated on hydrolysis of the metabolic product. The enzyme activity was found to be associated with the "microsomal" fraction, and with glucosylceramide as acceptor, 15 mM Mn^{2+} and 0.2% Cutscum gave maximum incorporation. A pH optimum of 6.7 was obtained with 3,3-dimethylglutarate as buffer, Tris buffer being less effective. A variety of sphingolipids were examined for their ability to stimulate the incorporation of [1-^{3}H]galactose into lipids, and except for glucosylceramide and lactosylceramide, none proved effective. Sphingosine, glucosyl-, galactosyl-, and lactosylsphingosine, ceramide, and galactosylceramide depressed the incorporation of [^{3}H]galactose. A K_m of 3.3×10^{-4} M was observed.[166] The formation of radioactive lactosylceramide was observed upon incubation of N-stearoyl[^{14}C]glucosylsphingosine in the presence of nonradioactive UDP-galactose.

The ability of embryonic chicken brain particles[134] and rat brain particles[55] to catalyze the synthesis of lactosylceramide has been reported. Both Mg^{2+} and Mn^{2+} appeared equally effective in stimulating this reaction with the chicken preparation. This preparation had a K_m for glucosylceramide of 1.7×10^{-4} M and this acceptor could be partially replaced by glucosylsphingosine but not by other sphingolipids. The K_m for UDP-[1-^{14}C]galactose was found to be 0.5×10^{-4} M. The identification of the product was well substantiated in the studied employing rat brain particles.

UDP-galactose   +   glucosylceramide   →   galactosylglucosylceramide
                                                                          (lactosylceramide) + UDP

The exact configuration of the galactosidic linkage as being $\beta 1 \rightarrow 4$ to the glucose moiety in the *in vitro* product was not documented. A soluble form of this galactosyltransferase has been partially purified from bovine milk.[166a]

Hydrolysis in Vitro.   The liberation of galactose from lactosylceramide was reported with a 20-fold enriched enzyme preparation from rat and calf brain.[141a] *p*-Nitrophenyl-β-D-galactoside was employed as substrate for this purification procedure; the pH optimum of the rat brain enzyme was found to be 3.1 and that of the calf brain, 4.2. The K_m for the rat preparation was 4×10^{-4} M and the K_i for galactonolactone, 1.8×10^{-4} M. Lactose and melibiose were poorer substrates than the *p*-nitrophenyl-β-D-galactoside. The ability of such enzyme preparations to cleave the galactose residue from lactosylceramide has been demonstrated.[142,143] A K_m of 2.2×10^{-5} M, a $V_{\max}$ of 0.3 μmole/mg per hr, a pH optimum of approximately 5.0, and a detergent requirement were reported, sodium taurocholate being most active for this hydrolysis. The preparation could not cleave galactosylceramide or galactosylsphingosine. Developmental studies with brain extracts have shown very small changes in specific activity for this hydrolytic activity from 4 to 320 days. In contrast, two β-galactosides were purified 500-fold and 60-fold from rat brain, and the major separation occurred with DEAE–cellulose column chromatography, where the more highly purified one was absorbed and the less highly purified was not. The ratio of activity toward 4-methylumbelliferyl- or *p*-nitrophenyl-β-D-galactoside and lactose was 1. Although activity toward lactosylceramide was present in early purification steps, the final preparation was devoid of this activity[167] and attempts at mixing fractions were not reported. Purified jack bean[168] and *E. coli*[169] β-galactosidase catalyze the release of the galactose from lactosylceramide.

galactosylglucosylceramide → glucosylceramide + galactose

There has been no unequivocal demonstration of a β-galactosidase specific for lactosylceramide. There are at least three or four chromatographically or electrophoretically distinct enzymes in most mammalian tissues capable of hydrolyzing the artificial β-galactoside substrates. There is the suggestion that two lactosylceramidases : β-galactosidases exist in tissues. A 250-fold purified lactosylceramidase II from human liver has been reported to hydrolyze lactosylceramide, GM$_1$, asialo GM$_1$, and 4-methylumbelliferyl-β-D-galactoside[168a] and is activated by either crude taurocholate or pure sodium taurodeoxycholate. A similar enzyme has been purified 400-fold from rabbit brain[168b] and 50-fold from rat brain[168c] It has been proposed that lactosylceramidase I is also a galactosylceramidase and is present in brain tissues. This form of the enzyme is optimally active in the presence of pure but not crude bile salts.[168d] A β-galactosidase has been obtained from human liver and claimed to possess β-D-galactosidase, β-D-glucosidase, β-D-fucosidase, β-D-xylosidase, and β-D-arabinosidase activities.[168e] However, the relationship of this panhydrolase to sphingolipid galactosidase is unknown.

2.7.2.2. Galactosylgalactosylglucosylceramide

In vivo studies specifically reporting the incorporation of radioactive tracers into this material have not been reported.

The *in vitro* biosynthesis of trihexosylceramide by rat spleen extracts[166] from lactosylceramide and UDP-[U-^{14}C]galactose has been demonstrated. This is the same enzyme preparation that had been previously employed to study the biosynthesis of lactosylceramide, and the majority of this enzyme activity resided in the "microsomal" subfraction. Product identification and assay relied upon cochromatography with standards on thin-layer chromatography. The endogenous amounts of lactosylceramide present in rat spleen, 10–20 μg/g wet wt tissue, were small compared to the quantity of exogenous acceptor added to the incubation tubes. The pH optimum was reported to be 6.1 with 3,3-dimethylglutarate as buffer; a K_m of 4×10^{-5} M for lactosylceramide, and stimulation both by Cutscum and by Mn^{2+} were observed. This activity, in contrast to the enzyme producing lactosylceramide, is stable to heating at 50°C for 5 min. Mixing experiments also suggest that these represent separate enzyme proteins.

More extensive examination of the reaction product obtained from these incubation mixtures containing [6-^{3}H]galactosylglucosylceramide as the acceptor and UDP-galactose as the donor indicated that a 1→3 linkage had been formed between the two galactose residues. Permethylation of a chromatographically purified trihexosylceramide fraction and methanolysis gave 2,4,6-trimethylgalactose as the sole radioactive species.[170] This indicated that the hydroxy group at C-3 was the site of attachment. Trihexosylceramide isolated from biological material possesses a 1→4 linkage joining the two galactoses, thus indicating differences with the *in vitro* product. Subsequent evidence indicated that rat kidney microsomes produce 1→3 and 1→4 positional isomers of the trihexosylceramide in the presence of UDP-galactose and lactosylceramide.[170a] Employing purified enzymes, it was shown that α-galactosidase from ficin but not β-galactoside from jack bean meal catalyzed the release of radioactive galactose from [U-^{14}C]galactosylgalactosylglucosyl- ceramide isolated from appropriate incubation mixtures.[171] The galactosyltransferases catalyzing the formation of trihexosylceramide and digalactosylceramide from lactosylceramide and galactosylceramide respectively may be identical.[171a]

Hydrolysis in Vitro. Galactosylgalactosylglucosylceramide was isolated from human kidney and subjected to general tritium labeling with ^{3}H$_2$ gas by the Wilzbach technique. During the course of this treatment, decomposition and side reactions occurred, and extensive purification of the trihexosylceramide was required. Upon hydrolysis, 4.3 times as much radioactivity was found in the lipid portion when compared to the carbohydrate portion; the theoretical value was calculated to be 3.7. The ratio of radioactivity in glucose was one-half that in galactose, as anticipated. This material was then employed as substrate for enzyme studies.[172] The highest activity was found in rat intestinal tissue and 220-fold purification was reported. Optimum activity occurred at pH 5.0, cholate at 2 mg/ml, 1 mg human serum albumin; a K_m of 3.7 × 10^{-4} M was calculated. The preparation did not hydrolyze glucosyl-, galactosyl-, or lactosylceramide, lactoside, *p*-nitrophenyl-β-D-galactoside, β-D-glucoside, or β-D-glucosaminide. Glucosylsphingosine was found to be a noncompetitive inhibitor with a K_i of 7.4 × 10^{-4} M. At the time of these studies, it had been erroneously assumed that the terminal galactose of trihexosylceramide was in β-linkage; however, subsequent investigations have indicated that this is an α-linkage. It probably is reasonable to assume that the purified enzyme was an α-galactosidase, although unfortunately *p*-nitrophenyl-α-galactoside was not examined as substrate in these investigations. The terminal galactose in this sphingolipid is susceptible to oxidation with galactose oxidase, and has been employed to prepare [6-^{3}H]galactosylgalactosylglucosylceramide,[169] which is a substrate for a α-galactosidases from ficin[168,169] and coffee bean.[173]

$$\text{GalGalGlcCer} \rightarrow \text{GalGlcCer} + \text{Gal}$$

The β-galactosidase from purified brain tissue[141a] was originally reported to hydrolyze the terminal galactose of trihexosylceramide.[174] This apparent discrepancy has been reconciled since it has been stated that the preparation has both α- and β-galactosidase activities.[175]

There appears to be a multitude of proteins in human plasma capable of cleaving the terminal galactose of trihexosylceramide. These enzyme activities were adsorbed onto an affinity chromatography column containing *p*-aminophenylmelibiose as the ligand. A total of five different forms of the enzyme were obtained from normal human plasma passed through such a column.[175a] Two of these activities, Form A-1 and Form A-2, have been characterized in great detail in terms of molecular weights, effectors, inhibitors, substrate specificities, and kinetics.[175b] The human serum enzymes have been subclassified into α-galactosidases of the A group, which are active at pH 5.4, and α-galactosidases of the B group, which are active at pH 7.2. Neuraminidase treatment of the acidic form (A) results in the production of a form electrophoretically similar to the basic form (B). Crude sialyltransferase was claimed to incorporate radioactivity from CMP-[^{14}C]-NeuAc to the acid form in the presence of the basic form (B) of the enzyme. Unfortunately, the hydrolytic activities of the products from these enzyme treatments were not documented nor the studies corroborated or extended.[175c] An enzyme has been purified 60-fold from human liver and has the capacity to hydrolyze the terminal galactose of trihexosylceramide and 4-methylumbelliferyl-α-D-galactoside.[175d] Myoinositol has been reported to be a competitive inhibitor with a K_i of about 10^{-2} M for trihexosylceramide hydrolysis by a rat intestinal preparation.[175e]

2.7.2.3. *N-Acetylgalactosaminylgalactosylglucosylceramide*

The available evidence for the formation of this material either *in vitro* or *in vivo* is limited. *In vitro,* a substrate as the carbohydrate donor was isolated from rats that had been administered [1-^{14}C]glucosamine. This biosynthetic material was a mixture of UDP-*N*-acetyl[^{14}C]glucosamine and UDP-*N*-acetyl[^{14}C]galactosamine in a ratio of 7 : 3 and had a specific activity of 0.3 Ci/μmole (5 × 10^4 cpm/0.1 μmole). The subcellular distribution of the transferase activity in rat brain indicated that the highest activity occurred with microsomes and mitochondria with a reported value of 7.0 pmoles/mg protein per hr. The particulate fraction possessed UDP-*N*-acetylglucosamine-4-epimerase activity, which allowed these workers to employ an impure substrate. The assay system was based upon the appearance of radioactivity in the lower phase after a Folch partitioning procedure. A variety of potential substrates were tested as acceptors for the amino sugar, glucosylceramide, galactosylceramide, and lactosylceramide being the most effective. Ceramide, galactosylgalactosylglucosylceramide, and GM$_3$ were the least effective. Investigations with brain particles from chick embryo have indicated that these two compounds were 10-fold better acceptors than lactosylceramide.[176] An approximate K_m of 1.2 × 10^4 M for the nucleotide-amino sugar was reported and both Triton X-100 and Mn^{2+} were reported to be stimulatory. The bulk of the radioactivity applied to a silicic acid column was eluted with chloroform–methanol 7 : 3 and found to cochromatograph with *N*-acetylgalactosaminylgalactosylglucosylceramide on thin-layer plates. Galactosamine was the sole radioactive material liberated by acid hydrolysis.[176a]

Further details about the occurrence of this reaction *in vitro* were also reported by DiCesare and Dain[177,178] in studies employing UDP-*N*-acetyl[^{14}C]galactosamine itself as the carbohydrate donor.

$$\text{GalGlcCer} + \text{UDP-GalNAc} \rightarrow \text{GalNAcGalGlcCer} + \text{UDP}$$

Hydrolysis in Vitro. Brain tissue has been reported to possess several types of hydrolytic activities for hexosaminides, and a particulate fraction was observed to have both galactosaminidase and glucosaminidase activities. Apparently such hydrolytic activity was able to survive solvent treatment, so that acetone powders of calf brain retain this capacity. A 63-fold purification of this enzyme was reported, employing either *p*-nitrophenyl-glucosamide or -galactosaminide as substrate.[179] The ratio of hydrolysis of both substrates remained reasonably constant during all stages of the purification procedures, suggesting a common protein catalyst. The K_m's for *p*-nitrophenyl-*N*-acetylglucosamide and *p*-nitrophenyl-*N*-acetylgalactosaminide were 0.8 and 0.54 mM while the calculated V_{max}'s were 19 and 3, respectively. The following K_i's (mM) were reported: with acetate buffer for *p*NP-*N*-acetylglucosaminide, 5.0; *p*NP-*N*-acetylgalactosaminide, 250; with free *N*-acetylgalactosamine; *p*NP-*N*-acetylglucosaminide, 0.5; *p*NP-*N*-acetylgalactosaminide, 2. These and other slight differences in inhibition by sulfhydryl-binding agents were interpreted as suggesting that different catalytic sites may be involved. Two other hex-osaminidase were found in the soluble portion of calf brain.[180] A 30,000g supernatant from a 25% homogenate prepared in 0.25 M sucrose was treated with protamine sulfate, which precipitated these activities. The hydrolytic activity was extracted with phosphate buffer and the bulk of the enzyme was precipitated with 25–40% saturated ammonium sulfate. In order to obtain "glucosaminidase" activity, dithiothreitol at 7×10^{-5} M was present in all solutions. The final purification was threefold and the ratio of glucosaminidase to galactosaminidase was 20 : 1. When the dithiothreitol was replaced by 10^{-3} M N-ethylmaleimide, a "galactosaminidase" was obtained with a threefold purification and this preparation had a glucosaminidase : galactosaminidase hydrolytic ratio of 0 : 25. The residual glucosaminidase activity still detectable was ascribed to the presence of residual "hexosaminidase" activity. The pH optimum was found to be 5.2 for "glucosaminidase," the K_m was 10^{-3} M, the K_i for acetate was 2.5×10^{-2} M, and N-acetylgalactosamine was not inhibitory. The pH optimum was found to be 5.5 for "galactosaminidase" and the K_m was 3×10^{-3} M. Addition of N-ethylmaleimide (10^{-3} M) to preparations previously not exposed to this reagent resulted in a 2- to 2.5-fold increase in "galactosaminidase" activity. This appears to be the sole report of success-ful separation of "glucosaminidase" and "galactosaminidase" activities. These enzymes have been termed "acetylhexosaminidases" to denote that they are nonspecific as to the nature of the N-acetyl amino sugar substrate. The role of these enzymes in the specific metabolism of sphingolipids possessing amino sugar residues is still unsettled. These matters will be illustrated in greater detail in the section on the metabolism of GM_2 (Section 2.7.2.6) and in the sections of Chapter 3 on the GM_2 gangliosidoses (Sections 3.6.1–3).

A radioactive substrate was prepared from ^{3}H-labeled asialo GM_1 by enzymatic degradation with β-galactosidase to yield N-acetyl[^{3}H]galactos-aminylgalactosylglucosyl(N-acyl) dihydrosphingosine. The formation of [^{3}H]lactosylceramide as a result of the enzymatic removal of the terminal N-acetylgalactosamine was monitored by thin-layer chromatography. Sodium taurocholate was found to stimulate the hydrolysis of sphingoglycolipids and to inhibit the hydrolysis of pNP-β-N-acetylgalactosamine.[181] Purified hex-osaminidase gave a K_m of 6×10^{-4} M, a V_{max} of 0.5 μmole/mg per hr, and a pH optimum of 3.8 with citrate–phosphate buffer. The two enzymes isolated from this supernatant, glucosaminidase and galactosaminidase, were unable to use this material as a substrate. Tay–Sachs' ganglioside itself, although not a substrate, was an effective inhibitor of N-acetylgalactosaminylgalactosyl-glucosylceramide hydrolysis. Hexosaminidases A and B have been purified approximately 4000- and 2000-fold respectively from human liver, employing p-NP-β-D-N-acetylglucosamine as the substrate. For hexosaminidase A with pNP-N-acetylglucosamine as substrate, the K_m was 0.67 mM and the V_{max} 117 μmoles/mg protein; with pNP-N-acetylgalactosamide as substrate, the K_m was 0.16 mM and the V_{max} 9.5 μmoles/mg protein. For hexosaminidase B with N-acetylgalactosaminylgalactosylglucosylceramide as substrate, the K_m was 0.2 mM and the V_{max} 1.1 μmoles/mg protein; with pNP-N-acetylglucosamine, K_m

0.67 mM and V_{max} 73 μmoles/mg protein; with pNP-N-acetylgalactosamine, K_m 0.15 mM and V_{max} 4.8, with N-acetylgalactosaminylgalactosylglucosylceramide, K_m 0.2 and V_{max} 1.1.[182] Thus, it appears that relatively nonspecific, highly purified β-hexosaminidase preparations have the capacity to hydrolyze the amino sugar from GA_2.

Site of Ganglioside Biosynthesis. The intracellular compartment containing the enzyme complement required for ganglioside production is not completely resolved. The Golgi apparatus of rat liver appears to contain the requisite glycosyltransferases.[182a] Synaptosomes have been found to possess these enzymes in embryonic chick brain.[182b]

2.7.2.4. Sialosylgalactosylceramide

Sialosylgalactosylceramide is a major constituent of the gangliosides of human white matter.[182c] This ganglioside can be produced by mouse brain particles in the presence of CMP-NeuAc and galactosylceramides with a K_m of 8.7×10^{-4} M for the acceptor. These same preparations produce hematoside (GM_3) when lactosylceramides replace the galactosylceramide with a K_m of 8.9×10^{-5} M. It appears that a single enzyme may be responsible for both of these reactions.[182d] Neuronal perikarya possess high sialosyltransferase activity for GM_4 biosynthesis.[182e]

$$\text{GalCer} + \text{CMP-NeuAc} \rightarrow \text{NeuAcGalCer} + \text{CMP}$$

2.7.2.5. Sialosylgalactosylglucosylceramide (GM₃)

The *in vitro* synthesis of this compound, which is the simplest ganglioside encountered in extraneural tissues, was originally described using particles isolated from rat brain.[183] The assay system was based upon the conversion of [1-^{14}C]-NeuAc from its CMP derivative to an organic solvent-soluble product. In addition, the product was further separated from the nucleotide sugar substrate with Sephadex G-24 column chromatography. This technique has been employed for the removal of water-soluble contaminants from lipid mixtures.[184] Unfortunately, the product of this reaction, hematoside, has peculiar distribution properties and it is found to partition into both organic and aqueous phases. The validity of this methodology as a suitable assay procedure for an accurate estimation of the *in vitro* kinetics was not examined. Lactosylceramide, which was employed as the acceptor, was prepared by acid hydrolysis of either total brain gangliosides or hematoside. The microsomes had the highest specific activity; the mitochondrial fraction was almost as efficient. Activation of the microsomal fraction was observed by disruption with deoxycholate, even though the enzyme activity was not "solubilized." The pH optimum was 6.8; MgCl was routinely present in the incubation mixtures although no information regarding this requirement or specificity was provided. Several galactosides were acceptors and their reported K_m's were: lactose 2.04×10^{-3} M, lactosylsphingosine 8×10^{-4} M, and lactosylceramide

1.7×10^{-4} M. Product identification as GM_3 was based solely on cochromatography with standards. A similar activity was reported to occur with particles from embryonic chick brain. Electrophoretic separation of the radioactive nucleotide sugar substrate from the sphingolipid products was the assay procedure employed in these investigations.[185] These studies indicated a requirement for a detergent mixture, Tween 80 and Triton CF-54 being the most effective, but no metal ion was needed. The stimulation of incorporation of $[^{14}C]$-NeuAc by tetrahexosylceramide (4950 cpm), lactosylceramide (850 cpm), GM_1 (1500 cpm), lactose (5544 cpm), N-acetyllactosamine (6942 cpm), and desialyzed fetuin (970 cpm) was reported with this particulate fraction. Heating these preparations at 50–60°C resulted in marked loss (66%) of tetrahexosylceramide and lactose acceptor activity with little effect on lactosylceramide stimulation. Lactose had little effect on lactosylceramide, tetrahexosylceramide, and GM_1 as stimulators for the formation of the corresponding sialyl-containing product. Only dihexosylceramide was found to be inhibitory for the lactose acceptor activity. The results of these competition and heat denaturation studies were interpreted as indicating the presence of a specific sialyltransferase for dihexosylceramide, which was different from the sialyltransferase for the other acceptors. The sialylated product obtained with tetrahexosylceramide as acceptor, although cochromatographing closely with GM_1 standards, was found to be neuraminidase-labile, thus indicating that it was not the presumed product, GM_1. The product obtained with dihexosylceramide as acceptor was analytically and chromatographically similar to the expected GM_3. Unfortunately, these workers have not published the details for this reaction or provided further evidence for the uniqueness of specificity of this enzyme.

Subsequent studies on the rat brain enzyme indicated that the "synaptosomal" and "light myelin" fractions had the highest specific activity with added lactosylceramide.[186] Developmental studies indicated a gradual increase in this enzyme activity with microsomes and osmotically shocked synaptosomes from 8 to 40 days of age.[187] The activity in intact synaptosomes, in contrast, gradually decreased during this period. A marked inhibition occurred when a concentrated $100,000g$ supernatant preparation was added to mitochondria (71%), microsomes (59%), or disrupted synaptosomes (74%). The addition of this inhibitory material increases the apparent K_m for lactosylceramide under the standard assay conditions. Preliminary studies indicate that the suppressing activity is not readily destroyed by heat, trypsin digestion, or organic solvents. It is nondialyzable and it has an apparent molecular weight of 70,000–80,000 by Sephadex G-100 column chromatography. Normal baby hamster kidney (BHK 21-C13) fibroblasts also have the capacity to catalyze this reaction. The enzyme activity is extremely low but is stimulated four- to sixfold by the addition of either cardiolipin or phosphatidylglycerol to the incubation mixtures.[188] Oligodendroglia possess high sialosyltransferase activity for GM_3 biosynthesis.[182e]

$$LacCer + CMP\text{-}NeuAc \rightarrow NeuAcLacCer\ (GM_3) + CMP$$

A variety of nonspecific neuraminidases have the capacity to cleave the NeuAc from several sialylated materials. One of the earliest indications of such enzyme activity in mammalian tissues employed neuraminlactose as substrate. The highest value was obtained with homogenates of lactating mammary gland followed by liver and brain.[189]

Hydrolysis in Vitro. Attempts to liberate neuraminic acid directly from calf brain particles were unsatisfactory; however, some purification was achieved starting with an acetone powder of this tissue. The starting material was subjected to two negative sodium cholate extractions, which still left most of the neuraminidase particle-bound. Triton X-100 extraction resulted in the liberation of approximately 35% of the activity into the 100,000g supernatant with a sevenfold purification.[190] This preparation also possessed β-glucosidase activity but was devoid of β-galactosidase and β-N-acetylhexosaminidase activities. Triton X-100 at 2–5 mg/ml was required for maximal activity even though it was present in the enzyme source; a pH optimum of 4.4 was observed with either acetate or citrate buffer (Table IX). This preparation catalyzed the release of neuraminic acid from mixed gangliosides, GT_1, GD_{1a}, and GM_3 almost equally well, while GD_{1b} was approximately 50% as effective as substrate. The following compounds were not substrates: GM_1, GM_2, sialosyllactose, and "glycoprotein." This enzyme is different from the earlier one[189] and indicated that except for GM_3, the sialic acid bound to the inner galactose of ganglioside is neuraminidase resistant. This presumably is due to this galactose being disubstituted with both GalNAc and NeuAc. Chemical degradation of GM_1 and subsequent treatment with neuraminidase has indicated that the NeuAc is bound by ketosidic linkage and is susceptible to hydrolysis after the GalNAc is removed.[191] A particulate fraction from rat liver sedi-

Table IX. Hydrolysis of Gangliosides by Calf Brain Neuraminidase[a,b]

Gangliosides	Sialic acid released (nmoles/mg/hr)
Mixed brain gangliosides	65
GT_{1b}	77
GD_{1a}	71
GD_{1b}	33
GM_{1a}	0
GM_2	0
GM_3	50
Sialosyllactose	0
Glycoprotein[c]	0

[a] Modified from Ref. 190.
[b] Incubation mixtures, in volumes of 0.2 mg, each contained 30 μmoles of acetate buffer (pH 4.4), 1 mg of Triton X-100, 0.2 mg of ganglioside, and 0.125 mg of enzyme. The tubes were incubated for 2 hr at 37°C and the sialic acid released was determined.
[c] "Sialic acid concentrate," Nutritional Biochemical Co.

menting at 900–30,000g was extracted with butanol to yield a "soluble" enzyme preparation.[192] GM$_3$ was subjected to partial tritium hydrogenation resulting in the introduction of label into the sphingosine moiety and was employed as substrate. The reaction was monitored for the formation of lactosylceramide by thin-layer chromatography. Mucin, mixed brain gangliosides, and sialo-syllactose were also substrates for these preparations.

An enzyme has been purified employing disialogangliosides as substrate from the 15,000g supernatants of pig brain homogenates some 600-fold through a series of ammonium sulfate and column chromatographic procedures.[193] Comparing the enzyme activity with different substrates, it was found to be most active with GQ$_1$, GD$_{1a}$, GT$_1$, 30–40% as active with GM$_3$, GD$_{1b}$, sialo-syllactose, and mucin, and inactive with GM$_2$ and GM$_1$. Therefore, it is somewhat similar to the calf brain particulate fraction. Similar studies have been carried out on the total particulate fraction from calf brain. The pH optima of neuraminidase varied for different substrates.[193a] Detergent effects upon the kinetics of hydrolysis of different gangliosides were studied.[193b] Species differences between the relative proportions of "soluble" and "particulate" neuraminidase for GD$_{1a}$ have been reported.[194] The removal of the NeuAc residue from hematoside has been reported to be more favored than that of the NeuGl residue from hematoside using a preparation from human brain.[195]

$$\text{NeuAcGalGlcCer} \rightarrow \text{GalClcCer} + \text{NeuAc}$$

2.7.2.6. *N-Acetylgalactosaminyl[sialosyl]galactosylglucosylceramide (GM$_2$, Tay–Sachs' Ganglioside)*

The information available regarding the *in vitro* biosynthesis of this material is not extensive. Three reports have appeared concerning the GM$_3$-stimulated incorporation of [^{14}C]-GalNAc from its UDP derivative into lipid by an embryonic chick brain particle system.[176,185,196] Of several sphingolipids employed, GM$_3$ appeared most effective, although some stimulation occurred with GM$_2$, the presumed product of the reaction, as well as with lactosylceramide. The nature of the sphingolipid synthesized or characteristics of the enzyme have not been well documented. The presence of this reaction was more satisfactorily investigated with rat brain preparations.[197] The total activity present in mitochondria and microsomal fractions was nearly equal, although the latter possessed a somewhat higher specific activity. Some of the general properties reported were: an Mn^{2+} requirement, a K_m for UDP[^{14}C]-GalNAc of 57 μM, and a K_m for GM$_3$ of 16 μM. Fractionation of the crude mitochondrial fraction indicated that although incorporation of tracer by endogenous acceptors occurred in nearly every subfraction, the major GM$_3$ stimulation was attained with the synaptic membrane-enriched fractions. The activity in the brains of 5-day-old animals appeared to be threefold higher than in 22-day-old rats, and peripheral organs in general possessed little enzyme. Studies comparing several acceptors indicated that GM$_3$ with NeuAc rather than with NeuGl was superior to the other gangliosides; slight activity was obtained with lactosylceramide. The product obtained, GM$_2$, was found to cochromatograph with authentic standards in several thin-layer chromatographic solvent systems. Mild acid hydrolysis under conditions to remove only NeuAc resulted in the production of a material that cochromatographed with GA_2.

$$NeuAcGalGlcCer (GM_3) + UDP\text{-}GalNAc \rightarrow$$

$$GalNAc[NeuAc]GalGlcCer (GM_2) + UDP$$

An early report concerning the *in vitro* formation of a ganglioside employed a rat kidney particulate system, which was fortified with UTP, UDP-glucose, UDP-galactose, and UDP-*N*-acetylglucosamine.[198] CMP-[14C]-NeuAc incorporation into ganglioside was measured. The ability of two asialosphingolipids to stimulate this incorporation was documented. Aminoglycolipids (*N*-acetylgalactosaminylgalactosylglucosylceramide) resulted in a threefold stimulation, whereas asialoganglioside (galactosyl-*N*-acetylgalactosaminylgalactosylglucosylceramide) was less than 50% stimulatory. Unfortunately, the evidence for the structures of these acceptors was poorly documented in these investigations. The assay was based upon the appearance of radioactivity in the upper phase of a Folch distribution. Under these assay conditions, only a trace of radioactivity was in the lower phase. This procedure would tend to rule out GM_2 or GM_3 as products since they would be expected to remain principally in the lower phase. The product cochromatographed with GM_2 and not GM_3 upon thin-layer examination. Although the radioactivity was released with mild acid hydrolysis, it was resistant to neuraminidase, also suggestive of GM_1 but not GM_3 or a disialoganglioside. Carrier dilution experiments proved that this was indeed a ganglioside. This observation has been questioned in a review article.[199] It was claimed that although there is a stimulation by *N*-triglycosylceramide, the products are GM_3 and [NeuAc]Gal[NeuAc]GalGlcCer. Although the data are not presented, the author is well respected for his contributions and experience with these compounds and may be correct in this observation. However, due to chromatographic, solvent partitioning, and neuraminidase susceptibility, suggested products would be easily distinguished from GM_2 as earlier reported.

Hydrolysis in Vitro. Neuraminidase activities capable of releasing the NeuAc residues from disialogangliosides and GM_3 were found ineffective with GM_1 and GM_2. The free sialic acid released was quantitated colorimetrically. It is possible that this analytical procedure lacked sufficient sensitivity for the small quantities of NeuAc that might have been released from these two monosialogangliosides. Radioactive gangliosides were prepared specifically labeled in the neuraminic acid residue. *N*-Acetyl[3H]mannosamine was injected intracerebrally into young rats and GM_2-[3H]-NeuAc was prepared through reasonably conventional techniques.[200] Employing this material as substrate, it was found that a variety of tissues possessed the capacity to catalyze the liberation of the radioactive sialic acid.[201] The highest specific activity found in the intestinal tissues occurred in the 600–22,000g particulate fraction. The pH optimum was found to be 5.0, a K_m of 5×10^{-4} M was reported, and the enzyme was found to be inhibited by detergents. Analysis of the products indicated that both free [^{3}H]-NeuAc and asialo GM$_2$ were formed during the hydrolysis of this substrate. The calf brain particulate sialidase previously described,[190] which hydrolyzed the disialogangliosides, did not cleave NeuAc from GM$_2$.

$$GalNAc[[^3H]-NeuAc]GalGlcCer \rightarrow GalNAcGalGlcCer + [^3H]-NeuAc$$

The hydrolysis of GM$_2$ was found to occur with a commercial preparation of *Clostridium perfringens* neuraminidase.[202] A pH optimum of 5.8 with acetate buffer and a requirement for a taurocholate[202a] were observed. The sialic acid released was measured colorimetrically and the asialo GM$_2$ detected by thin-layer chromatography after a 10-hr incubation. It is difficult to explain why this hydrolysis had not been detected by earlier workers; however, it should become a very useful tool for specifically degrading gangliosides produced in a variety of *in vitro* and *in vivo* systems. This neuraminidase, in the presence of purified hexosaminidase, has been demonstrated to produce lactosylceramide from GM$_2$.[202b] A similar neuraminidase activity has been reported with a brain lysosome-enriched fraction employing [^{3}H]-NeuAc GM$_2$.[203] The role of hexosaminidases for the hydrolysis of GM$_2$ is still unsettled. These enzymes have been operationally separated into hexosaminidase A, which is relatively acidic and is retained by DEAE-cellulose, and hexosaminidase B, which is less acidic and not retained by DEAE-cellulose. These will be dealt with in greater detail in relationship to the GM$_2$ gangliosidoses (Chapter 3).

Hexosaminidases A and B have been purified some 4000- and 2000-fold, respectively. They have similar molecular weights of about 100,000 and have different amino acid compositions. The carboxy-terminal amino acid in hexosaminidase A is serine and in hexosaminidase B, aspartic acid or asparagine. These investigators were unable to detect the hydrolysis of GM$_2$ by the purified enzymes, although crude placental tissue homogenates possessed this activity.[203a,b] This may be due to the removal of a heat-stable activator that stimulates the hydrolysis of several glycosphingolipids including GM$_2$ by hexosaminidases.[203c] The activator has been purified to homogeneity and its amino acid composition, molecular weight, stability, and immunological and stimulatory properties established.[203d] These observations are not compatible with a claim that 7000-fold purified human placental hexosaminidases A and B both cleave the amino sugar from GM$_2$.[203e] It is difficult to explain the discrepancy between these two laboratories. The simplest explanation would be the presence of "activator" in the more highly purified samples. A report using preparations of isoelectric-focused extracts of human liver of unspecified purification indicates that only hexosaminidase A hydrolyzes GM$_2$.[203f] The "activator" role is discussed in greater detail in Chapter 3 with respect to Tay–Sachs' ganglioside.

Hexosaminidase A has been purified by immunoaffinity,[203g] by affinity column containing a glycopeptide derived from bovine nasal system,[203h] and by conventional techniques from brain tissue.[203i] Homogeneous hexosaminidases A and B were obtained from beef spleen.[203j] The activity of these preparations, in common with many others described in the literature, toward naturally occurring substrates such as GM_2 was not reported. Indeed, a hexosaminidase from human liver has been described with N',N'-diacylchitobiosidase activity but devoid of activity toward 4MU-GlcNAc.[203k] Differing hydrolytic and transglycolytic activities toward several substrates have been noted for splenic hexosaminidases A and B.[203l]

As previously indicated, purified hexosaminidase from calf brain possessed the ability to hydrolyze the terminal GalNAc from GM_2 and under these conditions, Tay–Sachs ganglioside was found to be a poor substrate. However, GM_2 was reported to inhibit GA_2 hydrolysis.[181] Purified hexosaminidase A from human liver has been demonstrated to catalyze the hydrolysis of GM_2.[204] The activity was reported to be 0.1 nmole/mg protein per min for GM_2, 138.5 μmoles p-NP-N-acetylglucosaminide/mg protein per min, and 1.4 μmoles GA_2/mg protein per min. Rat brain lysosomes[203] apparently have the ability to catalyze a similar reaction; however, this activity appears to be at a rate approximately one-half that of sialidase for GM_2. Therefore, it appears that GM_2 may be a substrate for both hexosaminidase and neuraminidase enzymatic hydrolysis.

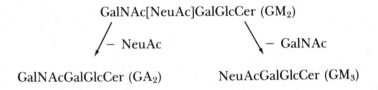

$$\text{GalNAc[NeuAc]GalGlcCer (GM}_2\text{)}$$

$$\diagup \, - \text{ NeuAc} \qquad\qquad \diagdown \, - \text{ GalNAc}$$

$$\text{GalNAcGalGlcCer (GA}_2\text{)} \qquad\qquad \text{NeuAcGalGlcCer (GM}_3\text{)}$$

The relative contribution of these two possible pathways for the catabolism of GM_2 in the intact animal is not established.

2.7.2.7. *Galactosyl-N-acetylgalactosaminyl[sialosyl]galactosylglucosylceramide (GM₁)*

The ability of brain particulate fractions to catalyze the GM_2-stimulated incorporation of galactose from UDP-[^{14}C]galactose has been well documented. Embryonic chick brain particles were shown to require a detergent mixture and $MnCl_2$.[205] This preparation also possessed galactosyltransferase activity with several glycoproteins as well as with N-acetylglucosamine. Competition experiments suggested that the activities for glycoproteins and gangliosides were catalyzed by different proteins. Heat inactivation studies also suggested that the transferase for glycoprotein and N-acetylglucosamine may be different from that for gangliosides. The hexosamine galactosyltransferase was found to possess the greatest activity and was nearly completely destroyed after 20 sec heating at 60°C, while the transferase for GM_2, which is only 40% as active, takes 40 sec to inactivate. The enzyme activity appears to be quite low in adult chickens. The product of the reaction, GM_1, was well characterized. In contrast to other animal species, the adult frog brain possesses appreciable GM_2 : galactosyltransferase activity.[206]

The capacity of rat brain particulates to catalyze this reaction increases in a linear manner from 5 days prior to birth and rapidly during the period of active myelination.[207] The K_m value observed for UDP-galactose was 12 μM and for GM_2, 94 μM. At higher concentrations of GM_2 of approximately 0.35 mM, which has been reported to be optimum for the chick brain system, inhibition occurs. UMPase has been reported to be rate limiting for this particular galactosyltransferase *in vitro*.[207a] The adult frog brain enzyme appears to be localized in the microsomal fraction.[208] The microsomal galactosyltransferase of 7-day-old rats has been solubilized by detergent treatment. Competition experiments suggest that both GM_2 and asialo GM_2 are acceptors for this enzyme.[178]

GalNAc[NeuAc]GalGlcCer (GM_2) + UDP-Gal →

GalGalNAc[NeuAc]GalGlcCer (GM_1) + UDP

A neuraminidase-labile ganglioside similar to GM_1 was formed in the presence of CMP-NeuAc and asialo GM_1.[208a] The structure of this product, termed GM_{1b}, has subsequently been determined as having the sialic acid at C-3 of the terminal galactose.[208b,c]

Hydrolysis in Vitro. A nonspecific β-galactosidase was purified from calf brain employing *p*-NP-β-D-galactoside as substrate. This preparation hydrolyzes the galactose of lactosylceramide and also hydrolyzes the terminal galactose of GM_1 as well as its asialo derivative.[174] The ability of rat and pig kidney particles to catalyze the stepwise degradation of tritiated asialoganglioside was monitored by thin-layer chromatographic analysis.[209] Evidence has appeared for the existence of two GM_1 β-galactosidases in human liver and brain tissues.[209a] Both forms were active toward artificial β-galactoside and GM_1. The "A" form of the liver enzyme has been purified to homogeneity. It hydrolyzed chromogenic and fluorogenic β-galactosides, lactose, *N*-acetyllactosamine, asialofetuin, β-D-fucoside, and α-L-arabinosides. Several other β-galactosides were resistant to hydrolysis including galactosylceramide and lactosylceramide.[209b] The activator that stimulates GM_2 hydrolysis also stimulates GM_1 hydrolysis by β-galactosidase.[203d] Lactosylceramidase II presumably hydrolyzes GM_1.[168a]

GalGalNAc[NeuAc]GalGlcCer (GM_1) → GalNAc[NeuAc]GalGlcCer (GM_2) + Gal

2.7.2.8. Disialogangliosides

A preliminary report appeared concerning the sialyltransferases that introduce a second NeuAc residue resulting in the formation of the polysialogangliosides.[210] The preparation employed may be identical to that catalyzing the formation of GM_3 from lactosylceramide and CMP-NeuAc.[185] If GM_1 is substituted as the acceptor, there is a transfer of [^{14}C]-NeuAc from CMP-NeuAc, which is dependent upon the presence of a detergent mixture. The general properties of the enzyme(s) are known; however, the presence of a protein for catalyzing this specific reaction is not extensively documented. The product formed was not well characterized but was claimed to be GD_{1a}, and it was suggested that the enzyme transferred the sialic acid to the terminal galactose moiety rather than the sialic acid.

GalGalNAc[NeuAc]GalGlcCer + CMP-NeuAc →

NeuAcGalGalNAc[NeuAc]GalGlcCer + CMP

These particles were also reported to catalyze an additional transfer of sialic acid from CMP-NeuAc involved with ganglioside formation. In the presence of GM_3 (the sample utilized in these studies contained NeuGl rather than NeuAc) and detergent, a disialoganglioside was produced. The addition of a histone caused a two- to threefold increase in this enzyme activity. Kinetic parameters relating strict proportionality with increasing protein concentration were difficult to obtain; however, the addition of boiled enzyme overcame this discrepancy. The material in the boiled preparation presumably responsible for this effect appeared to be lipoidal in character. Various lipids were tested and it was found that only phosphatidylethanolamine was capable of correcting the sigmoidal curve. Evidence was provided suggesting that the sialic acid added was linked to the sialic acid originally present in the acceptor GM_3.

NeuGlGalGlcCer (GM_3) + CMP-NeuAc → Gal[NeuAcNeuGl]GlcCer (GD_2) + CMP

A solubilized rat brain microsomal extract possesses an N-acetylgalactosaminyltransferase for gangliosides, which catalyzes the transfer of N-acetylgalactosamine from UDP-N-acetylgalactosamine to GM_3 resulting in GM_1 formation as described earlier.[185] In addition, GD_3 (sialosylsialosylgalactosylglucosylceramide) acts as an effective acceptor molecule with a K_m identical to that observed when GM_3 is employed as substrate. Attempts to differentiate these two transferases were equivocal; therefore, it is not possible to decide whether there are discrete enzyme proteins for each substrate.[211]

NeuNeuGalGlcCer + UDP-GalNAc → GalNAc[NeuNeu]GalGlcCer (GD_2) + UDP

A rat brain particulate fraction sedimenting at 900–20,000g was used for investigating the galactosyltransferases with several ganglioside acceptors. It was found that the presence of either GM_2 or GD_2 was effective in stimulating the incorporation of [^{14}C]galactose from UDP-[^{14}C]galactose.[212] The product with GM_2 as shown before was GM_1.[205] In an analogous manner, the product with GD_2 was shown to be GD_{1b}. Similarities of K_m's and the results of competitive studies suggest that the same galactosyltransferase may be responsible for catalyzing the synthesis of both GM_1 and GD_{1b}.

GalNAc[NeuNeu]GalGlcCer + UDP-Gal →

GalGalNAc[NeuNeu]GalGlcCer (GD_{1b}) + UDP

The coordinated activity of several membrane-embedded glycosyltransferases responsible for ganglioside formation have been discussed at some length. The discrepancies between *in vivo* and *in vitro* studies are also considered.[211a–c]

Hydrolysis in Vitro. As indicated in a previous section concerning the removal of NeuAc from GM_3, a variety of crude as well as highly purified neuraminidases have the capacity to remove the sialic acid residues from polysialogangliosides.[189–195] The product under these conditions, with mixed brain gangliosides as substrate, is GM_1. Earliest studies employed preparations from *C. perfringens*.[213] Neuraminidase is widespread in mammalian tissues, and depending upon species and treatment, varying amounts of "soluble" and "particulate" activity are detected.[194]

Studies to establish distinguishing characteristics of these two forms of neuraminidases have been carried out. The particulate enzyme activity is found associated with the "lysosomes"[214] in peripheral organs. Brain tissue is characterized by an enrichment of neuraminidase in gray matter vs. white matter and the activity appears concentrated in the synaptosomal or nerve-ending fractions.[215,216]

Rat heart muscle has been employed as the starting material for a neuraminidase that was purified 3500-fold. The release of radioactive sialic acid from suitably labeled GD_{1a} was the basic assay procedure used[216a]:

$$[NeuAc*]GalGalNAc[NeuAc]GalGlcCer \rightarrow$$

$$GalGalNAc[NeuAc]GalGlcCer + NeuAc*$$

The suggestion has been made that a "multiglycosyltransferase" system exists that operates in a highly integrated manner and is responsible for ganglioside biosynthesis.[217] Experimental support for such a concept has been offered from results of studies on products obtained with endogenous acceptors.[186,218] Similarly, a highly organized system may be responsible for the degradation of the gangliosides.[219]

2.8. Sphingomyelin

2.8.1. Biosynthesis in Vitro

This sphingolipid is ubiquitous in animal tissue and has been reported to occur in virtually every cell and membrane examined. The mode of biosynthesis of this compound at the *in vitro* level is somewhat problematical. The major publication reporting the formation of this compound employed a preparation of chicken liver microsomes and quantitated the incorporation of radioactivity from CDP-[1,2-^{14}C]choline onto an alkali-stable lipid as the measure of activity.[220] In the presence of particles, buffer, $MnCl_2$, Tween 20, and donor, negligible synthesis was observed (Table X). The addition of

Table X. Effect of Lipid Acceptors on Incorporation of Phosphorylcholine into
Alkali-Stable Lipid[a,b]

No.	Addition	Alkali-stable lipid synthesized (nmoles)
1	None	2
2	Sphingosine	1
3	Ceramide (obtained by cleavage of chicken liver sphingomyelin with lecithinase D)	1
4	Cerebroside-ceramide (obtained by hydrolysis of phrenosine with acetic–sulfuric acid by the method of Klenk)	66
5	Sphingosine (treated with acetic–sulfuric acid in the same manner as phrenosine)	120
6	Sphingosine (treated with propionic–sulfuric acid in the same manner as phrenosine)	540

[a] From Ref. 220.
[b] Each tube contained 0.25 ml of chicken liver particles; 20 μmoles of cysteine; 50 μmoles of Tris buffer, pH 7.4; 1 mg of Tween 20; 4 μmoles of lipid acceptor as shown; 10 μmoles of $MnCl_2$; and 0.8 μmole of CDP-[1,2-^{14}C]choline (16,000 cpm/μmole) in a total volume of 1 ml. The tubes were incubated at 37°C for 2 hr.

sphingosine or ceramide, which was obtained from naturally occurring sphingomyelin by the action of phospholipase D degradation, was ineffective in stimulating the system. Ceramide obtained from galactosylceramide as a result of acetic–sulfuric acid hydrolysis did support sphingomyelin formation. Similarly, sphingosine was active when treated with either an acetic–sulfuric acid mixture or a propionic–sulfuric acid mixture. N-Acetylsphingosine further purified through the crystalline triacetyl derivative was ineffective, which suggested that racemization may have occurred with the "active" material. Authentic samples of N-acetyl-DL-*threo*-sphingosine rather than the naturally occurring *erythro* derivative were found to be the active acceptor species (Table XI). Sphingosine bases were isolated from naturally occurring sphingomyelin, under hydrolytic conditions that would not result in isomerization, and were converted to an N-acetyl derivative. This product did not support sphingomyelin biosynthesis, unless it was treated with acetic–sulfuric acid. These observations agreed with the inactivity of the naturally occurring sphingosine base isomer, and the possibility of inversion of configuration after the enzymatic condensation occurred was also ruled out. The sphingomyelin synthesized with N-acetyl-DL-*threo*-sphingosine as acceptor was isolated and subjected to enzymatic hydrolysis with phospholipase D. The ceramide produced was found to be an active acceptor of phosphorylcholine, implying that it still retained the *threo* configuration. Mg^{2+} was required and an optimum pH of 7.5 was observed. A variety of *threo*-ceramides were synthesized with varying fatty acid composition. In the absence of detergent, the N-acetyl derivative was the most effective; however, in the presence of 5 mg/ml Tween 20, an eight-carbon fatty acid was most satisfactory. Presumably, longer-chain fatty acids were found to be inactive due to their solubility properties. Thus, N-

Table XI. Configuration of Active Ceramide Required
for Enzymatic Synthesis of Sphingomyelin[a,b]

Sample	Sphingomyelin synthesized (nmoles)
Experiment 1	
N-Acetyl-DL-*erythro-trans*-sphingosine	4
N-Acetyl-DL-*threo-trans*-sphingosine	105
Experiment 2	
N-Acetyl-DL-*erythro-trans*-sphingosine	5
N-Acetyl-DL-*threo-trans*-sphingosine	83
N-Acetyl-DL-*erythro-cis*-sphingosine	1
N-Acetyl-DL-*threo-cis*-sphingosine	1

[a] Modified from Ref. 220.
[b] Each tube contained 50 μmoles of Tris buffer (pH 7.4), 4 μmoles of MnCl$_2$, 20 μmoles of cysteine, 5 mg of Tween 20, 0.8 μmole of CDP-[1,2-^{14}C]choline, 4 μmoles of N-acetyl-DL-sphingosine isomer, and 0.25 ml of chicken liver particles in a final volume of 1 ml. The tubes were incubated for 2 hr at 37°C.

palmitoyl-*threo*-sphingosine was unreactive, while the N-oleoyl or N-linolyl derivative was reported to exhibit appreciable activity. A K_m of 7.5×10^{-4} M was observed for N-octonyl-*threo*-sphingosine. The reaction catalyzed

$$threo\text{-ceramide} + \text{CDP-choline} \rightarrow threo\text{-sphingomyelin} + \text{CDP}$$

although well documented, results in the formation of a sphingolipid with an unnatural base configuration. The possibility exists that although N-acetyl-DL-*erythro*-sphingosine was inactive, an N-oleoyl derivative might be effective. Short-chain N-acyl derivatives of sphingosine are rarely encountered in biological materials.

This observation was further documented in a study comparing the activity of both mitochondrial and microsomal chicken liver particles with several acceptors.[221] Naturally occurring ceramide and N-acetyl-*erythro*-sphingosine were somewhat more effective than N-acetyl-*threo*-sphingosine, employing mitochondria. Unfortunately, microsomal activity in the absence of an acceptor was not reported; however, it appeared that both N-acetyl-*erythro*- and N-acetyl-*threo*-sphingosine were equally active. This discrepancy with the earlier work indicating stimulation only with the *threo* derivative is difficult to rationalize or explain. The data presented could be interpreted as indicating a greater activity in mitochondria than in microsomes; however, since protein values were not provided, such comparisons are of little value.

Deoxycholate-treated mitochondria from chicken livers have the capacity to carry out the N-acetyl-*threo*-sphingosine-stimulated incorporation of choline into sphingomyelin. Under these conditions, N-acetyl-*erythro*-sphingosine is less than 10% as effective. The addition of intact microsomes inhibited the synthesis of sphingomyelin. There is no evidence that this is the result of enzymatic utilization of substrates in contrast to the presence of an inhibitor.[222a] The sulfhydryl reagent menadione, a vitamin K analog, has the property of causing the *erythro*-ceramide to become equally effective as the *threo* isomer. This reagent has no effect on the ability of N-acetyl-*threo*-sphingosine to act as acceptor.[222] Less stimulatory in this regard were N-ethylmaleimide, CoQ0, *o*-quinone, and flavin nucleotides, while dithiothreitol and cysteine were inactive. This effect of menadione does not occur with rat liver particles, in which oxidized CoA was claimed to be an activator. The unchanged ceramides, and the ceramides obtained after phospholipase treatment of the N-acylsphingosine phosphorylcholine synthesized, were isolated from the incubation mixtures. Analysis indicated that there was no inversion of the configuration. It has recently been observed that in the presence of ATP and CoA or several pantathine derivatives, N-acetyl-*erythro*-sphingosine can serve as an acceptor of phosphorylcholine from CDP choline in the presence of rat liver microsomes.[222b]

An alternative pathway for the formation of sphingomyelin has been proposed based upon the N-acylation of sphingosine phosphorylcholine. The substrate acceptor was obtained by alkaline treatment of sphingomyelin. The assay procedure depended upon the incorporation of radioactivity from [1-^{14}C]stearoyl CoA into a column chromatographic fraction containing principally sphingomyelin. Particles sedimenting both at 600–8400g and at 8400–100,000g possessed this activity, but only the activity in the former was rendered soluble with cholate extraction. Free stearic acid could not substitute for the thioester, and ATP was not required.[223] The preferential biosynthesis from *threo*- rather than *erythro*-sphingosine of sphingosine phosphorylcholine by chicken liver particles has been reported in a brief communication.[224] *Threo*- and *erythro*-sphingosine phosphorylcholine were found to be nearly equally effective acceptors, and palmitoyl CoA was the best donor with a mouse brain "mitochondrial fraction."[225]

The ability of several mammalian tissues to catalyze the *in vitro* incorporation of CDP-ethanolamine into ceramide phosphorylethanolamine has been reported. *Threo*- rather than *erythro*-ceramides, even in the presence of menadione, are the preferred acceptors, the activity with the N-octoyl derivative being greater than with the N-acetyl derivative. A pH optimum of 7.6 with Tris buffers and an Mn^{2+} requirement were found. The synthesis of ceramide phosphorylethanolamine was four times more effective than sphingomyelin under these experimental conditions.[226] The role of this reaction, if any, in sphingomyelin biosynthesis is unknown, but it is tempting to speculate that methylation of the ethanolamine moiety may occur.

Rat liver particles possessing the capacity to methylate dimethylphosphatidylethanolamine to lecithin were prepared. It was found that in the presence of S-adenosyl[C^{3}H$_3$]methionine, and Mn^{2+}- an ceramide phosphorylethanolamine-stimulated synthesis of the corresponding monomethyl derivative occurred. Under these conditions, ceramide phosphoryldimethylethanolamine conversion to sphingomyelin could not be detected.[227]

The results of these *in vitro* studies suggest:

ceramide + CDP-choline ⟶ ceramide phosphorylcholine

acyl CoA

sphingosine + CDP-choline ⟶ sphingosine phosphorylcholine   ↑SAM

ceramide + CDP-ethanolamine ⟶ ceramide phosphorylethanolamine

The importance of sphingosine phosphorylcholine and its subsequent acylation to sphingomyelin has been challenged.[227a] Radioactive substrates labeled in the base of choline portions were utilized for both *in vitro* and *in vivo* experiments. Sphingomyelin was not observed as a product under any experimental conditions employed.

An additional pathway has been offered from labeling studies on mammalian cells grown in tissue culture. A comparison of $^3H/^{32}P$ ratios in both lecithin and sphingomyelin after addition of the [^3H]choline and $^{32}PO_4$ or dual-labeled lecithin[228] to the growth medium as a function of time was undertaken. The kinetics were interpreted as indicating a transfer of intact phosphorylcholine from lecithin to sphingomyelin without the involvement of a nucleotide intermediate.[228a,b] A similar possibility has been offered for results obtained with mouse liver particles.[228b] Stoffel and Melzner[228c] have recently been challenged in studies representing the inability to reproduce these observations. However, their incubations contained lysolecithin as one of the routine ingredients, which has been shown to effectively inhibit phosphorylcholine transfer. A plasma membrane fraction of cultured cells was 50-fold enriched in the ability to incorporate the phosphorylcholine from exogenous lecithin into sphingomyelin.[228d] The availability of endogenous ceramide appears to limit the rate of this reaction.[228e] Certain phospholipases C from bacterial sources have been shown to catalyze the formation of sphingomyelin from lecithin. Only those preparations known to hydrolyze the phosphorylcholine moiety of lecithin and sphingomyelin were effective.[228f]

A transfer of the phosphorylcholine moiety from phosphatidylcholine to produce sphingomyelin was demonstrated with BHK cells.[228g]

2.8.2. Hydrolysis in Vitro

The hydrolysis of the polar phosphorylcholine portion from sphingomyelin has been reported to occur with preparations from a variety of biological sources. A 65-fold purification of a rat liver particulate enzyme catalyzing the release of phosphoryl[$^{14}CH_3$]choline from synthetic [$^{14}CH_3$]-sphingomyelin has been obtained. A pH optimum of 5.5 with acetate buffer and a K_m of 1.8×10^{-4} M were observed. Although lecithin was a competitive inhibitor with a K_i of 2×10^{-4} M, it was not a substrate.[229] The rat liver enzyme appears to be localized in the lysosomes.[230] The enzymes of both brain[231] and liver[232] utilize the naturally occurring *erythro* isomers preferentially to the corresponding *threo* isomer. The presence of this enzyme activity in rat intestine[233] has been reported. Examination of the physical state of the substrate has been undertaken.[232a–c] A 200-fold purification of an enzyme specific for sphingomyelin was obtained from the growth medium of *C. perfringens*. A stimulation was observed with Mg^{2+} and inhibition by Ca^{2+} and EDTA.[234] An increased activity of this enzyme in cerebral tissue during myelination has been observed.[235] Studies on the *in vivo* metabolism of dihydro[^{3}H]sphingomyelin have indicated the active conversion to dihydroceramide.[236]

A non-lysosomal neutral sphingomyelinase which requires Mg^{2+} or Mn^{2+} is enriched in rat liver plasma membranes.[236a]

$$\text{Sphingomyelin} \rightarrow \text{ceramide} + \text{phosphorylcholine}$$

The relationship to the dopamine system of a neutral sphingomyelinase of brain tissue has been contrasted to the acidic form of the enzyme.[236b,c] Lysosomal acid sphingomyelinase has been purified over 1000-fold but still possessed detectable mannosidase and galactosylceramide: β-galactosidase activities.[236d,e] The hydrophobic properties of highly purified placental[236f] and liver lysosomal[236g] sphingomyelinase have been described. Acid sphingomyelinase of human brain was purified to homogeneity and was active with several chromogenic and flurogenic analogues.[236h] A highly purified placental sphingomyelinase displayed complex kinetics with such "artificial" substrates.[236i] Sphingomyelinase activity of lysed human fibroblasts was inhibited by the addition of cholesterol.[236j]

2.9. Biosynthesis of Sphingolipids: Summary

A major problem concerning the proposed pathway of sphingolipid biosynthesis is that most of the individual reactions have been documented in relatively crude tissue extracts. Extensive purification of the specific enzymes proteins has not been accomplished. There are potentially five different galactosyl transferases involved in the formation of specific β-galactosidic bonds, as seen in Fig. 9. The general assumption has been that a distinct enzyme protein exists which is responsible for catalyzing the transfer of galactose residues to a specific acceptor molecule. The assumptions are based largely on the following observations: (a) the linkages formed are different, that is either 1→1, or 1→3 or 1→4, (b) the components to which the galactose becomes linked are different, such as sphingosine, glucose, or *N*-acetylgalactosamine, (c) competition experiments. Somewhat similar arguments have been employed for suggesting that there are at least 3 distinct sialyltransferases, two with a galactose moiety, and one with a sialic acid moiety as the acceptor.

The proposal for the existence of a "multiglycosyltransferase" complex which operates as an integrated unit for the production of sphingolipids is

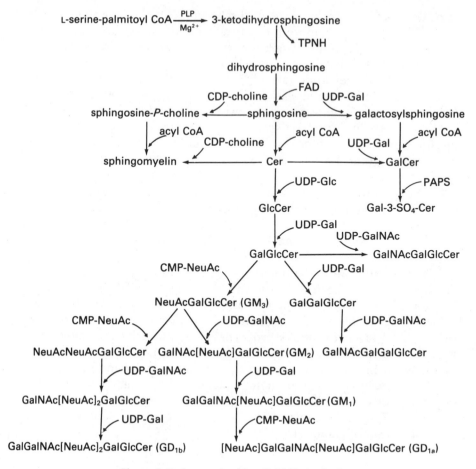

Figure 9. Pathways of sphingolipid biosynthesis.

reasonable. Conceptually it suggests that once the L-serine and palmitoyl CoA have condensed to produce 3-keto-dihydrosphingosine, the remaining reactions have been programmed into this architectural unit. The pathway presented in Fig. 9 has been derived from experiments in which the conversion of an appropriate radioactive substrate (usually a water soluble nucleotide derivative) to the anticipated product has been demonstrated. The ability of certain acceptor molecules to stimulate the endogenous activity is taken as an indication for the specificity for a given reaction. Two obvious problems arise with this approach. The first of these is the necessity of interfering with the inherent machinery of the particle to manufacture the sphingolipids efficiently. This is most markedly illustrated by the finding that N-acetyl-*threo*-sphingosine is the most effective exogenous acceptor to form N-acetyl-*threo*-sphingosine phosphorylcholine (sphingomyelin). Evidence for the natural occurrence of this particular isomer of sphingomyelin is nonexistent, since it characteristically possesses the *erythro* configuration for the base and long-chain fatty acids. However, the most generally accepted pathway for the *in vitro* formation of this sphingophospholipid is based upon the documented stimulation by this unnatural ceramide. In attempts to demonstrate the capacity of possible substrates to enhance the suspected enzyme activity, a variety of noxious and "unphysiological" materials are often added to reaction mixtures. Classically, these have been added in an attempt to solubilize a water-insoluble compound in order to render it suitable as a substrate for a given enzyme. It is extremely rare to find that detergents are omitted from *in vitro* incubation mixtures. The detergent specificity for these activations is difficult to understand and can be established only empirically. There is little rationale for deciding which material may be found most effective except for trial and error. Generally, one of the common commercially available ionic or nonionic detergents is effective. Mixtures of detergents are occasionally employed for maximal activations. The optimal concentration for a specific system varies, but amounts greater than 5 mg/ml are usually inhibitory. In addition to detergents, the stimulation by lipids coated on Celite, a diatomaceous earth, has been well documented, and amounts up to 50 mg/ml of this material have been employed. The peculiar aspect of this type of activation is that both the enzyme system and the lipid acceptor are bound to different particles. Thus, the transferase is usually present in microsomal particles and the lipid coated on Celite. The mechanism of proper interaction of these suspensions is unknown. The second problem is the minimum effort expended in investigating the reaction between the radioactive nucleotide sugar donor and the endogenous acceptor. If the concept of a "multienzyme complex" is correct, then this should be the most "physiological" state of the participants for studying in a particular reaction. The nature of these endogenous acceptors or the product formed has rarely been documented.

In a pathway involving several alternative directions for a given compound to travel, there are generally certain reactions termed "committed reactions." This concept implies that, depending upon the nature of the initial substrate modification, only certain very specific subsequent courses are then available. A typical illustration is the alternatives available for lactosylceramide as a substrate. If this compound reacts with the proper galactosyltransferase and UDP-galactose resulting in the formation of the α-galactosidic bond found in trihexosylceramide, then this lactosylceramide moiety can be incorporated *only* into the branch giving rise to "globosides." On the other hand, if the lactosylceramide becomes a substrate for a sialyltransferase, it can *only* become a portion of a ganglioside molecule. Thus, as a result of a specific transferase reaction, the lactosylceramide has been specifically committed to becoming *only* a portion of *one* or the other more complex molecules. In a similar manner, there must be "committed reactions" concerned with the primary hydroxy group of sphingosine, which result in a specific course that this amino alcohol will be destined to follow. If galactose is attached, only galactosylceramide, sulfatide, and digalactosylceramide will be formed. Sphingomyelin will be the *sole* product if phosphorylcholine is attached. However, if glucosylation occurs, then either a globoside or a ganglioside is the final product. The characteristic presence of hydroxy fatty acids in galactosylceramide and their absence in glucosylceramide and sphingomyelin suggest that the nature of the fatty acids in the ceramide acceptor may specify the carbohydrate moiety linked to the primary hydroxy group. However, this does not explain the presence of non-hydroxy fatty acid galactosylceramides. An appealing possibility is that a reaction between UDP-glucose and ceramide results in glucosylceramide formation, while galactosylceramide is produced via a galactosylsphingosine intermediate.

2.10. Hydrolysis of the Sphingolipids: Summary

It has generally been assumed that the specific enzymes catalyzing the hydrolysis of the individual bonds in the sphingolipids reside in the lysosomal subcellular fraction. This has been firmly established for several of these hydrolases and has been presumed to be true for others. Several sphingolipid hydrolases can be extracted employing similar techniques from a common crude brain particulate fraction. The existence of a common architectural unit possessing the necessary enzyme complement to degrade completely the most complex sphingolipid, GD_{1a}, to its individual components has been postulated.

The specificity for the substrate being tested is often open to question because only a few of the enzymes have been purified against a specific naturally occurring substrate. These enzymes have principally been the ceramidases, cerebrosidases, and sphingomyelinases for which radioactive substrates have been chemically prepared and their hydrolysis conveniently monitored. For several of these hydrolases, nonsphingolipid substrates have been employed. Enzyme preparations purified toward neuraminlactose can efficiently remove the sialic acid residues from the di- and trisialogangliosides. A very general β-galactosidase that catalyzes the cleavage of the terminal galactosidic bond of GM_1, GA_1, and lactosylceramide was purified with p-NP-β-D-galactoside as substrate. However, data from studies with tissue extracts obtained from GM_1 gangliosidosis patients suggest that different enzyme proteins specific for each of these β-galactosidic bonds must also be present in nature. In a similar manner, the N-acetylglucosaminidase and hexosaminidase A and B preparations that hydrolyze the terminal N-acetylgalactosamine bond of GA_2 were purified against artificial substrates. The question of enzyme specificity is somewhat unusual for the lysosomal acid hydrolases and the most satisfactory description is that they are often specifically nonspecific. A hydrolytic enzyme may be highly purified against an unnatural or "artificial" substrate; however, it often has the capacity to cleave a similar type of a bond in a naturally occurring biological macromolecule. These enzyme preparations are generally found more effective with the synthetic substrate than the naturally occurring compound. The possibility of "helper" or "specifier" proteins that invoke a degree of substrate specificity upon these general acid hydrolases has been suggested, but unequivocal evidence for their physiological importance is lacking. From the information available, it theoretically should be possible to mix a compound such as GD_{1a} with a purified lysosomal preparation and isolate the individual carbohydrates, the fatty acids, as well as the sphingosine bases from the reaction mixtures at the termination of a suitable incubation.

Two gray areas exist in the pathway presented in Fig. 10. The first centers on the "physiologically" more relevant removal of *N*-acetylgalactosamine in contrast to the sialic acid residue from GM$_2$. Both reactions have been reported to occur with lysosomal preparations, and this sequence is of some concern

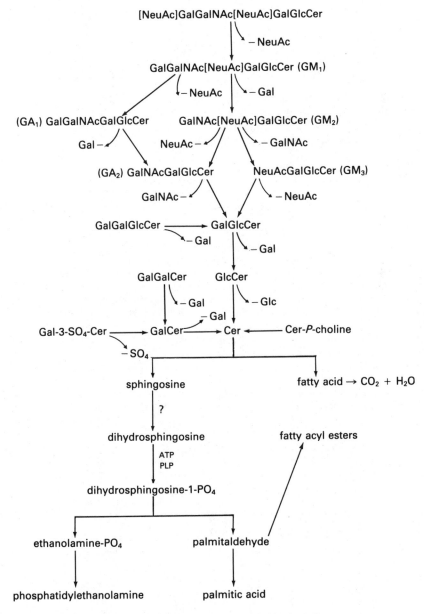

Figure 10. Pathways of sphingolipid hydrolysis.

in understanding the biochemical problems of Tay–Sachs' disease. The second area concerns the conversion of sphingosine produced by ceramidase action into dihydrosphingosine, which is presumed to be the actual substrate undergoing subsequent degradation. The gangliosides are the subject of two symposium volumes.[237,238]

2.11. References

1. Zabin, I., and Mead, J. F. *J. Biol. Chem.* **205,** 271 (1953).
2. Zabin, I., and Mead, J. F. *J. Biol. Chem.* **211,** 87 (1954).
3. Sprinson, D. B., and Coulon, A. *J. Biol. Chem.* **207,** 585 (1954).
4. Burton, R. M., In: *Lipids and Lipidoses* (Schettler, G., ed.), Springer-Verlag, New York (1967), p. 122.
5. Zabin, I. *J. Am. Chem. Soc.* **79,** 5834 (1957).
6. Brady, R. O., and Koval, G. *J. Biol. Chem.* **233,** 26 (1958).
7. Brady, R. O., Formica, J. V., and Koval, G. J. *J. Biol. Chem.* **233,** 1072 (1958).
8. Kanfer, J. N. *Chem. Phys. Lipids* **5,** 159 (1970).
9. Fujino, Y., and Zabin, I. *J. Biol. Chem.* **237,** 2069 (1962).
10. Weiss, B. *J. Biol. Chem.* **238,** 1953 (1963).
11. Gaver, R. C., and Sweeley, C. C. *J. Am. Chem. Soc.* **88,** 3643 (1966).
12. Menderhausen, P. B., and Sweeley, C. C. *Biochemistry* **8,** 2633 (1969).
13. Stoldola, F. H., Wickerham, L. J., Scholfield, C. R., and Dutton, H. J. *Arch. Biochem. Biophys.* **98,** 176 (1962).
14. Greene, M., Kaneshiro, T., and Law, J. H. *Biochim. Biophys. Acta* **98,** 582 (1965).
15. Braun, P. E., and Snell, E. E. *Proc. Natl. Acad. Sci. U.S.A.* **58,** 298 (1967).
16. Braun, P. E., and Snell, E. E. *J. Biol. Chem.* **243,** 3775 (1968).
17. Stoffel, W., LeKim, D., and Sticht, G. *Hoppe-Seyler's Z. Physiol. Chem.* **349,** 664 (1968).
18. Brady, R. N., DiMari, S. J., and Snell, E. E. *J. Biol. Chem.* **244,** 491 (1969).
19. DiMari, S. J., Brady, R. N., and Snell, E. E. *Arch. Biochem. Biophys.* **143,** 553 (1971).
20. Braun, P. E., Morell, P., and Radin, N. S. *J. Biol. Chem.* **245,** 335 (1970).
21. Kanfer, J. N., and Bates, S. *Lipids* **5,** 718 (1970).
22. Kurtz, D. J., and Kanfer, J. N. *J. Neurochem.* **20,** 963 (1973).
22a. Krisnanghura, K., and Sweeley, C. C. *J. Biol. Chem.* **251,** 1597 (1976).
23. Stoffel, W., LeKim, D., and Sticht, G. *Hoppe-Seyler's Z. Physiol. Chem.* **349,** 1637 (1968).
24. Stoffel, W. *Chem. Phys. Lipids* **5,** 139 (1970).
25. Carter, H., and Shapiro, D. *J. Am. Chem. Soc.* **75,** 5131 (1953).
25a. Shoyama, Y., and Kishimoto, Y. *Biochem. Biophys. Res. Commun.* **70,** 1035 (1976).
26. Stoffel, W., Sticht, G., and LeKim, D. *Hoppe-Seyler's Z. Physiol. Chem.* **349,** 1745 (1968).
27. Gatt, S., and Barenholz, Y. *Biochem. Biophys. Res. Commun.* **32,** 588 (1968).
28. Keenan, R. W., and Maxam, A. *Biochim. Biophys. Acta* **17,** 348 (1969).
29. Keenan, R. W., and Haegelin, B. *Biochem. Biophys. Res. Commun.* **37,** 888 (1969).
30. Hirschberg, C. B., Kisic, A., and Schroepfer, G. J. *J. Biol. Chem.* **245,** 3084 (1970).
31. Keenan, R. W. *Biochim. Biophys. Acta* **270,** 383 (1972).
32. Stoffel, W., Assman, G., and Binczek, E. *Hoppe-Seyler's Z. Physiol. Chem.* **351,** 635 (1970).
32a. Stoffel, W., Hellenbroich, B., and Heiman, G. *Hoppe-Seyler's Z. Physiol. Chem.* **354,** 1311 (1973).
32b. Louie, D. D., Kisic, A., and Schroepfer, G. J. *J. Biol. Chem.* **251,** 4557 (1976).
33. Stoffel, W., LeKim, D., and Sticht, G. *Hoppe-Seyler's Z. Physiol. Chem.* **350,** 1233 (1969).
34. Stoffel, W., and Assman, G. *Hoppe-Seyler's Z. Chem.* **351,** 1041 (1970).
35. Stoffel, W., and Assman, G. *Hoppe-Seyler's Z. Physiol. Chem.* **353,** 965 (1972).
35a. Akino, T., Shimojo, T., Miura, Y., and Schroepfer, G. J. *J. Am. Chem. Soc.* **96,** 939 (1974).
35b. Shimojo, T., Akino, T., Muria, Y., and Schroepfer, G. J. *J. Biol. Chem.* **251,** 4448 (1976).
36. Fujino, Y., and Nakano, M. *Agric. Biol. Chem.* **35,** 40 (1971).
37. Fujino, Y., and Nakano, M. *Agric. Biol. Chem.* **35,** 885 (1971).

38. Barkenhagen, L. F., Kennedy, E. P., and Fielding, L. *J. Biol. Chem.* **236**, 28 (1961).
39. Hübscher, G., Dils, R. R., and Pover, W. F. R. *Biochim. Biophys. Acta* **36**, 518 (1959).
39a. Ong, D. E., and Brady, R. N. *J. Biol. Chem.* **248**, 3884 (1973).
39b. Stoffel, W., and Bister, K. *Hoppe-Seyler's Z. Physiol. Chem.* **355**, 911 (1974).
40. Gal, A. E. *J. Labelled Compounds* **3**, 112 (1967).
41. Kanfer, J. N., and Gal, A. E. *Biochem. Biophys. Res. Commun.* **22**, 442 (1966).
42. Kanfer, J. N., and Gal, A. E. *J. Neurochem.* **14**, 1085 (1967).
43. Stoffel, W., and Sticht, G. *Hoppe-Seyler's Z. Physiol. Chem.* **348**, 941 (1967).
44. Barenholz, Y., and Gatt, S. *Biochem. Biophys. Res. Commun.* **27**, 319 (1967).
45. Stoffel, W., and Sticht, G. *Hoppe-Seyler's Z. Physiol. Chem.* **348**, 1561 (1967).
46. Stoffel, W., and Sticht, G. *Hoppe-Seyler's Z. Physiol. Chem.* **348**, 1345 (1967).
47. Stoffel, W., Sticht, G., and LeKim, D. *Hoppe-Seyler's Z. Physiol. Chem.* **350**, 63 (1968).
48. Stoffel, W., and Henning, R. *Hoppe-Seyler's Z. Physiol. Chem.* **349**, 1400 (1968).
49. Stoffel, W., Sticht, G., and LeKim, G. *Hoppe-Seyler's Z. Physiol. Chem.* **350**, 53 (1969).
50. Stoffel, W., LeKim, D., and Heyn, G. *Hoppe-Seyler's Z. Physiol. Chem.* **351**, 875 (1970).
50a. Stoffel, W., and Bister, K. *Hoppe-Seyler's Z. Physiol. Chem.* **354**, 169 (1973).
51. Burton, R. M., Sodd, M. A., and Brady, R. O. *J. Biol. Chem.* **233**, 1053 (1958).
52. Radin, N. S. In: *The Biology of Myelin* (Korey, S., ed.), Hoeber-Harper, New York (1959), p. 271.
53. Cleland, W. W., and Kennedy, E. P. *J. Biol. Chem.* **235**, 45 (1960).
54. Kanfer, J. N. *Chem. Phys. Lipids* **5**, 159 (1970).
55. Hildebrand, J., Stoffyn, P., and Houser, G. *J. Neurochem.* **17**, 403 (1970).
56. Miyatake, T., and Suzuki, K., *J. Biol. Chem.* **247**, 5398 (1972).
56a. Miyatake, T., and Suzuki, K. *J. Neurochem.* **22**, 231 (1974).
56b. Carter, T. P., and Kanfer, J. N. *Chem. Phys. Lipids* **13**, 340 (1974).
56c. Curtino, J. A., and Caputto, R. *Lipids* **7**, 525 (1972).
56d. Kanfer, J. N., Raghaven, S. S., and Mumford, R. A. *Biochem. Biophys. Res. Commun.* **391**, 129 (1975).
56e. Raghaven, S. S., Mumford, R. A., and Kanfer, J. N. *J. Lipid Res.* **15**, 484 (1974).
57. Brady, R. O. *J. Biol. Chem.* **237**, PC 2416 (1962).
58. Morell, P., Constantino-Ceccarini, E. C., and Radin, N. S. *Arch. Biochem. Biophys.* **141**, 738 (1970).
59. Fujino, Y., and Nakano, M. *Biochem. J.* **113**, 573 (1969).
60. Hammarström, S. *Biochem. Biophys. Res. Commun.* **45**, 459 (1971).
61. Hammarström, S. *FEBS Lett.* **21**, 259 (1972).
61a. Curtino, J. A., and Caputto, R. *Biochem. Biophys. Res. Commun.* **56**, 142 (1974).
62. Gatt, S. *Biochim. Biophys. Acta* **70**, 370 (1963).
63. Gatt, S. *J. Biol. Chem.* **238**, PC 3131 (1963).
63a. Stoffel, W., Krüger, E., and Melzner, I. *Hoppe-Seyler's Z. Physiol. Chem.* **361**, 77 (1980).
64. Sribney, M. *Biochim. Biophys. Acta* **125**, 542 (1966).
65. Gatt, S. *J. Biol. Chem.* **241**, 3724 (1966).
66. Dole, V. P. *J. Clin. Invest.* **35**, 350 (1956).
67. Yavin, E., and Gatt, S. *Biochemistry* **8**, 1692 (1969).
68. Zahler, W. L., Bardin, R. E., and Cleland, W. W. *Biochim. Biophys. Acta* **164**, 1 (1968).
68a. Sugita, M., Williams, M., Dulaney, J. T., and Moser, H. A. *Biochim. Biophys. Acta* **398**, 125 (1975).
69. Morell, P., and Radin, N. S. *J. Biol. Chem.* **245**, 342 (1970).
70. Radin, N. S. *Adv. Exp. Med. Biol.* **19**, 475 (1972).
71. Ullman, M. D., and Radin, N. S. *Arch. Biochem. Biophys.* **152**, 767 (1972).
71a. Singh, I., and Kishimoto, Y., *Biochem. Biophys. Res. Commun.* **82**, 1287 (1978).
71b. Singh, I., and Kishimoto, Y. *Arch. Biochem. Biophys.* **202**, 93 (1980).
71c. Kishimoto, Y., and Kawamura, N. *Mol. Cell. Biochem.* **23**, 17 (1979).
72. Radin, N. S., Martin, F. B., and Brown, J. R. *J. Biol. Chem.* **224**, 499 (1957).
73. Moser, H. M., and Karnovsky, M. L. *J. Biol. Chem.* **234**, 1990 (1959).
74. Burton, R. M., Garcia-Bunuel, L., Golden, M., and Balfour, Y. M. *Biochemistry* **2**, 580 (1963).
75. Hauser, G. *Biochim. Biophys. Acta* **84**, 212 (1964).

76. Mishimura, K., Ueta, N., and Yamakowa, T. *Jpn. J. Exp. Med.* **36,** 91 (1966).
77. Maker, H. S., and Hauser, G. *J. Neurochem.* **14,** 457 (1967).
78. Hajra, A. K., and Radin, N. S. *J. Lipid Res.* **3,** 327 (1962).
79. Hajra, A. K., and Radin, N. S. *J. Lipid Res.* **4,** 448 (1963).
80. Kishimoto, Y., Davies, W. E., and Radin, N. S. *J. Lipid Res.* **6,** 532 (1965).
81. Kishimoto, Y., Davies, W. E., and Radin, N. S. *J. Lipid Res.* **6,** 525 (1965).
82. Kanazawa, I., Ueta, N., and Yamakawa, T. *J. Neurochem.* **19,** 1483 (1972).
82a. Dhopeshwarkar, G. A., Subramanian, C., and Mead, J. F. *Biochim. Biophys. Acta* **248,** 41 (1971).
83. Kopaczyk, K. C., and Radin, N. S. *J. Lipid Res.* **6,** 140 (1965).
84. Morell, P., and Radin, N. S. *Biochemistry* **8,** 506 (1969).
85. Carter, H. E., Rathfus, J. A., and Gigg, R. *J. Lipid Res.* **2,** 228 (1961).
86. Fujino, Y., and Nakano, M. *Biochem. J.* **113,** 573 (1969).
87. Basu, S., Schultz, A. M., Basu, M., and Roseman, S. *J. Biol. Chem.* **246,** 4272 (1971).
88. Hammarström, S., and Samuelsson, B. *Biochem. Biophys. Res. Commun.* **41,** 1027 (1970).
89. Shah, N. *J. Neurochem.* **18,** 395 (1971).
90. Hammarström, S. *Biochem. Biophys. Res. Commun.* **45,** 468 (1971).
91. Behrens, N. H., Parodi, A. J., Lelair, L. F., and Krismon, C. R. *Arch. Biochem. Biophys.* **143,** 375 (1971).
92. Cabs, L., and Gray, G. M. *Biochim. Biophys. Acta* **38,** 520 (1970).
93. Gray, G. M. *Biochim Biophys. Acta* **239,** 494 (1971).
93a. Neskovic, N. M., Mandel, P., and Gatt, S. *Adv. Exp. Med. Biol.* **101,** 631 (1978).
93b. Constantino-Ceccarini, E., and Suzuki, K. *Arch. Biochem. Biophys.* **167,** 646 (1975).
93c. Constantino-Ceccarini, E., and Suzuki, K. *Brain Res.* **93,** 358 (1975).
93d. Chou, K. H., and Jungalwala, F. B. *Trans. Am. Soc. Neurochem.* **7,** 105 (1976).
93e. Constantino-Ceccarini, E., Cestelli, A., and De Vries, G. H. *J. Neurochem.* **32,** 1175 (1979).
93f. Sarlieve, L. L., Neskovic, N. M., Freysz, L., Mandel, P., and Rebel, G. *Life Sci.* **18,** 251 (1976).
93g. Neskovic, N. M., Sarlieve, L. L., and Mandel, P. *Biochim. Biophys. Acta* **429,** 342 (1976).
93h. Tatsumi, K., Murad, S., and Kishimoto, Y. *Arch. Biochem. Biophys.* **171,** 87 (1975).
93i. Akanuma, H., and Kishimoto, Y. *J. Biol. Chem.* **254,** 1050 (1979).
93j. Carter, T. P., and Kanfer, J. N. *J. Neurochem.* **23,** 589 (1974).
94. Anora, R. C., and Radin, N. S. *Biochim. Biophys. Acta* **270,** 254 (1972).
94a. Spik, G., Six, P., and Montreuil, J. *Biochim. Biophys. Acta* **584,** 203 (1979).
95. Brady, R. O., Gal, A. E., Kanfer, J. N., and Bradley, R. M. *J. Biol. Chem.* **240,** 3766 (1965).
95a. Kobayashi, T., and Suzuki, K. *J. Biol. Chem.* **256,** 7768 (1981).
96. Weinreb, N. J., Brady, R. O., and Tappel, A. L. *Biochim. Biophys. Acta* **159,** 141 (1967).
96a. Leese, H. J., and Semenza, G. *J. Biol. Chem.* **248,** 8170 (1973).
97. Hajra, A. K., Bowen, D. M., Kishimoto, Y., and Radin, N. S. *J. Lipid Res.* **7,** 379 (1966).
98. Agranoff, B. W., Radin, N. S., and Suomi, W. *Biochim. Biophys. Acta* **57,** 194 (1962).
99. Bradley, R. M., and Kanfer, J. N. *Biochim. Biophys. Acta* **84,** 210 (1964).
100. Bowen, D. M., and Radin, N. S. *Biochim. Biophys. Acta* **152,** 587 (1968).
101. Bowen, D. M., and Radin, N. S. *J. Neurochem.* **16,** 501 (1969).
101a. Takana, H., and Suzuki, K. *J. Biol. Chem.* **25,** 2324 (1975).
101b. Carter, T. P., and Kanfer, J. N. *J. Neurochem.* **27,** 53 (1976).
102. Radin, N. S., and Arora, R. C. *J. Lipid Res.* **12,** 256 (1971).
103. Arora, R. C., and Radin, N. S. *J. Lipid Res.* **13,** 86 (1972).
104. Arora, R. C., and Radin, N. S. *Lipids* **7,** 56 (1972).
104a. Arora, R. C., Lin Y.-N., and Radin, N. S. *Arch. Biochem. Biophys.* **156,** 77 (1973).
105. Radin, N. S., Brenkert, U., Arora, R. C., Sellinger, O. Z., and Flangas, A. L. *Brain Res.* **39,** 163 (1972).
106. Green, J. P., and Robinson, J. D., Jr. *J. Biol. Chem.* **235,** 1621 (1960).
107. Bakke, J. E., and Carnatzer, W. E. *J. Biol. Chem.* **236,** 653 (1961).
108. Davison, A. N., and Gregson, N. A. *Biochem. J.* **85,** 558 (1962).
109. Davison, A. N., and Gregson, N. A. *Biochem. J.* **98,** 915 (1966).
110. Hauser, G. *Biochim. Biophys. Acta* **84,** 212 (1964).

111. Pritchard, E. T. *J. Neurochem.* **13,** 13 (1966).
112. Lloyd, A. G., Large, P. J., James, A. M., and Dodgson, K. S. *J. Biochem.* **55,** 669 (1964).
113. Herschkowitz, N., McKhann, G. M., Saxena, S., and Shooter, E. M. *J. Neurochem.* **15,** 1181 (1968).
114. Zilversmit, D. B., Entenman, C., and Fishler, M. C. *J. Gen. Physiol.* **26,** 325 (1942).
115. Herschkowitz, N., McKhann, G. M., and Herndon, R. *J. Neurochem.* **16,** 1049 (1969).
115a. Nonaka, G., and Kishimoto, Y. *J. Neurochem.* **33,** 23 (1979).
116. Lipmann, F. *Science* **128,** 575 (1958).
117. McKhann, G. M., Levy, R., and Ho, W. *Biochem. Biophys. Res. Commun.* **20,** 109 (1965).
118. Balasubramanian, A. S., and Bachhawat, B. K. *Biochim. Biophys. Acta* **106,** 218 (1965).
118a. Bhandari, V. R., and Bachhawat, B. K. *Indian J. Biochem. Biophys.* **9,** 72 (1972).
118b. Sarlieve, L. L., Neskovic, N. M., Rebel, G., and Mandel, P. *J. Neurochem.* **26,** 211 (1976).
119. McKhann, G. M., and Ho, W. *J. Neurochem.* **14,** 717 (1967).
120. Cumar, F. A., Barra, H. S., Maccioni, H. J., and Caputto, R. *J. Biol. Chem.* **243,** 3807 (1968).
121. Stoffyn, P., Stoffyn, A., and Hauser, G. *J. Lipid Res.* **12,** 318 (1971).
122. Farrell, D. F., and McKhann, G. M. *J. Biol. Chem.* **246,** 4694 (1971).
122a. Farrell, D. F. *J. Neurochem.* **23,** 219 (1974).
123. Nussbaum, J. L., and Mandel, P. *J. Neurochem.* **19,** 1789 (1972).
123a. Fleischer, B., and Zombrano, F. *Biochem. Biophys. Res. Commun.* **52,** 951 (1973).
123b. Fleischer, B., and Zombrano, F. *J. Biol. Chem.* **249,** 5995 (1974).
124. Mehl, E., and Jatzkewitz, H. *Hoppe-Seyler's Z. Physiol. Chem.* **339,** 260 (1964).
125. Mehl, E., and Jatzkewitz, H. *Biochim. Biophys. Acta* **151,** 619 (1968).
126. Mehl, E., and Jatzkewitz, H. *Biochem. Biophys. Res. Commun.* **19,** 407 (1965).
127. Jatzkewitz, H., and Mehl, E. *J. Neurochem.* **16,** 19 (1969).
127a. Stenshoff, K., and Jatzkewitz, H. *Biochim. Biophys. Acta* **377,** 126 (1975).
128. Cuatrecasas, P., and Anfinsen, C. B. *Methods Enzymol.* **22,** 345 (1972).
128a. Lee, G. D., and Van Etten, R. L. *Arch. Biochem. Biophys.* **166,** 280 (1975).
128b. Shapira, E., and Nadler, H. L. *Arch. Biochem. Biophys.* **170,** 179 (1975).
128c. Nichol, L. W., and Roy, A. B. *J. Biochem.* **55,** 643 (1964).
128d. Stevens, R. L., Fluharty, A. L., Skokut, M. H., and Kihara, H. *J. Biol. Chem.* **250,** *2495 (1975).*
128e. Lee, G. D., and VanEtten, R. L. *Arch. Biochem. Biophys.* **171,** 424 (1975).
128f. Jerfy, A., and Roy, A. B. *Biochim. Biophys. Acta* **371,** 76 (1974).
128g. Graham, E. R. B., and Roy, A. B. *Biochim. Biophys. Acta* **329,** 88 (1973).
128h. Roy, A. B. *Biochim. Biophys. Acta* **377,** 356 (1975).
128i. Fluharty, A. L., Stevens, R. L., Miller, R. T., Shapiro, S. S., and Kihara, H. *Biochim. Biophys. Acta* **429,** 508 (1976).
128j. Yamato, K., Honda, S., and Yamakawa, T. *J. Biochem.* **75,** 1241 (1974).
128k. Fluharty, A. L., Stevens, R. L., Miller, R. T., and Kihara, H. *Biochem. Biophys. Res. Commun.* **61,** 348 (1974).
129. Breslow, J. L., and Sloan, H. R. *Biochem. Biophys. Res. Commun.* **46,** 919 (1972).
130. Clausen, J. *Eur. J. Biochem.* **7,** 575 (1969).
131. Kean, E. L. *J. Lipid Res.* **11,** 248 (1970).
132. Martensson, E. *Prog. Chem. Fats Other Lipids* **10,** 365 (1969).
133. Nishimura, K., Ueta, N., and Yamakawa, T. *Jpn. J. Exp. Med.* **36,** 91 (1966).
133a. Nishimura, K., and Yamakawa, T. *Lipids* **3,** 262 (1966).
133b. Metz, R. J., and Radin, N. S. *J. Biol. Chem.* **255,** 4463 (1980).
133c. Yamada, K., and Sasaki, T. *Biochim. Biophys. Acta* **687,** 195 (1982).
134. Basu, S., Kaufman, B., and Roseman, S. *J. Biol. Chem.* **243,** 5802 (1968).
134a. Basu, S., Kaufman, B., and Roseman, S. *J. Biol. Chem.* **248,** 1388 (1973).
135. Brenkert, A., and Radin, N. S. *Brain Res.* **36,** 183 (1972).
136. Hammarström, S. *Biochem. Biophys. Res. Commun.* **45,** 468 (1971).
136a. Vunnam, R. R., and Radin, N. D. *Biochim. Biophys. Acta* **573,** 73 (1979).
137. Kanfer, J. N., and Richards, R. L. *J. Neurochem.* **14,** 513 (1967).
138. Kanfer, J. N., Bradley, R. M., and Gal, A. E. *J. Neurochem.* **14,** 1095 (1967).

139. Shah, S. N., and Peterson, N. A. *Biochim. Biophys. Acta* **239,** 126 (1971).
140. Kanfer, J. N. *J. Biol. Chem.* **240,** 609 (1965).
141. Brady, R. O., Kanfer, J., and Shapiro, D. *J. Biol. Chem.* **240,** 39 (1965).
141a. Pentchev, P. G., Brady, R. O., Hibbert, S. R., Gal, A. E., and Shapiro, D. *J. Biol. Chem.* **248,** 5256 (1973).
141b. Kanfer, J. N., Mumford, R. A., Raghavan, S. S., and Byrd, J. *Anal. Biochem.* **60,** 200 (1974).
141c. Kanfer, J. N., Raghavan, S. S., and Mumford, R. A. *Biochim. Biophys. Acta* **391,** 129 (1975).
141d. Mumford, R. A., Raghaven, S. S., and Kanfer, J. N. *J. Neurochem.* **27,** 943 (1976).
141e. Ho, M. W. *Biochem. J.* **136,** 721 (1973).
141f. Ho, M. W., and Rigby, M. *Biochim. Biophys. Acta* **397,** 267 (1975).
141g. Ho, M. W. *FEBS Lett.* **53,** 243 (1975).
141h. Erickson, J. S., and Radin, N. S. *J. Lipid Res.* **14,** 133 (1973).
141i. Hyun, J. C., Misra, R. S., Greenblatt, D., and Radin, N. S. *Arch. Biochem. Biophys.* **166,** 382 (1975).
141j. Gatt, S., and Rapport, M. M. *Biochim. Biophys. Acta* **113,** 566 (1966).
142. Gatt, S., and Rapport, M. M. *Isr. J. Med. Sci.* **1,** 624 (1965).
143. Gatt, S. *Biochem. J.* **101,** 687 (1966).
144. Gatt, S. In: *Inborn Disorders of Sphingolipid Metabolism* (Aronson, S. M., and Volk, B. W., eds.), Pergamon Press, New York (1966), p. 261.
145. Suzuki, K. In: *Inborn Disorders of Sphingolipid Metabolism* (Aronson, S. M., and Volk, B. W., eds.), Pergamon Press, New York (1966), p. 215.
146. Vanier, M. T., Holm, M., Ohman, R., and Svennerholm, L. *J. Neurochem.* **18,** 581 (1971).
147. Suzuki, K. *J. Neurochem.* **12,** 969 (1965).
148. Burton, R. M. In: *Lipids and Lipidosis* (Schettler, G., ed.), Springer-Verlag, New York (1967), p. 122.
149. Kishimoto, Y., Davies, W. E., and Radin, N. S. *J. Lipid Res.* **6,** 525 (1965).
150. Kishimoto, Y., and Radin, N. S. *Lipids* **1,** 47 (1966).
151. Suzuki, K., and Korey, S. R. *Biochim. Biophys. Acta* **78,** 388 (1963).
152. Suzuki, K., and Korey, S. R. *J. Neurochem.* **11,** 647 (1964).
153. Suzuki, K., and Chen, G. C. *J. Neurochem.* **14,** 911 (1967).
154. Kanfer, J. N. *Methods Enzymol.* **14,** 660 (1969).
155. Suzuki, K. *J. Neurochem.* **14,** 917 (1967).
156. DeMarcioni, A. H. R., and Caputto, R. *J. Neurochem.* **15,** 1257 (1968).
157. Harzer, K., Jatzkewitz, H., and Sandhoff, K. *J. Neurochem.* **16,** 1279 (1969).
157a. Mack, S. R., Szuchet, S., and Dawson, G. *J. Neurosci. Res.* **6,** 361 (1981).
158. Kolodny, E. H., Brady, R. O., Quirk, J. M., and Kanfer, J. N. *J. Lipid Res.* **11,** 144 (1970).
159. Kanfer, J. N., and Ellis, D. A. *Lipids* **6,** 959 (1971).
160. DeVries, G. H., and Barondes, S. H. *J. Neurochem.* **18,** 101 (1971).
161. Maccioni, H. J., Arce, A., and Caputto, R. *Biochem. J.* **125,** 1131 (1971).
162. Holian, O., Dill, D., and Brunngrabier, E. G. *Arch. Biochem. Biophys.* **142,** 111 (1971).
163. Holm, M., and Svennerholm, L. *J. Neurochem.* **19,** 609 (1972).
163a. Orlando, P., Cocciante, G., Ippolito, G., Massace, P., Robert, S., and Tettamonte, G. *Pharmacol. Res. Commun.* **11,** 759 (1979).
164. Kampine, J. P., Martensson, E., Yankee, R. A., and Kanfer, J. N. *Lipids* **3,** 151 (1968).
165. Hauser, G. *Biochem. Biophys. Res. Commun.* **28,** 503 (1967).
166. Hildebrand, J., and Hauser, G. *J. Biol. Chem.* **244,** 5170 (1969).
166a. Bushway, A. A., and Keenan, T. W. *Biochim. Biophys. Acta* **572,** 146 (1979).
167. Jungalwala, F. B., and Robins, E. *J. Biol. Chem.* **243,** 4258 (1968).
168. Li, Y.-T. and Li, S.-C. *J. Biol. Chem.* **246,** 3769 (1971).
168a. Tanaka, H., Meisler, M., and Suzuki, K. *Biochim. Biophys. Acta* **398,** 452 (1975).
168b. Callahan, J. W., and Jerrie, J. *Biochim. Biophys. Acta* **391,** 141 (1975).
168c. Miyatake, T., and Suzuki, K. *J. Biol. Chem.* **250,** 585 (1975).
168d. Chester, M. A., Hultberg, B., and Öckerman, P. A. *Biochim. Biophys. Acta* **429,** 517 (1976).
168e. Wenger, D. A., Sattler, M., and Clark, C. *Biochim. Biophys. Acta* **409,** 297 (1975).
169. Clarke, J. T. R., and Wolfe, L. *J. Biol. Chem.* **246,** 5563 (1971).
170. Stoffyn, P., Stoffyn, A., and Hauser, G. *Trans. Am. Soc. Neurochem.* **3,** 125 (1972).

170a. Stoffyn, A., Stoffyn, P., and Hauser, G. *Biochim. Biophys. Acta* **360,** 174 (1974).
171. Hauser, G. Personal communication.
171a. Martensson, E., Öhman, R., Graves, M., and Svennerholm, L. *J. Biol. Chem.* **249,** 4132 (1974).
172. Brady, R. O., Gal, A. E., Bradley, R. M., and Martensson, E. *J. Biol. Chem.* **242,** 1021 (1967).
173. Bensaude, I., Callahan, J., and Philipart, M. *Biochem. Biophys. Res. Commun.* **43,** 913 (1971).
174. Gatt, S. *Biochim. Biophys. Acta* **137,** 192 (1967).
175. Gatt, S. *Adv. Exp. Med. Biol.* **25,** 141 (1972).
175a. Mapes, C. A., and Sweeley, C. C. *J. Biol. Chem.* **248,** 2461 (1973).
175b. Mapes, C. A., Suelter, C. H., and Sweeley, C. C. *J. Biol. Chem.* **248,** 2471 (1973).
175c. Mapes, C. A., and Sweeley, C. C. *Arch. Biochem. Biophys.* **158,** 297 (1973).
175d. Ho, M. W. *Biochem. J.* **133,** 1 (1973).
175e. Takenawa, T., Narunie, K., and Tsumita, T. *Jpn. J. Exp. Med.* **43,** 331 (1973).
176. Chien, J. L., Williams, T., and Basu, S. *J. Biol. Chem.* **248,** 1778 (1973).
176a. Handa, S., and Burton, R. M. *Lipids* **4,** 589 (1969).
177. DiCesare, J. L., and Dain, J. A. *Biochim. Biophys. Acta* **231,** 385 (1971).
178. DiCesare, J. L., and Dain, J. A. *J. Neurol.* **19,** 403 (1972).
179. Frohwein, Y. Z., and Gatt, S. *Biochemistry* **6,** 2775 (1967).
180. Frohwein, Y. Z., and Gatt, S. *Biochim. Biophys. Acta* **128,** 216 (1966).
181. Frohwein, Y. Z., and Gatt, S. *Biochemistry* **6,** 2783 (1967).
182. Sandhoff, K., and Wassle, W. *Hoppe-Seyler's Z. Physiol. Chem.* **352,** 1119 (1971).
182a. Keenan, T. W., Morre, D. J., and Basu, S. *J. Biol. Chem.* **249,** 310 (1974).
182b. Den, H., Kaufman, B., McGuire, E. J., and Roseman, S. *J. Biol. Chem.* **250,** 739 (1975).
182c. Ledeen, R. W., Yu, R. K., and Eng, L. F. *J. Neurochem.* **21,** 829 (1973).
182d. Yu, R. K., and Lee, S. H. *J. Biol. Chem.* **251,** 198 (1976).
182e. Stoffyn, A., Stoffyn, P., Farooq, M., Snyder, D. S., and Norton, W. T. *Neurochem. Res.* **6,** 1149 (1981).
183. Arce, A., Maccioni, H. F., and Caputto, R. *Arch. Biochem. Biophys.* **116,** 52 (1966).
184. Wells, M. A., and Dittmer, J. C. *Biochemistry* **2,** 1259 (1963).
185. Kaufman, B., Basu, S., and Roseman, S. In: *Inborn Errors of Sphingolipid Metabolism* (Aronson, S. M., and Volk, B. W., eds.), Pergamon Press, New York (1967), p. 193.
186. Arce, A., Maccioni, H. J., and Caputto, R. *Biochem. J.* **121,** 483 (1971).
187. Duffard, R. O., and Caputto, R. *Biochemistry* **11,** 1396 (1972).
188. Den, H., Schultz, A. M., Basu, M., and Roseman, S. *J. Biol. Chem.* **246,** 2721 (1971).
189. Carubelli, R., Trucco, R. E., and Caputto, R. *Biochem. Biophys. Res. Commun.* **60,** 196 (1962).
190. Leibovitz, Z., and Gatt, S. *Biochim. Biophys. Acta* **152,** 136 (1968).
191. Huang, R. T. C., and Klink, E. *Hoppe-Seyler's Z. Physiol. Chem.* **353,** 679 (1972).
192. Sandhoff, K., and Jatzkewitz, H. *Biochim. Biophys. Acta* **141,** 442 (1967).
193. Tettamonte, G., and Zambotti, V. *Enzymologia* **35,** 61 (1968).
193a. Preti, A., Lombardo, A., Cestaro, B., Zambotti, V., and Tettamonte, G. *Biochim. Biophys. Acta* **350,** 406 (1974).
193b. Tettamonte, G., Cestaro, B., Lombardo, A., Preti, A., Venderando, B., and Zambotti, V. *Biochim. Biophys. Acta* **350,** 415 (1974).
194. Tettamonte, G., Venerando, B., Preti, A., Lombardo, A., and Zambotti, V. *Adv. Exp. Med. Biol.* **25,** 161 (1972).
195. Öhman, R., Rosenberg, A., and Svennerholm, L. *Biochemistry* **9,** 3774 (1970).
196. Steigerwald, J. C., Kaufman, B., Basu, S., and Roseman, S. *Fed. Proc. Fed. Am. Soc. Exp. Biol.* **25,** 587 (1966).
197. DiCesare, J. L., and Dain, J. A. *Biochim. Biophys. Acta* **231,** 385 (1971).
198. Kanfer, J. N., Blacklow, R. S., Warren, L., and Brady, R. O. *Biochem. Biophys. Res. Commun.* **14,** 287 (1964).
199. Svennerholm, L. *Compr. Biochem.* **18,** 212 (1970).
200. Kolodny, E. H., Brady, R. O., Quirk, J. M., and Kanfer, J. N. *J. Lipid Res.* **11,** 144 (1970).
201. Kolodny, E. H., Kanfer, J. N., Quirk, J. M., and Brady, R. O. *J. Biol. Chem.* **246,** 1426 (1971).

202. Wenger, D. A., and Wardell, S. *Physiol. Chem. Phys.* **4**, 224 (1972).

202a. Wenger, D. A., and Wardell, S. *J. Neurochem.* **20**, 607 (1973).

202b. Wenger, D. A., Okada, S., and O'Brien, J. S. *Arch. Biochem. Biophys.* **153**, 116 (1972).

203. Tallman, J. F., and Brady, R. O. *J. Biol. Chem.* **247**, 7570 (1972).

203a. Srivastava, S. K., Awasthi, Y. C., Yoshida, A., and Beutler, E. *J. Biol. Chem.* **249**, 2043 (1974).

203b. Srivastava, S. K., Yoshida, A., Awasthi, Y. C., and Beutler, E. *J. Biol. Chem.* **249**, 2049 (1974).

203c. Li, Y.-T., Mazzotta, M. Y., Wan, C.-C., Orth, R., and Li, S.-C. *J. Biol. Chem.* **248**, 7512 (1973).

203d. Li, S.-C. and Li, Y.-T. *J. Biol. Chem.* **251**, 1159 (1976).

203e. Tallman, J. F., Brady, R. O., Quirk, J. M., Villalba, M., and Gal, A. E. *J. Biol. Chem.* **249**, 3489 (1974).

203f. Bach, G., and Suzuki, K. *J. Biol. Chem.* **250**, 132 (1975).

203g. Vladutiu, G. D., Carmody, P. J., and Rattazzi, M. C. *Prep. Biochem.* **5**, 147 (1975).

203h. Dawson, G., Propper, R. L., and Dorfman, A. *Biochem. Biophys. Res. Commun.* **54**, 1102 (1973).

203i. Aruna, R. M., and Basu, D. *J. Neurochem.* **25**, 611 (1975).

203j. Verpoorte, J. A. *J. Biol. Chem.* **247**, 4787 (1972).

203k. Sterling, J. L. *FEBS Lett.* **39**, 171 (1974).

203l. Werries, E., Neue, I., and Buddecke, E. *Hoppe-Seyler's Z. Physiol. Chem.* **356**, 953 (1975).

204. Sandhoff, K. *FEBS Lett.* **11**, 342 (1970).

205. Basu, S., Kaufman, B., and Roseman, S. *J. Biol. Chem.* **240**, PC 4115 (1965).

206. Yiamouyiannis, J. A., and Dain, J. A. *Lipids* **3**, 378 (1968).

207. Yip, G. B., and Dain, J. A. *Biochim. Biophys. Acta* **206**, 252 (1970).

207a. Dain, J. A., and Hitchener, W. R. *Adv. Exp. Med. Biol.* **101**, 649 (1978).

208. Yip, M. C. M., and Dain, J. A. *Biochem. J.* **118**, 247 (1970).

208a. Yip, M. C. M. *Biochem. Biophys. Res. Commun.* **53**, 737 (1973).

208b. Stoffyn, A., Stoffyn, P., and Yip, M. C. M. *Biochim. Biophys. Acta* **409**, 97 (1975).

208c. Stoffyn, P., and Stoffyn, A. *Carbohydr. Res.* **78**, 327 (1980).

209. Sandhoff, K., Pilz, H., and Jatzkewitz, H. *Hoppe-Seyler's Z. Physiol. Chem.* **338**, 281 (1964).

209a. Norden, A. G. W., and O'Brien, J. S. *Arch. Biochem. Biophys.* **159**, 383 (1973).

209b. Norden, A. G. W., Tennant, L. L., and O'Brien, J. S. *J. Biol. Chem.* **249**, 7969 (1974).

210. Kaufman, B., Basu, S., and Roseman, S. *J. Biol. Chem.* **243**, 5804 (1968).

211. Cumar, F. A., Fishman, P. H., and Brady, R. O. *J. Biol. Chem.* **246**, 5075 (1971).

211a. Maccioni, H. J. F., Arce, A., Landa, C., and Caputto, R. *Biochem. J.* **138**, 291 (1974).

211b. Mestrallet, M. G., Cumar, F. A., and Caputto, R. *Biochem. Biophys. Res. Commun.* **59**, 1 (1974).

211c. Caputto, R., Maccioni, H. J., and Arce, A. *FEBS Lett.* **4**, 97 (1974).

212. Cumar, F. A., Tallman, J. F., and Brady, R. O. *J. Biol. Chem.* **247**, 2322 (1972).

213. Burton, R. M. *J. Neurochem.* **10**, 503 (1963).

214. Mahadevan, S., Nduaguba, J. C., and Tappel, A. L. *J. Biol. Chem.* **242**, 4409 (1967).

215. Öhman, R. *J. Neurochem.* **18**, 89 (1971).

216. Schenguend, C. L., and Rosenberg, A. *J. Biol. Chem.* **245**, 6196 (1970).

216a. Tallman, J. F., and Brady, R. O. *Biochim. Biophys. Acta* **293**, 434 (1973).

217. Roseman, S. *Chem. Phys. Lipids* **5**, 270 (1970).

218. Maccioni, H. J. F., Arce, A., and Caputto, R. *FEBS Lett.* **23**, 136 (1972).

219. Gatt, S. *Chem. Phys. Lipids* **5**, 235 (1970).

220. Sribney, M., and Kennedy, E. P. *J. Biol. Chem.* **233**, 1315 (1958).

221. Fujino, Y., Nakano, M., Neigishi, T., and Ito, S. *J. Biol. Chem.* **243**, 4650 (1968).

222. Scribney, M. *Can. J. Biochem.* **49**, 306 (1971).

222a. Sribney, M. *Arch. Biochem. Biophys.* **126**, 954 (1968).

222b. Lyman, E. M., Knowles, C. L., and Sribney, M. *Can. J. Biochem.* **54**, 358 (1976).

223. Brady, R. O., Bradley, R. M., Young, O. M., and Kaller, H. *J. Biol. Chem.* **240**, 3693 (1965).

224. Fujino, Y., Negishi, T., and Ito, S. *Biochem. J.* **109**, 310 (1968).

225. Fujino, Y., and Negishi, T. *Biochim. Biophys. Acta* **152**, 428 (1968).

226. Muehlenberg, B. A., Stribney, M., and Duffe, M. K. *Can. J. Biochem.* **50,** 166 (1972).
227. Kanfer, J. N., and Papier, C. Unpublished observations.
227a. Stoffel, W., and Assman, G. *Hoppe-Seyler's Z. Physiol. Chem.* **353,** 65 (1972).
228. Marggraf, W. D., and Anderer, F. A. *Hoppe-Seyler's Z. Physiol. Chem.* **355,** 803 (1974).
228a. Diringer, H., Marggraf, W. D., Koch, M. A., and Anderer, F. A. *Biochem. Biophys. Res. Commun.* **47,** 1345 (1972).
228b. Ullman, M. D., and Radin, N. S. *J. Biol. Chem.* **249,** 1506 (1974).
228c. Stoffel, W., and Melzner, I. *Hoppe-Seyler's Z. Physiol. Chem.* **361,** 775 (1980).
228d. Wehrle, W. D., Anderers, A., and Kanfer, J. N. *Biochim. Biophys. Acta* **264,** 61 (1981).
228e. Marggraf, W. D., Zertani, R., Anderer, A., and Kanfer, J. N. *Biochim. Biophys. Acta* **710,** 314 (1982).
228f. Kanfer, J. N., and Spielvogel, C. *Lipids* **10,** 391 (1975).
228g. Voelker, D. R., and Kennedy, E. P. *Fed. Proc. Fed. Am. Soc. Exp. Biol.* **40,** 1805 (1981).
229. Kanfer, J. N., Young, O. M., Shapiro, D., and Brady, R. O. *J. Biol. Chem.* **241,** 1081 (1966).
230. Fowler, S. *Biochim. Biophys. Acta* **191,** 481 (1969).
231. Barenholz, Y., Roitman, A., and Gatt, S. *J. Biol. Chem.* **241,** 3731 (1966).
232. Heller, M., and Shapiro, B. *Biochem. J.* **98,** 763 (1966).
232a. Gatt, S., Herzl, A., and Barenholz, Y. *FEBS Lett.* **30,** 281 (1973).
232b. Yedgar, S., Barenholz, Y., and Cooper, V. G. *Biochim. Biophys. Acta* **363,** 98 (1974).
232c. Yedgar, S., Hertz, R., and Gatt, S. *Chem. Phys. Lipids* **13,** 404 (1974).
233. Nilsson, A. *Biochim. Biophys. Acta* **164,** 575 (1968).
234. Paston, I., Macchia, V., and Katzen, R. *J. Biol. Chem.* **243,** 3750 (1968).
235. Klein, F., and Mandel, P. *Biochemie* **54,** 371 (1972).
236. Schneider, P. B., and Kennedy, E. P. *J. Lipid Res.* **9,** 58 (1968).
236a. Hostetler, K. Y., and Yazaki, P. J. *J. Lipid Res.* **20,** 456 (1979).
236b. Rao, B. G., and Spence, M. W. *J. Lipid Res.* **17,** 506 (1976).
236c. Spence, M. W., Burgess, J. K., and Specker, E. R. *Brain Res.* **168,** 543 (1979).
236d. Yamaguchi, S., and Suzuki, K. *J. Biol. Chem.* **252,** 3805 (1977).
236e. Yamanaka, T., Handa, E., and Suzuki, K. *J. Biol. Chem.* **256,** 3884 (1981).
236f. Jones, C. S., Shankaran, P., and Callahan, J. W. *Biochem. J.* **195,** 373 (1981).
236g. Callahan, J. W., Gerrie, J., Jones, C. S., and Shankaran, P. *Biochem. J.* **193,** 275 (1981).
236h. Yamanaka, T. and Suzuki, K. *J. Neurochem.* **38,** 1753 (1982).
236i. Jones C. D., Davidson, D. J. and Callahan, J. W. *Biochem. Biophys. Acta* **701,** 261 (1982).
236j. Maziere, J. C., Wolf, C., Maziere, C., Mora, W., Bereziat, G., and Polonovski, J. *Biochem. Biophys. Res. Commun.* **100,** 1299 (1981).
237. Porcellatti, G., Ceccarelli, B., and Tettamanti, G. (eds.), *Advances in Experimental Medicine and Biology,* Vol. 71, Plenum Press, New York (1976).
238. Svennerholm, L., Mandel, P., Dreyfus, H., and Urban, P.-F. (eds.), *Advances in Experimental Medicine and Biology,* Vol. 125, Plenum Press, New York (1980).

Chapter 3

The Sphingolipidoses

Julian N. Kanfer

Heterogeneity exists in the clinical manifestation of most diseases or syndromes believed due to a deficiency of a single hydrolytic enzyme protein. Several of the enzymes responsible for the cleavage of the individual bonds present in the sphingolipids have been purified to homogeneity from human sources. Therefore, information concerning the physical properties, the amino acid composition and sequence for these proteins is limited. In some instances, there may be portions of the amino acid chain of a specific enzyme in which minor substitutions or alterations can occur resulting in only slight effect on the measured catalytic activity. Therefore, it may be impossible to distinguish one "normal" or "control" sample from another based solely upon the activity of the enzymes in crude cell-free assay systems. Experience in individual laboratories around the world has indicated that their "range for normals" for any specific enzyme may vary severalfold. This surprisingly large magnitude of difference may merely represent the "normal biological variation" among individuals. Alternatively, this might be ascribed to slight differences in the assay procedures. Limited studies with vertebrate enzymes suggest that heterozygosity may be inversely related to subunit number and directly to the chain length of the subunit. It is most likely that individual electrophoretic variations occur at general rather than catalytic sites and that there is greater heterozygosity in those enzymes with a greater number of general sites.[1] Perhaps it is more realistic merely to attempt to operationally distinguish between "pathological" and "nonpathological" tissue samples. The intriguing possibility might exist that persons with "normal" enzyme values, not identified as carriers, could have certain subclinical manifestations of the associated disease. In the adult form of Gaucher's disease, splenomegaly is associated with a β-glucosidase deficiency. An estimation of splenic mass and the activity of this enzyme in a "nonpathological" population may indicate that an inverse correlation exists. Another interesting possibility is that a couple with low detectable levels of a specific enzyme, who are not apparent carriers, may produce progeny with similarly reduced activity of the same enzyme.

The most extensive documentation on protein variations due to amino acid differences exists for the hemoglobins of the red blood cell where a multitude of different types have been reported. The methodology employed for the detection of these variations has largely been based on differences in the electrophoretic migrations either of the native proteins or of peptides produced by proteolytic enzyme hydrolysis. This suggests that the structural basis for these differences detected by this technique is principally due to changes in the charge on the molecule. Presumably, alterations of amino acid sequences that would not result in changes of electrophoretic migrations could go undetected.[1a,2–4]

The pathological conditions classified as lysosomal storage disorders are usually characterized by reduction in the amount of *detectable* enzyme activity. There is often some residual activity, which can represent 5–15% of that found in the "normal" control average. The complete absence of activity is rarely found and, in those instances where reported, might merely reflect the limitation in the sensitivity of the particular assay procedure employed.

The heterogeneity one may anticipate to exist in the "nonpathological" population has also been noted in many of these clinical conditions. The differences in clinical manifestations of a similar enzyme deficiency will be discussed under the individual disease.

The sphingolipidoses are presumably due to the reduced activity of an enzyme responsible for the cleavage of a specific bond. This results in the accumulation of the specific compound(s) within the cells of the affected individual. The enzyme proteins responsible for these hydrolyses are usually found in an intracellular organelle termed the lysosome. The discovery and early documentation of the compartmentalization of this group of hydrolytic enzymes was originally carried out in the laboratory of Dr. Christian De-Duve.[5] The principal criteria for the lysosomal nature of a given hydrolase are: (1) occurrence in a mixed mitochondrial–lysosomal population upon centrifugal subcellular fractionation, (2) acidic pH optimum, (3) structural latency. The lysosomes are usually found in the pellet obtained on centrifugation of a homogenate at 9–10,000g. However, this fraction also contains the bulk of the mitochondria and often fragments of miscellaneous other cellular organelles. Further lysosomal purification is usually accomplished by additional centrifugations through sucrose solutions of various densities. The ability of lysosomes to take up detergents, such as Triton WR 1339, has been exploited for these isolations. Fasted rats are administered this compound, the liver homogenate is centrifuged, and an opalescent turbid layer appears at the top of the tube. These are the "tritosomes" or Triton-filled lysosomes.

The lysosomal hydrolases have pH optima below neutrality, and this may vary from pH 3 for β-galactosidases to pH 5–6 for the cathepsins. Enzymes capable of hydrolyzing similar bonds are also encountered in other portions of the cells and are not exclusively associated with these particles. However, these "nonlysosomal" enzymes are easily distinguished since they generally possess neutral or alkaline pH optima. The property of the structural latency of these enzymes led to the discovery of lysosomes.[5] Increased detectable enzyme activity is usually observed when particle suspensions or whole tissue homogenates are subjected to treatment likely to cause lysis or fragmentation of these organelles. Repetitive cycles of freeze–thawing, osmotic shock, sonication, and detergent additions usually result in increased levels of detectable activity when compared to untreated samples.[6,7]

The individual lysosomal acid hydrolases are generally specific for the type of bond cleaved. The glycosidases are specific for the anomeric configuration at the glycosidic linkage as well as for the carbohydrate moiety. Proteases and esterases show similar specificities. The 4-methylumbelliferone (4MU) and p-nitrophenyl (pNP) derivatives are commercially available synthetic compounds that are commonly utilized as substrates for the hydrolases principally due to their convenience. Alkaline conditions cause these aglycones to be either fluorogenic (4MU) or yellow (pNP) and provide a simple and reliable enzyme assay system. The naturally occurring compounds that may be substrates for these enzymes are not readily accessible and usually more difficult to assay.

The ability of the lysosomal acid hydrolases to efficiently catalyze the liberation of these "unnatural" aglycones from their corresponding synthetic glycoside suggests a degree of enzymatic nonspecificity. It has been generally observed that for many hydrolytic activities the synthetic substrates are more rapidly cleaved than a naturally occurring material possessing the same glycosidic bond. The majority of information regarding the properties of the hydrolases and their individual purification has been obtained in studies where the pNP or 4MU derivatives were employed as substrate. Only a few reports on acid hydrolase purification have appeared utilizing a naturally occurring molecule as substrate throughout the isolation procedure. GM_1 ganglioside contains two distinct β-galactosidic bonds. Both the terminal galactose residue and the galactose residue linked to glucose are cleaved by several relatively nonspecific β-galactosidases. However, the galactose directly linked to the sphingosine moiety of ceramide in galactosylceramide appears to require a fairly specific enzyme.

The practical utility of these nonspecific artificial substrates is best illustrated by their reliability in the diagnosis of a variety of "inborn errors of metabolism." This applies to heterozygote and homozygote identification in addition to prenatal diagnosis with attendant therapeutic abortions.

From the biochemist's viewpoint, the simplest assumption is that the accumulation of a normally occurring macromolecule results from either an overproduction of the specific compound or its decreased catabolism. The accumulations found in "lysosomal disorders" are believed due to diminished catabolism. However, it should be noted that for many of the sphingolipidoses, the alternative possibility of overproduction has not been systematically or extensively investigated.

This chapter will not discuss each of the sphingolipidoses in an exhaustive manner. There have been several publications devoted to the body of information available concerning these diseases.[8–10] Individual articles are included on the specific diseases, written by researchers actively working in these areas. Also available are a variety of volumes reporting the current research efforts presented at symposiums.

The earliest proposals for the concept of "inborn lysosomal disease" appeared in a thoughtful publication by Hers.[11] This listed several of the criteria for the inclusion of any disease into this category.

3.1. Farber's Lipogranulomatosis

Farber's lipogranulomatosis was originally described in 1957, and since then approximately 15–20 cases have been reported.

There appear to be two clinical types, which differ from each other with respect to duration of the disease. The most prevalent type shows mortality by the age of 2 years. The other has been referred to as the "chronic" type since patients have lived into their teens. The "infantile" form has associated signs of central and peripheral nerve involvement, which are absent from the chronic type.

The early onset type is characterized by a hoarse, weak cry, irritability, swelling in the joints, severe motor and mental retardation, and death, usually from inanition and infection, often by the age of 2 years. The "chronic" variety is the less common form of the disease, which differs from the early onset type principally in chronology and lack of mental retardation.

A characteristic granulomatous lesion, in the center of which are cells distended by stored material, is seen in the subcutaneous tissues and joints. The nervous system shows evidence of neuronal lipid storage. These cells have a foamy appearance and are strongly positive with periodic acid–Schiff (PAS) reagent, suggesting a carbohydrate material. In the central nervous system, PAS-positive material distends the neuronal cytoplasm.[12] The combination of features of granulomatous lesion and the lipidoses suggested that they represented a bridge between a histiocytosis (a clinical pattern having an underlying theme of the development of a granulomatous lesion with a histiocytic proliferation) such as Hand–Schuler–Christian disease and a possible inborn metabolic error.[13] Routine laboratory tests provide no useful information regarding this disease.

3.1.1. Chemical Studies

The presence of PAS-positive foam cells in the characteristic lesions suggested that there might be a storage of carbohydrate material. Lipid extracts from peripheral organs were reported to contain elevated "hexose" that was stated not to be a cerebroside. This was claimed to be a part of a "lipoglycoprotein" of ill-defined composition.[14] No elevations in either tissue or urinary mucopolysaccharides have been observed except for two cases.

A preliminary report appeared concerning the elevation of N-acylsphingosine, ceramide, in a patient with Farber's disease. A more detailed publication reported the following elevations for ceramide: a 40-fold increase in liver, a 35-fold increase in lung, a 7-fold increase in kidney, and a 4-fold increase in cerebral white matter. A marginal elevation was observed in cerebral gray matter. Other changes reported were elevated hepatic and lung cholesterol, elevated hepatic glycolipid, and elevated hepatic, lung, femoral lymph node, and subcutaneous nodule "ganglioside sialic acid." In addition, these nodules contained large quantities of ceramide.[15] The presence of elevated amounts of ceramide has been confirmed.[16] Both hydroxy and nonhydroxy fatty acids are found in the ceramides isolated from brain and kidney of Farber's patients.[17]

3.1.2. Enzyme Defect

Attempts to demonstrate a ceramidase deficiency were initially unsuccessful.[15] A distinct ceramidase deficiency was subsequently demonstrated employing N-[^{14}C]oleoylsphingosine as substrate.[18] A striking reduction exceeding 95% in the hydrolysis of this substrate by homogenates of both kidney and cerebellar tissue was reported (Table I).[18] Normal values for neuraminidase, β-N-acetylhexosaminidase, β-galactosidase, and acid phosphatase were observed in Farber tissues, indicating the specificity of this deficiency. Ceramidase, in addition to the hydrolysis of ceramide, catalyzes the condensation of sphingosine and free fatty acid to form ceramide. This "synthetase" activity was similarly diminished in Farber tissues. This is believed to be a lysosomal enzyme activity with a pH optimum of 4.0. More recently, a ceramidase has been described that has an alkaline pH optimum of 9.0. There was no discernible decrease of this alkaline form in cerebellum from a patient with Farber's disease.[18a] Reduced levels of the "acid" ceramidase have been reported in cultured skin fibroblasts, suggesting that prenatal diagnosis is feasible.[18b] The uptake of radioactive ceramide was nearly identical for cultured skin fibroblasts from normal and Farber's patients. The decreased "acid" ceramidase resulted in the lysosomal ceramide accumulation.[18c]

Differential preference was reported of the neutral and acidic forms of ceramidase present in homogenized Farbers patients and controls cells based on the fatty chain length of the substrate.[18d]

The reduced "ceramidase" activity can readily explain the accumulation of ceramide in tissues of Farber patients. However, the problem of the reported elevations of other sphingoglycolipids has not been resolved.

3.2. Krabbe's Disease (Globoid Cell Leukodystrophy)

3.2.1. Clinical

Approximately 100 cases have been reported since Krabbe's original description in 1916. The general clinical course has been operationally divided into three stages.[19] The first stage is characterized by increased irritability and joint stiffness. The child appears normal during the first few postnatal months and then becomes hypersensitive to noise, to touch, or to visual stimuli and has episodes of spontaneous crying. Vomiting with attendant feeding problems and convulsive seizures may be early clinical manifestations. Cerebrospinal fluid protein levels are increased. The second stage is characterized by increased motor and mental degeneration. There is quadriparesis, and spasticity. The legs are often in extended and crossed posture, while the arms and head are flexed. Some toniclonic seizures may occur. The third stage has

Table I. Ceramidase–Enzyme Activities in Kidney and Cerebellum[a]

Diagnosis	Sex	Age at death	Time stored	Cerami-dase (nmole)	Neura-minidase (nmole)	β-Hexosa-minidase (μmole)	β-Galacto-sidase (μmole)	Acid phospha-tase (μmole)
Kidney								
Farber's disease	F	9 months	7 years	< 0.01	1.01	8.00	0.54	1.34
Biliary cirrhosis	M	2 years	4 years	2.48	0.77	3.66	0.96	1.51
Reye's syndrome	M	2 years	4 years	1.73	1.28	4.17	0.65	0.95
Familial microcephaly	F	4 years	4 years	2.28	0.70	3.02	0.67	1.03
Coarctation of aorta	F	2 months	5 years	2.26	1.52	7.99	0.96	1.69
Congenital heart disease	F	3 months	1 month	3.21	0.90	3.96	0.61	0.98
Burkitt's lymphoma	M	11 years	1 month	2.85	1.76	3.80	0.47	0.86
Metachromatic leukodystrophy	M	2 years	1 year	3.03		4.45	0.67	1.10
Infantile Gaucher's disease	M	2 years	11 years	0.66		4.20	0.72	1.61
Cerebellum								
Farber's disease	F	9 months	7 years	< 0.001		0.77	0.09	0.45
Congenital heart disease	F	1 month	1 day	1.34		0.68	0.16	0.64
Metachromatic leukodystrophy	M	2 years	1 year	0.68		0.89	0.11	0.62
B12 defect and homocystinuria	M	2 months	4 years	0.74		0.92	0.20	0.76
GM2 gangliosidosis, type AB	M	4 years	6 years	0.33		1.27	0.14	0.61
Hallervorden–Spatz's disease	F	24 years	5 years	0.44		0.33	0.14	0.49
Infantile Gaucher's disease	M	2 years	11 years	0.18		0.65	0.13	0.55

[a] Modified from Ref. 18.

been referred to as the "burned-out" stage and may last for several years, although death usually occurs by the second year. The child is decerebrate, blind, and unresponsive to external stimuli.

The symptoms and signs are restricted almost exclusively to the nervous system with lack of visceral involvement. Although there is peripheral neuropathy, this is often obscured by extensive damage to the central nervous system. A child who had been prenatally diagnosed as having Krabbe's disease displayed signs of peripheral neuropathy at 7 weeks of age.[19a] Routine blood analyses have not revealed any significant abnormality. Cerebrospinal fluid proteins are elevated and this is reflected principally by increased albumin and α_2-globulin fractions along with decreased β- and α-globulins.

Several other herediodegenerative diseases have a clinical course similar to Krabbe's disease, making the specific diagnosis difficult. However, this potential confusion is readily resolved by documentation of the specific enzyme defect.

3.2.2. Pathology

A uniform reduction in the size of the whole brain is observed. On sectioning, the white matter is grossly decreased in volume, with a firm rubbery consistency and a grayish-white color. There is a diffuse demyelination coupled with extensive astrocytic gliosis. The gray matter appears relatively normal except for reduced cortical thickness.[20]

Histologically, the white matter appears more abnormal than the gray matter. The characteristic findings are a large number of globoid cells,[21] a paucity of myelin, and an astrocytic gliosis. Two populations of globoid cells are seen, one type mononuclear and the other multinuclear. The globoid cells are the principal cellular type found in white matter, and it has been estimated that these cells may constitute 30–50% of the mass of this tissue. The globoid cell gives a positive PAS stain, usually interpreted as a demonstration of carbohydrate-containing materials, and a positive Sudan black stain, usually interpreted as a demonstration of lipids. This type of globoid cell appearance has been produced in experimental animals by the intracerebral injection of galactosylceramide[22] and lactosylceramide.[22a] A variety of other glycosphingolipids sphingolipids were not found to evoke this response.

A severe lack of myelin is characteristic of Krabbe's disease. A reciprocal relationship exists between the concentration of globoid cells and the quantity of demonstrable myelin. Thus, those areas with the least myelin have the greatest concentration of globoid cells. Globoid cells were absent in autopsy material from a 5-year-old boy with Krabbe's disease.[22b]

The globoid cells, when carefully examined under the electron microscope, have unusual membrane-bound inclusions found in the cytoplasm (Fig. 1).[23–25] These have been described as crystalloid tubules similar to those seen with pure galactosylceramide.[26] The presence of non-membrane-bound deposits in Krabbe tissues has been observed.[27]

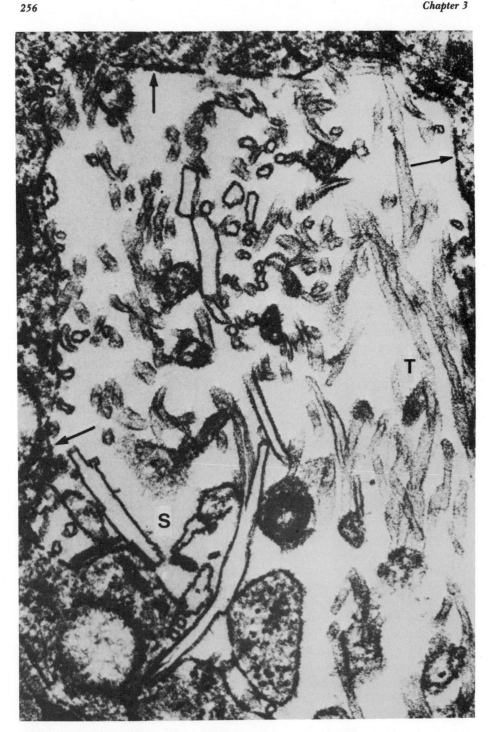

Figure 1. Portion of a globoid cell from Krabbe's patient. This sac contains both twisted (T) and straight (S) tubules. Note membrane (arrows). × 57,000. (From Ref. 23.)

3.2.3. Biochemistry

The problem that arose in the attempts at elucidation of the "stored material" in globoid leukodystrophy illustrates certain of the limitations inherent in the interpretation of data from quantitative analysis of tissue samples with extensive demyelination. Observations by pathologists suggested a close similarity between the appearance of the tubular inclusions found in the globoid cell and those seen in the "Gaucher" cell. Gaucher's disease is another sphingolipidosis characterized by glucosylceramide accumulation. Early studies designed to demonstrate a galactosylceramide accumulation in Krabbe's disease were equivocal.[28] However, subsequent analysis indicated that the total galactolipid content of the white matter was substantially lower than corresponding normal tissues.[29]

Evidence was presented reporting altered galactosylceramide/sulfatide ratios. Whereas the usual normal value for this ratio is 4, in Krabbe's disease this ratio was 5–10.[29,30] However, here also other pathological tissue samples have been reported to contain normal ratios. These observations suggested that the relationship between these two sphingolipids had been altered.

Initial studies on myelin isolated from globoid leukodystrophy tissues reported diminished sulfatide content. Subsequent investigations on isolated myelin indicated no quantitative analytical differences.[31] The procedures for myelin isolation employed by most investigators were originally devised for fresh rat and cattle brain tissues, and the product obtained is a reasonably well-purified preparation of myelin. Advantage is taken of the unique buoyant density properties of the lipid-enriched membrane for these isolation procedures. Therefore, abnormal myelin might not be expected to be recovered using this procedure. Indeed, well-documented reports on the isolation of grossly altered myelin from tissue obtained from patients with extensive demyelination are difficult to locate. These analytical studies on myelin from globoid leukodystrophy brain tissue suggested that the oligodendroglial cell in these patients is capable of producing a morphologically and chemically normal myelin.

The analytical reports of increased galactosylceramide with respect to sulfatide focused early attention on galactosylceramide-sulfotransferase as the possible enzyme defect in Krabbe's disease. A deficiency in this enzyme activity was demonstrated in kidney, cerebral cortex, and cerebral white matter of patients as compared to "normals" by Austin and co-workers in 1967.[32] The finding of normal levels of the sulfotransferase in metachromatic leukodystrophy, another white matter demyelinating disease, made this observation more meaningful. Subsequent investigations demonstrated a decreased sulfotransferase in both the white matter and the gray matter of five cases of Krabbe's disease. In this series of studies, a decreased transferase was observed in some, but not all, kidney samples from Krabbe patients.[33] The assay procedure utilized was dependent upon the presence of endogenous galactosylceramide in the tissue samples being analyzed to act as acceptor. Therefore, the decreased sulfotransferase activity observed could be due to either a reduced enzyme or a reduced sphingolipid acceptor. Mixing aliquots of normal and pathological white matter samples resulted either in questionable stimulation in two cases of a 32 to 82% inhibition in three other cases. Thus, it appears that a decreased galactosylceramide sulfotransferase is demonstrable in Krabbe's disease. However, this is not believed to be the primary enzyme defect.

The observations that galactosylceramide, when injected intracerebrally into animals, elicited the production of globoid-like cells and that the globoid cells obtained from pathological material have a significant quantity of this lipid suggested that a defect in the metabolism of this cerebroside might be responsible for Krabbe's disease. Reduced detectable activity of galactosyl-ceramide : β-galactosidase in homogenates of cerebral gray matter, cerebral white matter, liver, and spleen tissues from Krabbe's patients was reported (Table II).[34] A cerebroside was synthesized that was specifically labeled with tritium in the sixth carbon atom of the galactosyl moiety and this was employed as substrate. Levels of hydrolysis comparable to normals were observed with tissues obtained from patients with gangliosidosis, Gaucher's disease, Niemann–Pick's disease, and metachromatic leukodystrophy. This would suggest that a specificity exists for the finding with Krabbe's tissues. No changes were detected in the activity of several other lysosomal acid hydrolases. This enzyme deficiency has been demonstrated in serum, leukocytes, and skin fibroblasts grown in culture of Krabbe individuals.[35] Heterozygote detection appears possible,[35,36] but extensive studies have not been undertaken. However, at least one prenatal diagnosis has been accomplished and reported.[37]

A hypothesis has been offered attempting to correlate the biochemical and pathological features of this disease.[38] In the fetus and newborn, due to the small quantities of galactosylceramide present, the enzyme deficiency is of little consequence. There is a turnover of myelin as the child grows and myelination occurs. Galactosylceramide begins to accumulate due to the enzyme defect and this increase in its concentration within the brain tissue triggers the invasion by globoid cells. There is overgrowth and death of the oligodendroglial cells as the number of globoid cells increases, which results in the arrest of further myelination.

The deficiency in the hydrolysis of galactosylsphingosine (psychosine) by Krabbe's tissue samples has been reported.[39] The relationship of this observation to the development of the Krabbe's symptomatology is unclear. The presence of galactosylsphingosine has not been demonstrated in biological materials and its existence has been in question. Recent studies have reported an increase of this compound in Krabbe's cerebral tissue.[39a] The inability to cleave this compound may merely represent a nonspecificity of galactosylcer-amide : β-galactosidase for substrates. This interpretation receives support from the finding that Krabbe's tissue samples have reduced activity for cleaving the β-galactosidic bond in monogalactosyldiglyceride.[39b] The enzyme that is apparently missing in Krabbe's tissue appears to normally cleave β-galac-tosides that are linked directly to a lipoidal aglycone moiety. In common with most hydrolytic enzymes, rat brain galactosyl β-galactosidase was shown to catalyze a transgalactosylation reaction,[39c] which was absent in Krabbe's patient materials.[40]

Table II. Galactocerebroside : β-Galactosidase in White Matter[a]

Diagnosis	Age (years)	Galacto-cerebroside : β-galactosidase[b]	p-Nitrophenyl β-glycosidases[c]			
			β-Galactosidase	β-Glucosidase	N-Acetyl-β-glucos-aminidase	N-Acetyl-β-galactosaminidase
Krabbe's disease						
1	1.5	17.7	6.0	1.4	134	15.5
2	1	21.8	5.6	1.8	209	25.0
3	1.5	7.5	7.7	3.3	215	25.6
Schilder's disease	10	211	3.8	1.6	240	27.6
GM₁ gangliosidosis						
Early onset	2	335	—	—	—	—
Late onset	3	193	0.4	7.6	102	12.4
Tay–Sachs' disease	2	158	3.5	11.0	114[d]	12.5[d]
GM₂ gangliosidosis with total hexosaminidase deficiency	2.5	210	9.7	21.3	4.3	0.6
Hurler's syndrome	11	130	0.9	0.7	49.5	6.0
Gaucher's disease (infantile)	?	191	1.5	0.4	25.0	3.0
Niemann–Pick's disease	1	147	2.7	2.8	69.7	9.5
Metachromatic leukodystrophy	12	197	4.2	3.8	127	17.1
		Mean ± S.D. 197 ± 59				
Normal	0.3	236	2.7	1.8	19.4	3.2
Normal	21	254	2.2	1.1	24.1	3.3
Normal	57	161	—	—	—	—
Normal	72	143	1.8	0.8	24.1	3.3
		Mean ± S.D. 199 ± 55				

[a] Modified from Ref. 34.
[b] nmoles/hr/g.
[c] μmoles/hr/g.
[d] Hexosaminidase component A was deficient as demonstrated previously.

This aspect of broad specificity of lysosomal enzymes has been reinforced by the finding of reduced hydrolysis of the β-galactosidic bond present in lactosylceramide with tissues from patients with Krabbe's disease (Table III).[40a] The detection of both carriers and patients with Krabbe's disease appears to be more reliable employing lactosylceramide rather than galactosylceramide as the substrate (Table IV).[40b] Reduced levels of lactosylceramide : β-galactosidase activity of Krabbe's tissue have been corroborated.[40c,d] However, other workers were initially unsuccessful in demonstrating this deficiency.[40e] This unexpected observation resulted in a reassessment of the basic biochemical deficiency in this disease. The ability to demonstrate a reduced lactosylceramide : β-galactosidase in Krabbe's disease appears to depend upon the constituents present in the test tubes for the particular assay system employed. Of particular concern is the nature of the bile salt included as part of the incubation mixture. Thus, the ability to observe reduced lactosylceramide : β-galactosidase activity is dependent upon the presence of pure sodium taurocholate. Crude sodium taurocholate or other bile salts appear ineffective in this regard.[40f]

Table III. Galactosylceramide and Lactosylceramide :
β-Galactosidase Activity[a,b]

Substrate tissue	Galactosylceramide		Lactosylceramide	
	100,000g	11,500g	100,000g	11,500g
Brain				
Controls (3)	8.4–17.4	3.3–10.7	24.6–30.0	6.9–16.3
Tay–Sach	15.8	9.7	17.9	5.1
Metachromatic leukodystrophy	6.9	5.7	12.7	12.2
Generalized gangliosidosis	8.9	10.0	21.5	24.2
Fabry	27.8	7.4	50.0	15.8
Krabb (2)	0–0.05	0–0.03	0–0	0–0
Liver				
Controls (2)	0.57–2.43	0.68–0.84	2.03–6.2	2.2–12.9
Generalized gangliosidosis	2.2	4.2	6.5	2.9
Metachromatic leukodystrophy	1.3	1.3	1.8	2.4
Lipidoses (?)	3.9	5.2	3.9	5.2
Krabbe	0	0	0	0
Spleen				
Niemann-Pick type C	2.4	1.5	10.1	6.7
Fibroblasts	Total homogenate		Total homogenate	
Control	1.8		6.0	
Krabbe	0.02		0.04	

[a] Modified from Ref. 40a.
[b] Enzyme activities are expressed as nanomoles of galactose cleaved per milligram protein per hour at 37°C.

Table IV. Enzyme Activity in Cultured Skin Fibroblasts[a,b]

Subjects	Substrate	
	Galactosylceramide	Lactosylceramide
Controls		
1	5.97	18.25
2	2.26	10.07
3	—	10.83
4	—	13.87
Adult Gaucher's carrier	5.54	17.56
Lysosomal disease controls		
Sanfilippo A	2.83	11.11
Sanfilippo B	4.17	15.59
Juvenile GM$_2$ gangliosidosis	2.80	10.69
Infantile Gaucher's	4.64	18.17
Fucosidosis	7.70	31.44
Sudanophilic leukodystrophy	5.58	21.80
Niemann–Pick's type A	3.24	15.68
Niemann–Pick's type A	—	13.49
Niemann–Pick's (variant)	4.59	18.54
Average	4.48	16.22
Range	2.26–7.70	10.07–31.44
Standard deviation	± 1.64	± 5.65
Patients with Krabbe's disease		
1	0.12	1.07
2	0.21	0
3	0.08	0.13
4	0.30	0.77
Carriers of Krabbe's disease		
1	0.77	3.14
2	1.51	5.94
3	1.02	3.20

[a] Modified from Ref. 40b.
[b] Results are expressed as nanomoles of galactose released from either galactolipid per milligram protein per hour at 37°C.

In order to explain these perplexing observations, it has been suggested that most tissues contain two genetically distinct enzymes that have in common the capacity to remove the galactose moiety of lactosylceramide. It has been suggested that lactosylceramide : β-galactosidase I may be the predominant form in brain tissue. This form may be identical to galactosylceramide : β-galactosidase requiring pure sodium taurocholate for maximum activity and is reduced in Krabbe's disease. Lactosylceramide : β-galactosidase II may represent the predominant form in liver. This form may be identical to GM$_1$: β-galactosidase or the nonspecific acid β-galactosidase, requiring crude taurocholate for maximum activity and is reduced in GM$_1$ gangliosidosis.[40g] A highly purified human liver β-galactosidase was shown to be identical to lactosylceramidase II.[40g] This implies that there may be two distinct enzyme roteins with overlapping activity towards lactosylceramide. The ability to discriminate between these two activities in crude extracts of biological material appears to depend very greatly upon the constituents of the individual incubations. Evidence supporting the possible existence of separate β-galactosidases for the glycosphingolipids has appeared.[39c] A differential assay procedure has been devised for lactosylceramidase I and II.[40h] Separation of these β-galactosidases by Sephadex G-200 column chromatography has been reported with rat brain[40i] as well as with human brain and liver.[40j] Both β-galactosidases appeared to have the capacity of liberating galactose from lactosylceramide. One was also active towards galactosylceramide and the other towards GM_1 and "artificial" β-galactosides. Antibody has been prepared to galactosylceramide : β-galactosidase isolated from human placenta and used to demonstrate normal amounts of antigenically cross-reacting material in Krabbe's tissue samples. This antiserum did not react with neutral β-galactosidase or GM_1 : β-galactosidase. The enzyme has been partially purified from Krabbe liver tissue.[40k] Decreased metabolism was observed of labeled galactosylceramide added to cultures of Krabbe cells and only a slightly decreased metabolism of added lactosylceramide. This suggests that there probably is no functional abnormality in lactosylceramide metabolism in Krabbe's cells due to sufficient quantities of lactosylceramidase II.[40l]

The complexities of proper assessment of lactosylceramide : β-galactosidase activity are illustrated by the reports on lactosylceramidosis. Several papers appeared describing a condition of lactosylceramide accumulation concomitant with reduced lactosylceramide : β-galactosidase activity.[40m–o] Subsequently, it was shown that there was no such decrease, but that the patient probably was a variant of Niemann–Pick's disease.[40p] A deficiency of "neutral" β-glucosidase, which might also cleave lactosylceramide and 4MU-β-galactoside, has been reported for this particular "lactosylceramidase" patient.[40q,r]

Elevations of the human Krabbe's brain tissue activity of galactosylceramide : β-galactoside *in vitro* have been observed by the addition of *N*-decanoyl-2-amino-2-methylpropanol to incubation mixtures. This reagent has been claimed to specifically stimulate this hydrolytic enzyme activity.[41] Significant increases have been reported in the level of ceramide synthetase in Krabbe's brain samples.[42] The availability of an animal mutant for studying this disease should allow for significant *in vivo* and *in vitro* experiments designed to further understand this disease.[42a,b]

3.3. Metachromatic Leukodystrophy

3.3.1. Clinical

3.3.1.1. The Classical "Late Infantile" Form

Hagberg has conveniently divided this white matter disorder into four stages based upon increased motor involvement.[43] The first stage is defined as the time of detection of clinical symptoms and lasts an average of $1\frac{1}{4}$ years.

A loss of a previous ability to walk is often seen and this necessitates assistance for standing or walking. Deep tendon reflexes are diminished or lost, and the legs as well as the arms sometimes become flaccid. This stage may be misdiagnosed as "cerebral palsy" etc., until the progressive nature of the disease becomes apparent. In the second stage, the patient can no longer walk, but can still sit upright without assistance. Mental regression appears with a deterioration of speech. The speech disorder may reflect a combination of cerebral, brain stem, and peripheral nerve involvement. Ataxia is seen, and there may be intermittent bouts of pain in the extremities. This stage may last only 3–6 months. The child becomes quadriplegic and bedridden in the third stage. Dystonic, decerebrate or decorticate postures may be seen. There is particular difficulty in feeding and in maintaining clear airway passages. Mental defects are more marked and deep tendon reflexes are usually absent. This stage may last from $\frac{1}{4}$ to $3\frac{1}{2}$ years. In the fourth or terminal stage, the patient no longer responds to the environment. This stage is characterized by blindness, loss of speech and volitional movements. This period may last for long periods of time.

3.3.1.2. The "Adult" Form

These patients present a distinctly different clinical course than the classical infantile form. Often, some are residents of mental institutions for some time prior to the development of frank neurological disturbances. Loss of social inhibitions, euphoria, alcoholism, delusions of grandeur, poor memory, and dementia have often been cited as part of their clinical history. Except for a few cases, the diagnosis of metachromatic leukodystrophy has usually been determined based on postmortem tissue samples. A diagnosis of multiple sclerosis had occasionally been made on these individuals.[44]

These psychiatric manifestations could be considered characteristic of the disease. In addition, generalized seizures and slight incoordination are seen in later stages. In the terminal phase, the patient has seizures, is incontinent, mute, and generally deteriorated. The term "adult" indicates that the frank neurological manifestations occurred later than 21 years of age.

3.3.1.3. Variant with Multiple Sulfatase Deficiency

These children show signs suggestive of mucopolysaccharidosis in addition to those of leukodystrophy and may not show the usual psychomotor development that is characteristic of classical metachromatic leukodystrophy. They usually live only until 3–5 years of age, although one survived to 12. These patients are usually characterized by slow early development, as well as limited ambulatory and vocal skills. During the second year of life, signs of deterioration of existing skills are noted. Hepatosplenomegaly has been noted in some patients.[45] Heterozygote carriers for metachromatic leukodystrophy showed significant differences in some personality traits and spatial orientation.[45a]

3.3.2. Pathology

Deterioration of the central and peripheral nervous system white matter is the main pathological feature of this disease. Upon gross examination, the brain appears to be a wasted organ and the white matter has a grayish appearance.

Marked loss of both the myelin sheath and the oligodendrocytes is evident histologically. Granules or inclusions are found in macrophages, within the brain tissues, and within spared oligodendrocytes. The inclusions are observed to be metachromatic,[46] and it is for this staining property that the disease is named. These granules also stain positively for lipid with Sudan black and for carbohydrate with PAS. These staining properties are lost if the tissue is extracted with conventional lipid-solubilizing solvents and is consistent with the nature of the sphingoglycolipid, sulfatide, that is known to accumulate in these tissues. A morphologically heterogeneous and fibrous gliosis is found in the area of demyelination.

From electron microscopic studies, the inclusions appear to be lysosomal.[47]

These histological observations in the central nervous system are also evident in the peripheral nerves. Therefore, sural nerve biopsies are of diagnostic value in metachromatic leukodystrophy.

3.3.3. Biochemistry

An excessive quantity of sulfatide, ceramide galactoside-3-sulfate, was reported present in the white matter from metachromatic leukodystrophy patients almost simultaneously by Austin[48] and Jatzkewitz.[49] Sulfatide levels may be increased almost 10-fold, although the levels of cholesterol and galactosylceramide, two other myelin-associated lipids, are decreased. Excessive quantities of both sulfatides and cerebrosides have been observed in kidney tissues. Compositional studies have revealed no significant chemical differences between the sulfatide isolated from normal or metachromatic leukodystrophy tissue samples.[50,51] Decreases in the long-chain fatty acids containing from 21 to 26 carbon atoms have been reported for the cerebrosides and sphingomyelin obtained from metachromatic leukodystrophy white matter.[51a] The loss of these particular populations of sphingolipids may merely reflect the loss of myelin from this tissue.[52] There is a reported elevation of ceramide dihexoside sulfate in kidney tissue and in the urine of these patients.[53,54]

In addition to sulfatide, sulfated polysaccharides and sulfated sterols accumulate in tissues of patients with multiple sulfatase deficiencies.[55,56]

The metachromatic granules that are observed in histological sections of tissues have been isolated using classical centrifugal procedures. They appear to be 50% lipid on a dry weight basis, and have a molar ratio of cholesterol : cerebroside : sulfatide : phospholipid of 1 : 0.2 : 0.8 : 0.93. Gangliosides do not appear to be components of these granules.[57] These compositional data suggest a similarity between white matter, myelin, and the granules isolated from metachromatic leukodystrophy brain tissues. Due to their high protein content, they could be regarded as a lipoprotein matrix.

The sulfatases are a group of hydrolytic enzymes capable of removing sulfate ester groups and there appear to be two distinct populations differing in their intracellular location and pH optimum. The nature of their specific biological substrate, if there is any rigorous specificity, is not fully understood. Therefore, for the sake of convenience, they are often referred to as the arylsulfatases. The "C" type appears to be microsomal in location and has a pH optimum near neutrality. The types "A" and "B" are most likely lysosomal in origin and have acidic pH optimums. Austin and colleagues carried out a comparative study of nine different enzyme activities with brain, kidney, and liver samples from normals, globoid cell leukodystrophy, gargoylism, and metachromatic leukodystrophy. The sole consistent finding was a marked decrease of arylsulfatase A activity in the metachromatic leukodystrophy tissue samples (Table V).[58] These authors presented a very conservative interpretation of this observation, mainly due to the reported inability of arylsulfatase A preparations from human brain to cleave $^{35}SO_4$ from [^{35}S]sulfatide. This deficiency of arylsulfatase A, which was reported almost 20 years ago, is accepted as a reliable assessment of the basic enzymatic deficiency in this disease. Kidney tissue extracts from metachromatic leukodystrophy patients were subsequently shown to be unable to catalyze the release of the sulfate group from labeled sulfatide.[59] Further confirmation of the reliability of using arylsulfatase A activity as an operational estimation of this deficiency has appeared.

Highly purified arylsulfatase A was isolated from pig kidney and this preparation was found capable of catalyzing the liberation of sulfate from both nitrocatechol sulfate as well as from sulfatide. A heat-stable "complementary" fraction obtained from the kidney, which did not possess any hydrolase enzyme activity, caused a stimulation of sulfatide hydrolysis without having any effect upon nitrocatechol sulfate hydrolysis.[60] This has been purified from human liver and shown to be a protein.[60a] It appears to be

Table V. Specific Activities of Two Enzyme Systems Involved in Sulfate Metabolism[a]

Sample and diagnosis	Aryl sulfatase[b]			PAPS-degrading enzyme[b]		
	Brain	Liver	Kidney	Brain	Liver	Kidney
1. Normal	88.6	55.0	71.9	38.6	74.0	275
2. ML	1.1	10.4	3.9	21.4	96.4	85
3. Normal	33.5	29.7	56.4	106.4	59.1	46
4. GL	31.9	53.8	74.0	51.1	80.0	170
5. GARG	—	89.0	89.6	—	69.6	90
6. ML	3.1	6.9	—	39.6	44.6	—
7. Normal	52.1	—	—	33.0	—	—

[a] Modified from Ref. 58.
[b] Expressed in units per milligram protein.

localized in the lysosomal fraction of rat liver and possesses a pI of 4.1.[60b] Some of the interactions between the activator of the enzyme and the substrate have been described.[60c] The possibility arose that this material might be missing in metachromatic leukodystrophy tissues. Studies have indicated that similar quantities of this material are present in tissues from both normals and metachromatic leukodystrophy patients.[61] Rather than being a "complementary" fraction, it is currently believed to be an activator. Several publications have reported the reduction of both arylsulfatase A and cerebroside-3-sulfate sulfatase in material from metachromatic leukodystrophy patients.[62–64] Reduced hydrolysis of the sulfate esters present in sulfolactosylceramide,[64a] sulfogalactosylsphingosine,[64b] and sulfogalactosylglyceride[64c] has been reported for classical metachromatic leukodystrophy. This suggests that the enzyme that is decreased in this genetic disease is not absolutely substrate specific. Steroid sulfatase and arylsulfatase C activities do not appear impaired in classical metachromatic leukodystrophy. However, a reduction in testosterone-sulfate cleavage has been reported.[64d]

In vivo observations on the urinary excretion of radioactive sulfatide from patients receiving inorganic $^{35}SO_4$ support the proposed metabolic defect in sulfatide hydrolysis. Nonmetachromatic leukodystrophy individuals had an apparent turnover rate of 2–3 days, i.e., 5- to 10-fold lower than metachromatic leukodystrophy patients.[65]

The variant with multiple sulfatase deficiencies has, in addition to decreased arylsulfatase A, a diminution of arylsulfatases B and C.[66] The latter activity is believed to be localized in the microsomal fraction, whereas the former two are associated with the lysomal fraction. The hydrolysis of cholesterol sulfate and dehydroepiandrosterone sulfate is greatly diminished by tissues of these patients[56] as well as with cultured skin fibroblasts.[66a]

Few attempts have been undertaken to correlate the residual activity of the deficient enzyme, arylsulfatase A, with the onset of the clinical signs. Stumpf and Austin assayed a fraction precipitating between 1.8 and 2.2 M ammonium sulfate from the urine of normal individuals as well as from late infantile and juvenile metachromatic leukodystrophy patients. The specific activity of the measured residual arylsulfatase A for the late infantile patients was 8 nmoles/mg protein per hr; for the juvenile, 113 nmoles/mg protein per hr; and for the controls, 3.474 μmoles/mg protein per hr.[67] This is consistent with the possibility that the earlier the onset of the clinical manifestations in the late infantile form correlates with a more abnormal (?) enzyme protein. In a subsequent analysis employing solid tissue samples, no such correlation was found between the level of detectable enzyme, the level of sulfatide accumulation, and the onset of the disease.[62]

Antibodies have been raised in rabbits against a purified arylsulfatase A isolated from normal human liver. Precipitation lines against this antibody were obtained both with normal and with metachromatic leukodystrophy liver extracts using the technique of immunodiffusion. This suggests that the affected individual produces an immunologically active but catalytically inactive protein.[68] A technique has been developed that distinguishes subtle differences between those enzyme forms, differences that are not seen with usual immunodiffusion or immunoelectrophoresis.[68a] The effects on the activity of purified human liver arylsulfatase A by addition of this monospecific antibody have been reported. The antibody either stimulates or protects the enzyme activity under several specific experimental conditions.[69]

The usefulness of quantitation of arylsulfatase A enzyme activity, employing nitrocatechol sulfate as substrate, with a variety of biological tissue samples in order to detect the metachromatic leukodystrophy patient, is well documented.[70–72] Problems associated with the contribution of arylsulfatase B in these assays have been discussed.[72a–c] Heterozygote detection has been reported.[73,74] Due to the significant overlap with normals that has been observed, it may be of questionable reliability[75]; however, the 4MU substrate may be potentially more useful.[75a] Prenatal diagnosis of fetuses affected with metachromatic dystrophy has been reported.[76,77] Low arylsulfatase levels of cultured amniotic fluid cells of a fetus at risk were found. However, when sulfatide was employed as substrate in the assay, heterozygote values were obtained. This illustrates the potential problem of employing "artificial" substrates for diagnostic purposes.[77a]

The capacity of cultured MLD fibroblasts to convert exogenous [14C]cerebroside sulfate to cerebroside has been proposed as a useful diagnostic procedure.[77b] A patient with characteristic clinical and pathological signs of MLD had nearly normal arylsulfatase A and B with homogenates of cultured cells. However, these cells were unable to hydrolyze labeled sulfatide supplied in their culture medium.[77c]

Cultured skin fibroblasts derived from metachromatic leukodystrophy patients show enzyme deficiencies but no sulfatide accumulation.[78] When radioactive sulfatide is added to the culture medium, there is an accumulation of label in this sphingolipid with metachromatic leukodystrophy cells but not with normal cells.[79] This accumulation of sulfatide can be prevented by supplying a purified arylsulfatase A to the cells derived from these patients.[80]

Investigations employing fusion of human fibroblasts with Chinese hamster cells have provided data that have been interpreted as indicating that the gene for arylsulfatase A resides on chromosome 22 and that for arylsulfatase B on chromosome 5.[80a,b]

Attempts to improve the clinical state of patients by the administration of purified human urinary arylsulfatase A into the subarachnoid space[81] or the administration of a beef brain preparation by a combination of intravenous and intrathecal routes[82] have been unsuccessful.[83]

An oligodendroglia enriched fraction was obtained from autopsy tissue of a 3.8 year old MLD patient. These cells were maintained in culture, exposed to [3H]galactose and labeling of glycolipid and glycoprotein examined. The most striking difference was an observed increased amount of glycoprotein of MLD myelin capable of binding to wheat germ agglutinin.[83a]

3.4. Gaucher's Disease

This disease was the second sphingolipidosis to be defined clinically and pathologically and the first sphingolipidosis for which a specific lysosomal enzymatic deficiency was reported. The rationale for the biochemical approach to understanding Gaucher's disease provided the basic guideline for all subsequent elucidations of the enzymatic deficiency in the other sphingolipidoses.

3.4.1. Clinical

Approximately 1000 individuals with this disease have been described. There are certain general clinical manifestations although two or possibly three distinct variants have been reported. The useful classifications suggested will be employed.[84]

3.4.1.1. Type 1: Chronic Nonneuronopathic (Adult)

This is the most benign form of the disease, and the hallmark differentiating it from the other two forms is the absence of neurological or cerebral involvement. It may be recognized either in early childhood or in adulthood. The most common manifestation is an increased mass of the spleen and occasionally the liver. This splenic hypertrophy is usually associated with a variety of hematologic changes, which may be an anemia, a decrease in the circulating platelets (throbocytopenia), or a decrease in the number of circulating white blood cells (leukopenia). When the mass of the spleen is so enlarged that it causes discomfort, it is removed at elective surgery, and this results in improvement of the hematologic problem. Bone pain and pathological fractures caused by only mild injuries are not uncommon observations. These may be associated with, or be a result of vascular changes due to the glycolipid storage.

The prognosis for the individual with this type of Gaucher's disease is usually hopeful. In the past, these patients were often unable to combat intercurrent infections. With the advent of the modernday spectrum of antibiotics, this is rarely of great concern.

Many patients have brownish-bronzy skin pigmentation on exposed surfaces. These pigments are presumably a mixture of iron-containing deposits such as hemosiderin. There may be yellowish growths of connective-type tissues over the external part of the eye.

Gaucher's deposits were observed in the perivascular spaces of small blood vessels of the central nervous system of a type 1 patient. There was no storage seen in the neurons.[84a]

This particular form of Gaucher's disease is most prevalent in Ashkenazi Jews.

3.4.1.2. Type 2: Acute Neuronopathic

This has also been referred to as the acute, infantile, cerebral, or malignant form of Gaucher's disease. It is uniformly fatal and the affected infant rarely reaches 3 years of age.

The onset of the disease is usually rapid, and progressive neurological symptomatology is characterized by both cranial nerve and extrapyramidal tract dysfunction. The most common clinical features are the paralysis or incoordination of the extraocular muscles (strabismus), muscular spasticity and retroflexion of the head. Difficulty in opening the mouth (trismus), difficulty in swallowing (dysphagia), and increased deep tendon reflexes are less frequent. Terminally, most patients become apathetic, mentally retarded, and sometimes hypotonic.

In addition to these neurological defects, the clinical characteristics described above for the type 1 Gaucher's patient such as splenomegaly with associated hematologic changes, are also evident. This type appears to be panethnic in its occurrence.

3.4.1.3. Type 3: Subacute Neuropathic (Juvenile)

This classification includes older children, adolescents, and young adults with neurological involvement, splenomegaly, and characteristic pathology. Few reports of individual cases are present in the literature and only a single study of six patients in this category is available. This is a heterogeneous group of clinical descriptions, few of which have been unequivocally confirmed by specific biochemical techniques.[84b] The clinical description of 22 Norrbattnian type 3 Gaucher's patients has appeared.[84c]

3.4.2. Pathology

The most general pathological finding is the presence of a characteristic storage-cell, the "Gaucher's" cell, which in the light microscope has an appearance that has been described as "wrinkled tissue paper" or "crumpled silk."[85] These cells stain intensely with PAS reagents due to the storage of carbohydrates. In the electron microscope, a residual body or secondary lysosomes are seen containing a twisted tubular structure.[86] These deposits have been isolated and shown by X-ray diffraction to consist of a series of narrow twisted bilayers.[87] Tubules with similar appearance can be observed with pure preparations of glucosylceramide. These cells are observed in most tissues of the affected individual including blood, liver, adventitial cells of blood and lymph vessels. They are rarely encountered in the brain tissue itself, but often are present in the adventitia of the small cerebral vessels.

"Gaucher-like" cells have been observed in thalassemia[88] and chronic myelocytic leukemia.[89,90]

Scattered areas of neuronal loss in type 2 Gaucher's disease have been observed. Phagocytosis of the neurons (neuronophagia) is found in some brain areas. There is little unequivocal evidence for the storage of materials in the cell bodies of the neurons that stain positively with PAS, and this staining might merely reflect normal cellular constituents.

In addition to the presence of "Gaucher's" cells, an elevated serum acid phosphatase activity, which is not inhibited by the presence of tartrate, is usually considered to be characteristic of patients with this disease.[91] An improved assay procedure employing 4MU-phosphate as substrate has been developed[91a] for demonstrating the multiple forms of the plasma acid phosphatase of Gaucher patients.[91b] Elevated levels of serum angiotensin-converting enzyme have been reported for Gaucher patients.[91c] A provocative observation on the appearance of L-[1-^{14}C]glucosylceramide in the bile of experimental animals suggests this excretion as another route for removal of this compound from the body. Examinations were carried out on a restricted series of samples from three Gaucher's patients: one had elevated biliary and two had elevated hepatic cerebroside.[91d]

3.4.3. Biochemistry

Early attempts to understand the basis of this disease were obscured by the confusion surrounding the nature of the accumulating lipid. At that time, galactosylceramide had been identified as the major glycosphingolipid present in the nervous tissue, and only present in small quantities in peripheral organs. It was suggested that excess galactosylceramide was excreted from the brain into the blood and excessive quantities accumulated in the spleen of Gaucher's patients. This hypothesis became unacceptable when (1) only a small cerebroside excess was found in plasma and (2) the carbohydrate moiety was found to be glucose rather than galactose.[92]

In spleen tissue, there is an increase of glucosylceramide from approximately 0.17 mg/g wet wt in controls to levels ranging from approximately 5 to 30 mg/g wet wt in Gaucher's disease. The composition of the material appears to be nearly identical to the cerebroside usually present in this tissue.[93] Increased levels of hematoside (GM$_3$) have been observed in spleen tissue of both type 1 and type 2 patients.[94] Lactosylceramide but not glucosylceramide levels were found to be increased in leukocytes[94a] and elevated glucosylceramide levels in plasma but not erythrocytes of Gaucher patients.[94b]

The membranous deposits observed in Gaucher's tissues have been isolated and found to contain glucosylceramide, phospholipid, and cholesterol in a ratio of 12 : 3 : 2. The nonlipid residue accounts for 12% of the weight and includes a PAS-positive glycoprotein containing galactose, hexosamine, and sialic acid.[95]

Although the accumulation of glucosylceramide was documented in peripheral organs of type 1 Gaucher's tissues, information about this substance for the brain tissues of type 2 patients is conflicting. This difficulty may be due to the presence of galactosylceramide as the principal cerebroside of cerebral tissue. Most studies have observed a decrease of brain cerebrosides probably due to degeneration of the central nervous system. Several authors have been unable to demonstrate either the accumulation or the presence of glucosylceramide in brain tissues of Gaucher patients.[84a,96,97] In those instances where glucosylcerebroside has been found, there is a close compositional resemblance to the gangliosides present in the tissues. Stearic acid (C$_{18}$) represents the bulk (88 + %) of the acyl residues with traces (7–10%) of C$_{20}$ sphingosine base in addition to the most common C$_{18}$ base being observed. This is typical of the base and fatty acid composition of cerebral gangliosides.[98,99,99a]

Insight into the nature of the metabolic defect responsible for the accumulation of glucosylceramide was suggested two decades ago.[100] Slices of Gaucher and normal spleen tissues were incubated with either [14C]glucose or [14C]galactose and the labeled cerebrosides produced during the incubation were isolated. These were then hydrolyzed and the liberated glucose and galactose were recovered. There were no apparent differences between the ability of normal and Gaucher samples in converting either of the two radioactive carbohydrates into the glucose moiety of the cerebrosides. This observation allowed the authors to conclude that a deficiency in a hydrolytic enzyme was most likely responsible for the accumulation of glucosylceramide in Gaucher's tissues.

The documentation of this metabolic deficiency was reported almost simultaneously by workers on opposite sides of the North Atlantic Ocean. The group at the National Institutes of Health prepared [1-14C]glucosylceramide as substrate using strictly synthetic organic chemical techniques. Employing this material as substrate, it was observed that Gaucher's spleen tissues had a marked deficiency in the ability for the hydrolysis of the carbohydrate to a value only 10% of that observed in controls (Table VI).[101] Care was taken to correct for the dilution of the added substrate by the endogenous glucosylceramide present in patient material. It was therefore concluded from these studies that the accumulation appeared to be caused by a decreased glucosylceramide : β-glucosidase. Patrick, working in Great Britain, reported that a β-glucosidase deficiency was probably responsible for Gaucher's disease.[102] Gaucher's tissue possessed only 12% the capacity to cleave pNP-β-D-glucopyranoside when compared to control tissues. Nonradioactive glucosylceramide was also tested as substrate with normal and pathological tissue samples and the glucose released measured enzymatically with glucose oxidase. Zero

Table VI. Level of Glucocerebroside-Cleaving Enzyme in Human Spleen Preparations[a,b]

			Activity of glucocerebroside-cleaving enzyme	
Patients condition			Radioactivity in aqueous phase (cpm/mg protein)	Percent of non-Gaucher activity
Congenital hemolytic anemia	F	76	93.8	—
Congenital spherocytosis	M	4	72.1	—
Idiopathic thrombocytopenic purpura	F	37	77.0	—
Congenital hemolytic anemia	F	14	67.4	—
Average specific activity			77.6 ± 11.5[c]	
Gaucher's disease	F	33	13.9	17.9
	F	3	10.9	14.0
	M	13	8.4	10.8
Average specific activity			11.1 ± 2.24[d]	

[a] Modified from Ref. 101.
[b] The conditions of incubation and procedure for correcting for endogenous dilution of the labeled substrate are described in the text.
[c] Standard deviation.

activity was reported with Gaucher's spleen samples and detectable quantities with nonpathological samples. These studies led to the conclusion that the basic defect was a decreased β-glucosidase activity. [1-^{14}C]glucosylceramide hydrolysis in brain tissues from three infantile Gaucher (type 2) patients has been reported to be only 5% of that observed in control brain samples.[103] A 26,000-fold purification of human spleen glucosylceramide : β-glucosidase has been accomplished from both controls and a Gaucher patient. The control enzyme was about 10-fold more active than the mutant enzyme but most of the other properties were identical. Equivalent quantities of immunologically cross-reacting material were found in extracts of Gaucher tissues.[103a]

The usefulness of employing a synthetic or unnatural substrate in monitoring β-glucosidase activity for diagnostic purposes is well documented. Gaucher's patients as well as heterozygote carriers have been identified with peripheral leukocytes,[104,104a–e] solid tissues,[105] and skin fibroblasts[106] employing 4MU-β-D-glucoside as substrate. Difficulties in diagnosing Gaucher's disease employing liver and spleen samples with 4MU-β-glucosides were ascribed to residual "soluble" enzyme. The inclusion of 100 mM sodium chloride in the incubations overcame these difficulties, allowing unequivocal diagnosis.[106a] White blood cells from patients[107] and heterozygotes as well as skin fibroblasts[108] have also been identified employing [1-^{14}C]glucosylceramide as substrate. The prenatal diagnosis of an infantile Gaucher's fetus has been successfully accomplished.[109]

Comparable hydrolytic rates with either N-oleoyl- or N-stearoylglucosylceramide were observed with splenic extracts and fibroblasts from Gaucher's patients.[109a] Methodological improvements for identification of patients and carriers involving detergent,[109b] hydrolysis product quantitation,[109c] or the introduction of another unnatural substrate[109d] have been reported. Electrophoretic separation has been accomplished for the various forms of β-glucosidase.[109e] Attempts have been undertaken to study multiple forms of the membranous form of the enzyme.[109f,g]

Normal human liver β-galactosidase can be divided into three fractions on Sephadex G-150 columns and one of these fractions was found to have both β-glucosidase and β-D-xylosidase activities. Gaucher's liver samples subjected to a similar treatment show only the first two β-galactosidase peaks with an absence of glucosidase and xylosidase activity. This would suggest a commonality between one of the hepatic β-galactosidase and β-glucosidase and β-xylosidase.[110] This observation has been corroborated by the finding of decreased β-xylosidase activity with spleen, fibroblast, and leukocyte samples from Gaucher's patients[110a] as well as in the pharmacologically produced "Gaucher's" mouse.[110b] It is likely that particulate rodent liver β-glucosidase and β-xylosidase activity reside in the same enzyme.[110c]

It has been suggested that acid β-glucosidase and glucosylceramide : β-glucosidase activities consist of a soluble glycoprotein, present in elevated amounts in Gaucher's spleen, referred to as "P factor," and an insoluble component with low hydrolytic activity from normal spleen, referred to as "C factor."[111,112] When both of these factors were mixed together in the same incubation tubes, large increases were observed in the hydrolysis of both 4MU-β-D-glucoside and glycosylceramide. This would imply that the metabolism of these substrates requires both a catalytic hydrolytic protein and a "helper" nonhydrolytic glycoprotein. A model of glucosylcerebrosidase has been proposed that involves the catalytic protein, the "P factor" or effector, and phospholipids.[112a–d] This interpretation has been challenged and explained in terms of changes in the pH optimum for the hydrolysis of these substrates.[113] Highly purified glucosylceramide : β-D-glucosidase from placental tissue, which possesses only 4% activity towards the corresponding fluorogenic substrate, is not stimulated by a crude preparation of "P factor." These incubations are carried out at pH 6.0. However, when assayed at pH 4.0, under suboptimal conditions, there is enhanced activity, presumably due to stabilization of the enzyme.[114]

Highly purified "activators" have been isolated from both control and Gaucher's spleens. On a protein basis, the material from Gaucher's spleen is 16 times more active than from control spleen. The patient-derived material is a glycoprotein in contrast to the control material, and significant differences in the amino acid composition were also apparent.[114a] This activator appears to be associated with the lysosomal fraction of Gaucher cells.[114b] The β-glucosidase "stimulator," purified 32,000-fold from bovine spleen, decreased the K_m and increased the V_{max} of a partially purified glucosylceramide : β-glucosidase.[114c] Complex formation between the enzyme and the stimulator was prevented by phosphatidylserine. There was no complex formation between the stimulator and substrate.[114d]

Elevated levels of a glycopeptide have been reported to be present in Gaucher's spleen samples when compared to control tissues.[115] This may be related to the nonlipid component that has been observed to accumulate in Gaucher's lysosomes using microscopic techniques.[115a] This crude preparation has been further purified and fractionated into several components.[116] One of these fractions stimulated the hydrolysis of 4MU-D-glucoside by a calf spleen enzyme[117] purified by affinity column chromatography.[117a] The residual bodies isolated from Gaucher's spleen samples have been shown to possess β-glucosidase activity.[118]

In common with many hydrolytic enzymes, the glucosylceramide : β-glucosidase possess transglucolytic activity. Glucosylceramide is produced by the enzyme in the presence of ceramide acting as acceptor and several β-glucosides acting as glucosyl donors. This activity is absent in tissues from Gaucher's patients.[117] The reduced capacity for the hydrolysis of glucosylsphingosine by cultured skin fibroblasts as well as spleen tissue obtained from Gaucher's individuals was reported.[119] Evidence for the presence of glucosylsphingosine in the tissue samples was also presented. This material has been isolated and unequivocally identified as glucosylsphingosine.[119a] This observation has been supported by the presence of a similar compound in brain tissue of these patients.[120]

An inability to cleave steroid β-glucosides has been reported for Gaucher's tissues. The particulate fraction from both adult and infantile forms of the disease showed this reduced hydrolytic activity. In addition, decreased hydrolytic activity was obtained with the "soluble" fraction only from the infantile

Table VII. The Hydrolysis of 4MU-β-D-glucoside, Glucosylceramide, and 17α-Estradiol-3-β-D-glucoside by the Soluble and Particulate Fractions of Infantile and Adult Gaucher's Tissues[a,b]

	4MU-β-glucoside		Glucocerebroside		Steroid β-glucoside	
	Soluble	Particulate	Soluble	Particulate	Soluble	Particulate
Infantile control spleen (1)[c]	16	27.5	6.6	215.2	9.4	13.1
Infantile Gaucher spleen (2)	1.5	0.3	4.4	18.2	NDH[d]	NDH
Infantile Gaucher spleen (3)	NDH	0.4	2.6	0.9	1.6	0.8
Infantile control liver (1)	72.6	32.5	5.5	204.7	11.8	12.1
Infantile Gaucher liver (2)	51.6	2.6	13.2	27.1	29.7	1.7
Infantile Gaucher liver (3)	77.6	1.2	3.8	10.9	21.5	0.6
Infantile Gaucher liver (4)	NDH	1.0	4.9	17.3	13.8	0.6
Infantile control brain (1)	2.26	93.3	0	773.1	7.2	30.55
Infantile control brain (11)	3.8	27.9	0.7	679.4	7.5	27.1
Infantile Gaucher brain (2)	0	7.2	2.0	96.9	0.2	3.5
Infantile Gaucher brain (3)	0.7	2.1	2.5	24.1	2.1	1.2
Control adult spleen (5)	5.5	44.3	61.2	260.0	4.7	6.7
Control adult spleen (6)	32.5	43.4	9.5	248.1	9.3	25.5
Adult Gaucher spleen (7)	2.2	1.0	3.8	12.1	2.5	NDH
Adult Gaucher spleen (8)	5.5	3.3	9.2	33.7	9.5	1.6
Adult Gaucher spleen (9)	8.8	3.0	9.1	27.5	7.4	1.4
Adult Gaucher spleen (10)	12.9	5.2	6.3	51.8	15.3	3.1

[a] Modified from Ref. 120a.
[b] Activities expressed as nmoles cleaved/mg protein/unit time.
[c] Numbers in parentheses designate tissues from different individuals.
[d] NDH, no detectable hydrolysis.

patient material. This may represent the biochemical difference between these two forms of Gaucher's disease.[120a,b] These observations on the decreased hydrolytic activity were reinforced by finding reduced transglucolytic activity, i.e., decreased formation of steroid β-glucoside (Table VII). A deficiency of a relatively nonspecific soluble aryl glycosidase has been reported for liver and kidney of a juvenile Gaucher patient.[120c] This activity is probably identical to that that cleaves steroid β-glucosidases but is inactive with glucosylceramide.[120d] Membranes were prepared from control and Gaucher's spleen tissues and heated at 52.4°C for 3 hr. There was no loss in the ability to cleave 4MU-β-D-glucoside by the patient material but a 75% loss with the control preparation. The heat-stable β-glucosidase of control and patient membranes did not possess glucosylceramide : β-glucosidase activity.[120e] This heat-stable component may be related to the aryl glycosidase activity.

Exogenous [³H]glucosylceramide was not actively metabolized by cultured skin fibroblasts from Gaucher patients.[120f] The addition of N-hexyl-O-glucosylsphingosine to either cultured rat neuroblastoma[120g] or human skin fibroblasts resulted in elevated glucosylceramide levels. However, this material was ineffective in rats.[120h] Electrophoretic examination was performed on β-glucosidase of cultured normal and Gaucher skin fibroblasts. The residual activity present in the adult-type patient coincided with that of normals. The enzyme activity from the infantile form of the disease did not survive the electrophoresis.[120j] The addition of GM₁ ganglioside to homogenates of type 1 Gaucher fibroblasts increased the K_m but not the V_{max} for a β-glucosidase.[120k]

Conduritol β-epoxide was demonstrated to be a very effective inhibitor of purified β-glucosidase.[117] This inhibitor was employed in producing the "Gaucher" mouse.[120i] These animals have elevated tissue levels of glucosylceramide with a reduced β-glucosidase activity (Table VIII, IX). Restoration

Table VIII. Tissue Levels of Glucosylceramide in Experimental and Control Mice[a,b]

	Tissue	Glucosylceramide (nmoles/g wet wt.)
3-month-old mice		
Experimental	Spleen	138
Control	Spleen	74
Experimental	Liver	127
Control	Liver	75
Experimental	Brain	33
Control	Brain	6.9
Infant mice		
Experimental	Brain	60
Control	Brain	21
Experimental	Liver	49.3
Control	Liver	32.2

[a] Modified from Ref. 120g.
[b] Each value is the average of analysis of at least three tissue samples from separate animals.

Table IX. A Comparison of Tissue Levels of Several Hydrolytic Enzyme Activities from Experimental and Control Animals[a,b]

	α-Mannoside	β-Glucoside	β-Hexosaminide	β-Galactoside	α-Glucoside	β-Glucuronide
3-month-old mice						
Experimental liver	17.6	2.24	278	5.84	2.57	5.49
Control liver	20.1	67.8	287	6.6	3.22	8.24
Experimental brain	13.11	0.1	2016	17.8	4.45	5.4
Control brain	8.66	9.49	966	16.47	4.12	3.09
Infant mice						
Experimental spleen	86.2	1.49	1483	53.8	2.24	56.6
Control spleen	60.13	21.0	1858	55.3	5.79	80.2
Experimental brain	24.4	1.25	1896	22.26	1.52	15.18
Control brain	11.4	17.84	781	28.1	1.79	5.11

[a] Modified from Ref. 120g.
[b] All values are expressed as nmoles substrate hydrolyzed/mg protein/hr. Tissues from three different 3-month-old mice, eight experimental infant mice, or 16 control mice were assayed.

of tissue levels of glucosidase with reduction of accumulated glucosylceramide has resulted by cessation of drug administration.[120l] The appearance of characteristic morphological inclusion bodies was seen in the mice.[120m] A marked decrease of β-glucosidase activity occurred 5 hr after injection,[120n] and reduced β glucuronidase activity has been reported in these animals.[120o]

Liver transplantation has been carried out with a Gaucher patient[121] in an attempt at enzyme replacement. Erythrocyte ghosts have been incubated with a crude rat kidney glucosidase-enriched preparation. Uptake of β-glucosidase into the ghosts has been observed, suggesting a possible means of enzyme replacement therapy.[122] This interesting approach has the advantage of employing the patient's own red blood cells as the carrier for the enzyme. Highly purified placental glucosylceramide : β-glucosidase has been reported to substantially reduce hepatic glucosylceramide levels of Gaucher's patients.[122a] There was a maintenance at control levels of erythrocyte glucosylceramide for a period of 24 to 48 weeks postinjection (Fig. 2), which is unusual since others have not detected such elevations in erythrocytes.[94b] These authors have calculated that this treatment resulted in a reduction of hepatic glucosylceramide equivalent to 4 or 13 years' accumulation.[122b] Unfortunately, these workers were unable to reproduce these effects with other patients and indicate that successful enzyme replacement therapy has yet to be established.[122c] Attempts of enzyme replacement therapy with glucosylceramide : β-glucosidase entrapped in erythrocytes,[122d,e] monocytes,[122f] or lysosomes[122g] have provided equivocal success. A trimannosyl residue was

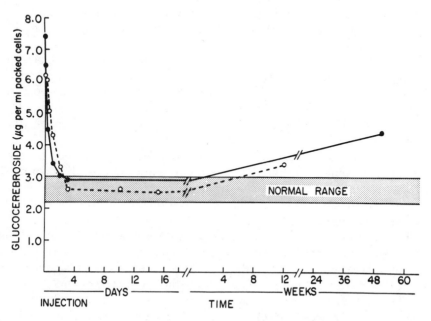

Figure 2. Effect of intravenous injection of purified human placental glucocerebrosidase on erythrocyte glucocerebroside in two patients with Gaucher's disease. (●) Patient No. 1; (○) patient No. 2.

chemically linked to purified glucosylcerebroside-β-glucosidase. Enhanced *in vitro* macrophage uptake and *in vivo* targeting of the enzyme to the reticularendothelial system was demonstrated with this altered enzyme.[122i]

3.5. Fabry's Disease (Angiokeratoma Corporis Diffusium)

Kuhnaii analyzed the heart muscle of a patient and noted an elevation of a poorly characterized phosphatide. He therefore concluded that Fabry's disease was a lipid storage disorder. Although the conclusion was correct, the material that accumulates is a glycosphingolipid rather than a phospholipid.

Until 1952, all reported patients had been males. Two half brothers, offsprings of the same mother but different fathers, both had this disease. This suggested the possibility that the disease is transmitted as a sex-linked characteristic. Studies of man–hamster hybridized cells have demonstrated that the gene expressing α-galactosidase is on the X chromosome.[122f] The female heterozygotes may have some of the typical clinical manifestations, but the disease does not lead to serious disability to these carriers.

3.5.1. Clinical

The earliest complaint usually relates to painful episodes in the extremities generally located over the joints. The discomfort is accentuated when the patient is subjected to extremes of temperature, either hot or cold. Further manifestations of this peripheral neuropathy involve bouts of tingling, burning, or numb sensations (paresthesias). Impaired sweat production (anhidrosis) is commonly seen.

The characteristic skin lesions appear as small dark-blue to black spots or papules distributed over the body, but are rarely found on the hands or feet. The spots are slightly overgrown by the horny layer of the epidermis (hyperkeratotic) and do not blanch with the application of pressure. These angiokeratomas do not bleed readily even after puncture, presumably due to restricted blood flow.[123]

In 30% of the patients, diffuse myocardial lesions have been observed. Coronary involvement, although it does occur, is not a uniformly encountered characteristic of Fabry's disease. Neurological involvement does occur in some of these patients. However, the diagnosis is difficult and complicated by the attendant widespread vascular lesion.

The accumulation of lipid over a protracted time course results in impaired renal function. Following gradual increases in blood urea nitrogen, frank uremia eventually occurs. Renal dialysis often satisfactorily controls this aspect; however, renal failure is commonly fatal to these patients.

The deposition of glycosphingolipids in the walls of the vasculature often results in cerebral vascular disease. Clot formation (thrombosis), seizures, paralysis (hemiplegia) or loss of tactile sensibility (hemianesthesia) on one side of the body, inability to communicate (aphasia), and cerebral hemorrhages are seen.[124] Death due to this disease may result from uremia or from strokes caused by cerebrovascular disease.

There is great variability in the extent and type of abnormalities found in the heterozygote. The most common and frequent manifestation is corneal involvement. Skin lesions, occasional pain in the extremities, or some signs of kidney malfunction are also found. The apparent severity of these signs increases with age. Death, when caused by the disease, often is due to cardiac or renal failure.[125]

3.5.2. Pathology

The principal changes observed are the thickening of the blood vessels and the presence in reticuloendothelial tissues of "storage" or "foam" cells, as well as storage material in the ganglion cell. The storage glycolipid appears as a birefringent Maltese cross upon histological examination.

The skin lesions observed are caused by dilation of blood vessels (angiectases) which gradually develop into the final telangiectases.

Lipoid deposition in the kidney is observed in cells of the loop of Henle and Bowman's capsule. In addition, the blood vessels within the kidney become involved. For some time the foam cells found in the glomeruli had been considered as pathognomonic of Fabry's disease; however, these do not appear to be specific since they are also observed in other unrelated renal diseases.

There is swelling of neurons in the central autonomic-nervous system and peripheral nervous system, in addition to the lipid deposit in the walls of the cerebral vasculature.

3.5.3. Biochemistry

The principal lipid that accumulates in Fabry's disease is trihexosylceramide (CTH) with smaller quantities of a digalactosylceramide (CDH); increases of 30-fold for both of these compounds were reported.[126] The structures originally proposed for these were:

$$Gal 1 \rightarrow 4Gal 1 \rightarrow 4GlcCer \text{ (CTH)}$$

$$Gal 1 \rightarrow 4GalCer \text{ (CDH)}$$

At the time of these particular studies, all of the glycosidic bonds present in the glycosphingolipids were believed to possess the β-configuration. These two sphingolipids are considered to be minor components and usually are not present in appreciable quantities in most mammalian tissues. Improved methodology and instrumentation have allowed for the reliable quantitation of these trace compounds. In all organs and body fluids from Fabry patients, increased levels of CTH have been observed. CDH appears to be present almost exclusively in the kidney of normal individuals.[127] This material has not been found to be elevated in the plasma of Fabry individuals although CTH may be increased 3- to 10-fold.

The earliest report concerning an enzymatic defect in Fabry's disease appeared in 1967.[128] These authors isolated CTH from mammalian tissues and subjected this product to general tritium labeling. The radioactive product was then employed as substrate and incubated with intestinal biopsy samples. The affected individual had a markedly decreased ability to catalyze the release of water-soluble radioactivity. The carrier was found to have an activity intermediate between the affected individual and the normal controls.

These studies demonstrated that the Fabry's individual lacked the capacity to catalyze the hydrolysis of CTH. It had been assumed that this was a reflection of a diminished β-galactosidase activity. Three years later, a brief report appeared concerning the inability of leukocytes prepared from Fabry's individuals to hydrolyze either 4MU- or pNP-α-D-galactoside (Table X).[129] Normal values were obtained for β-D-N-acetylhexosaminidase and β-D-galactosidase activities. These surprising observations suggested that the terminal galactose of CTH was in α- rather than β-glycosidic linkage, as proposed earlier. A flurry of activity by several sphingolipid chemists resulted in unequivocal demonstration that this was an α-linkage.[130–133] This is an interesting example of a diagnostic clinical observation providing a clue for a basic error in the fine structural details of a naturally occurring compound. The terminal linkage in the CDH accumulating in Fabry's kidney also possesses an α configuration.[134]

A terminal α-D-galactosidic bond is present in CTH, CDH, digalactosyldiglyceride, and the B-active blood group substances. A complex glycosphingolipid isolated from the pancreas of a Fabry's individual[135] possesses blood group B activity and has the following structure: Galα1→3[Fucα1→2]Galβ1→3(4)GlcNAcβ1→3Galβ1→4GlcCer.

Entensive quantitative data comparing the levels of this compound in normal individuals and Fabry's patients have not been reported. It seems reasonable to assume that elevated quantities of digalactosyldiglyceride may also be observed in tissues of Fabry's patients.[136]

Multiple forms of α-galactosidases that cleave 4MU-α-D-galactosidase have been observed in normal solid tissue.[136a] Skin fibroblasts and white blood cells appear to contain two electrophoretically and chromatographically distinct α-D-galactosidase activities. However, only one of these forms appears to be present in Fabry tissue samples.[137] The heat-labile component absent from all patient-derived materials has been operationally referred to as "α-galactosidase A," while the heat-stable component present in both normals and Fabry patients has been referred to as "α-galactosidase B."[138] The nature of the residual α-galactosidase activity present in Fabry's tissue has been investigated. The major component appears to resemble α-galactosidase B[138a,b] with smaller amounts of the A component.[138c] These two forms have been sufficiently purified from human placenta to be employed as antigens. The antibodies produced against either the "A" form or the "B" form of α-D-galactosidase do not cross-react upon double immunodiffusion assay.[139] Antiserum prepared against purified α-galactosidase A did not affect the α-galactosidase activity of Fabry tissue, suggesting that the B form is the sole form present.[139a] Myoinositol has been demonstrated to be an inhibitor of α-ga-

Table X. Enzyme Activities in Leukocytes of Control Subjects and of Patients with Fabry's Disease[a,b]

Subject	Kinship	Age (years)	β-Galactosidase (mean)	β-Acetyl-glucosaminidase (mean)	β-Acetyl-galactosaminidase (mean)	α-Galactosidase	
						Colorimetric (mean)	Fluorimetric (mean)
			Normal males (n = 15)				
		15–33	76 (27–153)	295 (139–446)	71 (45–128)	13 (6–25)	12 (7–17)
			Fabry's disease, affected males				
L.G.		31	110	433	105	<0.5	<0.2
Mo.A		16	71	485	114	<0.5	<0.2
Mu.A	Brother of Mo.A.	9	122	1220	173	<0.5	<0.2
			Fabry's disease, female carriers				
L.D.	Sister of L.G.	26	60	346	50	2.4	1.8
R.C.	Mother of L.G.	53	68	378	54	5.6	5.0
Fa.A.	Sister of Mo.A.	11	48	526	96	4.2	3.9
Na.A.	Sister of Mo.A.	6	56	1000	112	3.8	4.9

[a] Modified from Ref. 129.
[b] The values expressed are nanomoles of substrates hydrolyzed per hour per 10^6 cells. Numbers in parentheses indicate the range of values.

lactosidase activity of normal human fibroblasts. In contrast, this material does not affect the α-galactosidase of Fabry fibroblasts.[140] Purified human liver α-galactosidase B has been identified as an α-*N*-acetylgalactosaminidase.[140a,b] Three distinct forms of α-galactosidase have been found with normal human leukocytes. The residual activity in Fabry cells corresponds to the B form of the enzyme.[140c]

Human plasma has been applied to an affinity chromatographic column containing *p*-aminophenyl melibioside as the ligand. Five different forms of CTH : α-D-galactosidase and six forms of *p*NP-α-D-galactoside : α-D-galactosidase were isolated as separate fractions. Comparison of the column fractionation patterns from normal and Fabry serum samples indicated that three forms of the CTHase with a pH optimum of 7.2, two forms with a pH optimum of 5.4, and one of the general α-D-galactosidases were reduced in the samples obtained from these patients.[141] Readily available Cohn Fraction IV-1 of human plasma was used for a largescale isolation of α galactosidase A,[141a] and it was observed that the serum enzyme may be more highly sialylated than that from solid tissues.[141b]

Normal plasma has two separate protein activities towards CDH, and these activities are diminished in Fabry plasma.[142] The specific relationships between this complex of approximately 11 different α-galactosidic linkages present in CDH and CTH, as well as artificial substrates, are poorly understood. It will be interesting to learn if any of these enzyme fractions are specific for any given substrate. Investigations to answer these problems were pursued.[143,144]

Interconversions of purified CTH : α-D-galactosidase have been the subject of a preliminary report. The technique for monitoring these transformations is based upon the differences of electrophoretic mobility due to charge differences. When the acidic form of the enzyme was treated with neuraminidase to remove sialic acid, a series of electrophoretic bands were produced that appeared to possess less negative charges. A series of radioactive bands of increasing acidity was formed when a basic form of the enzyme was incubated with either CMP [^{14}C]sialic acid or UDP *N*-acetyl[^{14}C]glucosamine and a crude kidney sialyl transferase preparation. None of the products were reported to have any enzymatic activity.[145] Other investigators have reported that there are no detectable changes in α-galactosidase A upon treatment with neuraminidase.[145a] [6,6-^{2}H$_2$]-Glucose was administered to a patient with Fabry's disease and a control and the appearance in plasma glycosphingolipids measured. A twofold higher incorporation of label into lactosylceramide was found with the patient. A reasonable incorporation into CTH occurred with the control. However, there was no labeling of CTH in the Fabry's patient.[145b]

Fabry's disease has been a testing ground for most of the approaches of enzyme replacement therapy currently being undertaken. The initial attempt was infusion of plasma into two patients with this disease. Early transient increases in detectable CTHase activity coupled with decreases in circulating levels of CTH were reported.[146] Subsequent attempts by various laboratories have failed to obtain similar results. Several laboratories have reported salutary effects on some of the recipient patients.[147–149] Unfortunately, host rejection of the transplanted kidney occurred in most of these patients with death eventually ensuing. Highly purified human placental CTHase has been administered to a patient with Fabry's disease.[150] This enzyme rapidly disappeared from the plasma with a $t_{1/2}$ of about 10–12 min, and was apparently taken up the by the liver. Transient decreases in plasma CTH levels were seen with return within 2–3 days to the elevated preinfusion levels. α-Galactosidase from ficin was immobilized on an inert support and shown to hydrolyze CTH in the presence of buffer and sodium taurocholate. However, there was no reduction of CTH levels when the preparation was incubated with plasma from a Fabry's patient, suggesting that enzyme replacement therapy may be ineffective in this disease.[150a]

Heterozygote[128,151] and prenatal[152] detection have been successfully accomplished with a variety of tissues and fluids including tears.[152a]

3.6. The Gangliosidoses

Numerous review articles and clinical descriptions of the gangliosidoses have appeared[153–156] and therefore great detail will be avoided.

These can be conveniently divided into the GM_2 gangliosidoses and the GM_1 gangliosidoses on the basis of differences in both the chemical nature of the material stored and the separate enzymes affected. Subclassifications of these disorders are based upon either clinical or enzymatic variants.

3.6.1. GM₂ Gangliosidosis Type I (Classical Tay–Sachs' Disease)

This is the most commonly recognized sphingolipidosis and occurs in approximately 50 children born each year in the United States, of which 80% are to Jewish parents.

The earliest description of this disease is credited to Dr. Tay, an English ophthalmologist, who, in 1888, noted a cherry-red spot in the eye of a child. Over the next several years, Dr. Sachs, an American neurologist, described retardation and blindness in several infants. These patients with "amaurotic family idiocy" were the basis for the Tay–Sachs' disease of current interest.

The largest experience with such children in this country has been in Brooklyn, New York, at the Jewish Chronic Disease Hospital, where a ward had been established largely due to the efforts of Drs. B. Volk and S. Aronson. Much of the clinical material from Tay–Sachs children used by investigators throughout the world in order to further understanding of this disease has been made available from this particular service. This group of investigators has edited a monograph on the gangliosidoses.[156a]

Clinical symptoms are usually evident in the first year of life. Initial signs are not dramatic and present as enfeeblement, spasticity, and slow development. An exaggerated startle response to sound may be the most significant early sign of which a parent is aware.

As mentioned, the credit goes to an ophthalmologist for the earliest documented case of Tay–Sachs' disease. This is due to the characteristic and

striking ocular lesion with changes in the macular region leading to the presence of a cherry-red spot. A whitish space similar to a halo appears, due to
lipid accumulation in the ganglion cells that surround the macula.[157] Blindness occurs by the second year of life. Motor retardation progresses and, by
the beginning of the third year, spontaneous movement is absent. Seizures
and convulsions are evident in the second year and the neurological deterioration continues until a vegetative state is reached.

The appearance of the patient is not unpleasant with "doll-like" features,
pale clear skin, pink coloration, and fine hair.

No specific X-ray changes are noted. There is enlargement of the skull
but not of peripheral organs. Clinical differentiation from related disorders
is based on the absence of hepatosplenomegaly and the presence of a cherry-
red spot in the eye.

3.6.1.1. Pathology

The gray matter of the central nervous system is the site of the greatest
pathological changes. Cortical ganglion cells are "ballooned" or greatly distended, and the cytoplasm may have vacuoles that are filled with a substance
that reacts with only a few lipid stains. Electron microscopic examination of
tissue samples reveals the presence of numerous membranous cytoplasmic
bodies.[158] As the disease progresses, there is a gradual deterioration with
loss of healthy ganglion cells. Associated demyelination is a common feature.
Loss of Purkinje cells in the cerebellum is apparent.

This disease is transmitted as an autosomal recessive trait. Approximately
75+% of the documented cases are from parents with Ashkenazi Jewish
origins. These cases number in the thousands.

3.6.2. GM$_2$ Gangliosidosis Type II (Sandhoff–Jatzkewitz Variant)

The clinical description is nearly identical to that of Tay–Sachs' disease
and the major difference noted thus far is the non-Jewish parentage. Approximately one dozen cases have been documented since 1968. The pathological changes seen are similar to those found in Tay–Sachs' disease.[159]

3.6.3. GM$_2$ Gangliosidosis Type III (Juvenile)

The number of patients reported with this variant of Tay–Sachs' disease
is small, fewer than a dozen having been described.[159a] The onset of symptoms occurs between the second and sixth years of life. Loss of muscular
coordination associated with locomotion (ataxia) is frequently the earliest sign,
and blindness is present in the late stages of the disease. A decerebrate state
is gradually reached as the disease progresses. Although visual problems are
prevalent, there is no cherry-red spot in the macula of the eye.

Neuronal accumulation of lipoidal material with MCBs is similar to that
observed in GM$_2$ gangliosidosis tissue.[160]

3.6.3.1. Biochemistry

Klenk is credited with the earliest observation reporting excessive quantities of a lipid in Tay–Sachs' brain samples,[161] and all subsequent investigators have substantiated this initial finding. The structure of the accumulating material was described by Svennerholm[162] and, in greater detail, by Ledeen and Salsman.[163] This monosialoganglioside has been trivially referred to as "Tay–Sachs' " ganglioside, but using a more systematic abbreviation, it is commonly referred to as GM_2 ganglioside.[164] This material is a normal component of the ganglioside mixture that can be isolated from normal brain tissue, usually representing only a few percent of the total. In the brains of affected children, there is a substantial increase in total ganglioside, the overwhelming majority being GM_2 (Table XI). The structure of this material is formulated as $GalNAc\beta1\rightarrow4[NeuAc\alpha2,\rightarrow3]Gal\beta1\rightarrow4GlcCer$.

This large increase in ganglioside content is characteristic of the gray matter, since these sphingolipids are largely neuronal constituents.

Increased quantities of GD_2 and N-acetylgalactosamine GD_{1a} have been observed in Tay–Sachs brain tissue.[164a] GM_2 ganglioside accumulation has been seen with late-passage skin fibroblasts from Tay–Sachs and Sandhoff patients.[164b]

Elevated amounts of the asialoganglioside, that is, the product obtained through the removal of the N-acetylneuraminic acid, have been consistently observed in these cerebral tissues. This may comprise up to 3% of the total lipid in GM_2 type I and 12% in type II.[160] In addition to these increases, elevated amounts of glucosylceramide in the brain of Tay–Sachs children have been reported.[165,166] The structure of six neutral glycosphingolipids isolated from Tay–Sachs brain has been elucidated.[166a]

Increased quantities of GM_2 in type I patients have been observed for cerebrospinal fluid,[167] liver, and spleen.[165] The membranous cytoplasmic bodies have been isolated from pathological tissue samples. Ganglioside concentration is greatly increased when compared to similar bodies from metachromatic leukodystrophy and Niemann–Pick's disease.[168] These characteristic alterations in the content of the glycosphingolipids in Tay–Sachs tissue

Table XI. Gray Matter Ganglioside Mixture Composition^a

	Normal 4	GM_2 type I 15	GM_2 type II 16	GM_2 type III 8
Total ganglioside: (mg/100ml)	Ganglioside % distribution			
GM_2	7	85	85	40
GM_1	21	3	6	11
GD_{1a}	40	7	5	22
GD_{1b}	18	1	1	8
GT	17	1	—	7

^{a} From Ref. 160.

samples are well established and corroborated from similar findings in a number of well-documented reports. N-Acetylated hexosamines are present in other macromolecules such as the glycosaminoglycans and glycoproteins in addition to the glycosphingolipids. In most tissues, and especially in the brain, such materials are extremely heterogeneous and firmly membrane bound. To obtain the glycosaminoglycans and some glycoproteins, the lipid-free tissue sample is extensively treated with proteolytic enzymes, such as trypsin or papain to liberate the bound material. The solubilized products are then grossly separated into smaller-molecular-weight dialyzable and larger-molecular-weight nondialyzable fractions. A threefold increase in hexose and a nearly twofold increase in hexosamine have been observed in the dialyzable fraction prepared from Tay–Sachs gray matter employing this technique.[169] The hexoses were primarily mannose and glucose. There were no significant differences in the white matter. Extensive data are not available to indicate that this may be a finding applicable to all GM_2 gangliosidosis type I patients.

The enzymatic deficiency that is considered responsible for the deposition of GM_2 ganglioside in the organs of the type I patient is believed to be a specific β-N-acetylhexosaminidase species. Due to the technical difficulty in monitoring the hydrolysis of the accumulating substrate, GM_2, most of the studies have been carried out with a synthetic substrate such as a pNP- or 4MU-β-N-acetylglucosaminide or galactosaminide. Utilizing tissues from GM_2 type I patients, increases rather than reductions were observed in *total* β-D-N-acetylhexosaminidase activity employing these synthetic substances. In a now-classical publication, Okada and O'Brien observed that hexosaminidase A, one of the components of the total tissue hexosaminidase mixtures, is absent in GM_2 type I samples (Table XII).[170] These authors were aware of an earlier report describing the presence in human spleen of two distinct hexosaminidase components. One of these components has been referred to as hexosaminidase A and the other as hexosaminidase B.[171] Both hexosaminidases

Table XII. Activity of Hexosaminidase Components in Tay–Sachs' Disease[a,b]

Subjects	No.	β-N-Acetylglucosaminidase component		β-N-Acetylgalactosaminidase component	
		A	B	A	B
Cerebral cortex					
Controls	9	38.9 (77)	11.6 (23)	9.9	3.0
Tay–Sachs	4	0	157.9 (100)	0	98.6
Liver					
Controls	17	129.2 (40)	193.8 (60)	16.4	24.6
Tay–Sachs	3	0	106 (100)	0	14.4
Kidney					
Controls	2	123.6 (40)	185.4 (60)	18.0	27.1
Tay–Sachs	3	0	190 (100)	0	23.4

[a] From Ref. 170.
[b] Values expressed as nanomoles of substrate cleaved per milligram of wet tissue per hour. Values in parentheses are the percentages of the total hexosaminidase activity of each component.

have similar K_m values toward both glucosaminides and galactosaminides as well as similar subcellular distribution. The A form is more acidic than the B form and can be readily separated by electrophoresis and by anion-exchange column chromatography. These two forms of the enzyme can be operationally distinguished more conveniently by their differential heat stability at 50°C. Under these conditions, hexosaminidase A activity is destroyed while hexosaminidase B activity is stable. This property of differential heat stability has been the basis for a simple and reliable assay for the detection of both the affected individual and the heterozygote carrier state. Operationally, this involves duplicate incubations of the test tissue extract in which one sample is assayed directly for total hexosaminidase and this value represents the sum of the A and B forms. The duplicate sample is heated at 50°C for from 2 to 4 hr and then assayed and this value represents only the heat-stable or hexosaminidase B activity. Therefore, a simple calculation yields a value for percentage of hexosaminidase A in any given sample. The value for GM_2 type I patients approaches zero, while the value for carriers is approximately one-half the control value. This technique has been applied to assays of solid tissue, serum, urine, tears, skin fibroblasts, and amniotic fluid cells grown in culture.[172] The hexosaminidase deficiency has been confirmed by Kolodny *et al.*[173] in studies with heart muscle tissue in which the ability to remove either the N-acetylgalactosamine or N-acetylneuraminic acid residue from suitably labled GM_2 as substrate was investigated. Tissues from a GM_2 type I patient were unable to cleave the amino sugar residue but could adequately cleave the sialic acid residue, thus confirming that Tay–Sachs' disease is due to inability to catalyze the release of the galactosamine moiety from GM_2.

This type I GM_2 gangliosidosis with a specific deficiency of hexosaminidase A but no decrease in hexosaminidase B activity has also been termed the "variant B," denoting the presence of hexosaminidase B.[174]

In type II GM_2 gangliosidosis, the enzymatic deficiency is an absence of both forms of hexosaminidase,[175] and this type has also been referred to as the "variant 0" denoting an absence of both hexosaminidases A and B.[174] A juvenile form of this disease has been reported with somewhat higher hexosaminidase levels than those found in typical Sandhoff patients.[174a] In addition to increased GM_2 concentration, these patients have an accumulation of two other glycosphingolipids possessing a terminal N-acetylgalactosamine. These are the asialo derivative of GM_2, which is also trivially referred to as GA_2 (GalNAcβ1→4Galβ1→4GlcCer), and a globoside (GalNAcβ1→3Galα1→4Galβ1→4GlcCer).

Membranous cytoplasmic bodies from these patients contain elevated quantities of GA_2, 22% of total lipid, when compared to the bodies isolated from GM_2 type I patients.[176] In addition to the accumulation of hexosamine-containing sphingolipid, elevations in nonlipid hexosamine have been reported. A 3-fold elevation of hexosamine in aqueous extracts of such brain tissues has been found, suggesting accumulation of a glycoprotein material. This is further supported by observations of a 10- to 15-fold increased urinary excretion of free sialic acid, as well as oligosaccharides, which vary in the number of carbohydrate residues and contain principally mannose and N-

acetylglucosamine.[177] However, attempts to demonstrate elevated tissue gly-cosaminoglycans have been unsuccessful.[178] A heptasaccharide has been iso-lated from liver tissue of a variant 0 GM_2 gangliosidosis patient.[178a] The structure proposed was derived from a combination of degradative proce-dures and analysis of the components as well as by proton NMR spectroscopy (Fig. 3).

It had been generally assumed that most mammalian tissues contained two operationally distinct hexosaminidases, hexosaminidase A and hexos-aminidase B. A brief report appeared describing an additional activity, which was separable from hexosaminidases A and B by disc gel electrophoresis.

This new band was undetectable with 4MU-*N*-acetylgalactosaminide as substrate.[178b] This form, termed hexosaminidase C, was found to represent a larger proportion of total hexosaminidase in fetal tissue[178c] and immu-nologically distinct from hexosaminidase A.[178d] It has a pH optimum of about 7.0[178e] and appears to be the major form demonstrable in white blood cells obtained from Sandhoff patients[178f] (Fig. 4). Partial purification and prop-erties of hexosaminidase C of bovine[178g,m] and neonatal human[178h] brain have been investigated. The main isozyme present in the serum of pregnant humans, termed hexosaminidase P, has been highly purified and shown to be antigenically identical to hexosaminidase S but electrophoretically differ-ent.[178i] An additional isozyme termed hexosaminidase α, which appears to be distinct from either hexosaminidase A or C, has been obtained from the liver of Sandhoff patients. It is heat labile and cross-reacts strongly with anti-serum to hexosaminidase A and may represent two α_2 chains.[178j] This iso-zyme has also been purified from brain tissue.[178k] Hexosaminidase C was decreased or absent in brain tissue from four patients with Tay–Sachs' dis-ease.[178l]

The capacity to hydrolyze *N'*,*N*-diacetylchitobiose (*N*-acetylglucosaminyl-β-*N*-acetylglucosamine) is not reduced in liver tissue from either Tay–Sachs' or Sandhoff patients.[178i]

Patients with GM_2 gangliosidosis type III (shown in Table XI) do not have as dramatic an accumulation of GM_2 in the gray matter. In these patients, there is a decrease in detectable hexosaminidase A; however, this reduction is not as large as that observed in the other two forms.[179]

A single unusual patient with a decrease in neither hexosaminidase A

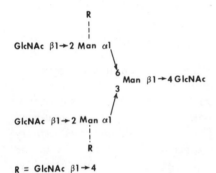

Figure 3. Structure of the major heptasaccharide (fraction 2) from the liver. The additional nonred-ucing GlcNAc is linked to either one of the 2-linked mannosyl residues. (From Ref. 178a.)

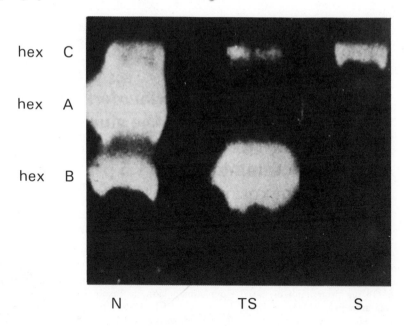

Figure 4. Electrophoretic separation of β-glucosaminidases extracted from white blood cells. N, control; TS, Tay–Sachs' disease; S, Sandhoff's disease. (From Ref. 178f.)

nor B has been classified as a GM_2 gangliosidosis variant type AB. The complete case description and pathological findings on the patient have not yet been published; however, elevated tissue levels of GM_2 and asialo GM_2 (GA_2) were reported.[180] The results of a study on an individual with similar symptoms have appeared.[180a] An adult "variant AB" GM_2 gangliosidosis patient was reported to have an eightfold elevation of GM_2 in the cerebral cortex with an 85% elevation of total hexosaminidases that is reflected in both forms of the enzyme.[180b] There were no differences compared to controls in the properties of hexosaminidase A purified from tissues of an AB variant.[180c] Stimulation of hexosaminidase A hydrolysis of GM_2 was obtained with heated extracts from controls, Tay–Sachs' disease (variant B), and Sandhoff (variant 0) kidney tissue (Fig. 5). There was no stimulation by extracts from the AB variant kidney, suggesting that this disease is not a result of a faulty enzyme activity but rather absence of an essential stimulatory factor.[180d] There was no observed decreased amount of a stimulatory factor in tissues of a different type of AB GM_2 gangliosidosis. This patient possessed a selective hexosaminidase A activity.[180e]

A patient has been described presenting with classical Tay–Sachs' disease but normal hexosaminidase A and B distribution and reduced detectible "activator" levels.[180f]

This emphasizes the problem that exists in understanding the biochemical basis for this disease. Convincing evidence directly demonstrating the cleavage of the N-acetylgalactosamine residue of GM_2 has been lacking. Highly purified hexosaminidase A from human liver has been incubated with radioactive GM_2

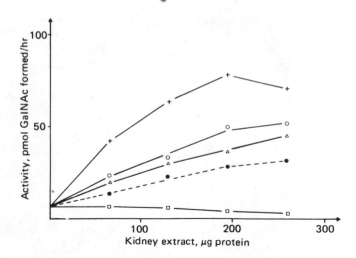

Figure 5. Absence of activating factor from unheated kidney extract of variant AB. Stimulation of the hexosaminidase A-catalyzed hydrolysis of [³H]-GM₂ was measured in the presence of extracts (13 mg protein/ml) prepared from human kidneys. Incubation mixtures containing increasing amounts of kidney extracts, 10 nmoles of [³H]-GM₂ and 65 mU of purified hexosaminidase A were analyzed for [³H]-GalNAc formed. Blanks were run for each value without additional hexosaminidase A and subtracted. ○, kidney extract, normal case 1; +, kidney extract, variant 0; △, kidney extract, variant B (Tay–Sachs' disease); □, kidney extract, variant AB; ●, mixture (1 : 1, vol/vol) of extracts from normal kidney (case 1) and variant AB kidney. (From Ref. 180d.)

and very little conversion to GM₃ was observed. Under the same conditions, when hexosaminidase B was employed, no cleavage was observed.[181,181a] However, both hexosaminidase A and hexosaminidase B were effective in catalyzing the removal of the terminal *N*-acetylgalactosamine of GA₂. Both hexosaminidases A and B can also hydrolyze the terminal amino sugar in globoside.[182] GM₂-hydrolyzing activity was found to be associated with the most acidic forms of crude liver extracts.[182a] In contrast, highly purified placental hexosaminidases A and B were claimed to be active towards GM₂.[184d] Both forms of hexosaminidase were highly purified from human liver by an affinity chromatographic procedure. In the presence of either taurodeoxycholate or stimulatory factor, hexosaminidase B was more effective in hydrolyzing GA₂ than was hexosaminidase A. Hexosaminidase A was superior in cleaving GM₂ in the presence of these additions, supporting the premise that GM₂ accumulation in Tay–Sachs' disease is due to the reduced hexosaminidase A activity. These workers also observed that hexosaminidase A but not hexosaminidase B hydrolyzed GalNAcβ1→4[NeuAcα2→3]Galβ1→4sorbitol.[182b] This observation is supported by the demonstration that hepatic human hexosaminidase A but not hexosaminidase B hydrolyzed either GlcNAcβ1→4GlcAβ1→3GlcNAc or GalNAcβ1→4GlcAβ1→3GalNAc prepared enzymatically from chondroitin-4-sulfate and hyaluronic acid.[182c]

The unsettled relationship between hexosaminidase A and GM₂ ganglioside hydrolysis appears to have been resolved. Li and co-workers have provided suggestive evidence for the role of a heat-stable, nondialyzable factor present in a crude preparation of human liver β-*N*-acetylhexosaminidase. This material stimulated aged or purified hexosaminidase A but not the hexosaminidase B form to catalyze the hydrolysis of GM_2. This effect was of such magnitude that the hydrolysis could be conveniently monitored by a thin-layer chromatographic separation of nonradioactive GM_2, the substrate, from GM_3, the product.[(183)] This material has been purified to apparent homogeneity. It has an apparent molecular weight of 21,800, is enriched in aspartate and glutamate, has a pI of 4.1, and appears to be a glycoprotein.[(183a)] In addition to stimulating GM_2 hydrolysis, it appears to stimulate the hydrolysis of GM_1 by β-galactosidase and of trihexosylceramide by α-galactosidase. The stimulatory activity for GM_2 and GM_1 have been obtained as separate entities.[(183b)] The effect of a highly purified protein activator of hexosaminidase A was examined. Its effect on the ability of both purified hexosaminidase A and B to hydrolyze several substrates was compared to the activation by sodium taurodeoxycholate. The activator stimulated hexosaminidase A hydrolysis of GM_2 gangliosides 107-fold, asialo GM_2 310-fold, and globoside 10-fold but had no effect on 4MU-β-GlcNAc hydrolysis. The activator caused a 5-fold increase of GM_2 and 70-fold increase of GA_2 hydrolysis by hexosaminidase B. Taurodeoxycholate caused a remarkable stimulation of the hydrolysis of the sphingolipids by both enzymes (Table XIII). These observations show that hexosaminidase A but not hexosaminidase B cleaves GM_2 and GA_2 and that the hydrolysis is stimulated by a protein activator.[(183c)] This activator protein was purified 10^5-fold from human liver and found to exert its maximal activity at a 1 : 1 molar ratio of enzyme.[(183d)] It is possible that previous efforts to demonstrate this hydrolysis may have been unsuccessful either due to the age of the sample or due to the absence of the stimulating factor from highly purified enzyme preparations. Therefore, one might suspect that the variant

Table XIII. Substrate Specificity of Hexosaminidases A and B in the Presence of Activator Protein and of Detergent[a,b]

		Stimulating agent		
Substrate	Hexosaminidase	None	Na TDC[c] (2 mM)	Activator protein[d] (13.5 AU/assay)
GM_2	A	0.009	0.40	0.97
	B	0.001	0.07	0.005
GA_2	A	0.008	6.3	2.48
	B	0.001	30.7	0.072
Globoside	A	0.004	14.4	0.042
	B	0.005	24.0	0.005
4-MU-GlcNAc	A	675	350	680
	B	1300	850	1390

[a] Modified from Ref. 183c.
[b] Values expressed as micromoles of substrate split per hour and milligram of enzyme protein.
[c] NA TDC, sodium taurodeoxycholate. Enzyme is inactivated during the incubation time. Values given were obtained with short incubation times (30 min).
[d] Glycolipid degradation rate depends on activator concentration, which was not saturating in this case.

AB, which had adequate amounts of both hexosaminidases, may be deficient in this factor since there is an accumulation of GM_2 in the tissues. The activator protein may act *in vitro* as an agent to transfer glycolipids between membranes.[183e]

The two major forms of hexosaminidases, A and B, have been obtained as fairly purified preparations, and apparently homogeneous proteins have been obtained from beef spleen.[184] The amino acid composition for the separated A and B forms was nearly identical. The greatest differences reported were a twofold larger quantity of both sialic acid and neutral hexose in hexosaminidase A compared to hexosaminidase B. Most of the kinetic and physical characteristics examined were identical.

The distinct differences in electrophoretic migrations and the ability of neuraminidase treatment of hexosaminidase A to form a hexosaminidase B-like product suggested that the B form merely represented the asialo derivative of the A form. This led to the proposal that the basic defect in Tay–Sachs' disease was the lack of a specific sialyltransferase.[184a] Independent observations demonstrated that this conversion was not dependent upon neuraminidase activity. The commercial neuraminidase samples employed had contained merthiolate as a preservative. This reagent has the ability to convert hexosaminidase A to a B-like form[184b,c] (Fig. 6).

Highly purified hexosaminidases A and B have been obtained from placental tissues. A B-like form was generated from the A form by carefully

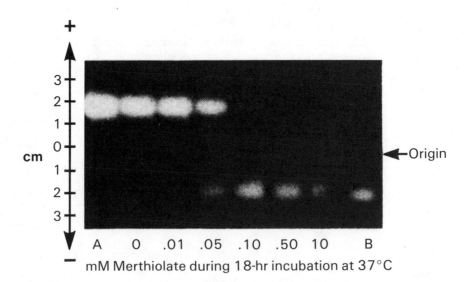

Figure 6. Electrophoresis of Hex A and Hex B on starch gel at pH 6.0. The channels marked A and B received purified placental Hex A and Hex B, respectively. The other channels received Hex A that had been incubated without merthiolate and with concentrations of merthiolate ranging from 0.01 to 10 mM as indicated. Initial Hex A activity in each incubated system was 1.8 U/ml but because of inactivation during incubation, between 6.4 and 83 mU of enzyme was placed into each slot. The B marker contained 5.6 mU of enzyme. (From Ref. 184b.)

controlled heating. This observation led this group to postulate that these two forms of hexosaminidase are merely conformers of one another.[184d]

An early model proposed speculated that there were subunit differences between the two forms of hexosaminidase.[184e] The enzymes were thought to be composed of two different polypeptide chains α and β. Hexosaminidase B would contain only β chains while hexosaminidase A would contain both α and β chains.[184d,f] Indirect evidence supporting this hypothesis has appeared.[184g,h] Preliminary experiments have been carried out on the *in vitro* translation of the human placental mRNA coding for the β hexosaminidase subunits.[184i]

Direct evidence for the subunit hypothesis has recently appeared in an elegant study on the chemical properties of highly purified placental hexosaminidases A and B.[184i–k] These authors found that sialic acid was present only in the A form. The B form contained twice as much cysteine + half-cystine as well as methionine (Table XIV). Further analysis indicated that there were no free sulfhydryl groups in hexosaminidase B and one or two in hexosaminidase A. As a result of this analysis, the model shown in Fig. 7 has been proposed. Thus, it would appear reasonable to assume that Tay–Sachs'

Table XIV. Amino Acid Analysis of Pure Hexosaminidases A and B[a]

	Hexosaminidase A		Hexosaminidase B	
	Mole%[b]	Moles of aa/mole enzyme[c]	Mole%	Moles of aa/mole enzyme
Lysine	5.38	47.8	5.65	54.7
Histidine	2.92	25.9	3.34	32.4
Arginine	4.64	41.2	4.57	44.3
Aspartic acid	9.76	86.7	9.54	92.4
Threonine	5.08	45.1	5.98	57.9
Serine	7.67	68.1	8.55	82.8
Glutamic acid	10.96	97.3	9.09	88.1
Proline	6.55	58.2	6.13	59.4
Glycine	7.48	66.4	6.52	63.2
Alanine	5.51	48.9	5.51	53.4
Cysteine + half-cystine[d]	1.43	12.7	2.54	24.6
Valine	5.99	53.2	4.99	48.3
Methionine	1.34	11.9	2.06	20.0
Isoleucine	3.86	34.3	4.07	39.4
Leucine	10.46	92.9	9.15	88.6
Tyrosine	4.21	37.4	5.05	48.9
Phenylalanine	5.24	46.5	5.53	53.6
Tryptophan	1.48	13.1	1.70	16.5
Sialic acid		1.65		UD[e]

[a] Modified from Ref. 184i.
[b] Expressed as moles of amino acid per 100 moles of total residues in hydrolysate.
[c] Expressed as the number of each residue in hexosaminidases A and B; the molecular weights taken for the calculations were 100,000 and 110,000, respectively.
[d] Determined as cysteic acid.
[e] UD, undetermined.

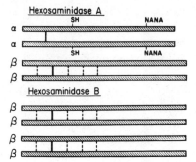

Figure 7. Suggested molecular structure of hexosamin-idases A and B. Solid lines connecting two chains represent S–S bridge; broken lines represent either intra- or interchain S–S bridges. In the α chains of hexos-aminidase A, in addition to the interchain S–S bridge, one sulfhydryl group (SH) and one sialic acid residue (NANA) are present. (From Ref. 184i.)

disease results in the inability to produce the α subunit and Sandhoff's disease the inability to produce the β subunit. Desialylated purified human hexosa-minidases A and B were substrates for a sialotransferase.[184l]

Immunological evidence supports the close chemical similarities of these two hexosaminidases, since antibodies raised against either purified hexos-aminidase A or B cross-react with both forms of the enzyme.[185] The anti-bodies raised against purified hexosaminidase A or B from normal human liver reacted with comparable fractions prepared and isolated from Tay–Sachs tissues.[186] No cross-reacting material was found present in the variant 0 of GM$_2$ gangliosidosis, which suggests that neither enzymatically nor immuno-logically active protein is produced in Tay–Sachs tissues. It has been suggested that there is a common antigenic site that is shared by both hexosaminidases A and B. An additional specific antigenic site is believed to be present in hexosaminidase A but absent from the B form.[186a] This may be the α subunit.

A low-molecular-weight (around 20,000) protein was found that cross-reacts with antibody toward hexosaminidase A and B in liver tissue of control, Tay–Sachs', and Sandhoff's patients.[186b] Radial immunodiffusion and ra-dioimmunoassay techniques have been developed for hexosaminidase A and B determinations. The absence of detectable hexosaminidase A has been confirmed using these methodologies.[186c] Antibodies were raised against glu-taraldehyde-cross-linked purified placental hexosaminidase A. This antibody precipitated cross-reacting material from Tay–Sachs tissues.[186d]

Hybridization experiments between cells from rodent and man have been employed to examine certain genetic aspects of the GM$_2$ gangliosidosis.[187] Early studies revealed that the genes for hexosaminidase A and B behave as independent markers.[187a,b] The gene responsible for hexosaminidase B has been assigned to chromosome 5,[187c,d] whereas hexosaminidase A is syntenic with the gene locus for mannose phosphate isomerase and pyruvate kinase-3 expression.[187d] This supports the observation that Tay–Sachs' disease and Sandhoff's disease are the result of separate genetic defects. Further evidence for distinct genetic loci is derived from complementation studies. Fibroblasts derived from Tay–Sachs and Sandhoff cells were fused. Heterokaryons ob-tained in these experiments were found to produce a form of hexosaminidase that was absent from both parental strains. This appeared to have the usual heat inactivation, electrophoretic, and immunological properties of hexosaminidase A.[187e–g]

Determination of hexosaminidase A levels in biological materials has been useful for the identification of heterozygote carriers and affected individuals of GM_2 gangliosidosis type I.[188,189] Recent efforts in this area appear to be largely technical and oriented toward improving existing methodology. Heterozygote detection of GM_2 type II based on hexosaminidase levels has been accomplished.[190]

The diagnostic usefulness of cultured skin fibroblasts was demonstrated as a prerequisite for the use of cultured amniotic fluid cells. Prenatal diagnosis of affected GM_2 gangliosidosis type I[191,192] and type II[193] fetuses has been successfully accomplished. The abortuses in these particular cases were examined and found to have the characteristic biochemical abnormalities of the disease.

Neuropathological examination of several of these abortuses obtained as early as 16 weeks *in utero* has revealed the presence of characteristic multilaminar cytoplasmic bodies.[194,194a,b] This observation suggests that if meaningful enzyme replacement therapy is to be attempted, it should be begun either *in utero* or immediately after birth of the affected individual.[194c]

A complication of the diagnostic values of hexosaminidase A activities has been revealed by the observed absence of detectable hexosaminidase A in a normal woman.[195] In addition, four members of a single sibship were reported to have no detectable hexosaminidase A activity.[196] These patients appeared to be comparable to Tay–Sachs' heterozygotes when the hydrolysis of GM_2 was examined employing leukocyte preparations.[196a] Low levels of hexosaminidase A were also demonstrated in these individuals by using a radioimmunoassay.[196b] The simplest conclusion, although not necessarily correct, is that hexosaminidase A is of questionable importance in GM_2 ganglioside metabolism. These individuals may have been aborted if a prenatal diagnosis had been undertaken on their mother. The most important practical problem raised by these observations is the usefulness of mass screening of Tay–Sachs' carriers,[197] which has recently been automated.[198]

The father of two children who died with the symptoms of Tay–Sachs' disease was found to have low total levels of hexosaminidase in serum and white blood cells. When assays were carried out with GA_2 as substrate, this individual appeared to be a carrier for Sandhoff-type GM_2 gangliosidosis.

Unusual hexosaminidase A mutations have been seen in two separate families; one is characterized by reduced GM_2 but unchanged 4MU-GlcNAc hydrolyses, and the other is characterized by the reverse situation.[196b,c] A "classical" Tay–Sachs child with reduced hexosaminidase A activity and an unusually thermolabile hexosaminidase B activity.[196d,e] Heat-labile forms have been found in the serum of two healthy adults.[196f] Serum hexosaminidase deficiencies with adequate ability to cleave GM_2 and 4MU-*N*-acetyl glucosaminide have been reported in clinically normal individuals.[196g] A systematic nomenclature has been proposed relating the several hexosaminidase A and B phenotypes to the genotypes described by the postulated subunit structures.[196g]

Enzyme replacement therapy through plasma infusion of GM_2 gangliosidosis type I[160] and GM_2 gangliosidosis type II[199] has revealed a rapid removal of the hexosaminidases from the serum. This rapid clearance from the circulation into the liver would not be expected to alleviate the symptoms in this disease. Radiolabeled hexosaminidase was rapidly cleared from the serum of rats by a recognition system present on hepatic sinusoidal cells.[199a] It was concluded that intrathecal injections of pure hexosaminidase A were not beneficial for Tay–Sachs patients.[199b] Another approach for the administration of hexosaminidase has been proposed. Hexosaminidase A was incorporated into liposomes that had been coated with human IgG. These were then incubated with leukocytes from a Tay–Sachs patient and enzyme uptake was observed.[199c] Con A facilitated the uptake of hexosaminidase A into brain cells maintained in culture. There was a significant reduction in GM_2 of cells from Tay–Sachs patients.[199d] These cells had decreased UDP-galactose : GM_2 galactosyl transferase activity, which is required for GM_1 formation.[199e]

GM_2 gangliosidosis with reduced levels of both hexosaminidases A and B has been reported in cats[199f,g] and pigs.[199h] The cat model of this disease has led to experiments on enzyme replacement therapy. Blockade of hepatic uptake of injected hexosaminidase was accomplished with mannan. This treatment in combination with hyperbaric oxygen resulted in the appearance of the injected enzyme in the animals' nervous tissue.[199i,j] Hypertonic mannitol modification of the blood–brain barrier allowed the appearance of ^{125}I human hexosaminidase A in experimental animals.[199k]

The biochemistry and genetics of the GM_2 gangliosidoses have been reviewed.[199i,j] Precursor forms of hexosaminidase may be produced in cultured human skin fibroblasts.[199l]

3.6.4. GM₁ Gangliosidosis Type I (Pseudo-Hurler's, Landing's Syndrome; Neurovisceral Lipidosis)

The clinical manifestations appear early in life and may be apparent even at birth. Life expectancy is short with death usually prior to the third year. The physical development is slow due to small birth weight, poor feeding and appetite. There are a variety of gross physical abnormalities, including enlarged tongue (macroglossia), protuberant forehead (frontal bossing), depressed nasal bridge, a soft downy hair on the face (hirsutism), and a cherry-red spot on the macula in one-half of the patients. The presence of clear corneas sets this apart from Hurler's disease. A rhythmical shakiness of the eyeballs (nystagmus), and nonparallelism in focus of the eyes (strabismus) are occasionally seen.

A distended abdomen due to an enlarged spleen and liver is evident in the first year of life. Curvature of the spine, both laterally and backwards (kyphoscoliosis), is usual. Hyperactive reflexes, poor muscle strength, and decreased muscle tone are evident. Clonic tonic convulsions and rapid deterioration occur during the second year. Decerebrate rigidity, with a "frog-like" position, and quadriplegia are seen in the terminal stages.[200]

3.6.4.1. Pathology

Extensive vacuolization of the reticuloendothelial system with deposition of Sudanophilic and PAS-positive material is seen. Neuronal ballooning is observed and the material stored in the neurons is Sudanophilic and weakly PAS positive, suggesting a lipid accumulation. The opposite is true for the glial cell and systemic organs, suggesting a polysaccharide accumulation.[201] Fine structure of the brain reveals membranous cytoplasmic bodies similar to those observed in type I GM_2 gangliosidosis.

Bony abnormalities are evident upon X-ray examination. These skeletal changes observed are most characteristic of GM_1 gangliosidosis type I and are quite reminiscent of the abnormalities found in Hurler's syndrome.[202] The GM_1 gangliosidosis type I patient is readily distinguishable from GM_2 gangliosidosis patients due to the presence of hepatosplenomegaly and bony changes. It differs from Hurler's disease because of the very rapid neurological deterioration of the patients.

3.6.5. GM_1 Gangliosidosis Type II (Juvenile GM_1 Gangliosidosis; Derry's Syndrome)

These patients appear normal with respect to all the usual milestones during the first year of life. The earliest manifestation involves locomotion with awkward gait and frequent falling. Loss of coordinative manipulation and speech is frequently seen and early neurological examination is asymptomatic. Gradual deterioration of mental and motor functions is evident. Severe progressive nervous system dysfunction, which varies with the individual patient, develops over the course of life.

There is no cherry-red spot in the macula of the eye and visceromegaly is absent, thus distinguishing these patients from GM_1 gangliosidosis type I.

A third type, not well defined clinically, is seen in adults having an acid galactosidase deficiency. This has been termed the adult or variant form of GM_1 gangliosidosis. Some of these patients may be mildly retarded, some may have bony abnormalities, and some may have macular cherry-red spots.[202a–c]

3.6.5.1. Pathology

Radiological examination reveals only minimal bony abnormalities. Characteristic changes are the vacuolation of cells of visceral organs, and the demonstrable lipidosis of neurons. Membranous cytoplasmic bodies similar to those found in the type I patient are evident.

3.6.5.2. Biochemistry

The brain samples of patients with GM_1 gangliosidosis have an approximate doubling of total ganglioside content. Approximately 75% of the gangliosides present in this mixture are the GM_1 type whereas in normals this usually represents about 20%. In addition, there is a 10- to 20-fold elevation in tissue levels of the corresponding asialo derivative, GA_1. Analysis of peripheral organs reveals a 20- to 50-fold increase in hepatic GM_1 levels.[203]

In addition to accumulation of GM_1, several incompletely characterized heteropolysaccharides are present in excessive quantities, as described in a thorough analysis of tissues from five patients.[204] One of these macromolecular materials is a mucopolysaccharide composed mainly of galactose and glucosamine referred to as an undersulfated keratan sulfate due to its low sulfur content. The structural details of this material have not been provided. Both chondroitin sulfate and heparatin sulfates are elevated in Hurler's patients. An increase in a "sialomucopolysaccharide" has also been observed in GM_1 gangliosidosis and this has been obtained as a single electrophoretic band. Galactose and glucosamine in a 1 : 1 ratio were the major products obtained upon acid hydrolysis. However, sialic acid was present at only approximately one-tenth the amount of these carbohydrates. The following structure has been proposed for the material isolated from liver tissue of an affected individual[204a]:

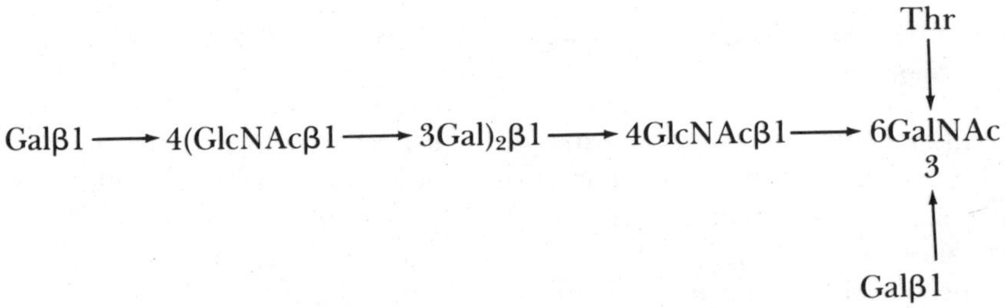

$$\text{Thr}$$
$$\downarrow$$
$$\text{Gal}\beta1 \longrightarrow 4(\text{GlcNAc}\beta1 \longrightarrow 3\text{Gal})_2\beta1 \longrightarrow 4\text{GlcNAc}\beta1 \longrightarrow 6\text{GalNAc}$$
$$3$$
$$\uparrow$$
$$\text{Gal}\beta1$$

This formulation was arrived at after both chemical analysis and enzymatic degradation.

Further studies of these compounds obtained from liver tissues have revealed the presence of yet another polysaccharide, which upon hydrolysis yielded galactose, mannose, and N-acetylglucosamine in roughly equimolar quantities; small amounts of sialic acid were also present.[205] Further purification of this material suggests it is a mixture of highly branched structures and is apparently unrelated to keratan sulfates. A structural relationship to the erythrocyte MN substrate has been noted. An oligosaccharide was isolated from the liver of a GM_1 gangliosidosis type I patient.[205a] Its structure is believed to be:

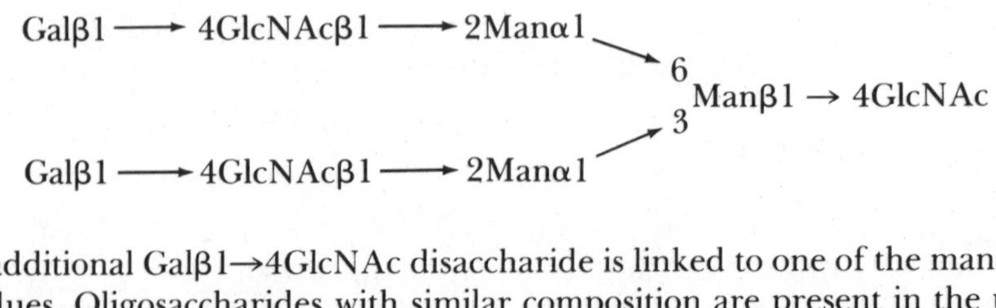

$$\text{Gal}\beta1 \longrightarrow 4\text{GlcNAc}\beta1 \longrightarrow 2\text{Man}\alpha1 \searrow$$
$$6$$
$$\text{Man}\beta1 \rightarrow 4\text{GlcNAc}$$
$$3$$
$$\text{Gal}\beta1 \longrightarrow 4\text{GlcNAc}\beta1 \longrightarrow 2\text{Man}\alpha1 \nearrow$$

An additional Gal$\beta1{\rightarrow}4$GlcNAc disaccharide is linked to one of the mannosyl residues. Oligosaccharides with similar composition are present in the urine of type I and II patients.[205b] These materials appear to be identical to those found in GM_2 gangliosidosis[178a] except that a β-galactosyl residue is present. There have been observations concerning accumulation of a similar material in brain tissue.[206]

The accumulation of the gangliosides and these heteropolysaccharides is presumably due to a decrease in lysosomal β-galactosidase activity. GM_1 ganglioside was biosynthetically prepared with [^{14}C]galactose in the terminal position. When this material was incubated with tissues from GM_1 gangliosidosis patients, a greatly reduced capacity to catalyze the release of free galactose from this substrate was found (Table XV).[207] Decreased general β-galactosidase with the *p*NP-β-D-galactoside was also seen. In tissues from GM_1 gangliosidosis patients, impaired galactose release from both mucopolysaccharides and glycoprotein has also been observed.[208]

The pH optimum obtained with the residual β-galactosidase present in the type II patient tissues is approximately 6, while in normals the major peak of β-galactosidase activity is at pH 4.5 with only a shoulder at 6. Electrophoresis revealed that several bands were present in control samples, while in the type I patient sample these were absent. In type II patient samples, one band was still detectable.[200] Differences in the heat stability and the pH optimum have been observed for the residual β-galactosidase activity with cultured skin fibroblasts[208a] and liver[208b] from type I and II patients. Variant forms of the GM_1 gangliosidosis have been reported. A patient with the total absence

Table XV. Ganglioside GM_1 β-Galactosidase Activity[a,b]

Age	Diagnosis	Enzyme activity
	Cerebral gray matter	
3 years	Cardiac anomaly	46
10 years	Ataxia-telangiectasia	23
59 years	Adenocarcinoma of lung	25
81 years	Myocardial infarction	36
2 years	Generalized gangliosidosis	1
	Liver	
2 days	Meconium peritonitis	77
1 month	Gastroenteritis	73
1 month	Multiple congenital anomalies	52
2 years	Lymphangioma	65
32 months	Chronic renal disease	51
8 months	Generalized gangliosidosis	4
2 years	Generalized gangliosidosis	2

[a] From Ref. 207.
[b] Activity is expressed as [^{14}C]galactose released (cpm) in 60 min by galactosidase, purified by the method of Gatt and Rapport from 10 mg of tissue (wet weight) at 37°C. Each sample contained labeled GM_1 (4500 cpm, 276 mCi/mmole) added to 0.1 ml of purified β-galactosidase in acetate buffer (pH 5.0) to make a final volume of 0.2 ml. Radioactive assays were done in duplicate. Nonradioactive GM_1 was added to all control tissue homogenates before β-galactosidase purification in concentrations equivalent to those found in the patients' tissues (0.5% of wet weight).

of all forms of β-galactosidase activity has been referred to as the juvenile GM₁ gangliosidosis type.[208c] A 14-year-old girl has been seen without neurological involvement and drastically reduced acid β-galactosidase activity.[208d] A typical type II patient was found to possess residual β-galactosidase activity that was 10- to 50-fold greater than that seen with other patients. The residual activity also electrophoresed differently and appeared to have increased antigenicity.[209]

There is a terminal β-galactosidic linkage present in three of the glycosphingolipids. These are in galactosylceramide, the compound that accumulates in Krabbe's disease, galactosylglucosylceramide, and GM₁ ganglioside. Tissues from GM₁ gangliosidosis patients have no deficit in the ability to hydrolyze either galactosylceramide or galactosylglucosylceramide, suggesting that a "specific" β-galactosidase isozyme exists that is involved in GM₁ hydrolysis.[210] Attempts to separate these β-galactosidase isozymes and assignment of their relationship to GM₁ gangliosidosis are not extensive.[211,212] The discussion of lactosidases I and II should be referred to in the earlier section on Krabbe's disease.

The β-galactosidase possessing lactosylceramide : β-galactosidase II activity has been purified 54,000-fold from human placenta tissue. Two peaks of enzyme activity of molecular weights 420,000 and 220,000 were obtained from gel exclusion columns. The purified enzyme was devoid of lactosylceramide : galactosidase I activity.[212a]

Cell hybridization experiments have been carried out with fibroblasts from type I and type II patients and there was no genetic complementation observed.[212b] However, complementation was seen when infantile type I and adult type IV cells were fused.[212c,d] The molecular genetics of the various forms of the GM₁ gangliosidoses have been thoroughly discussed with respect to the various enzyme forms and substrates.[212e]

Immunologically cross-reacting material to purified "acid" β-galactosidase has been demonstrated in liver tissue from a GM₁ gangliosidosis type I patient.[212f]

Bovine testicular β-galactosidase "cures" the storage of sulfated mucopolysaccharides when added to cultured GM₁ gangliosidosis fibroblasts.[212g]

The occurrence of GM₁ gangliosidosis in Siamese cats has been documented.[213,214,214a,b] The availability of such an animal model should be useful for experimentation with a variety of potential therapeutic approaches. Thus, enzyme replacement, organ transplantation, *in utero* intervention could be first attempted in the cats prior to human trial.

Prenatal diagnosis of the GM₁ gangliosidosis fetus and therapeutic abortions have been successfully accomplished. In both instances, pathological changes characteristic of the disease were observed.[215,216] Successful prenatal diagnosis of a type II GM₁ gangliosidosis fetus has been accomplished.[216a] The genetic material for the expression of *E. coli* β-galactosidase has been successfully incorporated into GM₁ fibroblasts by using a specialized transducing phage λplac. Increased levels of detectable β-galactosidase were demonstrated.[216b] This approach may ultimately be useful for enzyme replacement therapy via "genetic engineering."

The lysosomal β-galactosidase of humans has been assigned to chromosome 3.[216c,d]

3.7. Niemann–Pick's Disease

This was the second sphingolipidosis for which a specific enzymatic deficiency was discovered. The earliest clinical description of this disease is credited to Niemann, who had a young patient with clinical features similar to the then recently reported Gaucher's disease. However, due to the rapid decline in the course of this patient, he felt it was atypical of Gaucher's disease. Pick, after having seen several other patients, chose to call this a "lipoid cell splenomegaly" in order to distinguish it from Gaucher's disease.[217]

Niemann–Pick patients, as with several of the other sphingolipidoses, have been classified into several types due to distinct clinical differences. The system that has received general acceptance is that originally proposed by Crocker.[218]

3.7.1. Clinical

3.7.1.1. Type A (Acute Neuronopathic Form)

This represents the classical form of the disease and is characterized by both visceral and nervous system involvement shortly after birth. Enlarged spleen may be evident in the second month, followed by feeding difficulties. This failure to thrive coupled with hepatosplenomegaly results in a patient having an enlarged abdomen and thin arms and legs. A discoloration of the skin is often seen. Loss of intellectual and motor functions are usual neurological signs. Death usually occurs by the fourth year.

This is the most common form of the disease.

3.7.1.2. Type B (Chronic Form without Nervous System Involvement)

Visceral signs of the disease may occur in infancy, as seen in the type A patient; however, there is no evidence for central nervous system involvement. Patients of this type have lived into the second decade and appear free of neurological problems.

3.7.1.3. Type C (Subacute or Juvenile)

These are similar in principal to the type A patient except that the onset and duration are more extended. Symptoms may not become apparent until the second year or as late as the sixth. Neurological involvement is the basis for differentiation from the type B. The visceromegaly is less pronounced than in type A or B. A continuous deterioration of mental and motor functions is seen during the life of the patient.

3.7.1.4. Type D (Nova Scotia Variant)

This group represents those patients with ancestral origin in Nova Scotia having a clinical course similar to the type C patient.

Spanish-American children have been seen with a variant form of this disease.[218a] as well as a family possessing a heat-labile form of sphingomyelinase.[218b]

3.7.2. Pathology

Lipid-laden "foam cells" present in the reticuloendothelial system are not exclusive for this disease; however, these cells can be differentiated from the Gaucher's cell by the experienced pathologist. There is a distribution of discrete lipoidal deposits throughout the cytoplasm, causing a "mulberry"-like appearance in the cells. They give a positive reaction with most lipid strains such as Sudan black and a negative reaction with most carbohydrate stains such as PAS.[219] These bodies are presumed to be the products of fusion of secondary lysosomes and the accumulating lipid.

A pigment referred to as ceroid or lipofuscin is also found in the foam cells. This material is present in most cells upon aging; however, the chemical composition of these deposits is not well defined. A component of this "aging pigment" is speculated to result from peroxidation of cellular lipids.

A variety of changes are found in the quantity of circulating formed elements of the blood which is presumably correlated with the deposition of lipid in the cells of the hemopoietic system. Microcytic anemia, leukopenia, and thrombocytopenia usually develop.

The liver is firm, has a yellowish cast, and possesses vacuolated Kupffer cells. Jaundice is common in the type D patient. Liver function tests are usually normal except for some changes occasionally seen in the type B individuals.

Splenic mass may increase as much as 10-fold and the tissue is infiltrated with foam cells throughout the splenic pulp. Although the kidney has lipoidal deposition, there is no evidence of interference with renal function.

The pathological changes seen in the central nervous system have an irregular pattern. Pyramidal cells often become ballooned, lose their characteristic shape, and develop a vacuolated cytoplasm with a loss of Nissl substance. Membranous cytoplasmic bodies are readily observable under the electron microscope. Extensive gliosis is usually apparent. There have been no systematic attempts to correlate the pattern, or extent of changes observed in the central nervous system with the different types of Niemann–Pick's disease.

3.7.3. Biochemistry

The lipid that accumulates in Niemann–Pick's disease was correctly identified as sphingomyelin nearly 50 years ago by Klenk.[220] An attempted correlation between the lipid concentration and the type of disease is presented in Table XVI.[221]

Table XVI reveals an elevation of splenic and hepatic sphingomyelin, which usually is most marked in type A and B patients. An increased quantity

Table XVI. Tissue Sphingomyelin and Cholesterol Concentration in the Different Clinical Types of Niemann–Pick's Disease[a]

| Type | Liver[b] | | | Spleen[b] | | |
	A Sphingomyelin	B Cholesterol	B/A	A Sphingomyelin	B Cholesterol	B/A
A	230.5	160	0.69	271	266	0.98
B	162	207	1.27	344	103	0.30
C	37.2	64.7	1.74	119	145	1.22
D	19.4	142	7.3	71.4	220	3.08
Controls	12.9	51	3.9	19.4	64	3.34

[a] From Ref. 221.
[b] Values are μmoles/g tissue.

of tissue cholesterol is also evident. It has been claimed that this increase is of such a magnitude that sphingomyelin accounts for 2–5% of the total body weight of type A patient.[221] There is a larger amount present in the spleen than liver, and a concomitant rise in cholesterol content is usually observed. The reason for the sterol accumulation has not been explained although there has been speculation.

Early claims of unusual fatty acid composition in sphingomyelin have been revised. These reflect the characteristic differences of the sphingomyelin isolated from white matter, the average fatty acid chain length of which is C_{24}, or from gray matter, the average fatty acid chain length of which is C_{18}.

More recently, another lipid, lysobisphosphatidic acid, was reported to be increased in Niemann–Pick's tissues[222,223]:

$$
\begin{array}{ccc}
& \overset{\displaystyle O}{\overset{\displaystyle \|}{\text{CH}_2\!-\!\text{OCR}}} & \overset{\displaystyle O}{\overset{\displaystyle \|}{\text{CH}_2\!-\!\text{OCR}}} \\
\text{HO}\!-\!\text{CH} & & \text{HO}\!-\!\text{C}\!-\!\text{H} \\
& \overset{\displaystyle O}{\overset{\displaystyle \|}{}} & \\
\text{H}_2\text{C}\!-\!\text{O}\!-\!\text{P}\!-\!\text{O}\!-\!\text{CH}_2 & & \\
& \overset{}{\underset{\displaystyle \text{O}^-}{|}} &
\end{array}
$$

This compound and free cholesterol have been found to accumulate in both patients and experimental animals that have received the drug 4,4'-diethyl aminothoxyhexestrol.[224] These observations on the effect of this drug have not been extensively exploited for studying the pathobiology of Nie-mann–Pick's disease. Attempts should be undertaken to correlate the levels of lysobisphosphatidic acid and the different types of Niemann–Pick's disease. Significant elevations may be found in types C and D, where sphingomyelin levels are not as greatly increased as they are in types A and B.

Elevated levels of plasma[224a] and cerebral cortex[224b] glycosphingolipids have been reported for Niemann–Pick's disease. Elevations of gangliosides,

Table XVII. Level of Sphingomyelin-Cleaving Enzyme in Human Tissue Preparations[a]

Tissue	Source	Patient	Diagnosis	Age (years)	Sex	Enzyme activity (U/mg protein)[b]
Liver	Fresh (necropsy)	NH	Rheumatic heart disease	55	M	4.7
		RD		45	M	4.4
		NK	Neuroblastoma	64	M	10.3
		CD	Tetralogy of fallot	10	F	4.9
		JD	Mitral stenosis	48	M	11.1
		NN	Rheumatic heart disease	44	F	4.4
	Biopsy	DK	Normal	63	F	9.5
	Frozen, 3 years	KS	Heart disease	5	F	8.9
		KW	Septal defect	7	F	4.4
	1 year	JV	A_β-lipoproteinemia	11	M	4.5
	4 years	ZK	Gaucher's disease	31	M	4.6
	3 years	WC	Infantile Gaucher's disease	1	M	7.0
						Mean 6.6 ± 2.6[c]
Kidney		KW	Septal defect	7	F	4.4
Liver	2 years	KP	Niemann–Pick's disease[d]	2	F	0.0
	3 years	NW		3	F	0.54
	½ year	PW		3	M	0.61
	⅔ year	ND		3	M	0.26
	Biopsy	DB		9 mo.	M	0.46
		ES		9 mo.	M	0.89
						Mean 0.46 ± 0.31[c]
Kidney	Frozen, 2 years	KP		2	F	0.0

[a] Modified from Ref. 227.
[b] One unit of enzyme activity is defined as the amount of enzyme required to catalyze the hydrolysis of 1 nmole of sphingomyelin per hour using the conditions of incubation described in the text.
[c] Standard deviation.
[d] Clinical details covering the first four patients with Niemann–Pick's disease appear elsewhere in a review (Ref. 8) coded as Nos. 163 (KP), 70 (NW), 71 (PW), and 164 (ND). The biopsy sample on DB, an infant with history entirely compatible with the infantile form of the disease, was generously provided by Dr. Janet Cuttner, Mt. Sinai Hospital, New York. The biopsy sample on ES was supplied by Dr. Robert Kaye, Children's Hospital, Philadelphia. The proportion of total liver phospholipid phosphorus represented by sphingomyelin in these six patients was 60, 59, 55, 49, 51, and 62%, respectively. The normal value is less than 10%.

glycoproteins, and glycosaminoglycan have been demonstrated for an infantile Niemann–Pick's patient.[224c] Unfortunately, this individual was not fully biochemically characterized since the activity of sphingomyelinase was not reported.

Membranous cytoplasmic bodies and myelin have been isolated from the brain of a type A patient and only minimal differences were found in the composition of myelin when compared to normals. The membranous cytoplasmic bodies were 90 + % lipid, cholesterol represented approximately 20% and sphingomyelin 50–60% of total lipid content.[225]

The rationale for biochemically approaching the enzymatic abnormality in Niemann–Pick's disease rested on an early observation made by Crocker and Mays.[226] Slices of liver and spleen tissue from two patients with this disease were incubated with ^{32}P and the specific activity of the sphingomyelin produced during the course of the experiment was compared to that obtained with slices from control tissues. The values obtained with patient tissues were lower than those obtained with control tissues. These authors concluded that sphingomyelin accumulation in Niemann–Pick's disease probably was not due to overproduction of this compound.

For subsequent studies, the assumption was made that a deficiency in an enzyme responsible for hydrolysis of sphingomyelin results in the lipid accumulation. This decreased sphingomyelinase activity was demonstrated in studies employing radioactive sphingomyelin prepared by organic synthesis methods (Table XVII).[227] This reaction results in sphingomyelin cleavage with the release of choline phosphate and ceramide. Limited data are available correlating the level of sphingomyelinase activity with the specific clinical type of Niemann–Pick's patient (shown in Table XVIII).

The sphingomyelinase deficiency appears to be most marked in type A and B patients, with no demonstrable decrease in sphingomyelinase activity with samples from types C and D. These observations should be correlated with the lack of sphingomyelin accumulation in the tissues of type C and D patients (Table XVI). If the biochemical definition of this disease includes

Table XVIII. Sphingomyelin Hydrolysis by Tissue Obtained from Various Types of Niemann–Pick's Patients[a]

Type	Liver[b]	Spleen[c]	Fibroblasts[d] Bone	Fibroblasts[d] Skin
A	0.55	1.2	0.3	0.13
B	1.13	4.6	3.2	1.75
C	8.8	158	30.6	—
D	6.25	36	30.1	—
Controls	4–11	65–200	52 ± 10	35 ± 6

[a] From Ref. 221.
[b] U/mg protein.
[c] μU/mg protein.
[d] U/10⁶ cells.

both an elevated tissue sphingomyelin level and a concomitant decrease in its hydrolysis, then the types classified as C and D should not be considered as Niemann–Pick's disease patients. The acid sphingomyelinase has been partially purified from the brain of a type C Niemann–Pick individual. Its properties and activities were found to be similar to those prepared from control tissues, suggesting this enzyme is not involved in this form of the disease.[227a]

An Mg^{2+}-dependent sphingomyelinase activity with a pH optimum of 7.4 has been reported to be unaltered in Niemann–Pick type C spleen samples.[228] This activity appears to be different from the major lysosomal sphingomyelinase. Preliminary observations concerning the properties of a microsomal, Mg^{2+}-dependent sphingomyelinase with a pH optimum of 7.4 have appeared.[228a] This activity is found principally in the brain and is enriched in the gray matter.[228b] This activity was present in the brain tissue of a patient with "infantile" Niemann–Pick's disease at a level claimed to be within the range for normals.[228c] The contribution of this extralysosomal enzyme to brain sphingomyelin metabolism is unknown. Presumably it would not be accessible to the storage vesicles containing the excessive quantities of sphingomyelin that occur in this disease. An additional sphingomyelinase, which is Mg^{2+} independent and has a pH optimum of 7.0,[228d] may be associated with rat brain myelin. The Mg^{2+}-dependent, neutral sphingomyelinase of rat liver is present in the plasma membrane subfraction.[228e]

The inability of Niemann–Pick type C patients to catalyze the cleavage of bis-pNP-phosphate was reported at a meeting.[229] This activity required Mg^{2+}, cysteine, and had a pH optimum around neutrality, suggesting that it may represent a general phosphodiesterase deficiency. Full and careful documentation of this preliminary observation will be needed before this can be generally accepted. Decreased hydrolytic activity towards bis-4MU-pyrophosphate diester has been observed with fibroblasts from Niemann–Pick types A and B with only marginal decrease with the type C patients. This deficiency was greatest at pH 5.0 and not evident at pH 7.2.[229a] The addition of Triton X-100 stimulates the activity towards this substrate at acidic pH values. Under these conditions, a decrease was seen with fibroblasts from type C patients; however, it is not clear that this was due to the presence of detergent.[229b]

Sphingomyelinase isozymes obtained by isoelectric focusing of tissue extracts from normals and patients with type C Niemann–Pick's disease have been compared.[228f,g] Normals have two major and three minor components. An apparent loss of the component referred to as Peak II was seen in liver (Fig. 8) and brain tissue from two patients. Although these observations were reproduced, it was alleged that they were merely artifacts created by the particular extraction procedure employed.[228h] A comparative study of the sphingomyelinase isozymes of brain and liver, separated by gel filtration, is in accord with tissue-specific forms of the disease.[228i] Extracts of cultured fibroblasts from type A, B, and E Niemann–Pick's patients were electrofocused and the activities of the various isozymes compared to controls. Reduced activity of all isozyme species was seen with type A and B cell extracts. The reductions observed were more marked with type A cells than with type B cells.[228j] This is in accord with the information presented in Table XVIII.

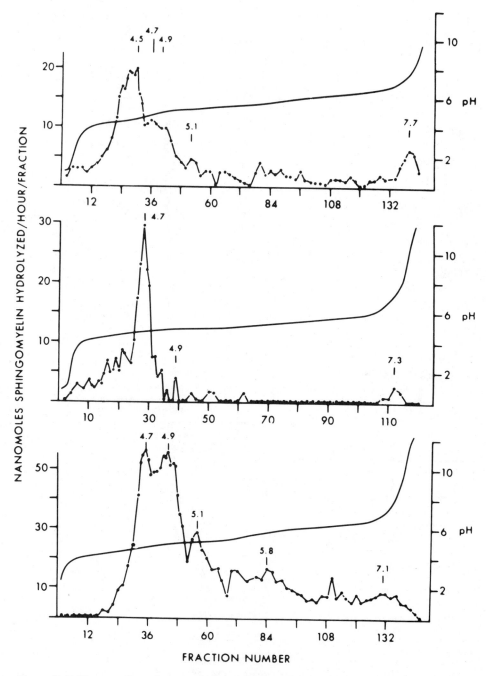

Figure 8. Sphingomyelinase isozymes in normal liver and in type C Niemann–Pick liver. The extracts were prepared from 1 g frozen liver and analyzed separately. Normal liver (lower panel) is compared with liver from case 1 (upper panel) and from case 2 (middle panel). The isoelectric points of the major peaks are indicated. Recovery was 85, 65, and 78% for the normal, case 1, and case 2, respectively. The pH gradient is the solid line. (From Ref. 228g.)

The most reliable diagnostic procedure relies on the cleavage of radioactive sphingomyelin. Heterozygote detection employing leukocytes and cultured fibroblasts has been achieved.[108,230,230a,b] Prenatal diagnosis of an affected fetus, based upon the inability to cleave sphingomyelin by fibroblasts propagated in culture, has been documented.[231] The fetal brain and liver, obtained after therapeutic abortion, were also unable to cleave this substrate. It is of historical interest to point out that Niemann–Pick's disease was the first lipid storage disease diagnosed *in utero*. Based upon quantitation of sphingomyelin of cultured amniotic fluid cells, it was correctly predicted that the offspring would be affected.[232]

A chromogenic substrate, 2-*N*-(hexadecanoyl)-amino-4-nitrophenyl-phosphorylcholine hydroxide, for sphingomyelinase quantitation has been synthesized.[232a] The product of enzymatic cleavage is yellow under alkaline conditions, providing a convenient assay procedure. The usefulness of this substrate for diagnostic purposes has been examined[232b] with solid tissues (Table XIX) and cultured fibroblasts. It appears that this material may be as reliable as sphingomyelin for both homozygous affected and heterozygote identifications.[232c] A chromogenic analog of sphingomyelin, *N*-ω-trinitrophenylaminolaurylsphingosylphosphorylcholine, has been synthesized and shown to distinguish between normal and Niemann–Pick fibroblasts.[232d]

Table XIX. Hydrolysis of 2-Hexadecanoylamino-4-nitrophenyl-phosphorylcholine (HNP), Shingomyelin, and Glucocerebroside by Various Enzyme Preparations[a]

Source of enzyme	HNP	[^{14}C]Sphingomyelin	[^{14}C]Glucocerebroside
		(nmoles/mg protein /hr)	
Purified placental sphingomyelinase[c]	3942.0	4062.0	0.0
C. welchii phospholipase C[d]	0.0	0.0	0.0
B. cereus phospholipase C[d]	0.0	0.0	0.0
Human liver control[e]	78.0	38.0	32.0
Liver from patient with Tay–Sachs' disease (Case 1)[e]	62.0	31.0	38.0
Liver from patient with Tay–Sachs' disease (Case 2)[e]	53.0	27.0	32.0
Liver from patient with Gaucher's disease[e]	34.0	21.0	2.6
Liver from fetus with type A Niemann–Pick's disease[e]	0.0	0.1	58.0

[a] Modified from Ref. 232b.
[b] Mixtures incubated for 1 hr at 37°C under conditions described in text.
[c] 500 μg of human serum albumin added to incubation mixtures to stabilize enzyme.
[d] Under these conditions, the *C. welchii* and *B. cereus* phospholipase C preparations hydrolyzed 177 and 235 nmoles of phosphatidylcholine, respectively.
[e] The human liver specimens were obtained at postmortem examination from a 61-year-old man who died of a myocardial infarction (control), two patients with Tay–Sachs' disease each 2½ years old at the time of death, a 4-month-old child with Gaucher's disease, and a 20-gestational-week fetus with Niemann–Pick's disease. Previous studies have shown that extracts of liver tissue from a fetus without Niemann–Pick's disease at this stage contained 13.1 nmoles of sphingomyelinase activity/mg protein per hr.

The term "activator" has been used in association with sphingomyelinase. This term should refer to those materials that are not detergents but are heat-stable factors derived from vertebrates that stimulate the hydrolysis of lipids. This describes the activator of hexosaminidase[183] or glucosylceramide.[114a] A tissue extract has been prepared that is much less effective than Triton X-100 in supporting sphingomyelin hydrolysis. Unfortunately, data were not provided on this basal enzyme activity in the absence of either "activator" and detergent.[232e] Triton X-100 at high concentrations inhibits a solubilized preparation of the Mg^{2+}-dependent pH 7.4 sphingomyelinase. A heat stable-factor that is equated with "activator" present in brain and other tissues prevents this inhibition.[232f] In neither case has an "activator," as defined previously, of sphingomyelinase been demonstrated.

Morphological examination of a 19-week abortus with Niemann–Pick's disease revealed the presence of characteristic lipoidal cytoplasmic inclusions. Therefore, the pathological changes characteristic of the affected individual are probably already developing *in utero.*[233] Prenatal diagnosis of Niemann–Pick's disease has been the subject of several reports.[233a,b] The activity of the lysosomal sphingomyelinase was reduced but the activity of the neutral, Mg^{2+}-dependent form of the enzyme was unaffected.[233b] Accumulation of sphingomyelin in liver, spleen, and placenta but not brain was seen with a type A Niemann–Pick abortus. Bislysophosphatidic acid was also seen in the tissues.[233c] Tissue from a presumed type C fetus had slightly decreased acid sphingomyelinase activity.[233d]

Some caution must be used in such attempts at prenatal diagnosis. They should be considered only when the phenotype of a classical type A patient is present in the family. It is not currently possible to distinguish between the enzymic deficiencies of types A and B. The second group of patients can live a useful life that is not severely affected by the disease.[234]

Sea-blue histiocytosis syndrome may be identical to Niemann–Pick type B.[234a] A family described as having sea-blue histiocytosis has an unusually heat-sensitive sphingomyelinase present in fibroblast extracts. It was proposed to create a new category, Niemann–Pick type F, to accommodate these individuals.[234b]

The possible occurrence of Niemann–Pick's disease in Siamese cats has been reported.[235,235a]

3.8. GM₃ Gangliosidosis

This disease differs from the other diseases discussed since it is not correlated with the decrease of a lysosomal acid hydrolase.

Only a single patient has been thoroughly characterized thus far. The child had generalized seizures, feeding difficulty, and was lethargic shortly after birth. At 1 month, its appearance suggested infantile GM_1 gangliosidosis. He was hypotonic, unresponsive, had coarse facies, macroglossia, and was hirsute. Enlarged spleen and liver were evident. The patient died by $3\frac{1}{2}$ months.[236] The features distinguishing this from GM_1 gangliosidosis have been reviewed.[237]

3.8.1. Pathology

A pervading vacuolation or sponginess of the white matter was the primary histological finding. There was no evidence for the accumulation of storage materials. There were no ultrastructural abnormalities in vascular elements or oligodendroglial cells.[238]

3.8.2. Biochemistry

A distinctly altered ganglioside pattern was seen with brain tissue from the patient (Table XX). The concentration of GM_3 and GD_3 and relative percent of total ganglioside were dramatically increased. There was no reduction in lysosomal acid hydrolase activity. This increased GM_3 is apparently not the result of reduced neruaminidase activity, but rather a consequence of reduced UDP-GalNAc : GM_3 N-acetylgalactosaminyl transferase activity.[239] This enzyme results in GM_2 production, which in turn is converted to higher gangliosides.

3.9. General Comments

All of the sphingolipidoses are inherited as autosomal recessive characteristics, except for Fabry's disease, which is transmitted as an X-linked recessive trait.[240]

The comparative studies cited concerning the metabolism of glucosylceramide by Gaucher's tissues[100,101] and sphingomyelin by Niemann–Pick tissues[226,227] provided suggestive evidence for a catabolic defect in these two diseases. Similar information on both the anabolic and the catabolic activities

Table XX. Ganglioside composition of Brain from the Patient and an Age-Matched Control[a]

	Patient		Control[b]	
	nmoles of sialic acid/mg protein	% of total	nmoles of sialic acid/mg protein	% of total
GM_3	1.6	67	0.46	3.6
GM_2	0.1	—	0.41	3.3
GM_1	0.1	—	2.6	21
GD_3	0.8	33	0.34	2.6
GD_{1a}	0.1	—	3.1	25
GD_{1b}	0.1	—	2.9	23
GT_1	0.1	—	2.8	22
Total gangliosides	2.4	—	12.6	

[a] From Ref. 236.
[b] Values for control corresponded well with those obtained by Suzuki except for slightly higher figures for GM_3 and GM_2, which were quantitatively recovered by the isolation procedure used in our experiments, but were probably retained in the chloroform phase and not determined by Suzuki.

is not available for the other sphingolipidoses. This is most likely due to the priority and emphasis given by investigators for examining hydrolytic enzymes as the probable biochemical lesion in these groups of diseases.

The patient with metachromatic leukodystrophy has been postulated to form an unstable myelin due to sulfatide accumulation. This may be due to a surplus of negative charges, which tend to repulse one another.[241] There are few theories currently available attempting to explain the basic pathophysiological events and damage to the function of an organ as a result of the accumulation of sphingolipids. This probably relates to the general deficit in understanding the basic physiological role of the sphingolipids.

The progress of our understanding of the sphingolipidoses provides an excellent example of the importance of basic biochemical research. The early workers in this field could not have predicted or foreseen the developments that have been built upon their contributions. The sequence of events follows an extremely reasonable and logical course of events. This involved: elucidation of the structure of the accumulating material; purification and establishment of the properties of the enzyme responsible for its hydrolysis; demonstration of quantitative reduction in this hydrolytic enzyme in patients; heterozygote detection; amniocentesis and prenatal diagnosis; mass carrier identification screening program; and the current attempts at enzyme re-

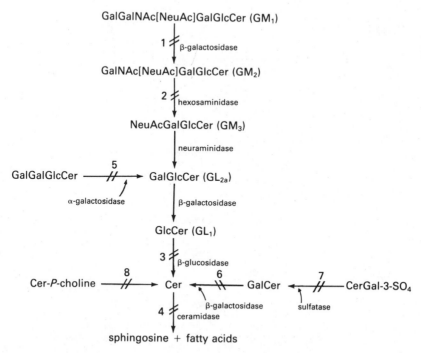

Figure 9. Lysosomal acid hydrolase deficiencies associated with the sequential cleavage of bonds present in sphingolipids. 1, GM_1 gangliosidosis; 2, GM_2 gangliosidosis; 3, Gaucher's disease; 4, Farber's disease; 5, Fabry's disease; 6, Krabbe's disease; 7, metachromatic leukodystrophy; 8, Niemann–Pick's disease.

placement therapy.[242,243] These events have contributed to the dramatic growth and development of biochemical genetics, genetic counseling, public health and health care delivery systems.

The enzymatic defects associated with these diseases are summarized in Fig. 9. It is apparent that there is a specific enzyme deficiency associated with the cleavage of individual linkages in the sphingolipids, except for neuraminidase. The relationship of the recently reported neuraminidase deficiency in I cell disease (mucolipidosis II) fibroblasts[244] to ganglioside metabolism is not yet established. *N*-Acetylneuraminic acid is an important constituent of many glycoproteins, including many lysosomal acid hydrolases, hormones, and cell membrane constituents. It might be assumed that a deficiency in the activity of such an enzyme would be catastrophic to the function of cells and therefore would be a lethal mutation.

There are a number of books available for those requiring further information about the sphingolipidoses. Individual chapters devoted to each of the sphingolipidoses are contained in the several editions of *The Metabolic Basis of Inherited Diseases*.[245] Recurring symposia focusing on research about the sphingolipidoses have been published.[246–248] Summaries of the individual sphingolipidoses, which may be mainly of historical interest, are available.[249,250] Volumes have appeared focusing upon the gangliosidoses[251,252] or diagnostic procedures.[253]

3.10. References

1. Ward, R. D. *Biochem. Genet.* **16,** 799 (1978).
1a. Childs, B., and Der Kalavstain, V. M. *N. Engl. J. Med.* **279,** 1205, 1267 (1968).
2. Mellman, W. J. *J. Pediatr.* **72,** 727 (1968).
3. Harris, H. *The Principle of Human Biochemical Genetics,* American Elsevier, New York (1970).
4. Harris, H. *Triangle* **10,** 41 (1971).
5. DeDuve, C. In: *Lysosomes in Biology and Pathology* (Dingle, J. T., and Fell, H. B., eds.), North-Holland, Amsterdam (1969), p. 3.
6. Tappel, A. L. In: *Lysosomes in Biology and Pathology* (Dingle, J. T., and Fell, H. B., eds.), North-Holland, Amsterdam (1969), p. 207.
7. Barrett, A. J. In: *Lysosomes in Biology and Pathology* (Dingle, J. T., and Fell, H. B., eds.), North-Holland, New York (1969), p. 245.
8. Schettler, G. (ed.). *Lipids and Lipidosis,* Springer-Verlag, New York (1967).
9. Stanbury, J., Wynngaarden, J., and Fredrickson, D. *The Metabolic Basis of Inherited Diseases,* 4th ed., McGraw-Hill, New York (1978).
10. Hers, H. G., and Von Hoof, F. (eds.), *Lysosomes and Storage Diseases,* Academic Press, New York (1973).
11. Hers, G. H. *Gastroenterology* **48,** 625 (1965).
12. Crocker, A. C., Cohen, J., and Farber, S. In: *Inborn Disorders of Sphingolipid Metabolism* (Aronson, S., and Volk, B. W., eds.), Pergamon Press, Oxford, England (1967), p. 485.
13. Farber, S. *Am. J. Dis. Child.* **84,** 899 (1952).
14. Farber, S., Cohen, J., and Uzman, L. L. *J. Mt. Sinai Hosp.* **24,** 817 (1957).
15. Moser, H. M., Prensky, A. L., Wolfe, H. J., and Rosmann, N. P. *Am. J. Med.* **47,** 869 (1969).
16. Samulssen, K., and Zetterstrom, R. *Scand. J. Clin. Lab. Invest.* **27,** 393 (1971).
17. Sugita, M., Connolly, P., Dulaney, J. T., and Moser, H. W. *Lipids* **8,** 401 (1973).
18. Sugita, M., Dulaney, J. T., and Moser, H. W. *Science* **178,** 1100 (1972).
18a. Sugita, M., Williams, M., Dulaney, J. T., and Moser, H. W. *Biochim. Biophys. Acta* **398,** 125 (1975).

18b. Dulaney, J., Moser, H. W., Sidbury, J., and Milunsky, A. *Adv. Exp .Med. Biol.* **68,** 403 (1976).
18c. Chen, W. W., Moser, A. B., and Moser, H. W. *Arch. Biochem. Biophys.* **208,** 244 (1981).
18d. Momoi, T., Ben-Yoseph, Y., and Nadler, H. L. *Biochem. J.* **205,** 419 (1982).
19. Hagberg, B. *Acta Pediatr. Scand.* **52,** 213 (1963).
19a. Lieberman, J. S., Oshtory, M., Taylor, R. G., and Dreyfus, P. M. *Arch. Neurol.* **37,** 446 (1980).
20. Krabbe, K. *Brain* **39,** 74 (1916).
21. Collier, J., and Greenfield, J. G. *Brain* **47,** 489 (1924).
22. Austin, J. H., and Lehfeldt, D. *J. Neuropathol. and Exp. Neurol.* **24,** 265 (1965).
22a. Suzuki, K., Tanaka, H., and Suzuki, K. *Adv. Exp. Med. Biol.* **68,** 99 (1976).
22b. Dunn, H. G., Dolman, C. L., Farrell, D. F., Tischler, B., Hasinoff, C., and Woolf, L. I. *Neurology* **26,** 1035 (1976).
23. Yunis, E. J., and Lee, R. E. *Lab. Invest.* **21,** 415 (1969).
24. Schochel, S. S., Hardman, J. M., Lampert, P. W., and Earle, K. M. *Arch. Pathol.* **88,** 305 (1969).
25. Suzuki, K. *Lab. Invest.* **23,** 612 (1970).
26. Yunis, E. J., and Lee, R. E. *Science* **169,** 64 (1970).
27. Blinzinger, K., and Anzil, P. *Experientia* **28,** 780 (1972).
28. Austin, J. H. *J. Neurochem.* **10,** 921 (1963).
29. Austin, J. *Arch. Neurol.* **9,** 207 (1963).
30. Svennerholm, L. In: *Brain Lipids and Lipoproteins and the Leukodystrophies* (Folch-Pi, J., and Bauer, H., eds.), Elsevier, Amsterdam (1963), p. 104.
31. Eto, Y., Suzuki, K., and Suzuki, K. *J. Lipid Res.* **11,** 473 (1970).
32. Bachhawat, B. K., Austin, J., and Armstrong, D. *Biochem. J.* **104,** 15c (1967).
33. Austin, J., Suzuki, K., Armstrong, D., Brady, R. O., Bachhawat, B. K., Schlenker, J., and Stumpf, D. *Arch. Neurol.* **23,** 502 (1970).
34. Suzuki, K., and Suzuki, K. *Proc. Natl. Acad. Sci. U.S.A.* **66,** 302 (1970).
35. Suzuki, Y., and Suzuki, K. *Science* **171,** 73 (1971).
36. Farrell, D. F., Percy, A. K., Kaback, M. M., and McKhann, G. M. *Am. J. Hum. Genet.* **25,** 604 (1973).
37. Suzuki, K., Schneider, E. L., and Epstein, C. J. *Biochem. Biophys. Res. Commun.* **45,** 1363 (1971).
38. Suzuki, K., and Suzuki, Y. In: *The Metabolic Basis of Inherited Disorders* (Standbury, J. B., Wyngaarden, J. B., and Fredrickson, D. S., eds.), McGraw-Hill, New York (1972), p. 760.
39. Miyatake, T., and Suzuki, K. *Biochem. Biophys. Res. Commun.* **48,** 538 (1972).
39a. Svennerholm, L., Vanier, M. T., and Mansson, J. E. *J. Lipid Res.* **21,** 53 (1980).
39b. Wenger, D. A., Sattler, M., and Markey, S. P. *Biochem. Biophys. Res. Commun.* **53,** 680 (1973).
39c. Carter, T. P., and Kanfer, J. N. *J. Neurochem.* **27,** 53 (1976).
40. Carter, T. P., Beblowski, D. W., Savage, M. H., and Kanfer, J. N. *J. Neurochem.* **34,** 189 (1980).
40a. Wenger, D. A., Sattler, M., and Heate, W. *Proc. Natl. Acad. Sci. U.S.A.* **71,** 854 (1974).
40b. Wenger, D. A., Sattler, M., Clark, C., and McKelvey, H. *Clin. Chim. Acta* **56,** 199 (1974).
40c. Svennerholm, L., Hakansson, G., and Vanier, M. T. *Acta Paediatr. Scand.* **64,** 649 (1975).
40d. Tanaka, H., and Suzuki, K. *Arch. Biochem. Biophys.* **175,** 332 (1976).
40e. Suzuki, Y., and Suzuki, K. *J. Biol. Chem.* **249,** 2098, 2105 (1974).
40f. Wenger, D. A., Sattler, M., and Clark, C. *Biochim. Biophys. Acta* **409,** 297 (1975).
40g. Tanaka, H., and Suzuki, K. *J. Biol. Chem.* **250,** 2324 (1975).
40h. Tanaka, H., and Suzuki, K. *Clin. Chim. Acta* **75,** 267 (1977).
40i. Miyatake, T., and Suzuki, K. *J. Biol. Chem.* **250,** 585 (1975).
40j. Tanaka, H., and Suzuki, K. *Arch. Biochem. Biophys.* **175,** 332 (1976).
40k. Ben-Yoseph, Y., Hungerford, M., and Nadler, H. L. *Am. J. Hum. Genet.* **30,** 644 (1978); *Arch. Biochem. Biophys.* **196,** 93 (1979).
40l. Tanaka, H., and Suzuki, K. *J. Neurol. Sci.* **38,** 409 (1978).
40m. Dawson, G., and Stein, O. *Science* **170,** 556 (1970).
40n. Dawson, G. *J. Lipid Res.* **13,** 207 (1972).
40o. Dawson, G., Matalon, R., and Stein, A. O. *J. Pediatr.* **79,** 423 (1971).

40p. Wenger, D. A., Sattler, M., Clark, C., Tanaka, H., Suzuki, K., and Dawson, G. *Science* **188,** 1310 (1975).

40q. Miller, A. L., Frost, R. G., and O'Brien, J. S. *Biochem. J.* **165,** 591 (1977).

40r. Burton, B. K., Ben-Yoseph, Y., and Nadler, H. L. *Clin. Chim. Acta* **88,** 483 (1978).

41. Radin, N. S., Arora, R. C., Ullman, M. D., Brenkert, A. L., and Austin, J. H. *Res. Commun. Chem. Pathol. Pharmacol.* **3,** 637 (1972).

42. Suzuki, Y., Austin, J., Armstrong, D., Suzuki, K., Schlenker, J., and Fletcher, T. *Exp. Neurol.* **29,** 65 (1970).

42a. Suzuki, Y., Miyatake, T., Fletcher, T. F., and Suzuki, K. *J. Biol. Chem.* **249,** 2109 (1974).

42b. Costantino-Ceccarine, E., Fletcher, T. F., and Suzuki, K. *Adv. Exp. Med. Biol.* **68,** 127 (1976).

43. Hagberg, B. In: *Brain Lipids and Lipoproteins and the Leucodystrophies* (Folch-Pi, J., and Bauer, H., eds.), Elsevier, Amsterdam (1963), p. 134.

44. Miller, D., Pilz, H., and Meulen, V. T. *J. Neurol. Sci.* **9,** 567 (1969).

45. Austin, J. In: *Lysosomes and Storage Diseases* (Hers, H. G., and Von Hoof, F., eds.), Academic Press, New York (1973), p. 412.

45a. Christomanou, H., Martinius, J., Jaffe, S., Betke, K., and Förster, C. *Hum. Genet.* **55,** 103 (1980).

46. Hirsch, T., and Peiffer, J. *Arch. Psychiatr. Nervenkr.* **194,** 88 (1955).

47. Resibios, A. *Acta Neuropathol. (Berlin)* **13,** 149 (1969).

48. Austin, J. H. *Proc. Soc. Exp. Biol. Med.* **100,** 361 (1959).

49. Jatzkewitz, H. *Z. Physiol. Chem.* **311,** 279 (1958).

50. Malone, M. J., Stoffyn, P., and Moser, H. *J. Neurochem.* **13,** 1033 (1966).

51. O'Brien, J. S. *Biochem. Biophys. Res. Commun.* **15,** 484 (1964).

51a. Pilz, H., and Heipertz, R. *Z. Neurol.* **206,** 203 (1974).

52. Ställberg-Stenhagen, S., and Svennerholm, L. *J. Lipid Res.* **6,** 146 (1965).

53. Martensson, E., Percy, A., and Svennerholm, L. *Acta Paediatr. Scand.* **55,** 1 (1966).

54. Philippart, M., Sarlieve, L., Meurant, C., and Mechler, L. *J. Lipid Res.* **12,** 434 (1971).

55. Bischel, M., Austin, J., and Kemeny, N. *Arch. Neurol.* **15,** 13 (1966).

56. Murphy, J. V., Wolfe, H. J., Balags, E. A., and Moser, H. W. In: *Lipid Storage Diseases* (Bernsohn, J., and Grossman, H. J., eds.), Academic Press, New York (1971), p. 67.

57. Suzuki, K., Suzuki, K., and Chen, G. C. *J. Neuropathol. Exp. Neurol.* **26,** 537 (1967).

58. Austin, J. H., Balasubramanian, A. S., Pattabiraman, T. N., Saraswathi, S., Rosie, D. K., and Bachhawat, B. K. *J. Neurochem.* **10,** 805 (1963).

59. Mehl, E., and Jatzkewitz, H. *Biochem. Biophys. Res. Commun.* **19,** 407 (1965).

60. Mehl, E., and Jatzkewitz, H. *Biochem. Biophys. Acta* **151,** 619 (1968).

60a. Fischer, G., and Jatzkewitz, H. *Hoppe-Seyler's Z. Physiol. Chem.* **356,** 605 (1975).

60b. Mraz, W., Fischer, G., and Jatzkewitz, H. *Hoppe-Seyler's Z. Physiol. Chem.* **357,** 1191 (1976).

60c. Fischer, G., and Jatzkewitz, H. In: *Enzymes of Lipid Metabolism* (Gatt, S., Freysz, L., and Mandel, P., eds.), Plenum Press, New York (1978), p. 573.

61. Jatzkewitz, H., and Stenshoff, K. *FEBS Lett.* **32,** 129 (1973).

62. Jatzkewitz, H., and Mehl, E. *J. Neurochem.* **16,** 19 (1969).

63. Percy, A. K., Farrell, D. F., and Kaback, M. M. *J. Neurochem.* **19,** 233 (1972).

64. Harzer, K., Stinshoff, K., Mraz, W., and Jatzkewitz, H. *J. Neurochem.* **20,** 279 (1973).

64a. Harzer, K., and Benz, H. A. *Hoppe-Seyler's Z. Physiol. Chem.* **355,** 744 (1974).

64b. Eto, Y., Wiesmann, U., and Herschkowitz, N. N. *J. Biol. Chem.* **249,** 4955 (1974).

64c. Yamaguchi, S., Aoki, K., Handa, S., and Yamakawa, T. *J. Neurochem.* **24,** 1084 (1975).

64d. Iwamori, M., Moser, H. W., and Kishimoto, Y. *J. Neurochem.* **27,** 1389 (1976).

65. Moser, H. W., Moser, A. B., and McKhann, G. M. *Arch. Neurol.* **17,** 494 (1967).

66. Austin, J., Armstrong, D., Shearer, L., and McAfee, D. *Arch. Neurol.* **14,** 259 (1966).

66a. Eto, Y., Wiesmann, U. N., Carson, J. H., and Herschkowitz, N. N. *Arch. Neurol.* **30,** 153 (1974).

67. Stumpf, D., and Austin, J. *Arch. Neurol.* **24,** 117 (1971).

68. Stumpf, D., Neuwelt, E., Austin, J., and Kohler, P. *Arch. Neurol.* **25,** 427 (1971).

68a. Neuwelt, E., Kohler, P. F., and Austin, J. *Immunochemistry* **10,** 767 (1973).

69. Neuwelt, E., Stumpf, D., Austin, J., and Kohler, P. *Biochim. Biophys. Acta* **236**, 333 (1971).
70. Percy, A. K., and Brady, R. O. *Science* **161**, 594 (1968).
71. Leroy, J. G., Dumon, J., and Radermecker, J. *Nature (London)* **226**, 553 (1970).
72. Thomas, G. H., and Howell, R. R. *Clin. Chim. Acta* **36**, 99 (1972).
72a. Shapira, E., and Nadler, H. L. *Clin. Chim. Acta* **65**, 1 (1975).
72b. Kolodny, E. H., and Mumford, R. A. *Adv. Exp. Med. Biol.* **68**, 239 (1976).
72c. Dubois, G., Turpin, J. C., and Baumann, N. *Adv. Exp. Med. Biol.* **68**, 233 (1976).
73. Boss, N., Witmar, E. J., and Dreifuss, F. E. *Neurology* **20**, 52 (1970).
74. Kaback, M. M., and Howell, R. R. *N. Engl. J. Med.* **282**, 1336 (1970).
75. Kihara, H., Porter, M. T., Fluharty, A. L., Scott, M. L., de la Flor, S., Trammell, J. L., and Nakamura, R. N. *Am. J. of Ment. Defic.* **77**, 389 (1973).
75a. Christomanou, H., and Sandohoff, K. *Neuropaediatrie* **9**, 385 (1978).
76. Leroy, J. G., Van Elsen, A. F., Martin, J. J., Dumon, J. E., Hulet, A. E., Okada, S., and Navarro, C. *N. Engl. J. Med.* **288**, 1365 (1973).
77. Van der Hagen, C. B., Borresen, A. L., Molne, K., Oftedal, G., Bjoro, K., and Berg, K. *Clin. Genet.* **4**, 256 (1973).
77a. Booth, C. W., Chen, K. K., and Nadler, H. L. *J. Pediatr.* **86**, 560 (1975).
77b. Kadoh, T., and Wenger, D. A. *J. Clin. Invest.* **70**, 89 (1982).
77c. Hahn, A. F., Gordon, B. A., Feleki, V., Hinton, G. G., and Gilbert, J. J. *Ann. Neurol.* **12**, 33 (1982).
78. Porter, M. T., Fluharty, A. L., and Kihara, H. *Proc. Natl. Acad. Sci. U.S.A.* **62**, 887 (1969).
79. Porter, M. T., Fluharty, A. L., Harris, S. E., and Kihara, H. *Arch. Biochem. Biophys.* **138**, 646 (1970).
80. Porter, M. T., Fluharty, A. L., and Kihara, H. *Science* **172**, 1264 (1971).
80a. DeLuca, C., Brown, J. A., and Shaws, T. B. *Proc. Natl. Acad. Sci. U.S.A.* **76**, 1957 (1979).
80b. Bruns, G. A. P., Mintz, B. J., Leary, A. C., Regina, V. M., and Gerald, P. S. *Cytogenet. Cell Genet.* **22**, 182 (1978).
81. Austin, J. In: *Inborn Disorders of Sphingolipid Metabolism* (Aronson, S., and Volk, B., eds.), Pergamon Press, New York (1966), p. 359.
82. Greene, H. L., Hug, G., and Schubert, W. K. *Arch. Neurol.* **20**, 147 (1969).
83. Austin, J. *Birth Defects: Orig. Artic. Ser.* **9**, 125 (1973).
83a. Poduslo, S. E., Miller, K., and Jang, Y. *Acta Neuropathol.* **57**, 13 (1982).
84. Fredrickson, D. S., and Sloane, H. S. In: *Inherited Basis of Metabolic Diseases* (Stanbury, J., Fredrickson, D. S., and Wyngaarden, J., eds.), McGraw-Hill, New York (1972), p. 730.
84a. Soffer, D., Yamanaka, T., Wenger, D. A., Suzuki, K., and Suzuki, K. *Acta Neuropathol.* **49**, 1 (1980).
84b. Miller, J. D., McCluer, R., and Kanfer, J. N. *Ann. Intern. Med.* **78**, 883 (1973).
84c. Dreborg, S., Erikson, A., and Hagborg, B. *Eur. J. Pediatr.* **133**, 197 (1980).
85. Block, M., and Jacobson, L. O. *Acta Haemat.* **1**, 165 (1948).
86. Lee, R. E. *Proc. Natl. Acad. Sci. U.S.A.* **61**, 484 (1968).
87. Lee, R. E. *Arch. Biochem. Biophys.* **159**, 259 (1973).
88. Zaino, E. C., Rossi, M. G., Pham, T. D., and Azai, H. A. *Blood* **38**, 457 (1971).
89. Albrecht, M. *Blut* **13**, 169 (1966).
90. Kattlove, H. E., Williams, J. C., Gaynor, E., Spivack, M., Bradley, R. M., and Brady, R. O. *Blood* **33**, 379 (1969).
91. Tuchman, L. R., Goldstein, G., and Clyman, M. *Am. J. Med.* **27**, 959 (1959).
91a. Chambers, J. P., Aquino, L., Glew, R. H., Lee, R. H., and McCafferty, L. R. *Clin. Chim. Acta* **80**, 67 (1977).
91b. Chambers, J. P., Peters, S. P., Glew, R. H., Lee, R. E., McCafferty, L. R., Mercer, D. W., and Wenger, D. A. *Metabolism* **27**, 801 (1978).
91c. Lieberman, J., and Beutler, E. *N. Engl. J. Med.* **294**, 1442 (1976).
91d. Pentchev, P. G., Gal, A. E., Wong, R., Morrone, S., Neumeyer, B., Massey, J., Kanter, R., Sawitsky, S., and Brady, R. O. *Biochim. Biophys. Acta* **665**, 615 (1981).
92. Aghion, A. Thesis, University of Paris (1934).
93. Suomi, W. D., and Agranoff, B. W. *J. Lipid Res.* **6**, 211 (1965).

94. Philippart, M., and Menkes, J. *Biochem. Biophys. Res. Commun.* **15,** 551 (1964).
94a. Klibansky, C., Ossimi, Z., Mathoth, Y., Pinkhas, J., and De Vries, A. *Clin. Chim. Acta* **72,** 141 (1976).
94b. Dawson, G., and Oh, J. H. *Clin. Chim. Acta* **75,** 149 (1977).
95. Glew, R. H., and Lee, R. E. *Arch. Biochem. Biophys.* **156,** 626 (1973).
96. Inose, T., Sakai, M., Tano, T., and Kaneko, Y. *Yokahama Med. Bull.* **18,** 215 (1967).
97. French, J. H., Brotz, M., and Poser, C. M. *Neurology* **19,** 81 (1969).
98. Svennerholm, L. In: *Inborn Errors of Sphingolipid Metabolism* (Aronson, S. M., and Volk, B. W., eds.), Pergamon Press, New York (1967), p. 169.
99. Kubota, M. *Jpn. J. Exp. Med.* **42,** 513 (1972).
99a. Sudo, M. *J. Neurochem.* **29,** 379 (1977).
100. Trams, E. G., and Brady, R. O. *J. Clin. Invest.* **39,** 1546 (1960).
101. Brady, R. O., Kanfer, J. N., and Shapiro, D. *Biochem. Biophys. Res. Commun.* **18,** 221 (1965).
102. Patrick, A. D. *Biochem. J.* **97,** 17c (1965).
103. Brady, R. O. In: *Neurosciences Research,* Vol. 2 (Ehrenpreis, S., and Solnitzky, O. C., eds.), Academic Press, New York (1969), p. 301.
103a. Pentchev, P. G., Brady, R. O., Blair, H. E., Bretton, D. E., and Sarrell, S. H. *Proc. Natl. Acad. Sci. U.S.A.* **75,** 3970 (1978).
104. Beutler, E. and Kuhl, W. *J. Lab. Clin. Invest.* **76,** 747 (1970).
104a. Klibansky, C., Hoffmann, J., Pinkhas, J., Algom, D., Dintzman, M., Ben-Bassat, M., and de Vries, A. *Eur. J. Clin. Invest.* **4,** 101 (1974).
104b. Peters, S. P., Lee, R. E., and Glew, R. H. *Clin. Chim. Acta* **60,** 391 (1975).
104c. Sengers, R. C. A., Lamers, K. J. B., Bakkeren, J. A. J. M., Schretlen, E. D. A. M., and Trijbels, J. M. F. *Neuropaediatrie* **6,** 377 (1975).
104d. Beutler, E., Kuhl, W., Matsumoto, F., and Pangalis, G. *J. Exp. Med.* **143,** 975 (1976).
104e. Raghavan, S. S., Topol, J., and Kolodny, E. H. *Am. J. Hum. Genet.* **32,** 158 (1980).
105. Hultberg, B., and Ockerman, P. A. *Clin. Chim. Acta* **28,** 169 (1970).
106. Beutler, E., Kuhl, W., Trinidad, F., Teplitz, R., and Nadler, H. *Am. J. Hum. Genet.* **23,** 62 (1971).
106a. Butterworth, J., and Broadhead, D. M. *Clin. Chim. Acta* **87,** 433 (1978).
107. Kampine, J. P., Brady, R. O., Kanfer, J. N., Feld, M., and Shapiro, D. *Science* **155,** 86 (1967).
108. Brady, R. O., Johnson, W. G., and Uhlendorf, B. W. *Am. J. Med.* **51,** 423 (1971).
109. Schneider, E. L., Ellis, W. G., Brady, R. O., McCulloch, J. R., and Epstein, C. J. *J. Pediatr.* **81,** 1134 (1972).
109a. Peters, S. P., Aquino, L., Nascarato, W. F., Gilbertson, J. R., Diven, W. F., and Glew, R. H. *Biochim. Biophys. Acta* **575,** 27 (1979).
109b. Raghavan, S. S., Topol, J., and Kolodny, E. H. *Am. J. Hum. Genet.* **32,** 158 (1980).
109c. Choy, F. Y. M., and Davidson, R. G. *Am. J. Hum. Genet.* **32,** 670 (1980).
109d. Johnson, W. G., Gal, A. E., Miranda, A. F., and Pentchev, P. G. *Clin. Chim. Acta* **102,** 91 (1980).
109e. Shafit-Aagardo, B., Devine, E. A., and Desnick, R. J. *Biochim. Biophys. Acta* **614,** 459 (1980).
109f. Carroll, M. *J. Inher. Metab. Dis.* **4,** 11 (1981).
109g. Yagoob, M., and Carroll, M. *Biochem. J.* **185,** 541 (1980).
110. Ockerman, P. A. *Biochim. Biophys. Acta* **165,** 59 (1968).
110a. Chiao, Y., Peters, S. P., Diven, W. F., Lee, R. E., and Glew, R. H. *Metabolism* **28,** 56 (1979).
110b. Stephens, M. C., Bernatsky, A., Legler, G., and Kanfer, J. N. *J. Neurochem.* **32,** 969 (1979).
110c. Stephens, M. C., Bernatsky, A., Legler, G., and Kanfer, J. N. *Biochim. Biophys. Acta* **571,** 70 (1979).
111. Ho, M. W., and O'Brien, J. S. *Proc. Natl. Acad. Sci. U.S.A.* **68,** 2810 (1971).
112. Ho, M. W., O'Brien, J. S., Radin, N. S., and Erickson, J. S. *Biochem. J.* **131,** 173 (1973).
112a. Ho, M. W., and Light, N. D. *Biochem. J.* **136,** 821 (1973).
112b. Ho, M. W. *FEBS Lett.* **53,** 243 (1975).
112c. Ho, M. W., and Rigby, M. *Biochim. Biophys. Acta* **397,** 267 (1975).
112d. Dale, G. L., Villacorte, D. G., and Beutler, E. *Biochem. Biophys. Res. Commun.* **71,** 1048 (1976).

113. Pentchev, P. G., and Brady, R. O. *Biochim. Biophys. Acta* **297,** 491 (1973).
114. Pentchev, P. G., Brady, R. O., Hibbert, S. R., Gal, A. E., and Shapiro, D. *J. Biol. Chem.* **248,** 5256 (1973).
114a. Peters, S. P., Coyle, P., Coffee, C. J., Glew, R. H., Kuhlenschmidt, M. S., Rosenfeld, L., and Lee, Y. C. *J. Biol. Chem.* **252,** 563 (1977).
114b. Chiao, Y. B., Chambers, J. P., Glew, R. H., Lee, R. H., and Wenger, D. A. *Arch. Biochem. Biophys.* **186,** 42 (1978).
114c. Berent, S. L., and Radin, N. S. *Arch. Biochem. Biophys.* **208,** 248 (1981).
114d. Berent, S. L., and Radin, N. S. *Biochim. Biophys. Acta* **664,** 572 (1981).
115. Kanfer, J. N., Stein, M., and Spielvogel, C. In: *Sphingolipids, Sphingolipidoses and Allied Diseases* (Volk, B. W., and Aronson, S. M., eds.), Plenum Press, New York (1972), p. 225.
115a. Elleder, M., and Smid, F. *Virchows Arch. B* **26,** 133 (1977).
116. Friedman, R., and Kanfer, J. N. *Biochim. Med.* **9,** 327 (1974).
117. Kanfer, J. N., Raghavan, S. S., and Mumford, R. A. *Biochim. Biophys. Acta* **391,** 129 (1975).
117a. Kanfer, J. N., Mumford, R. A., Raghavan, S. S., and Byrd, J. *Anal. Biochem.* **60,** 200 (1974).
118. Glew, R. H., Christopher, A. R., and Schnure, F. W. *Arch. Biochem. Biophys.* **160,** 163 (1974).
119. Raghavan, S. S., Mumford, R. A., and Kanfer, J. N. *Biochim. Biophys. Res. Commun.* **54,** 256 (1973).
119a. Raghavan, S. S., Mumford, R. A., and Kanfer, J. N. *J. Lipid Res.* **15,** 484 (1974).
120. Nilsson, O., and Svennerholm, L. *J. Neurochem.* **39,** 709 (1982).
120a. Kanfer, J. N., Raghavan, S. S., Mumford, R. A., Labow, R. S., Williamson, D. G., and Layne, D. S. *Biochem. Biophys. Res. Commun.* **67,** 683 (1975).
120b. Kanfer, J. N., Raghavan, S. S., Mumford, R. A., Sullivan, J., Spielvogel, C., Legler, G., Labow, R. S., Williamson, D. G., and Layne, D. *Adv. Exp. Med. Biol.* **68,** 77 (1976).
120c. Chiao, Y. B., Hoyson, G. M., Peters, S. P., Lee, R. E., Diven, W., Murphy, J. V., and Glew, R. H. *Proc. Natl. Acad. Sci. U.S.A.* **75,** 2448 (1978).
120d. Kanfer, J. N., Munford, R. A., and Raghavan, S. S. *Can. J. Biochem.* **55,** 140 (1977).
120e. Yaqoob, M., and Carroll, M. *Biochem. J.* **185,** 541 (1980).
120f. Barton, N. W., and Rosenberg, A. *J. Biol. Chem.* **250,** 3966 (1975).
120g. Dawson, G., Stoolmiller, A. C., and Radin, N. S. *J. Biol. Chem.* **249,** 4638 (1974).
120h. Radin, N. S., Warren, K. R., Arora, R. C., Hyren, J. C., and Misra, R. S. In: *Modification of Lipid Metabolism* (Perkins, E. G., and Witting, L. A., eds.), Academic Press, New York (1975), p. 87.
120i. Kanfer, J. N., Legler, G., Sullivan, J., Raghavan, S. S., and Mumford, R. A. *Biochem. Biophys. Res. Commun.* **67,** 85 (1975).
120j. Dale, G. L., Gudas, J., Woloszyn, W., and Beutler, E. *Am. J. Hum. Genet.* **31,** 518 (1979).
120k. Mueller, O. T., and Rosenberg, A. *J. Biol. Chem.* **254,** 3521 (1979).
120l. Stephens, M. C., Bernatsky, A., Barachinsky, V., Legler, G., and Kanfer, J. N. *J. Neurochem.* **30,** 1023 (1978).
120m. Adachi, M., and Volk, B. W. *Arch. Pathol. Lab. Med.* **101,** 255 (1977).
120n. Hara, A., and Radin, N. S. *Biochim. Biophys. Acta* **582,** 412 (1979).
120o. Hara, A., and Radin, N. S. *Biochim. Biophys. Acta* **582,** 423 (1979).
121. Groth, C. G., Blomstrand, R., Hagenfeldt, L., Ockerman, P. A., Samuelsson, K., and Svennerholm, L. In: *Sphingolipids, Sphingolipidoses and Allied Disorders* (Volk, B. W., and Aronson, S. M., eds.), Plenum Press, New York (1972), p. 633.
122. Ihler, G. M., Glew, R. H., and Schnure, F. W. *Proc. Natl. Acad. Sci. U.S.A.* **70,** 2663 (1973).
122a. Brady, R. O., Pentchev, P. G., Gal, A. E., Hibbert, S. R., and Dekaban, A. S. *N. Engl. J. Med.* **291,** 989 (1974).
122b. Pentchev, P. G., Brady, R. O., Gal, A. E., and Hibbert, S. R. *J. Mol. Med.* **1,** 73 (1975).
122c. Pentchev, P. G., Barranger, J. A., Gal, A. E., Furbish, F. S., and Brady, R. O. *ACS Symp. Ser.* **80,** 150 (1978).
122d. Beutler, E., Dale, G. L., Guinto, E., and Kuhl, W. *Proc. Natl. Acad. Sci. U.S.A.* **74,** 4620 (1977).
122e. Dale, G. L., Kuhl, W., and Beutler, E. *Proc. Natl. Acad. Sci. U.S.A.* **76,** 473 (1979).
122f. Humphreys, J. D., and Ihler, G. *J. Lab. Clin. Med.* **96,** 682 (1980).
122g. Braidman, I., and Gregoriadas, G. *Biochem. Soc. Trans.* **4,** 259 (1976).

122h. Rebourcet, R., Weil, D., Van Cong, N., and Frezal, J. *C. R. Acad. Sci.* **278,** 3379 (1974).

122i. Doebbert, T. W., Wu, M. S., Bugianesi, R. L., Ponpipom, M. M., Furbish, F. S., Barranger, J. A., Brady, R. O., and Shen, T. Y. *J. Biol. Chem.* **257,** 2193 (1982).

123. Kahlke, W. In: *Lipids and Lipidoses* (Schettler, G., ed.), Springer-Verlag, New York (1969), p. 32.

124. Wise, D., Wallace, H. J., and Jellinck, E. H. *Q. J. Med.* **31,** 177 (1962).

125. Burda, C. D., and Winder, P. R. *Am. J. Med.* **42,** 293 (1967).

126. Sweeley, C. C., and Klionsky, B. *J. Biol. Chem.* **238,** 3148 (1963).

127. Mortensson, E. *Biochim. Biophys. Acta* **116,** 296 (1966).

128. Brady, R. O., Gal, A. E., Bradley, R. M., Martensson, E., Warshaw, A. L., and Laster, L. *N. Engl. J. Med.* **276,** 1163 (1967).

129. Kint, J. A. *Science* **167,** 1268 (1970).

130. Clarke, J. T. R., Wolfe, L. S., and Perlin, A. S. *J. Biol. Chem.* **246,** 5563 (1971).

131. Hakomori, S., Seddiqui, B., Li, Y.-T., Li, S. C., and Hellerquist, C. G. *J. Biol. Chem.* **246,** 2271 (1971).

132. Handa, S., Ariga, T., Miyatake, T., and Yamakawa, T. *J. Biochem. (Tokyo)* **69,** 625 (1971).

133. Li, Y.-T., and Li, S. C. *J. Biol. Chem.* **246,** 3769 (1971).

134. Li, Y.-T., Li, S. C., and Dawson, G. *Biochim. Biophys. Acta* **260,** 88 (1972).

135. Whenett, J. R., and Hakomori, S. I. *J. Biol. Chem.* **248,** 3046 (1973).

136. Kint, J. A., and Huys, A. In: *Glycolipids, Glycoprotein and Mucopolysaccharides of the Nervous System* (Zambotti, V., Tettamanti, G., and Arrigoni, M., eds.), Plenum Press, New York (1972), p. 273.

136a. Kano, I., and Yamakawa, T. *J. Biochem.* **75,** 347 (1974).

137. Ho, M. W., Beutler, S., Tennant, L., and O'Brien, J. S. *Am. J. Hum. Genet.* **24,** 256 (1972).

138. Beutler, E., and Kahl, W. *Am. J. Hum. Genet.* **24,** 237 (1972).

138a. Kano, I., and Yamakawa, T. *Chem. Phys. Lipids* **13,** 283 (1974).

138b. Riétra, P. J. G. M., Van Den Bergh, F. A. J. T. M., and Tager, J. M. *Clin. Chim. Acta* **62,** 401 (1975).

138c. Romeo, G., D'Urso, M., Pisacane, A., Blum, E., De Falco, A., and Ruffilli, A. *Biochem. Genet.* **13,** 615 (1975).

139. Beutler, E., and Kuhl, W. *J. Biol. Chem.* **247,** 7195 (1972).

139a. Riétra, P. J. G. M., Molenaar, J. L., Hamers, M. N., Tager, J. M., and Borst, P. *Eur. J. Biochem.* **46,** 89 (1974).

140. Crawhall, J. C., and Benfalvi, M. *Science* **177,** 527 (1972).

140a. Dean, K. J., Sung, S. J., and Sweeley, C. C., *Biochem. Biophys. Res. Commun.* **77,** 1411 (1977).

140b. Dean, K. J., and Sweeley, C. C. *J. Biol. Chem.* **254,** 9,994, 10,001 (1979).

140c. Salvayre, R., Maret, A., Negre, A., and Douste-Blazy, L. *Eur. J. Biochem.* **100,** 377 (1979).

141. Mapes, C. A., and Sweeley, C. C. *J. Biol. Chem.* **248,** 2461 (1973).

141a. Bishop, D. F., Wampler, D. E., Sgouris, J. T., Bonefeld, R. J., Anderson, D. K., Hawley, M. C., and Sweeley, C. C. *Biochim. Biophys. Acta* **524,** 109 (1978).

141b. Bishop, D. F., and Sweeley, C. C. *Biochim. Biophys. Acta* **525,** 399 (1978).

142. Mapes, C. A., and Sweeley, C. C. *Biochem. Biophys. Res. Commun.* **53,** 1317 (1973).

143. Mapes, C. A., Svelter, C. H., and Sweeley, C. C. *J. Biol. Chem.* **248,** 2471 (1973).

144. Ho, M. W. *Biochem. J.* **133,** 1 (1973).

145. Mapes, C. A., and Sweeley, C. C. *Arch. Biochem. Biophys.* **158,** 297 (1973).

145a. Romeo, G., Di Matteo, G., D'Urso, M., Li, S. C., and Li, Y.-T. *Biochim. Biophys. Acta* **391,** 349 (1975).

145b. Vance, D. E., Krivit, W., and Sweeley, C. C. *J. Biol. Chem.* **250,** 8119 (1975).

146. Mapes, C. A., Anderson, R. L., and Sweeley, C. C. *Science* **169,** 987 (1970).

147. Philippart, M., Franklin, S. S., and Gordon, A. *Ann. Intern. Med.* **77,** 195 (1972).

148. Clarke, J. T. R., Guttmann, R. D., Wolfe, L. S., Beaudain, J. G., and Morehouse, D. D. *N. Engl. J. Med* **287,** 1215 (1972).

149. Desnick, R. J., Allen, K. Y., Simmons, R. L., Woods, J. E., Anderson, C. F., Najacian, J. S., and Krivet, W. *Birth Defects: Orig. Artic. Ser.* **9,** 88 (1973).

150. Brady, R. O., Tallman, J. F., Johnson, W. G., Pal, A. E., Leaky, W. R., Quirk, J. M., and Dekeban, A. S. *N. Engl. J. Med.* **289,** 9 (1973).

150a. Schrom, A. W., Hamers, M. N., Oldenbroek-Haverkamp, G., Strijland, A., DeJonge, A., Van Den Bergh, F. A. T. M., and Tager, J. M. *Biochim. Biophys. Acta* **527,** 456 (1978).
151. Desnick, R. J., Allen, K. Y., Desnick, S. J., Ramon, M. K., Bernlohr, R. W., and Krivet, W. *J. Lab. Clin. Med.* **81,** 157 (1973).
152. Brady, R. O., Uhlendorf, B. W., and Jacobson, C. B. *Science* **172,** 174 (1971).
152a. Johnson, D. L., DelMonte, M. A., Cotlier, E., and Desnick, R. J. *Clin. Chim. Acta* **63,** 81 (1975).
153. O'Brien, J. S., Okada, M. W., Fillereys, D. L., Veath, M. L., and Adams, K. *Fed. Proc.* **30,** 956 *Fed. Am. Soc. Exp. Biol.* (1971).
154. Schneck, L., Volk, B. W., and Saifer, A. *Am. J. Med.* **46,** 245 (1969).
155. O'Brien, J. S. *N. Engl. J. Med.* **284,** 893 (1971).
156. O'Brien, J. S. *Fed. Proc. Fed. Am. Soc. Exp. Biol.* **32,** 191 (1973).
156a. Volk, B. W., and Schneck, L., *The Gangliosidoses*, Plenum Press, New York (1975).
157. Greenfield, J. G. *Proc. R. Soc. Med.* **44,** 686 (1951).
158. Terry, R. D., and Weiss, M. *J. Neuropathol. Exp. Neurol.* **22,** 18 (1963).
159. Sandhoff, K., and Harzer, K. In: *Lysosomes and Storage Diseases* (Hers, H. G., and Van Hoof, F., eds.), Academic Press, New York (1973), p. 345.
159a. Menkes, J. H., O'Brien, J. S., Okada, S. et al., *Arch. Neurol.* **25,** 14 (1971).
160. O'Brien, J. S. In: *Lysosomes and Storage Diseases* (Hers, H. G., and Van Hoof, F., eds.), Academic Press, New York (1973), p. 323.
161. Klenk, E. *Ber. Dtsch. Chem. Ges.* **75,** 1632 (1942).
162. Svennerholm, L. *Biochem. Biophys. Res. Commun.* **9,** 436 (1962).
163. Ledeen, R., and Salsman, K. *Biochemistry* **4,** 2225 (1965).
164. Svennerholm, L. *J. Lipid Res.* **5,** 145 (1964).
164a. Iwamori, M., and Nagai, Y. *J. Neurochem.* **32,** 767 (1979).
164b. Prellarkat, R. K., Reha, H., and Berates, N. G. *Biochem. Biophys. Res. Commun.* **92,** 149 (1980).
165. Olofsson, O. E., Kristensson, K., Laurander, P., and Svennerholm, L. *Acta Paediatr. Scand.* **55,** 546 (1966).
166. Suzuki, K., and Chen, G. C. *J. Lipid Res.* **8,** 105 (1967).
166a. Inagawa, A., Maketa, A., Gasa, S., and Arashima, S.-I. *J. Neurochem.* **33,** 369 (1979).
167. Bernheimer, H. *Klin. Wochenschr.* **46,** 258 (1968).
168. Suzuki, K., and Suzuki, K. In: *Handbook of Neurochemistry, Vol. 7 (Lajtha, A., ed.), Plenum Press, New York (1972), P. 131.*
169. Brunngraber, E. G., Berra, B., and Zambotti, V. *Brain Res.* **38,** 151 (1972).
170. Okada, S., and O'Brien, J. S. *Science* **165,** 698 (1969).
171. Robinson, D., and Stirling, J. L. *Biochem. J.* **107,** 321 (1965).
172. Kabach, M. *Methods Enzymol.* **28,** 862 (1972).
173. Kolodny, E. H., Brady, R. O., and Volk, B. *Biochem. Biophys. Res. Commun.* **37,** 526 (1969).
174. Sandhoff, K., and Jatzkewitz, H. In: *Sphingolipids, Sphingolipidoses and Allied Disorders* (Volk, B. W., and Aronson, S. M., eds.), Plenum Press, New York (1972), p. 305.
174a. Wood, S., and MacDougall, B. G. *Am. J. Hum. Genet.* **28,** 489 (1976).
175. Sandhoff, K., Andreae, U., and Jatzkewitz, H. *Pathol. Eur.* **3,** 278 (1968).
176. Suzuki, Y., Jacob, J. C., Suzuki, K., Kutty, K. M., and Suzuki, K. *Neurology* **21,** 313 (1971).
177. Strecker, G., and Montreul, J. *Clin. Chim. Acta* **33,** 395 (1971).
178. Applegarth, D. A., and Bozoian, J. *Clin. Chim. Acta* **39,** 269 (1972).
178a. Kin, N. M. K. N. Y., and Wolfe, L. S. *Biochem. Biophys. Res. Commun.* **59,** 837 (1974).
178b. Hooghwinkel, G. J. M., Veltkamp, W. A., Overdijk, B., and Lisman, J. J. W. *Hoppe-Seyler's Z. Physiol. Chem.* **353,** 839 (1972).
178c. Poenaru, L., and Dreyfus, J. C. *Clin. Chim. Acta* **43,** 439 (1973).
178d. Poenaru, L., Weber, A., Vibert, M., and Dreyfus, J. C. *Biomedicine* **19,** 538 (1973).
178e. Braidman, I., Carroll, M., Dance, N., and Robinson, D. *Biochem. J.* **143,** 297 (1974).
178f. Penton, E., Poenaru, L., and Dreyfus, J. C. *Biochim. Biophys. Acta* **391,** 162 (1975).
178g. Overdijk, B., Van der Kroef, W. M. J., Veltkamp, W. A., and Hooghwinkel, G. J. M. *Biochem. J.* **151,** 257 (1975).
178h. Besley, G. T. N., and Broadhead, D. M. *Biochem. J.* **155,** 205 (1976).

178i. Geiger, B., Calef, E., and Arnon, R. *Biochemistry* **17,** 1713 (1978).

178j. Sterling, J. L. *Biochem. J.* **141,** 597 (1974).

178k. Potier, M., Teitelbaum, J., Melancon, S. B., and Dallaire, L. *Biochim. Biophys. Acta* **566,** 80 (1979).

178l. Minami, R., Nakamura, F., Oyanagi, K., and Nagao, T. *Tohoku J. Exp. Med.* **133,** 175 (1981).

178m. Overdijk, B., Van Der Kroef, W. M. J., Van Steijn, G., and Lesman, J. J. W. *Biochim. Biophys. Acta* **659,** 255 (1981).

179. O'Brien, J. S. *Lancet* **1,** 805 (1969).

180. Sandhoff, K., Harzer, K., Wassle, W., and Jatzkewitz, H. *J. Neurochem.* **18,** 246 (1971).

180a. DeBalcque, C. M., Suzuki, K., Rapin, I., Johnson, A. B., Whethers, D. L., and Suzuki, K. *Acta Neuropathol.* **33,** 207 (1975).

180b. O'Neell, B., Butler, A. B., Young, E., Falk, P. M., and Bass, N. H. *Neurology* **28,** 1117 (1978).

180c. Conzelmann, E., Sandhoff, K., Nehrkorn, H., Geiger, B., and Arnon, R. *Eur. J. Biochem.* **84,** 27 (1978).

180d. Conzelmann, E., and Sandhoff, K. *Proc. Natl. Acad. Sci. U.S.A.* **75,** 3979 (1978).

180e. Li, S.-C., Hirabayashi, Y., and Li, Y.-T. *Biochem. Biophys. Res. Commun.* **101,** 479 (1981).

180f. Hechtman, P., Gordon, B. A., and Kin, N. M. K. N. Y. *Pediatr. Res.* **16,** 217 (1982).

181. Sandhoff, K., and Wassle, W. *Hoppe-Seyler's Z. Physiol. Chem.* **352,** 1119 (1971).

181a. Sandhoff, K., Conzelmann, E., and Nehrokorn, H. *Hoppe-Seyler's Z. Physiol. Chem.* **358,** 779 (1977).

182. Wenger, D. A., Okada, S., and O'Brien, J. S. *Arch. Biochem. Biophys.* **153,** 116 (1972).

182a. Bach, G., and Suzuki, K. *J. Biol. Chem.* **250,** 1328 (1975).

182b. Sandhoff, K., Conzelmann, E., and Nehrkorn, H. *Hoppe-Seyler's Z. Physiol. Chem.* **358,** 779 (1977).

182c. Bearpork, T. M., and Stirling, J. L. *Biochem. J.* **173,** 997 (1978).

183. Li, Y. T., Mazzotta, M. Y., Wan, C. C., Orth, R., and Li, S. C. *J. Biol. Chem.* **248,** 7512 (1973).

183a. Li, S. C., and Li, Y.-T. *J. Biol. Chem.* **251,** 1159 (1976).

183b. Li, S.-C., Nakamura, T., Ogamo, A., and Li, Y.-T. *J. Biol. Chem.* **254,** 10592 (1979).

183c. Conzelmann, E., and Sandhoff, K. *Hoppe-Seyler's Z. Physiol. Chem.* **360,** 1837 (1979).

183d. Li, S.-C., Hirabayashi, Y., and Li, Y.-T. *J. Biol. Chem.* **256,** 6234 (1981).

183e. Conzelmann, E., Burg, J., Stephan, G., and Sandhoff, K. *Eur. J. Biochem.* **123,** 455 (1982).

184. Verpoorte, J. A. *J. Biol. Chem.* **247,** 4787 (1972).

184a. Goldstone, A., Konecny, P., and Koenig, H. *FEBS Lett.* **13,** 68 (1971).

184b. Carmody, P. J., and Rattazzi, M. C. *Biochim. Biophys. Acta* **371,** 117 (1974).

184c. Beutler, E., Villacorte, D., Kuhl, W., Gunito, E., and Srivastava, S. *J. Lab. Clin. Med.* **86,** 195 (1975).

184d. Tallman, J. F., Brady, R. O., Quirk, J. M., Villalba, M., and Gal, A. E. *J. Biol. Chem.* **249,** 3489 (1974).

184e. Srivastava, S. K., and Beutler, E. *Nature (London)* **241,** 463 (1973).

184f. Beutler, E. Kuhl, W., and Comings, D. *Am. J. Hum. Genet.* **27,** 628 (1975).

184g. Beutler, E., and Kuhl, W. *Nature (London)* **258,** 262 (1975).

184h. Srivastava, S. K., Wiktorowicz, J. E., and Awasthi, Y. C. *Proc. Natl. Acad. Sci. U.S.A.* **73,** 2833 (1976).

184i. Douglas, G. C., and Mackenzie, A. *Biochem. Soc. Trans.* **6,** 1072 (1978).

184j. Geiger, B., and Arnon, R. *Biochemistry* **15,** 3485 (1976).

184k. Freeze, H., Geiger, B., and Miller, A. L. *Biochem. J.* **177,** 749 (1979).

184l. Joziasse, D. H., Van Den Eijnden, D. H., Leiman, J. J. W., and Hooghwinkel, G. J. M. *Biochim. Biophys. Acta* **660,** 174 (1981).

185. Srivastava, S. K., and Beutler, E. *Biochem. Biophys. Res. Commun.* **47,** 753 (1972).

186. Carroll, M., and Robinson, D. *Biochem. J.* **131,** 91 (1973).

186a. Bartholomew, W. R., and Rattazzi, M. C. *Int. Arch. Allergy Appl. Immunol.* **46,** 512 (1974).

186b. Carroll, M., and Robinson, D. *Biochem. J. 137,* 217 (1974).

186c. Geiger, B., Navon, R., Ben-Yoseph, Y., and Arnon, R. *Eur. J. Biochem.* **56,** 311 (1975).

186d. Srivastava, S. K., Ansari, N. H., Hawkins, L. A., and Wiktorowicz, J. E. *Biochem. J.* **179,** 657 (1979).

187. Thompson, J. N., Stoolmiller, A. C., Matalon, R., and Dorfman, A. *Science* **181,** 866 (1973).

187a. Van Someren, H., and van Henegouwen, H. B. *Humangenetik* **18,** 171 (1973).

187b. Lolley, P. A., Rattazzi, M. C., and Shows, T. B. *Proc. Natl. Acad. Sci. U.S.A.* **71,** 1569 (1974).

187c. Boldecker, H. J., Mellman, W. J., Tedesco, T. A., and Croce, C. M. *Exp. Cell Res.* **93,** 468 (1975).

187d. Gilbert, F., Kucherlapati, R., Cregan, R. P., Murnane, M. J., Darlington, G. J., and Ruddle, F. H. *Proc. Natl. Acad. Sci. U.S.A.* **72,** 263 (1975).

187e. Thomas, G. H., Taylor, H. A., Miller, C. S., Axelman, J., and Migeon, B. R. *Nature (London)* **250,** 580 (1974).

187f. Galjaard, H., Hoogeveen, A., de Wit-Verveek, H. A., Reuser, J. J., Keijzer, W., Westerveld, A., and Bootsma, D. *Exp. Cell Res.* **87,** 444 (1974).

187g. Rattazzi, M. C., Brown, J. A., Davidson, R. G., and Shows, T. B. *Am. J. Hum. Genet.* **28,** 143 (1976).

188. O'Brien, J. S., Okada, S., Chen, A., and Fillerup, D. L. *N. Engl. J. Med.* **283,** 15 (1970).

189. Suzuki, Y., Berman, P. H., and Suzuki, K. *J. Pediatr.* **78,** 643 (1971).

190. Kolodny, E. H. In: *Sphingolipids, Sphingolipidoses and Allied Disorders* (Volk, R. W., and Aronson, S., eds.), Plenum Press, New York (1972), p. 321.

191. O'Brien, J. S., Okada, S., Fillerup, D. L., Veath, G. L., Adornato, B., Brenner, P. H., and Leroy, J. G. *Science* **172,** 61 (1971).

192. Schneck, L., Valent, C., Amsterdam, D., Friedland, J., Adachi, M., and Volk, B. W. *Lancet* 582 (1970).

193. Desnick, R. J., Kribit, W., and Sharp, H. L. *Biochem. Biophys. Res. Commun.* **51,** 20 (1973).

194. Schneck, L., Adachi, M., and Volk, B. W. *Pediatrics* **49,** 342 (1972).

194a. Adachi, M., Schneck, L., and Volk, B. W. *Lab. Invest.* **30,** 102 (1974).

194b. Navon, R., Geiger, B., Yoseph, Y. B., and Rattazzi, M. C. *Am. J. Hum. Genet.* **28,** 339 (1976).

194c. Cutz, E., Lowden, J. A., and Conen, P. E. *J. Neurol. Sci.* **21,** 197 (1974).

195. Vidgoff, J., Buist, N. R. M., and O'Brien, J. S. *Am. J. Hum. Genet.* **25,** 372 (1973).

196. Navon, R., Padek, B., and Adam, A. *Am. J. Hum. Genet.* **25,** 287 (1973).

196a. Tallman, J. F., Brady, R. O., Navon, R., and Padek, B. *Nature (London)* **252,** 254 (1974).

196b. Dreyfus, J. C., Poenaru, L., and Svennerholm, L. *N. Engl. J. Med.* **292,** 61 (1975).

196d. O'Brien, J. S., Tennant, L., Veath, M. L., Scott, C. R., and Buchnall, W. E. *Am. J. Hum. Genet.* **30,** 602 (1978).

196e. Momoi, T., Sudo, M., Tanioka, K., and Nakao, Y. *Pediatr. Res.* **12,** 77 (1978).

196f. Thomas, G. H., Raghavan, S., Kilodny, E., Frisch, A., Neufeld, P., O'Brien, J. S., Reynolds, L., Miller, C., Shapiro, J., Kazazima, H. H., and Heller, R. H. *Pediat. Res.* **16,** 232 (1982).

196g. Hechtman, P., and Rowland, A. *Am. J. Hum. Genet.* **31,** 428 (1979).

196h. O'Brien, J. S. *Am. J. Hum. Genet.* **31,** 672 (1979).

197. Kaback, M. M., and Zeiger, R. S. In: *Sphingolipids, Sphingolipidoses and Allied Diseases* (Volk, B. W., and Aronson, S., eds.), Plenum Press, New York (1972), p. 613.

198. Lowden, J. A., Skomorowski, M. A., Henderson, F., and Kaback, M. M. *Clin. Chem.* **19,** 1345 (1973).

199. Johnson, W. G., Desnick, R. J., Lang, D. M., Sharp, H. L., Krevit, W., Brady, B., and Brady, R. O. *Birth Defects: Orig. Artic. Ser.* **9,** 120 (1973).

199a. Steer, C. J., Kusiak, J. W., Brady, R. O., and Jones, E. A. *Proc. Natl. Acad. Sci. U.S.A.* **76,** 2774 (1979).

199b. Von Specht, B. U., Geiger, B., Arnon, R., Passwell, J., Keren, G., Goldman, B., and Padek, B. *Neurology* **29,** 848 (1979).

199c. Cohen, C. M., Weissmann, G., Hoffstein, S., Awasthi, Y. C., and Srivastava, S. K. *Biochemistry* **15,** 452 (1976).

199d. Brooks, S. E., Hoffman, L., Adachi, M., Amsterdam, D., and Schneck, L. *Acta Neuropathol.* **50,** 9 (1980).

199e. Baser, M., Presper, K. A., Basu, S., Hoffman, L. M., and Brooks, S. C. *Proc. Natl. Acad. Sci. U.S.A.* **76,** 4270 (1979).

199f. Cork, L. C., Munnell, J. F., Lorenz, M. D., Murphy, J. V., and Rattozgi, M. C. *Science* **196,** 1014 (1977).

199g. Cork, L. C., Mannell, J. F., and Lorenz, M. D. *Am. J. Pathol.* **90,** 723 (1978).

199h. Kosanke, S. D., Pierce, K. R., and Bay, W. W. *Vet. Pathol.* **15,** 685 (1978).

199i. Rattazzi, M. C., Lanse, S. B., McCullough, R. A., Nester, J. A., and Jacobs, E. A. *Birth Defects Orig. Artic. Ser.* **16,** 179 (1981).

199j. Rattazzi, M. C., Appel, A. M., Baker, H. J., and Nester, J. In: *Lysosomes and Lysosomal Storage Diseases* (Callahan, J., and Lowden, J. A., eds.), Raven Press, New York (1981), p. 405.

199k. Neuwelt, E. A., Barranger, J. A., Brady, R. O., Pagel, M., Furbish, F. S., Quirk, J., Mook, G. E., and Frankel, E. *Proc. Natl. Acad. Sci. U.S.A.* **78,** 5838 (1981).

199l. Frisch, A., and Neufeld, E. F. *J. Biol. Chem.* **256,** 8242 (1981).

200. O'Brien, J. S. In: *The Metabolic Basis of Inherited Diseases* (Stanbury, J. B., Wyngaarden, J. L., and Fredrikson, D. S., eds.), McGraw-Hill, New York (1972), p. 639.

201. Gonatas, N. K., and Gonatas, J. *J. Neuropathol. Exp. Neurol.* **24,** 318 (1965).

202. Landing, B. H., Silverman, F. N., Craig, M. M., Jacoby, M. D., Lahey, M. E., and Chadwick, D. L. *Am. J. Dis. Child.* **108,** 503 (1964).

202a. O'Brien, J. S., Geiger, E., Giedion, A., Weessman, U., Herschhoritz, N., Meier, C., and Leroy, J. *Clin. Genet.* **9,** 495 (1976).

202b. Arbisser, A. D., Donnelly, K. A., Scott, C. I., Di Ferrante, N., Singh, J., Stevenson, R. E., Aylesworth, S., and Howell, R. R. *Am. J. Med. Genet.* **1,** 195 (1977).

202c. Wenger, D. A., Sattler, M. O., Mueller, O. T., Myers, G. G., Schneiman, R. S., and Nixon, G. W. *Clin. Genet.* **17,** 323 (1980).

203. O'Brien, J. S., Stern, M. B., Landing, B. H., O'Brien, J. K., and Connell, G. N. *Am. J. Dis. Child.* **109,** 338 (1965).

204. Suzuki, K., Suzuki, K., and Kamoshita, S. *J. Neuropathol. Exp. Neurol.* **28,** 25 (1969).

204a. Tsay, G. C., Dawson, G., and LI, Y.-T. *Biochim. Biophys. Acta* **385,** 305 (1975).

205. Wolfe, L. S., Clarke, J. T. R., and Senior, R. G. In: *Sphingolipids, Sphingolipidoses and Allied Diseases* (Volk, B. W., and Aronson, S., eds.), Plenum Press, New York (1972), p. 373.

205a. Wolfe, L. S., Senior, R. G., and Kin, N. M. K. N. Y. *J. Biol. Chem.* **249,** 1828 (1974).

205b. Kin, N. M. K. N. Y., and Wolfe, L. S. *Biochem. Biophys. Res. Commun.* **66,** 123 (1975).

206. Brunngraber, E. G., Berra, B., and Zambotti, V. *FEBS Lett.* **34,** 350 (1973).

207. Okada, S., and O'Brien, J. S. *Science* **160,** 1002 (1968).

208. MacBrinn, M. C., Okada, S., Ho, M. W., Hu, C. C., and O'Brien, J. S. *Science* **163,** 947 (1969).

208a. Pinsky, L., Miller, J., Shanfield, B., Watters, G., and Wolfe, L. D. *Am. J. Hum. Genet.* **26,** 563 (1974).

208b. Yutaka, T., Okada, S., Mimaki, K., Sugita, T., and Yabuuchi, H. *Clin. Chim. Acta* **59,** 283 (1975).

208c. Lowden, J. A., Callahan, J. W., Norman, M. G., Thain, M., and Prichard, J. S. *Arch. Neurol.* **31,** 200 (1974).

208d. O'Brien, J. S., Geigler, E., Giedion, A., Wiessman, U., Herschowitz, N., Meier, C., and Leroy, J. *Clin. Genet.* **9,** 495 (1976).

209. Norden, A. G. W., and O'Brien, J. S. *Proc. Natl. Acad. Sci. U.S.A.* **72,** 240 (1975).

210. Brady, R. A., O'Brien, J. S., Bradley, R. M., and Gal, A. E. *Biochem. Biophys. Res. Commun.* **210,** 193 (1970).

211. Ho, M. W., Cheetham, P., and Robinson, D. *Biochem. J.* **136,** 351 (1973).

212. Norden, A. G. W., and O'Brien, J. S. *Arch. Biochem. Biophys.* **159,** 383 (1973).

212a. Lo, J., Mukerji, K., Awasthi, Y. C., Handa, E., Suzuki, K., and Srivastava, S. K. *J. Biol. Chem.* **254,** 6710 (1979).

212b. Galjaard, H., Hoogeveen, A., Keijzer, W., de Wit-Verbeek, H. A., Reuser, A. J. J., Ho, M. W., and Robinson, D. *Nature (London)* **257,** 60 (1975).

212c. De Wit-Verbeek, H. A., Hoogeveen, A., and Galjaard, H. *Exp. Cell Res.* **113,** 215 (1978).

212d. Hoeksema, H. L., Van Diggelen, O. P., and Galjaard, H. *Biochim. Biophys. Acta* **566,** 72 (1979).

212e. O'Brien, J. S. *Clin. Genet.* **8,** 303 (1975).

212f. Meisler, M., and Rattazzi, M. C. *Am. J. Hum. Genet.* **26,** 683 (1974).

212g. Distler, J., Hieber, V., Schmickel, R., Myerowitz, R., and Jourdian, G. W. *Birth Defects Orig. Artic. Ser.* **11,** 311 (1975).

213. Blakemore, W. F. *J. Comp. Pathol.* **82,** 179 (1972).

214. Farrell, D. J., Baker, H. J., Herndon, R. M., Lindsey, J. R., and McKhann, G. M. *J. Neuropathol. Exp. Neurol.* **32,** 1 (1973).

214a. Holmes, E. W., and O'Brien, J. S. *Biochem. J.* **175,** 945 (1978).

214b. Holmes, E. W., and O'Brien, J. S. *Am. J. Hum. Genet.* **30,** 505 (1978).

215. Lowden, J. A., Catz, E., Conen, P. E., Rudd, N., and Doran, T. A. *N. Engl. J. Med.* **288,** 225 (1973).

216. Kaback, M. M., Sloan, H. R., Sonneborn, M., Herndon, R. M., and Percy, A. K. *J. Pediatr.* **82,** 1037 (1973).

216a. Booth, C. W., Gerbie, A. B., and Nadler, H. L. *Pediatrics* **52,** 521 (1973).

216b. Horst, J., Kluge, F., Beyreuther, K., and Gerok, W. *Proc. Natl. Acad. Sci. U.S.A.* **72,** 3531 (1975).

216c. Burns, G. A. P., Leary, A. C., Regina, V. M., and Gerald, P. S. *Cytogenet. Cell Genet.* **22,** 177 (1978).

216d. Shows, T. B., Scrafford-Wolf, L., Brown, J. A., and Meisler, M. *Cytogenet. Cell Genet.* **22,** 219 (1978).

217. Pick, L. *Med. Klin.* **23,** 1483 (1927).

218. Crocker, A. C. *J. Neurochem.* **7,** 69 (1961).

218a. Wenger, D. A., Barth, G., and Githens, J. H. *Am. J. Dis. Child.* **131,** 955 (1977).

218b. Schneider, E. L., Pentchev, P. G., Hibbert, S. R., Sawitsky, A., and Brady, R. O. *J. Med. Genet.* **15,** 307 (1978).

219. Crocker, A. C., and Farber, S. *Medicine* **37,** 1 (1958).

220. Klenk, E. *Hoppe-Seyler's Z. Physiol. Chem.* **229,** 161 (1934).

221. Fredrickson, D. S., and Sloan, H. R. In: *Inherited Basis of Metabolic Disorders* (Stanbury, J. B., Wyngaarden, J. B., and Fredrickson, D. S., eds.), McGraw-Hill, New York (1972), p. 783.

222. Rouser, G., Kretchevsky, G., Yamamoto, A., Knudson, A. G., and Simon, G. *Lipids* **3,** 287 (1968).

223. Seng, P. N., Debuch, H., Witter, B., and Wiedemann, H. R. *Hoppe-Seyler's Z. Physiol. Chem.* **352,** 280 (1971).

224. Yamamoto, A., Adachi, S., Ishikawa, K., Yokomura, T., Kitani, T., Nasu, T., Imoto, T., and Nishikawa, M. *J. Biochem.* **70,** 775 (1971).

224a. Dacremont, G., Kint, J. A., Carton, D., and Cocquyt, G. *Clin. Chim. Acta* **52,** 365 (1974).

224b. Kannan, R., Tjiong, H. B., Debuch, H., and Wiedemann, H. R. *Hoppe-Seyler's Z. Physiol. Chem.* **355,** 551 (1974).

224c. Brungraber, E., Berra, B., and Zombotti, V. *Clin. Chim. Acta* **48,** 173 (1973).

225. Kamoshita, S., Aron, A. M., Suzuki, K., and Suzuki, K. *Am. J. Dis. Child.* **117,** 379 (1969).

226. Crocker, A. C., and Mays, V. B. *Am. J. Clin. Nutr.* **9,** 63 (1961).

227. Brady, R. O., Kanfer, J. N., Mock, M. B., and Fredrickson, D. S. *Proc. Natl. Acad. Sci. U.S.A.* **55,** 366 (1966).

227a. Muller, H., and Harzer, K. *J. Neurochem.* **34,** 446 (1980).

228. Schneider, P. B., and Kennedy, E. P. *J. Lipid Res.* **8,** 202 (1967).

228a. Gatt, S. *Biochem. Biophys. Res. Commun.* **68,** 235 (1976).

228b. Rao, B. G., and Spence, M. W. *J. Lipid Res.* **17,** 506 (1976).

228c. Gatt, S., Dinur, T., and Kopolovic, J. *J. Neurochem.* **31,** 547 (1978).

228d. Yamaguchi, S., and Suzuki, K. *J. Biol. Chem.* **253,** 4090 (1978).

228e. Hostetler, K. Y., and Yazaki, P. J. *J. Lipid Res.* **20,** 456 (1979).

228f. Callahan, J. W., Khalil, M., and Gerrie, J. *Biochem. Biophys. Res. Commun.* **58,** 384 (1974).

228g. Callahan, J. W., Khalil, M., and Philippart, M. *Pediatr. Res.* **9,** 908 (1975).

228h. Harzer, K., Anzil, A. P., and Schuster, I. *J. Neurochem.* **29,** 1155 (1977).

228i. Yamaguchi, S., and Suzuki, K. *Biochem. Biophys. Res. Commun.* **77,** 999 (1977).

228j. Callahan, J. W., and Khalil, M. *Pediatr. Res.* **9,** 914 (1975).

229. Callahan, J. W., Lassila, E. L., and Philippart, M. *Biochem. Med.* **11,** 262 (1974).

229a. Fensom, A. H., Benson, P. F., Babarik, A. W., Grant, A. R., and Jacobs, L. *Biochem. Biophys. Res. Commun.* **74,** 877 (1977).

229b. Besley, G. T. N. *Clin. Chim. Acta* **90,** 269 (1978).

230. Brady, R. O. *Semin. Hematol.* **9,** 273 (1972).

230a. Besley, G. T. N. *J. Inher. Metab. Dis.* **1,** 29 (1978).

230b. Zitman, D., Chazan, S., and Klibansky, C. *Clin. Chim. Acta* **86,** 37 (1978).

231. Epstein, C. J., Brady, R. O., Schneider, E. L., Bradley, R. M., and Shapiro, D. *Am. J. Hum. Genet.* **23,** 533 (1971).

232. Uhlendorf, B. W., Holtz, A. I., Mock, M. B., and Fredrickson, D. S. In: *Inborn Errors of Sphingolipid Metabolism* (Aronson, S. M., and Volk, B. W., eds.), Pergamon Press, Oxford (1967), p. 443.

232a. Gal, A. E., and Fash, F. J. *Chem. Phys. Lipids* **16,** 71 (1976).

232b. Gal, A. E., Brady, R. O., Hibbert, S. R., and Pentchev, P. G. *N. Engl. J. Med.* **293,** 632 (1975).

232c. Gal, A. E., Brady, R. O., Barranger, J. A., and Pentchev, P.G. *Clin. Chim. Acta* **104,** 129 (1980).

232d. Gatt, S., Dinur, T., and Barenholz, Y. *Biochim. Biophys. Acta* **530,** 503 (1978); *Clin Chem.* **26,** 93 (1980).

232e. Barton, G., and Revol, A. *Clin. Chim. Acta* **76,** 339 (1977).

232f. Gatt, S., Dinur, T., and Liebovitz-Bera Gershon, Z. *Biochim. Biophys. Acta* **531,** 206 (1978).

233. Schneider, E. L., Ellis, W. G., Brady, R. O., McCulloch, J. R., and Epstein, C. J. *Pediatr. Res.* **6,** 720 (1972).

233a. Chazan, S., Zitman, D., and Klibansky, C. *Clin. Chim. Acta* **86,** 45 (1978).

233b. Wenger, D. A., Wharton, C., Sattler, M., and Clark, C. *Am. J. Med. Genet.* **2,** 345 (1978).

233c. Klibansky, C., Chazan, S., Schoenfeld, A., and Abramovici, A. *Clin. Chim. Acta* **91,** 243 (1979).

233d. Harzer, K., Schlote, W., Peiffer, J., Benz, H. U., and Anzil, A. P. *Acta Neuropathol.* **43,** 97 (1978).

234. Lowden, J. A., and LaRamee, M. A. In: *Sphingolipids, Sphingolipidoses and Allied Diseases* (Volk, B. W., and Aronson, S. M., eds.), Plenum Press, New York (1972), p. 257.

234a. Fried, K., Beer, S., Krespin, H. I., Leiba, H., Djaldetti, M., Zitman, D., and Klibonsky, R. *Eur. J. Clin. Invest.* **8,** 249 (1978).

234b. Schneider, E., Pentchev, P. G., Hibbert, S. R., Sawitsky, A., and Brady, R. O. *J. Med. Genet.* **15,** 370 (1978).

235. Chrisp, C. E., Ringler, D. H., Abrams, G. D., Radin, N. S., and Brenkert, A. *J. Am. Vet. Med. Assoc.* **156,** 616 (1970).

235a. Wenger, D. A., Sattler, M., Kadoh, T., Snyder, S. P., and Kingston, R. S. *Science* **208,** 1471 (1980).

236. Max, S. R., Maclaren, N. K., Brady, R. O., Bradley, M. S., Rennels, M. B., Tanaka, J., Garcia, J. H., and Cornblath, M. *N. Engl. J. Med.* **291,** 929 (1974).

237. Maclaren, N. K., Max, S. R., Cornblath, M., Brady, R. O., Ozand, P. T., Campbell, J., Rennels, M., Wolfgang, J. M., and Garcia, J. H. *Pediatrics* **57,** 106 (1976).

238. Tanaka, J., Garcia, J. H., Max, S. R., Viloria, J. E., Kamijyo, Y., Maclaren, N. K., Cornblath, M., and Brady, R. O. *J. Neuropathol. Exp. Neurol.* **34,** 249 (1975).

239. Fishman, P. H., Max, S. R., Tallman, J. F. *Science* **187,** 68 (1974).

240. Kolodny, E. H. *Semin. Hematol.* **9,** 251 (1972).

241. O'Brien, J. S., and Sampson, E. L. *Science* **150,** 1613 (1965).

242. Tager, J. M., Hooghwinkel, G. J. M., and Daems, W. T. (eds.). *Enzyme Therapy in Lysosomal Storage Diseases,* North-Holland/American Elsevier, New York (1974).

243. Desnick, R. J., Thorpe, S. R., and Fiddler, M. B. *Physiol. Rev.* **56,** 57 (1976).

244. Thomas, G. H., Tiller, L. W., Reynolds, L. W., Miller, C. S., and Bace, J. W. *Biochem. Biophys. Res. Commun.* **71,** 188 (1976).

245. Stanbury, J. B., Wyngaarden, J. B., and Fredrickson, D. S. (eds.). *The Metabolic Basis of Inherited Diseases,* McGraw-Hill, New York (1978).

246. Aronson, S. M., and Volk, B. W. (eds.). *Inborn Disorders of Sphingolipid Metabolism,* Pergamon Press, London (1967).

247. Volk, B. W., and Aronson, S. M. (eds.). *Sphingolipids, Sphingolipidoses and Allied Diseases,* Plenum Press, New York (1972).

248. Volk, B. W., and Schneck, L. (eds.). *Current Trends in Sphingolipidoses and Allied Disorders,* Plenum Press, New York (1976).

249. Schettler, G. (ed.). *Lipids and Lipidoses,* Springer-Verlag, New York (1967).

250. Thannhauser, S. J. *Lipidoses, Diseases of the Cellular Lipid Metabolism,* 2nd ed., Oxford Medical Publications, New York (1950).

251. Kaback, M. M. (ed.). *Tay Sachs Disease Screening and Prevention,* Alan R. Liss, New York (1977).

252. Volk, B. W., and Schneck, L. (eds.). *The Gangliosidoses,* Plenum Press, New York (1975).

253. Glew, R. H., and Peters, S. P. (eds.). *Practical Enzymology of the Sphingolipidoses,* Alan R. Liss, New York (1977).

Chapter 4

Glycosphingolipids in Cellular Interaction, Differentiation, and Oncogenesis

Sen-itiroh Hakomori

4.1. Introduction

The idea that glycosphingolipids (or, briefly, glycolipids) are ubiquitous components of plasma membrane and display cell type-specific patterns perhaps stemmed from the classical studies on glycolipids of erythrocyte membranes.[1,2] Subsequently, plasma membranes of various animal cells were successfully isolated and analyzed; all were characterized by their much higher content of glycolipid than was found in intracellular membranes.[3–8] It is generally assumed that glycolipids are present at the outer leaflet of the plasma membrane bilayer, although this assumption is based only on experiments with surface-labeling by galactose oxidase–NaB[^{3}H]$_4$ of intact and lysed erythrocyte membranes and inside-out vesicles.[9,10] Obviously, further extensive studies of the organization of glycolipid in other eukaryotic cell membranes are necessary. Interestingly, the majority of neutral glycolipids present in plasma membranes are cryptic (see Section 4.2.1).

The high content of glycolipid at the outer half of the lipid bilayer suggests two possibilities: (1) Glycolipids may contribute to the structural rigidity of the surface leaflet (see Table I, item 1). Ceramides confer more structural rigidity than glycerides, because ceramides contain both a hydrogen acceptor (amide carbonyl) and a donor (hydroxy) to form stable hydrogen bonds. In contrast, glycerides have only a hydrogen acceptor (ester carbonyl).[11–13] Ceramides with sugar may confer even higher stability.[13] The higher rigidity of sphingolipid liposomes as compared to glyceride liposomes has been demonstrated in studies with nuclear magnetic and electron spin resonance (ESR) spectroscopy[11,12] as well as in kinetics of complement-dependent lysis of liposomes by antibodies.[14] (2) Glycolipids at the outer leaflet of a bilayer are well suited to interact with exogenous ligands through their carbohydrate chains. Various functional speculations about glycolipids in cellular interactions, differentiation, and immune response and their deficiency in trans-

Table I. Functions of Glycosphingolipids

1.	Constitutive component of outer leaflet of lipid bilayer; conferring structural rigidity[11–13]
2.	Cell-surface markers and antigens (see Chapter 5)
3.	Cell–cell interaction and recognition
	a. Cell contact response and contact inhibition in general[55–58,106]
	b. Neuroglial interaction[59]
	c. Retinotectal recognition[60,61]
	d. Neuromuscular recognition[62]
4.	Differentiation marker
	a. Crypt-to-villus differentiation[90,91]
	b. Erythrocyte differentiation (glycolipid branching and i to I)[93–95]
	c. Myogenic differentiation[103,104]
	d. Neuroblastoma[101,102]
	e. Teratocarcinoma[80,83,85,89]
	f. Lymphoid cell differentiation and mytogenesis[105,113,114,116]
5.	Cell growth regulation and oncogenesis (see Table VII)
6.	Interaction with bioactive factors: bacterial toxins, glycoprotein hormones, and viruses (implication as receptor) (see Chapter 6)

formed cells can all be viewed from a common standpoint, i.e., glycolipids as cell-surface interactants and transducers (see Table I, items 2–6).

During the last decade, various new methods for the separation and characterization of glycolipids have been introduced. Thus, a large number of glycolipids with new structures (about 50–70) have been added to our list (see Chapter 1). We now visualize a much more complex pattern of glycolipid composition for each type of cell than previously. About 6 or 7 glycolipids from human erythrocyte membranes were known some ten years ago. We have now separated and characterized 15 neutral glycolipids and 9 monosialosyl gangliosides from the same cell membranes.[15–17] Perhaps 10 more neutral glycolipids and 15 more gangliosides remain to be characterized. Polyglycosylceramides[18,19] with more than 30 sugar residues are not included in this list. Such complex glycolipid patterns reinforce the concept that glycolipids are receptors for a large number of specific signals and that they serve as a variety of specific antigens on the cell surface (for reviews, see Marcus and Schwarting,[20] Hakomori and Young,[21] and Chapters 5 and 6).

This chapter will focus upon the following four topics: (1) glycolipid organization in membranes; (2) glycolipids in cellular interaction and differentiation; (3) the role of glycolipids in cell growth control and oncogenic transformation; and (4) glycolipid response on immune recognition.

4.2. Organization and Dynamic State of Glycolipids in Membranes

4.2.1. Crypticity and Organization of Glycolipids

The ceramide moiety of glycolipids is assumed to be inserted into the outer leaflet of the lipid bilayer and the carbohydrate moieties may end toward the external environment.[9,10] However, the bulk of some glycolipids were cryptic, and their organization is unknown. A striking feature of glycolipid at the plasma membrane is the crypticity and a change of the crypticity associated with the change of cellular functions. The availability of GbOse$_4$Cer is greatly enhanced at the G$_1$ phase of the cell cycle.[22,23] Similarly, the cell-surface reactivity of both GM$_3$ and GgOse$_4$Cer to their respective antibodies was greatly increased at the G$_1$ phase of the cell cycle.[24] There was a less pronounced increase in the reactivity of GM$_3$ at the G$_2$ phase (see Fig. 1). Despite such remarkable changes in cell-surface glycolipids, the chemical quantity of glycolipids in membranes was unchanged during the cell cycle. Thus, the crypticity of membrane glycolipids must be coupled to the functional change of membrane. The cryptic glycolipids became highly exposed when cells were transformed[23] (Table II). The method of galactoseoxidase–NaB[^{3}H]$_4$ labeling, however, does not clearly indicate the chemical quantity of the oxidized glycolipids, i.e., those exposed to the surface. Galactose oxidase treatment followed by reduction with NaBH$_4$ resulted in deuterated glycolipids that can be determined by mass spectrometry. Only 10–15% of GbOse$_4$Cer in human erythrocytes was oxidized by this method (Matsubara and Hakomori, unpublished data). The majority of GbOse$_4$Cer was cryptic and was not labeled from both sides of the plasma membrane. Galactose oxidase-oxidized cerebroside can be readily separated from unoxidized cerebroside by high-performance liquid chromatography. The majority of cerebroside (over 98%) of the myelin sheath membrane was not oxidized.[25] These data indicate the possibility that a large portion of membrane-bound glycolipids could be masked due to an extensive association with integral membrane components, such as an as yet unknown "cofactor" of membrane enzymes or functional proteins. Sulfatide has been claimed as the K$^+$-selective cofactor site of the Na$^+$-K$^+$-dependent adenosine triphosphatase.[26] Gangliosides are implicated as a Ca^{2+}-binding cofactor in synaptic transmission.[27] They greatly activated the Mg^{2+}-dependent adenosine triphosphatase, but not the Na$^+$-K$^+$-dependent enzyme of brain microsomes,[28] and they may be a cofactor of membrane adenylate cyclase.[29] A type of glycolipid organization described as "annular glycolipid" surrounding a functional protein, correlated with the conformational change of such a protein,[13] could be in the same category as cryptic glycolipid. An idealized version of the organization of the cryptic glycolipid is shown in Fig. 2.

4.2.2. Association of Membrane Proteins and Glycolipids

The association of glycolipids with a specific membrane protein has been suggested by the specific requirement of glycolipids for membrane enzyme as described above. The possible association between glycolipids and specific proteins was further indicated by cross-linking studies. In a model experiment, the coliphage M13 membrane protein incorporated into a lipid bilayer was preferentially cross-linked with the exogenously added photosensitive glycolipid probe.[30] The association of GbOse$_4$Cer with a membrane protein has recently been demonstrated by cross-linking studies with the heterobifunctional methyl-4-azidobenzoimidate with or without anti-GbOse$_4$Cer antibodies.[31] A low-molecular-weight amphipathic polypeptide possessing Paul–

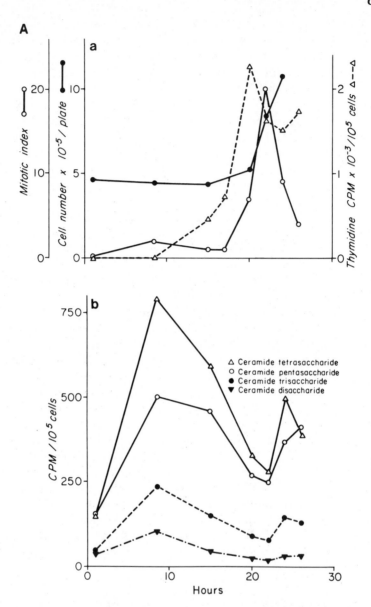

Figure 1. Cell cycle-dependent changes of the surface-exposed glycolipids. (A) Variation of the surface-label activities of glycolipids in synchronized hamster NIL cells. Synchronization started from the confluent, trypsinized cells. (a) Changes of mitotic index, cell number, and thymidine incorporation into DNA; (b) label activities of various glycolipids at times corresponding to those in (a). It is clear that the S phase occurred at 16–20 hr and mitosis at 22–26 hr after seeding. Note that a pronounced peak of the label occurred at the G_1 phase, decreased at the S phase, and increased again after mitosis. A similar glycolipid label was obvious in synchronized cells with the double thymidine block method. (B) Variation of the surface-label activities of glycolipids in synchronized hamster NILpy cells. Synchronization was followed after thymidine block of cell growth at the G–S interphase. (a) Changes of mitotic index and cell numbers; (b) changes of glycolipid label at time cor-

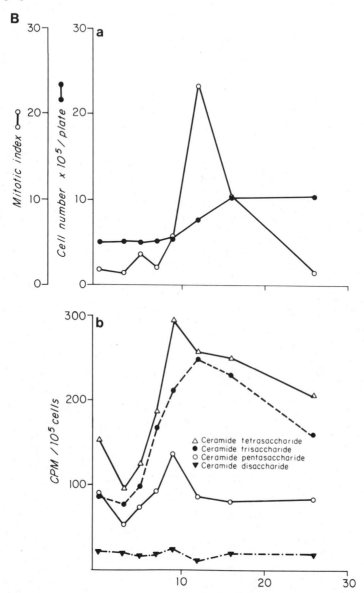

responding to those in A. (C) Cell cycle-dependent variation in cell-surface reactivity of NIL cells to anti-GM$_3$ (*N*-acetylhematoside) (○) and to antigloboside (●) antibodies. ■, Thymidine incorporation (cpm/cell × 10^2); △, cell numbers. (D) Cell cycle-dependent variation in cell-surface reactivity of NILpy cells to anti-GM$_3$ (○) and to antigloboside (●). Both NIL and NILpy cells were synchronized by the mitotic selection method,[236] which is better than trypsinization or thymidine block. The obvious S phase is present at 7–10 hr after plating, and the M phase started 15 hr after plating. The reactivity to the antiglycolipid antibody was determined by iodinated protein A. Note the presence of a remarkable peak before S (corresponding to G$_1$) that is higher in NILpy cells. The chemical quantity of these glycolipids at the G$_1$ phase is unchanged. (Data from Refs. 22–24.)

(continued)

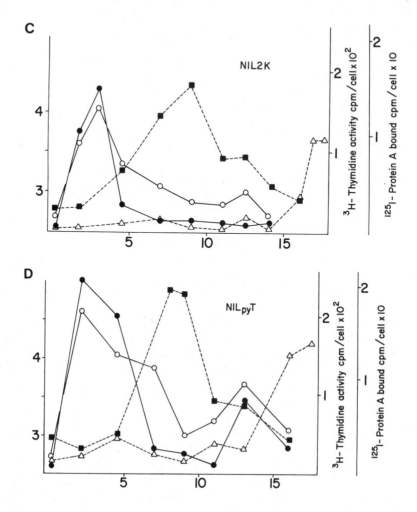

Figure 1. (*continued*)

Table II. Surface-Labeled Glycolipid Activity of Normal and Transformed Cells as Compared to Chemical Quantity of Glycolipids[23]

	GL5	GL4	GL3	GL2
Cell line	Radioactivity of glycolipid (GL) by surface label with galactose oxidase–NaB[^{3}H]$_4$ (cpm/10^5 cells)			
NIL	89	196	163	20
NILpy	178	480	288	43
	Specific activity (cpm/mole glycolipid $\times$ 10^{-3})			
NIL	261	297	543	36
NILpy	4944	3265	1811	44

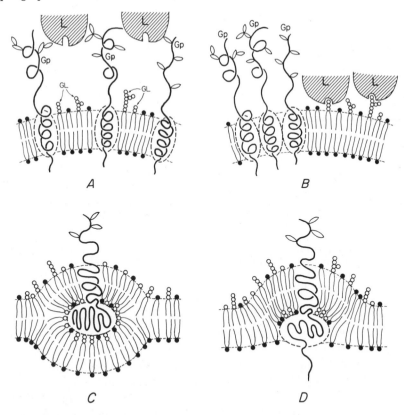

Figure 2. Idealized version of the organization of glycolipids [○ (GL)] with proteins or glycoproteins (Gp) and phospholipid (●) at the cell-surface membranes (see Section 4.2.1). Glycolipids are present at the outer half-leaflet of the lipid bilayer, but rarely at the inner leaflet. However, under normal conditions, a large proportion of glycolipids that must be present in a cryptic form were not labeled. Although we do not know the exact organization of the cryptic form of glycolipids, they may be hidden among "branches" of a glycoprotein "tree" (A). Under certain conditions (the G_1 phase of the cell cycle or membrane perturbation caused by trypsin or sialidase treatment), glycoproteins may cluster so as to make a space for glycolipids accessible to exogenous ligands (L) (galactose oxidase or antibodies) (B). Some membrane proteins are extensively associated with glycolipids and phospholipids. The glycolipids with a strong tendency to associate with a specific type of protein (or glycoprotein) must be cryptic, which may control the function of such a membrane protein. Such a membrane protein ("glycolipid-associated protein") could be a part of a membrane enzyme system, as discussed in the text, or an unknown system that may control membrane fluidity and mobility. C, D, Possible organization of cryptic glycolipid associated with membrane protein. This cryptic glycolipid must be the same as the "annular glycolipid" discussed in the text.

Bunnell (P-B) antigen activity was isolated from bovine erythrocyte membranes and was shown to have a specific affinity for GM_3.[32] Interestingly, the P-B antigenicity at the cell surface was greatly decreased by sialidase treatment of intact cells, but the antigenicity of the purified peptide did not contain sialic acid and was not sialidase-sensitive. The gangliophilic antigen polypeptides (see Table III) may be associated with gangliosides at the cell

Table III. Affinity of Paul–Bunnell Antigen Polypeptides to GM₃ Ganglioside
Demonstrated by Coprecipitation[a]

Expt. no.	GM$_3$		Globoside		Sialyl paragloboside	
	With P–B antigen	Without P–B antigen	With P–B antigen	Without P–B antigen	With P–B antigen	Without P–B antigen
1	7800	750	450	365	540	750
2	7600	430	315	310	ND	ND
3	3500	150	250	150	130	128
4	3250	1100	6500	1560	ND	ND

[a] ^{3}H-labeled glycolipids were dissolved in chloroform–methanol (2 : 1) and mixed with P-B antigen polypeptides in the same solvent. The mixture was evaporated, and the dried residue was dissolved in phosphate-buffered saline and precipitated with anti-P-B antiserum with the second antibody. Results are expressed as the radioactivity (cpm/μg) found in the precipitate formed by anti-P-B and anti-human IgM antibodies. The specific activities of ^{3}H-labeled glycolipids were in a relatively narrow range (15,000–25,000 cpm/μg). Note that GM$_3$ was coprecipitated with P-B antigen in all the experiments. Experiment 4 was carried out with higher-molecular-weight P-B antigen in which globoside was also coprecipitated with P-B antigen. ND, not done. For experimental details see Watanabe *et al.*[32]

surface, and such a complex displays its maximum antigenic activity. Thus, ganglioside could regulate the antigenicity of a polypeptide.[32] These results support the idea that proteins or polypeptides could be associated with a specific type of glycolipid, although their specificity may not be absolute. It is reasonable to assume that such proteins may interact with the cytoskeletal system as well. Exogenously added glycolipids in media are taken up quickly by cells.[33–35] A kinetic study of GM$_1$ uptake by lymphocytes indicated the presence of a specific binding site at the cell surface. The binding was Ca^{2+}-dependent, and not competitive with LacCer or other gangliosides[36]; therefore, it may represent a gangliophilic protein as discussed above.

4.2.3. Dynamic Behavior of Glycolipids in Membranes

The organization of carbohydrate heads and their dynamic state in phospholipid bilayers and in cultured cell membranes were studied by ESR spectrometry with spin-labeled GM$_1$ (labeled at the sialosyl tail). The results indicate that GM$_1$ moves rapidly on the lipid bilayer. However, head-group mobility decreased with increasing bilayer concentration of unlabeled GM$_1$, perhaps as a result of cooperative head-group interactions.[12,37] A physiological concentration of Ca^{2+} caused a head-group cross-linking that led to decreased mobility in solution as well as in lipid bilayers. Interaction between glycophorin and GM$_1$ was observed by the spectral change, especially in the presence of Ca^{2+}.[38] Spin-labeled GM$_1$ at a concentration lower than the critical micelle concentration was rapidly incorporated into the intact cell membrane and was not released by trypsinization. The head-group mobility was essentially the same as that in phospholipid bilayers. Surprisingly, the mobility of the head group increased abruptly on addition of a very small quantity of wheat germ lectin, which binds to sialic acid; it decreased steadily on further addition of the lectin[39] (see Fig. 3). It seems that perturbation of

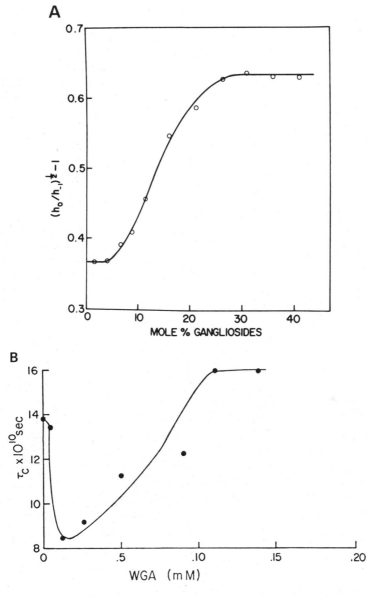

Figure 3. Dynamic behavior of the GM_1 head group revealed by ESR spectra. A, Effect of increasing amount of total ganglioside on the head-group mobility of spin-labeled gangliosides (see Chapter 1, Fig. 18) in fluid liquid vesicles. Egg lecithin vesicles contain 1.5 moles% spin-labeled ganglioside and variable amounts of unlabeled ganglioside. The head-group mobility is inversely related to $[(h_o/h_{-1})^{1/2}-1]$ (for this fraction, see Chapter 1, Fig. 19). A dramatic increase of this factor with increasing concentration of ganglioside indicates a drastic decrease of the head-group mobility. B, Effect of a specific binding event with wheat germ lectin [wheat germ agglutinin (WGA)] on ganglioside head-group mobility. The egg lecithin bilayers contained 4.8 moles% spin-labeled ganglioside in pH 7.4 buffer without Ca^{2+} or Mg^{2+} and added with varying amounts of WGA. Interestingly, an initial dip (increased head-group mobility) was observed when less than 0.1 mM WGA was added. The mobility decreased on further addition of WGA. This initial increase of mobility occurs at a stoichiometry such that a few lectin molecules cause increasing mobility in the majority of ganglioside head groups. This suggests the initial presence of ganglioside clusters in the lipid bilayer.

the entire GM_1 head group was induced when only a small population of GM_1 was bound to wheat germ lectin. Interestingly, the mobility of GM_1 as compared with that of GM_3, GM_2, GD_{1a}, and GD_{1b} was the most restricted in micelles and in mixed dispersions with phospholipid as measured by fluorescence anisotropy and the nanosecond time-dependent change of the fluorescence using 1,6-diphenyl-1,3,5-hexatriene as the motion probe.[40] The cooperativity of phospholipids in liposomes was greatly reduced by GD_{1a} inclusion, as shown by broadening of the endothermal curve in differential scanning colorimetry. The endothermal change in phospholipid–GD_{1a} vesicles was much greater in the presence of 10 mM $CaCl_2$. The Ca^{2+} effect was not observed without GD_{1a}.[41]

These various physical parameters point to the following features of ganglioside behavior in lipid bilayers: (1) head-group structure may define the intrinsic mobility of each ganglioside; (2) the mobility of one species of ganglioside is greatly reduced by increasing the concentration of the same ganglioside species through a possible aggregation; and (3) ligand-induced mobility increased greatly when a small population was bound to ligands, but the mobility was reduced on saturation with ligands.

Cap formation of the choleragen receptor (assumed to be GM_1) was detected with anticholeragen antibody and a fluorescein-labeled second antibody.[42] Dansyllabeled GM_1 incorporated into lymphocytes was capped on addition of cholera toxin. The capping was inhibited by treatment with colchicine and trypsin.[43] Patching followed by capping of GM_1 in thymocytes was induced by anti-GM_1 antibodies. The capping was inhibited by cytochalasin B.[44] Interestingly, anti-GM_1 induced a remarkable mitogenesis.[44] There are three possible explanations for the capping of GM_1: (1) capping may be caused by a GM_1-associated transmembrane protein; (2) glycolipids on membranes are continuously motile according to continuously oriented motion of membranes (the continuous lipid-flow model), cross-linkage by ligands then resulting in trapping at the rear "pole" of cells[45]; or (3) the same carbohydrate structure as GM_1 could be present on glycoprotein ("ganglioprotein")[46] and the observed capping was, in fact, a capping of "ganglioprotein" (see Sections 4.2.4 and 4.7).

4.2.4. Common Carbohydrate Chain in Glycolipids and Glycoproteins

Blood group ABH and Ii determinants are present both in glycolipids and in Bands 3 and 4.5 glycoproteins of human erythrocyte membranes (for a review, see Hakomori[47]). The repeating Galβ1→4GlcNAcβ1→3Gal structure (polylactosamime), the carrier of the blood group determinants, has been found in simpler glycolipids[48–50] and in polyglycosylceramides,[18,19] as well as in the carbohydrates of Bands 3 and 4.5 glycoproteins,[51] which are all susceptible to endo-β-galactosidase of *Escherichia freundii*.[52] Thus, cell-surface labeling by galactose oxidase–$NaB[^3H]_4$ followed by endo-β-galactosidase treatment revealed the presence of the common structure shared between glycolipids and glycoproteins[52] (see Fig. 4). The cell-surface labeling of human erythrocytes, membrane isolation, detergent solubilization, double immune precipitation with antigloboside antibodies followed by sodium dodecyl sulfate–polyacrylamide gel electrophoresis (SDS-PAGE) detected globoside and a few major glycoproteins with the same migration as "PAS IV" ("globoprotein").[46] The same approach with anti-GM_1 antibodies applied to 3T3 cells detected a few glycoproteins reactive to anti-GM_1 ("ganglioprotein")[46] (see Fig. 5). Methylation analysis of brain glycoprotein detected the sugar residue NeuAc2→8NeuAc2→3Gal, similar to that present in ganglioside.[53] Thus, the presence of a common carbohydrate sequence in glycolipids and glycoproteins is increasingly apparent, and many of the same carbohydrate structures as in glycolipids must be present in the peripheral region of glycoprotein carbohydrates, although the core structure of the carbohydrates in glycoproteins is entirely different from that of glycolipids (for a brief review, see Järnefelt *et al.*[54]).

The biological significance of the presence of a common carbohydrate chain in glycolipids and in glycoproteins in not known. However, a two-step binding of ligands from a glycoprotein to glycolipid is possible.

4.3. Glycolipids in Cellular Interaction and Differentiation

4.3.1. Cellular Interaction

Certain lines of study indicate that glycolipids on cell surfaces could be recognized by other cell surfaces and vice versa. This is an important basis of cellular recognition and histogenesis. A few examples and a possible mechanism are discussed below.

4.3.1.1. Ganglioside Changes and Neuroglial Recognition

Cell contact induces enhanced synthesis of a particular glycolipid prior to "contact inhibition" between homologous cells.[55–58] A similar glycolipid response was even more pronounced in heterologous neuroglial cell contact. Pronounced induction of GD_{1b} synthesis was detected by coculturing glial and neuronal cell lines, although neither cell line contained GD_{1b} initially. Similarly, GT_{1b} and an unidentified GT appeared to be synthesized when two different glial cell lines, M1 and MT16, were cocultivated, although both cell lines lacked GT species initially.[59]

4.3.1.2. Retinotectal Recognition

A glycolipid marker possibly involved in the preferential adhesion of chick neuronal retinal cells to the surfaces of intact optic tectum was studied. The adhesion of ventral retina to dorsal tecta depends on recognition between β-GalNAc residues on dorsal tecta and a specific protein located on ventral

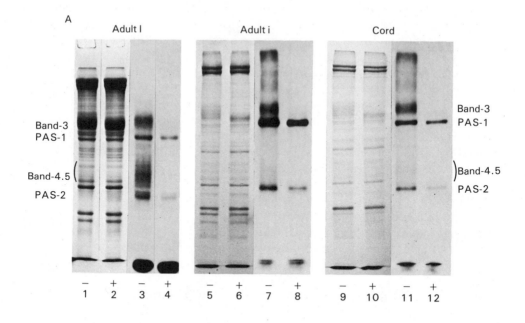

A

Adult I   Adult i   Cord

Band-3
PAS-1

Band-4.5

PAS-2

Band-3
PAS-1

Band-4.5

PAS-2

− + − +   − + − +   − + − +
1 2 3 4   5 6 7 8   9 10 11 12

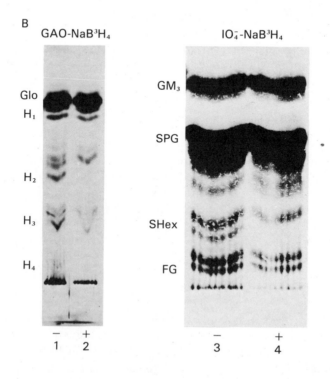

B   GAO-NaB³H₄   IO₄⁻-NaB³H₄

Glo
H₁

H₂

H₃

H₄

GM₃

SPG

SHex

FG

−   +   −   +
1   2   3   4

retina. A reverse recognition was observed between dorsal retina and ventral tecta; i.e., a GalNAc residue on dorsal retina is recognized by a specific protein on ventral retina. The specific GalNAc residue present on dorsal tecta or dorsal retina was not affected by protease, and liposomes containing GM_2 (NeuAcGgOse$_3$Cer) inhibited these adhesions.[60] Consequently, the retinotectal adhesion was found to be inhibited by the affinity-purified antibody directed to GM_2, but not by preimmune serum. Furthermore, the oligosaccharides released from GM_2 but not from other glycolipids affected the adhesive specificity.[61]

4.3.1.3. Neuromuscular Recognition

The formation of neuromuscular junctions is based on a highly specific recognition process that is now believed to involve glycolipids. In culture, the formation of the neuromuscular junction was specifically inhibited by Gb-Ose$_4$Cer at high concentrations (0.25–0.5 mM) and was somewhat stimulated at low concentrations (8–63 μM) of the same glycolipid.[62]

4.3.1.4. Cell-Surface Recognition Sites for Glycolipids and a Possible Mechanism for Cellular Interactions through Glycolipids

In all the phenomena described above, it is assumed that glycolipids on one type of cell surface are recognized by the surfaces of the complementary cells involved. The following two, basically different mechanisms can be considered: (1) The order and the number of glycolipids organized at the cell surface are complementary to the same order and number of glycolipids on the counterpart cell surface (self–self interaction).[63] (2) The molecules that could recognize or interact with glycolipids would be expected to be present; they could be a lectinlike protein[64] or enzymes.[65–67] A "lock-and-key" in-

---

Figure 4. Presence of common carbohydrate chains that are susceptible to endo-β-galactosidase of glycolipids and glycoproteins at the erythrocyte membranes. A, SDS-PAGE of erythrocyte membrane proteins of adult blood group I cells (lanes 1–4), adult blood group i cells (lanes 5–8), and cord erythrocytes (lanes 9–12). Cells were surface-labeled by the galactose oxidase–NaB[^{3}H]$_4$ method. −, +, Indicate that cells were not or were treated with endo-β-galactosidase. Lanes 1, 2, 5, 6, 9, and 10 were stained by Coomassie blue. Lanes 3, 4, 7, 8, 11, and 12 were autoradiographed (fluorograms of gels 1, 2, 5, 6, 9, and 10, respectively). Note that the major surface-labeled carbohydrates locating at Bands 3 and 4.5 were lost after treatment with endo-β-galactosidase. B, Fluorography pattern of thin-layer chromatography of neutral glycolipids (lanes 1 and 2) and gangliosides (lanes 3 and 4). The upper neutral glycolipids from adult erythrocytes are shown before (lane 1) and after (lane 2) treatment with endo-β-galactosidase. Gangliosides of erythrocytes are shown before (lane 3) and after (lane 4) treatment with endo-β-galactosidase. Note that H$_3$, H$_4$, sialosylhexaosylceramide (SHex), and fucoganglioside (FG) were greatly diminished or disappeared after endo-β-galactosidase treatment. These data clearly indicate that a common carbohydrate (termed polylactosaminoglycan) is present both in Bands 3 and 4.5 glycoproteins and in various fractions of glycolipids that are all susceptible to endo-β-galactosidase. (Data from Ref. 52.)

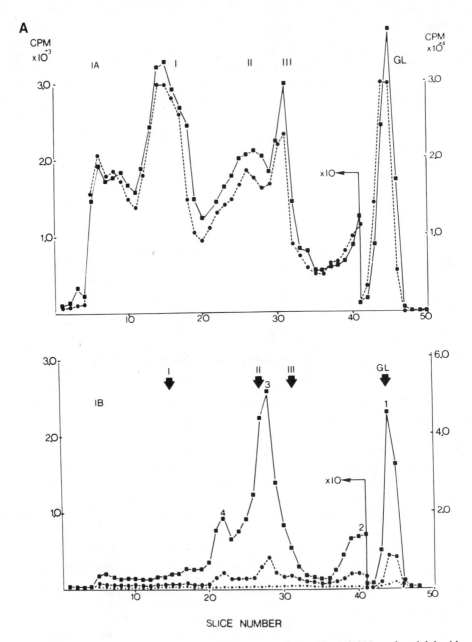

Figure 5. Demonstration of the glycoproteins that cross-react with anti-GM$_1$ and antigloboside antibodies (glanglioproteins and globoproteins). The presence of glycoproteins that cross-react with antiglycolipid antibodies was best demonstrated by cell-surface labeling and specific immune precipitation, followed by SDS-PAGE. A—*Top:* SDS-PAGE pattern of the surface-labeled human erythrocyte membranes. ■, Activities solubilized in SDS on heating; ●, Empigen–Tris extract; GL, position of glycolipid. *Bottom:* SDS-PAGE pattern of the fraction reactive to antigloboside and antiparagloboside isolated from the Empigen–Tris extract of the surface-labeled membranes. ■, Activity with antigloboside; ●, activity with antiparagloboside; peak I, globoside. The major peak corresponds to the PAS-II band, which is termed "globoprotein." B—*Top:* SDS-PAGE pattern of the fraction reactive to anti-GM$_1$ ganglioside antibodies (■) and that with anti-asialo GM$_2$ antibody (●) isolated from 3T3 cells. *Bottom:* SDS-PAGE pattern of the same fraction as above, but isolated from 3T3KiMSV cells.

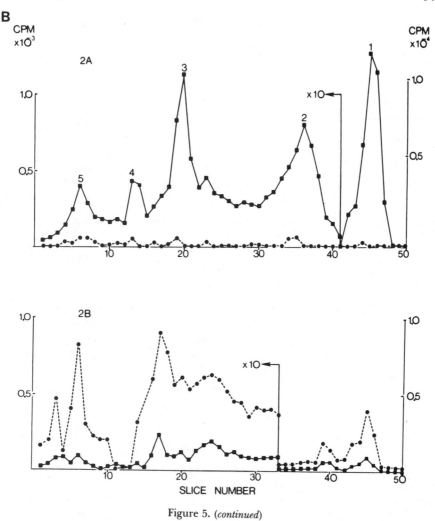

Figure 5. (*continued*)

teraction is possible through these sugar-binding proteins (see Fig. 6). The presence of a specific binding protein to glycolipid at the cell surface was suggested by the specific adherence of glycolipid liposomes containing fluorescein labels.[68] The specific binder at the cell surface could be either a lectin or an enzyme. Animal lectins have been isolated from various tissues and cells, but their presence at the cell surface and their involvement in cellular interactions are not clear (for reviews, see Barondes and Rosen[64] and Frazier and Glazer[69]). A protein in retinotectal recognition was suggested to be a GalNAc-transferase.[60] While a classical proposal of glycosyltransferase at the cell surface[65] aroused considerable debate,[66,67] the possibility of the presence of glycosylhydrolases at the cell surface has now been raised.[58,70–74] In fact, specific cell adhesion can be induced on solid substrates of β-galactosidase or sialidase at physiological pH, under which condition the hydrolytic–catalytic

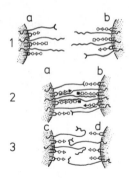

Figure 6. A model for cell–cell interaction through some glycolipids and ectoproteins at the cell surfaces. 1, Growing cells: Certain proportions of glycolipids and proteins (or glycoproteins) are arranged in a certain order. The structures and the organization (order) of carbohydrates and proteins should be complementary. 2, Confluent cells: When cells a and b meet, the carbohydrates and proteins are linked through complementary structure and some carbohydrates can extend their chains for better linkages between complementary proteins. These linkages should be noncovalent, and are possibly mediated by bivalent cations, since cells are dissociable by EDTA. 3, Transformed malignant cells: Carbohydrate chains are incomplete; consequently, no complementary structures were found between glycolipids and proteins. Intercellular linkages were therefore not formed. This model is based on the decrease of reactivity of cells to antiglycolipids and glycosyl extension of some glycolipids when cells contact cells. A similar model proposed by Roseman[65] is based on the presence of glycosyltransferase on cell surfaces. In his model, glycosyltransferases are the complementary proteins that can be "sites" for binding surface glycans of counterpart cells. Recently, some evidence has been presented that the ectoprotein could be cell-surface lectins or glycohydrolases (see the text).

activity was suppressed but the binding activity remained.[73] Cell-surface hydrolase, if present, may well function as cell-surface lectin. A cellular protein released by ethidium bromide could induce growth inhibition[75] and had affinity for glycolipids, particularly GM_3. One component was adsorbed on a GM_3 column, but was difficult to elute.[76] A possible role of glycolipid-associated protein[31,72] in cellular interaction remains to be studied.

4.3.2. Glycolipids as Differentiation Markers

Embryonic or histogenetic differentiation is the orderly process whereby the zygote generates a large diversity of cell types. A continuous change in cell-surface structure encoded by a genetic program is instrumental to perform this orderly differentiation. A remarkable phase-dependent change of glycolipid composition and synthesis was observed during differentiation of various cell systems, indicating that glycolipids may be involved in cellular recognition during differentiation. The phase-dependent glycolipid may represent an "area-code molecule"[77] that may define cell–cell interactions and migrations to specific tissues. A few lines of study are described below.

4.3.2.1. Carbohydrate Markers at Early Embryonic Stages and in Teratocarcinoma

The changes in cell-surface molecules during embryonic development have been detected by immunological methods and have been described as developmental antigens. These include blood group ABH,[78] Forssman,[79,80] Ii,[81,82] F9,[83,84] TerC,[85] and stage-specific embryonic antigen-1 (SSEA-1).[86] In each of these cases, the antigenic molecules were found to be a glycolipid or carbohydrate moiety of glycoprotein and were expressed at the defined stage of embryonic development. A stage-specific expression of F9 antigen has been well documented; it was expressed most strongly at the morulae and blastocyte stage and disappeared on further development.[83] Interestingly, anti-F9 antibody Fab reversibly prevents compaction of mouse morulae.[87] The F9 antigen could be a glycoprotein partially susceptible to endo-β-galactosidase.[84] A similar surface antigen, SSEA-1, recognized by the monoclonal antibody to F9 cells, appeared at the 8-cell embryonic stage, was maximally expressed at the morulae stage, and disappeared at the blastocyte stage[86] (see Fig. 7). The antigen was recently identified as a group of glycolipids having type 2 chain (Galβ1→4GlcNacβ1→3Galβ1→R) with a Le^x determinant (Galβ1→4[Fucα1→3]GlcNacβ1→R) at the terminus (Table IV). These glycolipid antigens are unbranched and susceptible to endo-β-galactosidase. The antigen is therefore closely related to i rather than I (see Fig. 8).[240] The reactivity of the monoclonal antibody SSEA-1 with a meconium glycoprotein was specifically inhibited by lacto-N-fucopentaose III and a synthetic trisaccharide (Galβ1→4[Fucα1→3]GlcNac), but not by lacto-N-fucopentaose II.[241,242] The carbohydrate structures recognized by monoclonal anti-I (Ma), anti-I (Step), and anti-i (Dench) were shown to be expressed on early postimplantation mouse embryo and teratocarcinoma cells. Undifferentiated stem cells were rich in surface-associated and cytoplasmic I antigen, whereas i antigen appeared when cells were differentiated into primary endoderm.[82] The Ii antigens are also known as the differentiation marker of later-stage erythroid cells.[81] The surface antigen, "TerC," of mouse testicular teratoma was recently identified as glycolipid.[85] Glycolipids of teratocarcinoma cell lines were altered by the induction of differentiation. The slow-migrating glycolipids, corresponding to GM$_1$ and GD$_{1a}$, appeared on differentiation.[89]

Table IV. Structures of SSEA-1 Active Glycolipids

Glycolipid	Structure
Y2 glycolipid[a]	Galβ1–4GlcNAcβ1–3Galβ1–4GlcNAcβ1–3Galβ1–4Glcβ1–1Cer 3 / Fucα1
Z1 glycolipid[a]	Galβ1–4GlcNAcβ1–3Galβ1–4GlcNAcβ1–3Galβ1–4GlcNAcβ1–3Galβ1–4Glcβ1–1Cer 3 / Fucα1
Z2a glycolipid[a]	Galβ1–4GlcNAcβ1–3Galβ1–4GlcNAcβ1–3Galβ1–4GlcNAcβ1–3Galβ1–4Glcβ1–1Cer 3 3 / / Fucα1 Fucα1
Le^x glycolipid[b]	Galβ1–4GlcNAcβ1–3Galβ1–4Glcβ1–1Cer 3 / Fucα1

[a] Kannagi, Nudeluran and Hakomori.[240]

[b] Glycolipid previously isolated and characterized from human adenocarcinoma tissue.

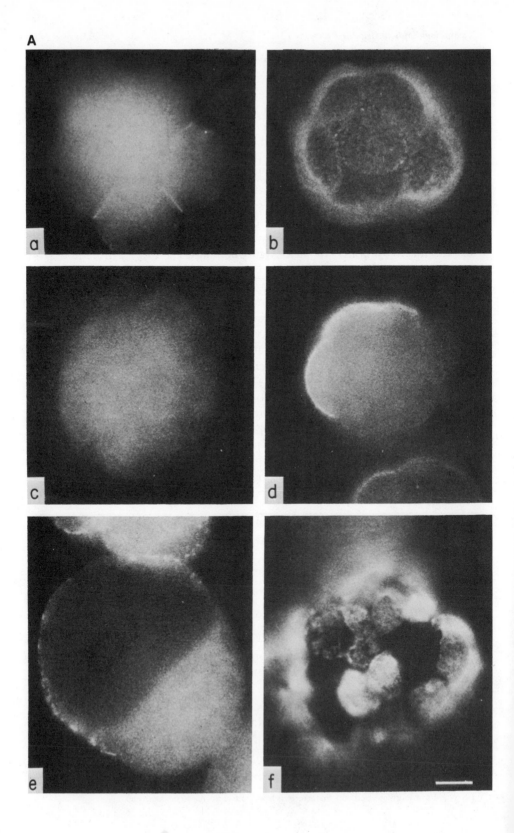

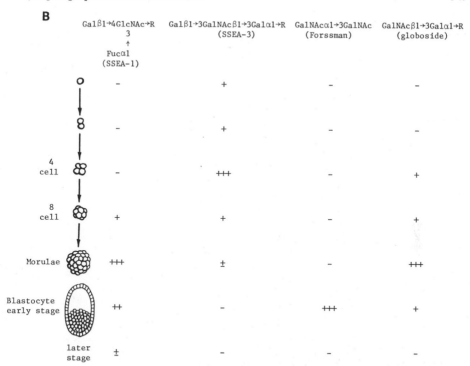

Figure 7. Stage-specific expression of glycolipid haptens in mouse pre-implantation embryo. A, The change of reactivity in mouse embryo directed to monoclonal SSEA-1 antibody. (a) Eight-cell stage, two positive blastomeres; (b) 8-cell stage, all but one blastomere are positive; (c) completely negative morulae; (d) partially positive morulae; (e) blastocyte, few trophoblastic cells positive; (f) inner cell mass grown *in vitro* for 3 days, patches of positive and negative endoderm cells are seen. Scale bar represents 20 μm. B, A schematic drawing of a clear stage-specificity expressed by each carbohydrate determinant. SSEA-1 and globoside are maximally expressed at morulae, SSEA-3 at 4-cell stage, and Forssman at blastocyte. Results summarized from Solter and Knowles,[86] Shevinsky *et al.*,[262] and Willison *et al.*[263]

4.3.2.2. Glycolipids in Intestinal Epithelial Differentiation

The undifferentiated "crypt" cells of intestinal epithelia undergo progressive maturation within 12 hr as they migrate up the villus and develop a well-defined brush-border cell that contains a variety of digestive enzymes. GM_3 was virtually absent from crypt cells, and the enzyme required for sialic acid addition (CMP-NeuAc : CDH-sialyltransferase) was also absent from the crypt-cell fraction, whereas the villus cell was characterized by a high level of GM_3 and the sialyltransferase. GM_3 synthesis is therefore correlated with rapid differentiation of intestinal epithelia.[90,91] Similarly, undifferentiated basal or parabasal germinating layers of buccal epithelia were stained by monoclonal antibodies to the precursor chains for blood group ABH determinants, in contrast to the differentiated spinous cells and flat epithelial cells which were

A   Development of Carbohydrate Chain in Intestinal
Mucosal Epithelia (Bouhour and Glickman)

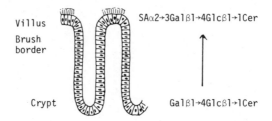

Villus
Brush
border   SAα2→3Galβ1→4Glcβ1→1Cer

Crypt   Galβ1→4Glcβ1→1Cer

B   Development of Carbohydrate Chain in Buccal
Mucosal Epithelia (Dabelsteen and Hakomori)

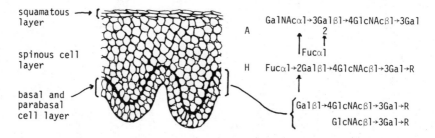

squamatous
layer

spinous cell
layer

basal and
parabasal
cell layer

A   GalNAcα1→3Galβ1→4GlcNAcβ1→3Gal
                               2
                               ↑
     ↑Fucα1

H   Fucα1→2Galβ1→4GlcNAcβ1→3Gal→R
     ↑

     { Galβ1→4GlcNAcβ1→3Gal→R
     { GlcNAcβ1→3Gal→R

Figure 8. Glycolipid changes associated with epitheliogenesis. A, Glycolipid changes from crypt cells to villus cells in intestinal epithelia. Data taken from Bouhours and Glickman.[91] B, The change of blood group carbohydrate chain from germinating basal cell layers to squamatous epithelial cells in buccal epithelia. Data taken from Dabelsteen *et al.*[243]

stained by antibodies directed to blood group ABH determinant.[243] Thus, the chain elongation reaction of blood group determinant may account for differentiation of epithelial tissue. The differentiation-dependent glycolipid changes in epitheliogenesis are illustrated in Fig. 8.

4.3.2.3. Erythrocyte Differentiation and Glycolipid Changes

Agglutinability by anti-GbOse$_4$Cer was much higher with fetal erythrocytes as compared with newborn and adult erythrocytes. This change was ascribed to the process of "masking" rather than the chemical change of GbOse$_4$Cer.[92] In contrast, a remarkable change of agglutinability to anti-I and anti-i antibodies after birth is based on a chemical change of the carbohydrate chain. The i-antigen, which is present in fetal erythrocytes, was identified as a linear repeating N-acetyllactosamine structure (Galβ1→ 4GlcNAcβ1→3Gal) that is represented by lacto-N-*nor*-hexaosylceramide.[93] I antigen, which is lacking in fetal erythrocytes but predominantly present in adult erythrocytes, is now identified as a branched structure represented by lacto-N-*iso*octaosylceramide.[94,95] Various monoclonal anti-i antibodies may recognize various domains of structure within an unbranched linear lacto-N-*nor*-hexaosylceramide (Chapter 5). A linear repeating N-acetyllactosamine structure was found in fetal polyglycosylceramide, while the branched structure was found in adult polyglycosylceramide[19] as well as in a glycopeptide fraction derived from Band 3 glycoprotein.[51] The branching enzyme, which could define a crucial step in erythrocyte differentiation, remains to be elucidated. Interestingly, the time for switching synthesis from fetal hemoglobin to adult hemoglobin coincides with the time for i-to-I conversion.[96] Endo-β-galactosidase of *Escherichia freundii*[97] hydrolyzes repeating N-acetyllactosamine as described above, but a linear unsubstituted chain was degraded much more readily than the branched structure,[98] and Ii activity *in situ* was destroyed.[52] The major Ii determinants were present in the Bands 3 and 4.5 glycoproteins besides the lactoglycosyl series glycolipid.[52] The branching process associated with erythrocyte differentiation is a general phenomenon not only referred in Ii antigens but also related to ABH determinants.[99] (see Figs. 9 and 10). By cell-surface labeling and with reaction with anti-branched H (anti-H₃) antibodies, the branched structures were found to be much less predominant in fetal than in adult erythrocytes. A weak fetal erythrocyte reaction to an antiglobulin test with anti-A,B immunoglobulin G (IgG) was correlated to the lack of bivalent A,B antigen in fetal erythrocytes.[100]

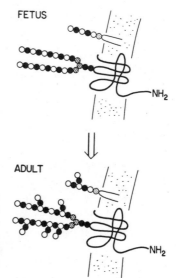

Figure 9. Idealized version of the changes in carbohydrate chains (polylactosaminoglycan) associated with erythrocyte development. A single unbranched chain linked to a ceramide and multiple unbranched chains linked to a core oligosaccharide of Band 3 protein are converted to the branched chain during development from fetal to adult erythrocytes. ○, Gal; ●, GlcNAc; ○, Man; ○, Glc.

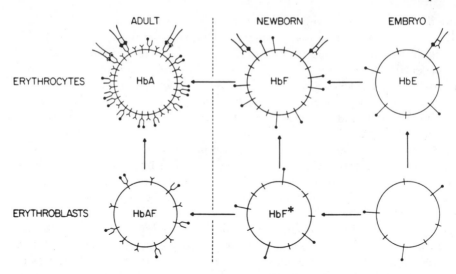

Figure 10. Schematic representation of blood group ABH and Ii determinants in embryo, new-born, and adult erythrocytes. Blood group ABH determinants are carried by the long unbranched carbohydrate chain in erythrocytes or erythroblasts of the embryo or newborn. These determinants bind only weakly with IgG anti-A, -B, or -H because they are lacking in bivalent binding sites. These cells also have the precursors of unbranched determinants, which are indicated as short lines (—) and are now identified as i antigen. The quantity of both ABH (●—) and i (—) increases when cells develop from embryo to newborn and when erythroblasts develop into erythrocytes. Blood group determinants (●) in adult erythrocytes and erythroblasts are carried by branched carbohydrate chains (⊃—), thus forming the bivalent binding sites. These structures will bind IgG anti-A or -B much more strongly than unbranched determinants. Adult erythrocytes and erythroblasts also contain their precursors (>—), which are now identified as I antigen. Erythroblasts may also contain the branched determinants, but in much smaller numbers than mature erythrocytes. The hemoglobin differs at each stage of differentiation, embryonic (HbE), fetal (HbF), and adult (HbA), and is distinguishable in embryo, newborn, and adult erythrocytes. Newborn erythroblasts have a variant of HbF (HbF*), whereas adult erythroblasts contain both HbA and HbF.

4.3.2.4. Other Differentiation Systems

A slow-migrating ganglioside increased when neuroblastoma cell lines were induced to differentiate. Increased activities of galactosyltransferase and sialyltransferase in an adrenergic clone of neuroblastoma was found on addition of dibutyryl cyclic AMP.[101,102] Enhanced synthesis of GD_{1a} of L6 muscle cells at the aligning stage shortly before cell fusion was observed.[103] However, what was described as GD_{1a} was recently identified as GD_3.[104] In addition, a still-unidentified slow-migrating neutral glycolipid (corresponding to ceramide heptasaccharide) was present in fused mixed culture but not in bromodeoxyuridine-inhibited muscle cultures.[104] The presence of asialo GM_1 ($GgOse_4Cer$) in a mature T-cell population was recently observed.[105] Mouse myeloid leukemia Ml cells differentiate into macrophages by various agents such as the protein factor in the conditioned medium of mouse lung fibroblasts, steroids, polyanions, and lipopolysaccharides (LPS). Associated with this induction of differentiation, adhesiveness and phagocytotic capacity were enhanced, and Fc and C3 receptors appeared at the cell surface. Recently, gangliosides of Ml cells and their dramatic changes during the induction of differentiation have been reported. A severalfold increase of GM_{1b} ganglioside with an inconsistent decrease in disialoganglioside was observed.[235] More recent studies have indicated that differentiation of Ml cells to macrophagelike cells, accompanies a series of glycolipid changes. Undifferentiated cells are characterized by the presence of ganglio-series glycolipids, $GgOse_3Cer$, $GgOse_4Cer$, and GM_{1b}. In the early stage of differentiation, synthesis of Gg-series declines and synthesis of lacto-series is enhanced, resulting in the conversion of i to I antigen. In the later stage of differentiation, synthesis of globo-series increases and $GbOse_3Cer$ (P^k antigen) becomes the major cell surface component in differentiated macrophage-like cells.[259] This is a typical example of the shifting of the synthetis of one series of glycolipids to another series associated with differentiation. The typical glycolipid changes associated with differentiation are shown in Fig. 11. Thus, it is increasingly apparent that either ontogenetic development or histogenetic differentiation (such as hematopoiesis or epitheliogenesis or spermatogenesis) or both are mediated by glycolipid changes.

4.4. Role of Glycolipids in Cell Growth Control

4.4.1. Cell Contact Response of Glycolipids as Related to 'Contact Inhibition'

An unknown mechanism operates to enhance glycolipid synthesis of the contact-inhibitable cells when they contact each other at high cell population density. The phenomenon is called "glycolipid contact response." The possible regulation of cell growth through glycolipids was suggested by the observation that "glycolipid contact response" was closely related to "contact inhibition" of cell growth. Synthesis of a certain glycolipid, such as $GbOse_3Cer$ in BHK,[55] $GbOse_3Cer$, $GbOse_4Cer$, $GbOse_5Cer$, and GM_3 in NIL,[56,57] GD_{1a} in 3T3,[58] and GD_3 and GM_1 in human fibroblasts 8166,[55] was greatly enhanced prior to density-dependent growth inhibition (see Fig. 12). The enzymatic basis of the response can be at least partially measurable[106,107] (see Table V), and the response is based on cell contact, rather than other factors.[107] The loss of this response is closely related to the loss of growth control in oncogenic transformants.[55–58] Chemically transformed mouse embryo cells showed a glycolipid pattern very similar to that of nontransformed cells, but they lacked the density-dependent response of GD_{1a}.[108] However, the lack of this response may not be an absolute criterion expressing tumorigenicity, since there are some exceptions; GM_3 or $GbOse_5Cer$ in some clones of transformed NIL cells[109] and GD gangliosides of Ehrlich carcinoma cells also showed a density-dependent enhancement.[110]

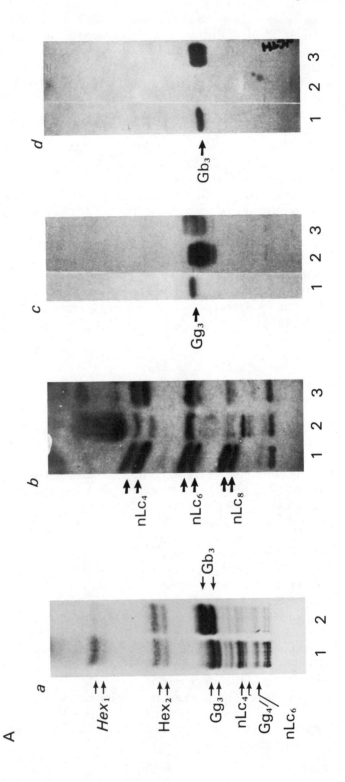

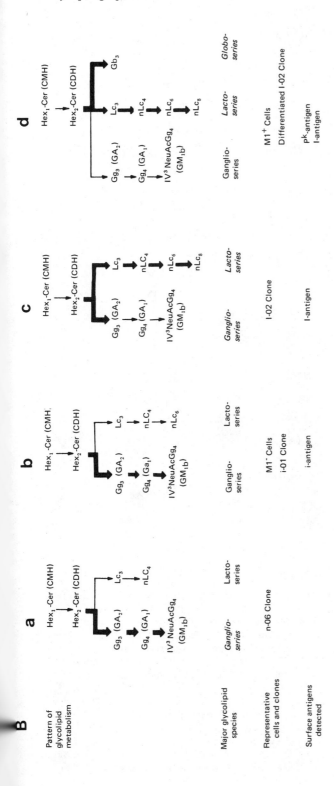

Figure 11. A, Glycolipid changes of mouse myelogeneous leukemia M1 cells during differentiation as revealed by immunostaining. TLC (a), and immunostaining pattern on TLC with monoclonal anti-Galβ1→4GlcNAc (b), anti-gangliotriaosylceramide (c), and anti-Pᵏ (d), antibodies of M1 cell neutral glycolipids. In (a), lane 1, neutral glycolipids prepared from undifferentiated M1 cells; lane 2, from differentiated M1 cells (72 hr culture with 10% conditioned medium); visualized with orcinol reagent. For immunostaining with monoclonal antibodies (b–d), M1 cell glycolipids were chromatographed on an HPTLC plate and reacted with anti-Galβ1→4GlcNAc (for b, mouse IgM);[260] anti-GgOse₃Cer (for c, rat IgM);[261] antibody followed by a second antibody and ^{125}I labeled protein A solution. Autoradiography was carried out using Kodak X-ray films. Lane 1, control glycolipids; lane 2, neutral glycolipids prepared from undifferentiated M1 cells; lane 3, from differentiated M1 cells (72 hr). The undifferentiated M1 cell glycolipid shows a strongly stained spot of Gg₃ (lane 2c), whereas staining for the differentiated M1 cell glycolipids is faint (lane 3c). In (d), only globotriaosylceramide (Gb₃) is stained with this antibody. No glycolipids were stained for the undifferentiated cells (lane 2d), and a strongly positive spot was detected for the differentiated cells (lane 3d). B, The figure illustrates a shifting of glycolipid synthesis from ganglio-series to lacto-series, and finally to globo-series during differentiation of M1 cells. Cells having undifferentiated characteristics can be divided into two stages. In (a), synthesis of ganglio-series is predominate and that of lacto-series is almost absent. These cells are i⁻I⁻ (shown in the undifferentiated clone N-06). In stage (b), the synthesis of ganglio-series is still predominant, but the synthesis of a linear unbranched lacto-series has been initiated and cells become i⁺. In stage (c), the synthesis of ganglio-series declines and that for lacto-series is greatly enhanced, resulting in the appearance of I antigen. In the differentiated macrophage-like cells (d), synthesis of ganglio-series has almost ceased and that of GbOse₃Cer appears. (Data from Ref. 259.)

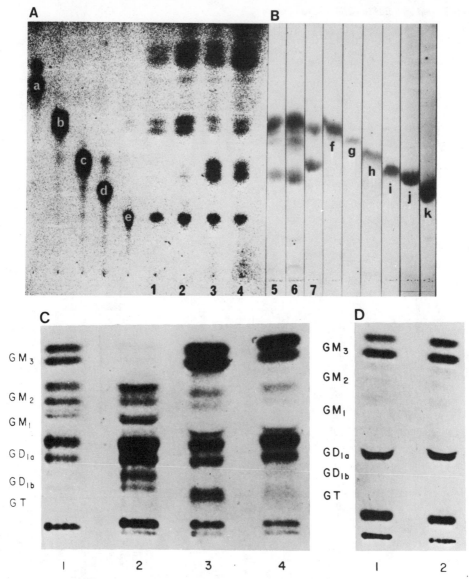

Figure 12. Cell-population-density-dependent changes of glycolipid composition as revealed by thin-layer chromatography. A, BHK cells and their transformants. Lanes: 1, BHKpy cells at confluency; 2, BHKpy cells at low density; 3, BHK cells at confluency; 4, BHK cells at sparse culture. Spots a–e are standards for CM, CD, CT (GbOse$_3$Cer), globoside, and GM$_3$. Note that CT (Gb-Ose$_3$Cer) (c) greatly increased at high cell population density, which is completely absent in transformed BHKpy cells. B—Lanes: 5, Sparse culture of human fibroblast 8166; 6, confluent culture of the same cells; 7, SV40-transformed human fibroblasts. Spots: f, N-acetylhematoside; g, N-glycolylhematoside; h, GD$_3$ (N-acetyl); i, GM$_2$; j, GD$_3$ (N-glycolyl); k, GM$_1$. C, Autoradiograms of the gangliosides from the three cell lines metabolically labeled with [^{14}C]glucosamine. Lanes: 1, Exponentially growing nontransformed C3H/10T1/2 cells; 2, same cells at confluency; 3, exponentially growing dimethylbenzanthracene-transformed TCL1 cells; 4, same cells at confluency. D, Similar autoradiograms of the gangliosides from 3-methylcholanthrene-transformed TCL15 cells. Lanes: 1, Exponentially growing cells; 2, confluent cells. Note that a great increase of the band corresponding to GM$_1$ and GD$_{1a}$ occurred at confluency of nontransformed cells, but the glycolipid pattern was almost identical between two different growth states of transformed cells. Plates A and B reproduced with permission from Ref. 55; C, D reproduced through the courtesy of Dr. R. Langenbach from Ref. 108.

Table V. Activities of UDP-Gal : Glycolipid α- and β-Galactosyltransferase of "Fraction P-3" Prepared from Normal and Transformed BHK and NIL Cells

Cells	Cell population densities	Glycolipid synthesized (μmoles/mg P-3 protein/hr) (complete system)	
		CTH synthesis[a]	CDH synthesis[a]
BHK	Sparse ($\leqslant 5 \times 10^4$/cm^2)	109 (4)	122
	Confluent ($>10^5$/cm^2)	385 (5)	167
BHKpy	Low ($\leqslant 10^5$/cm^2)	55	
	High ($>10^5$/cm^2)	48 (4)	121
NIL-2E	Sparse ($\leqslant 5 \times 10^4$/cm^2)	385 (2)	247 (2)
	Confluent ($>10^5$/cm^2)	1052 (2)	257 (2)
NILpy	Low ($\leqslant 10^5$/cm^2)	26	400 (2)
	High ($>10^5$/cm^2)	38 (2)	206

[a] Numbers in parentheses, represent the mean value of multiple experiments. (Data from Ref. 106.)

4.4.2. Cell Cycle and Mitogenesis

Changes associated with the cell cycle have also been observed in the exposure of glycolipids at the cell surface[22,23] (see Figs. 1 and 2) and the metabolic incorporation of precursor sugars[111,112] into glycolipids. Lectin-induced mitogenesis of sheep lymphocytes is accompanied by enhanced synthesis of ceramide trihexoside and a change of ganglioside composition.[113] Other studies on murine lymphocytes have indicated that synthesis of all glycolipids was enhanced with a maximum peak 30 hr after addition of lectin.[114,115] Moreover, mitogen-activated T and B cells could be distinguished by glycolipid pattern.[116]

To test the possibility that cell-surface glycolipids may control cell proliferation, three approaches have been undertaken: (1) exogenous addition of glycolipids; (2) application of antibodies directed to glycolipids; and (3) measurement of glycolipid changes caused by differentiation inducers.

4.4.3. Glycolipid Addition in Cell Culture

Addition of glycolipid antigens can cause conversion of cellular antigenicity; e.g., Le^{a-} erythrocytes converted to Le^{a+} erythrocytes by incubation with Le^a glycolipids. Thus, added glycolipids may well function like those synthesized *in situ*. Incubation with glycoprotein antigen did not change the antigen status.[33] Hamster NIL cells cultured in media containing GbOse$_4$Cer showed more than 2-fold enhancement of the GbOse$_4$Cer levels at the plasma membrane. Consequently, cell adhesiveness increased, morphology of NILpy cells resembled that of normal NIL cells, and the prereplicative period, particularly the G$_1$ phase, became twice as long as before treatment[34] (see Fig. 13). Similarly, the addition of several gangliosides to culture media reduced both the growth rate and the saturation density of simian virus 40 (SV40)-transformed and untransformed 3T3 cells.[35] No apparent cytotoxicity was demonstrated even with the most effective GM$_1$, GD$_{1b}$, and GT$_{1b}$; free ceramide and GD$_{1a}$

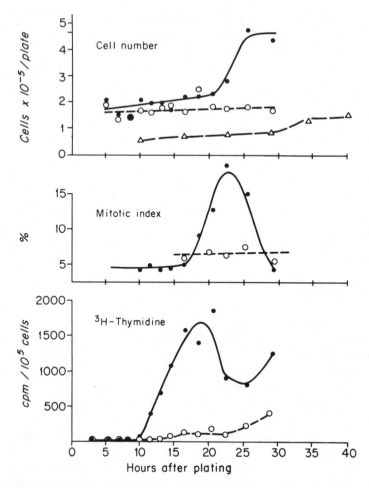

Figure 13. Extension of the prereplicative phase of NIL cells by addition of exogenous globoside. The effect of globoside medium on synchronized NIL cell culture is shown. *Top:* Change of cell numbers. ●, Normal Eagle's medium; [, △, globoside-enriched Eagle's medium (8 × 10⁻⁴ mol). *Middle:* Change of mitotic index. ●, Normal Eagle's medium; ○, globoside-enriched Eagle's medium. *Bottom:* [³H]Thymidine incorporation determined after the culture was incubated with medium containing 4 × 10⁵ cpm [³H]thymidine for 30 min. ●, Normal Eagle's medium; ○, globoside-enriched Eagle's medium. (Data from Ref. 34.)

were also ineffective. In these studies, gangliosides added to the culture media were rapidly accumulated by the cells as constituents of the plasma membrane. However, the exogenously supplied gangliosides could be distinguished from those synthesized endogenously by the lability of the former to sialidase.[35] While some of the added glycolipids appear to be removable by trypsin treatment,[117] some are certainly oriented with the hydrophobic ceramide moiety embedded in the lipid bilayer.[38,43,117] This approach is important to establish the function of membrane components as related to cellular growth behavior. In chemically defined medium, the effect of gangliosides exogenously added to the culture medium is much more remarkable than when they are added in the conventional serum-containing medium. Furthermore, a selective effect by a specific ganglioside is apparent in the chemically-defined medium. Recent studies with this approach strongly suggest that GM_3, and not GM_1, in hamster fibroblasts may regulate the function of the receptor for fibroblast growth factor (FGF).[237] A similar approach has been applied, not only for glycolipids but for fibronectin and other cell surface molecules as well. Cell growth behavior, motility, and morphology can be modified significantly when fibronectin is added to the culture medium.[118,119]

4.4.4. Modification by Antiglycolipid Antibodies

The second approach to studying the function of specific glycolipids on cell surfaces is to observe the functional changes following the addition of specific antiglycolipid antibodies.[24] While lectins such as concanavalin A (Con A) and their monovalent or divalent derivatives have been used to affect cell growth behavior,[120] lectins are cytotoxic and their specificity is too broad to specify the function of a specific carbohydrate. However, affinity-purified antiglycolipid antibodies[121] should be suitable for this purpose. Anti-GM_3, but not anti-$GbOse_4Cer$, markedly inhibited the growth of hamster NIL and mouse BALB/c fibroblasts, but not that of their transformed derivatives. In subconfluent cell cultures, the interaction led to a 1.5-fold stimulation of [^{14}C]-Gal incorporation specifically into GM_3. However, glycolipid synthesis in confluent cell cultures was not stimulated on addition of anti-GM_3 antibody. Thus, the interaction with antiglycolipid mimicked those cellular interactions that stimulate synthesis of density-dependent glycolipid and may lead to cessation of cell growth ("mimic contact inhibition").[24] Significant growth inhibition of astrocytoma cells was induced by anti-GM_1 and anti-GM_3 whole antibodies. The induced growth inhibition was accompanied by increased adenylate cyclase and decreased guanylate cyclase activity.[59] Incubation of BALB/c–3T3 cells with anti-GM_1 antibodies inhibited murine sarcoma virus (MSV)-dependent expression of transformed phenotype. Rat kidney NRK cells transformed with temperature-sensitive avian sarcoma virus LA23 or LA25, preincubated with anti-GM_3 antibodies at a nonpermissive temperature, completely inhibited the expression of transformed phenotype such as morphology change, and growth in soft agar, when cells were exposed to the permissive temperature.[122] A block of the differentiation process of early mouse embryo by anti-F9 antibodies was mentioned previously.[87] Studies on the mechanism by which antiglycolipids affect cell growth, transformation, and differentiation may provide clues for understanding the function of glycolipids as "transducer."

4.4.5. Glycolipid Changes Caused by Differentiation Inducers

Shorter-chain fatty acids, particularly *n*-butyric acid, added in culture media remarkably reduce cell saturation density and restore contact inhibition and are regarded as differentiation inducers.[123] Associated with these changes in cell growth behavior is a greatly increased chemical level of GM_3 due to an enhanced GM_3 synthetase (CMP-sialic acid : CDH sialyltransferase[124,125]; the reactivity to cholera toxin was also enhanced due to the synthesis of GM_1.[126] Butyrate is now recognized as a potent differentiation inducer for various cell systems, particularly for erythroid cells.[127] The tumor promoter phorbol-12-myristate-13-acetate for mouse skin was found to induce differentiation of human melanoma cells, inducing melanin synthesis and increasing GM_3 synthesis similar to the effect of butyrate.[128]

Other powerful differentiation inducers are the retinoid compounds (retinol, retinoic acids, and their analogs). Transformed cells tend to restore their normal growth behavior and show density-dependent growth inhibition *in vitro* on addition of retinoids (20–100 nmoles/ml),[129] and chemical and viral carcinogenesis *in vivo* were prevented or inhibited by retinoid administration.[130] Enhanced synthesis of GM_3 in NIL cells and enhanced cell contact response of gangliosides in 3T3 cells have been observed.[131] Restored contact inhibition on feeding retinoid compounds may well be related to restored contact response of glycolipid synthesis. Although the mechanism by which butyrate, phorbol derivatives, and retinoids induce functional changes is unknown, the effect may be due in part to the change of membrane gangliosides.

4.5. Glycolipid Changes in Oncogenic Transformation: Deficiency in Glycolipid Function

The idea that glycolipid composition in tumor tissue could be different from that in normal tissue was suggested by the classical studies on lipid haptens by Witebsky and Hirszfeld 50 years ago. Some of these haptens were later identified as the chemically well-defined glycolipids and termed "cytolipins" by Rapport and his colleagues (see Chapter 5).

However, unequivocal demonstration of glycolipid changes associated with oncogenic transformation became obvious only after knowledge and technology that allowed transformation of the cultured cells *in vitro* were established during the early part of the 1960s. The first clear demonstration of the change of glycolipid composition associated with oncogenic transformation was described in 1968 using hamster fibroblast BHK cells and their polyoma virus transformants,[132,133] and subsequently with mouse 3T3 and NALN cells transformed by SV40.[134,135] Since then, a number of studies have been carried out employing established fibroblast cell lines or primary cultures transformed by various tumor viruses, chemical carcinogens, and *in vivo* tumors including human cancers. These results are summarized in Table VI. Since a certain mechanism for cell growth regulation and cellular interaction is lost or aberrant in many transformed cells, a distinct change of glycolipid composition and metabolism associated with the transformation has received a great deal of interest in the hope of discovering a possible function of glycolipids at the cell surface and in cell growth regulation and cellular interaction. In fact, many studies directed toward glycolipid function at the

Table VI. Chemical Changes in Glycolipid Composition and Synthesis of Glycolipids in Various Transformed Cells

Change and cell type	Authors	Description of change
A. Transformation by DNA viruses		
Hamster		
BHK/BHKpy	Hakomori and Murakami[132]	Decreased GM_3, increased GL2
BHK/BHKpy	Hakomori et al.[133]	Decreased GM_3
Mouse		
3T3sw/3T3sv; 3T3svpy	Hakomori et al.[133]	Decreased GM_3
3T3sw/3T3sv; 3T3svpy	Mora et al.[134]	Decreased GM_1, GD_{1a}
NALN/SVS ALN	Mora et al.[134]	Decreased GM_1, GD_{1a}
Hamster		
NIL/NILpy	Robbins and Macpherson,[136] Sakiyama et al.,[56,57] Kijimoto and Hakomori[106]	Loss of contact-dependent synthesis of GL3, GL4, GL5
Hamster		
C12TSVS-R/C12TSV3-S	Nigam et al.[137]	Deletion of higher ganglioside (GD_{1a})
Human		
8166/transformed SV60	Hakomori[55]	Deletion of GM_1 ganglioside, increased GM_2
Mouse		
STU/STUSV40	Diringer et al.[138]	Decreased GM_3
STU/STU Friend		
3T3/py-6; SV-479	Yogeeswaren et al.[139]	Deletion of GM_1, GD_{1a}
3T3/SV-A26; SVCE56	Yogeeswaran et al.[139]	Decreased GM_3, increased GD_{1a}
B. Transformation by RNA viruses		
Hamster BHK/BHK-RSV	Hakomori et al.[133]	Decreased GM_3
Chicken embryo fibroblasts/RSV	Hakomori et al.[140]	Decreased GM_3, GD_3
Hamster NIL/hamster sarcoma virus	Sakiyama et al.[56]	Decreased GL3, GL4, GL5
Mouse embryo cells		
MSV	Mora et al.[141]	Deletion of GD_{1a}
3T3/Kirsten MSV	Fishman et al.[142]	Deletion of GM_1, increased GM_2

(continued)

Table VI. (Continued)

Change and cell type	Authors	Description of change
C. "Revertant" clones of transformed cells		
Pollack's flat revertant	Mora et al.[143]	GD_{1a}, GM_1 level reverted to normal level of 3T3
3T3sv (FL²SV101)		
3T3py (Fl py 11)		
Sachs' revertant	Den et al.[144]	GD_{1a}, GM_1 did not revert to the level of normal hamster embryo cells
Golden hamster embryo cells transformed by py and dimethylnitrosamine		
Hamster cells/SV40	Nigam et al.[137]	GM_2 level parallel to normal phenotype
D. Transformation by RNA or DNA virus with thermosensitive mutants, or transformed cells showing thermosensitive growth behavior		
BHKpy ts3	Hammarström and Bjursell[145]	CTH level did not revert
BHKpy ts3	Gahmberg et al.[146]	CTH level did not revert CDH and GM_3 levels reverted
CEF RSV ts7	Warren et al.[147]	CM_3 activity did not revert
CEF RSV ts	Hakomori et al.[148]	GM_3 chemical level reverted
Mouse fibroblast C3H Sv (ts A58, B_1 tsB11)	Onodera et al.[149]	No qualitative difference, but CDH level greatly increased at permissive temperature
Swiss 3T3/Svts	Itaya and Hakomori[150]	GM_2 synthesis greatly decreased at permissive stage
ts mutant of dimethylnitrosamine-transformed cells	Buehler and Moolten[151]	Glycolipid synthesis (CTH) inhibited at permissive temperature
E. Chemically transformed cells *in vitro* and *in vivo*		
Rat hepatocytes and rat Morris hepatoma *in vitro*	Brady et al.[152]	Deletion of GD_{1a}, increased GM_3
Rat liver and rat Morris hepatoma	Siddiqui and Hakomori[153]	Deletion of GT, increased GD_{1a}, GM_1
Rat liver and rat Morris hepatoma	Cheema et al.[154]	Deletion of GT, increased GD_{1a}, GM_1
Plasma membranes of rat liver and Morris hepatoma	Dnistrian et al.[155]	Deletion of GT, accumulation of GD_{1a}, GM_1, and tetraosylceramide

Cell/tissue	Reference	Observation
Regenerating rat liver, rat hepatoma 27, and rat liver	Dyatlovitskaya et al.[156]	Deletion of GT; accumulation of GD_{1b} only in hepatoma, not in regenerating liver
Rat liver and Novikoff hepatoma	Van Hoeven and Emmelot[157]	Deletion of GD, GT
Rat mammary tissue and rat mammary tumor	Keenan and Morré[158]	Deletion of GD_{1a}
Rat embryo cells transformed by 3-methylcholanthrene *in vitro* with or without combination of Rauscher leukemia virus	Langenbach[159]	Deletion or decrease of GM_2, increased GM_3, new synthesis of GT
Mouse C2H embryo cells transformed by dimethylbenzanthracene (CDMBA-TCL-1), 3-methyl-cholanthrene (MCA-TCL-5)	Langenbach and Kennedy[108]	Loss of density-dependent increased GM_1 and GD_{1a}
Rat hepatocarcinogenesis by *N*-2-fluorenylacetamide	Morré and associates[160–162]	Inhibition of disialoganglioside pathway, enhancement of monosialoganglioside pathway
Rat hepatocyte, hepatoma HTC, and H-35 cells	Baumann et al.[163]	Induction of fucoganglioside and neutral fucolipids
F. Human cancer		
Human leukocytes and leukemic leukocytes	Hildebrand et al.[164,165]	Deletion of higher gangliosides
Human gastrointestinal tumor	Hakomori and associates[166,167,168,181,238,239]	Accumulation of various fucolipids including di- or trifucosylated type 2 chain glycolipids and fucosylceramide
Human brain tumor	Seifert and Uhlenbruck[169] Kostic and Buchheit[170] Kanazawa and Yamakawa[174]	Accumulation of GM_3, GD_3, CDH, CMH, and deletion of higher gangliosides
Human lung tumor	Gasa and Makita[172] Yoda et al.[173]	Enhanced Forssman antigen
Human seminoma	Ishizuka and Yamakawa[175]	Deletion of seminolipids
Human kidney cancer	Karlsson et al.[171]	Deletion of higher glycolipids

cell surface, as described in this chapter, have originated from observations on transformation-dependent glycolipid changes.

4.5.1. Common Features of Glycolipid Changes

The following important features of glycolipid changes associated with oncogenic transformations have been noticed: (1) Such changes can be observed in a large variety of transformed cells caused by tumor viruses and chemical carcinogens and also in spontaneous tumors including various human cancers (Table VII). Namely, the change must be the most common membrane phenotype associated with oncogenic transformation. (2) The variety of changes that affect glycolipid composition and metabolism is considerable, depending on the combination of causative agents and the type of cells, but the two kinds of changes described below are basically observable.

One type of change is the blocked synthesis of complex glycolipids associated with accumulation of their precursors.[132–158] The second type of change is the induction of new glycolipid synthesis that is absent in progenitor cells.[181–188] This includes the synthesis of Forssman glycolipid antigen in tumors derived from Forssman-negative tissue,[183,187] A-like antigen in tumors derived from blood group O or B individuals,[166,184,188] the appearance of P_1 antigen in the tumor derived from the p individual (Mrs. J., with genotype *pp*)[185,186] fucosylceramide in human cancers,[181] and the induction of various fucolipids and fucogangliosides in rat hepatomas.[163,182] The accumulation of fucolipids with Le^x determinant and those with di- and poly-

Table VII. Various Types of Glycolipid Changes Associated with Oncogenic Transformation

A. Block of synthesis

Type 1.   Decrease or deletion of GM_3, GD_3; LacCer or GlcCer increased at later stage e.g., BHKpy,[132,133] STV/STVSV40,[138] CEF/RSV,[140] BHK/RSV[133]

Type 2.   Decrease or deletion of GT, GD_{1a}, GD_{1b}, GM_1, GM_3, GM_2 (increase of GM_3, GM_2 often not obvious) (e.g., 3T3sw/SV, 3T3SVpy[134]; ALN/SV[134]; hamster C12TSVS-R/C12TSV3[137]; human fibroblast 8166/SV[55]; 3T3/py, 3T3/SV-A26; SVCE56[139]; rat hepatoma[152–158]

Type 3.   Accumulation of "asialo core" of ganglioside that is normally absent (e.g., 3T3/KiMSV,[176] NIL/py,[177] mouse lymphoma L5178Y,[178] human acute lymphatic leukemia[179]

Type 4.   Decrease or deletion of $GbOse_3Cer$, $GBOse_4Cer$, $GbOse_5Cer$, and other neutral glycolipids (e.g., NIL/py[56,136]; A9HT, B82HT[180]

B. Induction of Synthesis

Type 5.   Increase or induction of GD_{1a} or GD_{1b}, GT; deletion of GM_3 (e.g., 3T3/SV-A26,[139] rat embryo cells/3-methylcholanthrene[159]

Type 6.   Induced synthesis of fucoganglioside (e.g., fucosylceramide synthesis in human colonic carcinoma[181]; fucoganglioside and other fucolipid induction in 2-*N*-fluorenylacetamide-induced hepatoma[182]; hepatoma HTC, H-35 cells[163]

Type 7.   New synthesis of $GbOse_5Cer$ in F^- host[183]

Type 8.   New synthesis of incompatible blood group antigens foreign to the host (e.g., A-like antigen in tumors of host blood group B or O[166,184,188,264] or PP_1 antigen in the pp individual[185,186,265] fucosylated structures (poly-Le^x or poly-X) has recently been found in a number of human adenocarcinoma.[238,239] These glycolipid changes will be discussed as glycolipid tumor antigens in Chapter 5. These chemical changes are listed in Table VII, in which types 1–4 are blocked synthesis and types 5–8 are induction of new synthesis of glycolipids. Typical glycolipid changes associated with oncogenic transformation are shown in Fig. 14. Both the incomplete synthesis that results in accumulation of the precursor and the induction of new glycolipid synthesis may lead to accumulation of glycolipids that are absent in progenitor cells or normal tissues. These can be recognized as tumor-associated antigens or tumor-distinctive cell-surface markers.

The major findings up to 1975–1977 were previously reviewed.[189–194] Recently, the following lines of work have been developed: (1) studies with temperature-sensitive *(ts)* mutants of oncogenic viruses or with tumor cells showing temperature-sensitive growth behavior; (2) comparison of tumor cells with different degrees of tumorigenicity and metastatic properties; (3) studies of glycosphingolipid changes with chemical carcinogens, with particular emphasis on the precancerous state; (4) studies on glycosphingolipids of various types of human cancer. These are described in the following sections.

4.5.2. Studies with Temperature-Sensitive Mutants

The type 1 change of glycolipids (Table VII) was observed in chick embryo fibroblasts (CEF) transformed with *ts* mutants of avian sarcoma viruses LA23, LA24, and LA25 at permissive temperature (35°C).[148] The type 1 change with accumulation of LacCer in BHK cells transformed with polyoma virus Ts4 was remarkable at the permissive temperature (32°C), and the change disappeared at the nonpermissive temperature (41°C).[146] Similarly, the level of lactosylceramide of the mice fibroblasts infected with the *ts* mutant of adenovirus at the permissive temperature was much higher than at the nonpermissive temperature, although the level of gangliosides was unchanged.[149] A *ts* mutant of BHK cells transformed with dimethylnitrosamine exhibits transformed behavior at 38.5°C, but not at 32°C. Synthesis of all neutral glycolipids was unimpaired at 32°C, but was impaired at 38°C; the observed change was similar, but not identical, to type 3 in Table VII.[151] All these studies indicate that membrane transformation can result from a functional mutation affecting glycolipid synthesis. The results with CEF/LA23-25 (see Table VIII) indicated that GM$_3$ synthesis could in some way be controlled through the *src* gene, if not directly.

4.5.3. Comparison of Tumor Cells with Different Degrees of Tumorigenicity, Malignancy, and Metastatic Capability

Glycolipids of low-tumorigenic mouse L-cell lines resistant to 8-azaguanine (A9) or bromodeoxyuridine (B82) and their high-malignant isolates (A9HT or B82HT), obtained through *in vivo* passage, and their various hybrids with high or low tumorigenicity have been compared.[180] The highly tumorigenic L-cell sublines were characterized by the absence or decrease of the long-

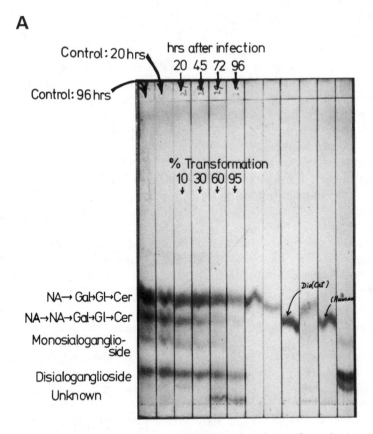

A

Control: 20 hrs

hrs after infection
20 45 72 96

Control: 96 hrs

% Transformation
10 30 60 95

NA→ Gal→Gl→Cer

NA→NA→Gal→Gl→Cer

Monosialoganglio-
side

Disialoganglioside

Unknown

Did(Cat)

(Human)

Figure 14. Typical changes of glycolipids associated with oncogenic transformation as revealed by thin-layer chromatography (TLC). A, Ganglioside changes of CEF by Rous sarcoma virus (RSV). Primary culture of CEF was infected with RSV, and aliquots were extracted at different times as indicated. Ganglioside patterns were compared at 20, 45, 72, and 96 hr after infection. Almost complete transformation was achieved at 96 hr. The approximate rate of transformation is indicated (10, 30, 60, and 95% of the total population was transformed at 20, 45, 72 and 96 hr after infection). Note that GM_3 and GD_3 were drastically decreased and almost completely disappeared at 96 hr. The unknown slow-migrating component appeared at the stage of complete transformation. This is a typical type 1 change of transformation. (Data from Ref. 140.) B, Autoradiogram of TLC pattern of glycolipids isolated from 3T3 and 3T3 cells transformed by MSV of Kirsten strain (3T3KiMSV). a, Ganglioside fraction of 3T3 cells; b, ganglioside fraction of 3T3KiMSV cells; c, neutral glycolipid fraction of 3T3 cells; d, neutral glycolipid fraction of 3T3KiMSV cells. Note the deletion of Bands 4 (GD_{1a}) and 3 (GM_1) and increase of Bands 2 (GM_2) and 8 (asialo GM_2) ($GgOse_3Cer$) in KiMSV cells. (Data from Ref. 176.) C, Fucolipid changes in two hepatomas as compared to rat liver. Fucolipids were metabolically labeled by [³H]fucose, and neutral fucolipid fractions were isolated from rat liver hepatoma H35 and HTC cells and separated by TLC followed by autoradiography. Lanes: 1, ³H-Labeled globoside; 2, neutral fucolipids from normal rat liver; 3, neutral fucolipids of H35 hepatoma cells; 4, neutral fucolipids from hepatoma HTC cells. (Data from Ref. 163.) A, B, Developed with chloroform–methanol–water–concentrated ammonia (60 : 35 : 7 : 1); C, developed with chloroform–methanol–water (65 : 25 : 4).

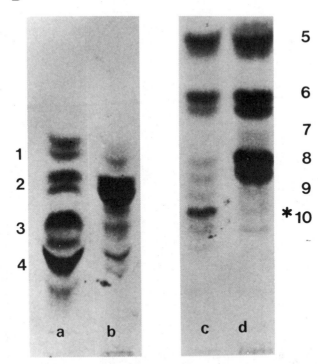

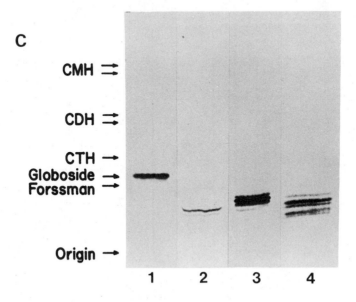

Table VIII. Glycolipids of Chick Embryo Fibroblasts Infected with Temperature-Sensitive Mutants of Rous Sarcoma Virus[a]

		Virus and temperatures[b]									
		Control CEF		wt PR-A		LA334		LA24		LA25	
Glycolipid[c]		35°C	41°C	35°C	41°C	35°C	41°C	35°C	41°C	35°C	41°C
		Chemical quantity (μ/10 mg protein)									
Hematoside	SL-1	25.5	36.8	8.5	12.0	7.5	19.5	8.0	45.2	7.8	20.1
Gangliosides	SL-2	8.3	8.5	2.0	2.5	3.2	3.1	2.0	12.0	2.2	3.5
	SL-3	15.2	10.2	1.5	1.0	3.0	2.5	3.0	2.5	3.0	3.2
	SL-4	ND	ND	2.0	2.0	3.2	2.0	1.5	2.0	1.0	1.2
		Radioactivity (cpm/10 μg protein)[d]									
Hematoside	SL-1	2580	2850	230	280	250	2780	350	3010	380	1500
Gangliosides	SL-2	320	450	200	150	580	1150	200	1500	210	220
	SL-3	1850	2580	180	130	330	620	300	250	260	450
	SL-4	2520	2660	200	350	500	520	150	330	160	120

[a] Data from Ref. 148.
[b] ND, Not determined.
[c] SL, Sialosyl lipid.
[d] Cells were metabolically labeled with [^{14}C]glucosamine. Numbers express the activity incorporated into each SL.

Table IX. *Glycolipid Composition of Normal Rat Liver and of Two Forms of Rat Ascites Hepatomas*[a]

	Normal rat liver		Ascites hepatoma			
			Island form		Free form	
Glycolipid	Adult	Neonatal	AH130	AH7974	AH130 FN	AH7974F
	Sugar residue (μg/mg protein)					
GlcCer (CMH)	0.53	0.23	0.38	0.68	1.63	2.57
LacCer (CDH)	0.25	0.21	0.76	1.01	1.78	3.56
GbOse$_3$Cer (CTH)	—	—	0.44	0.21	—	—
GgOse$_3$Cer (asialo GM$_2$)	—	—	—	—	1.52	2.74
GgOse$_4$Cer (asialo GM$_1$)	—	—	—	—	0.96	2.47
GbOse$_4$Cer (globoside)	—	—	0.73	0.34	—	—
GM$_3$	0.19	0.12	0.58	2.26	—	—
Fucolipid	—	—	—	—	—	4.60

[a] Data from Refs. 195–198.

chain neutral glycolipid or GM$_3$ compared to low-tumorigenic parent cells. Thus, the level of GM$_3$ and the long-chain neutral glycolipid reflected tumorigenicity (Table VII, types 1 and 4 changes). The most impressive correlation between the glycolipid pattern, tumor cell growth behavior, and enzymatic activities was demonstrated with two types of ascites hepatomas[195–198]: the "island" form and the "free" form; the "island" form aggregates and displays lower tumorigenicity and metastasis as compared to the "free" form, which grows as a free, separate form and displays higher malignancy. Glycolipids and their synthesis in five sublines of each island and free form have been studied. All of the island forms contain GbOse$_3$Cer, GbOse$_4$Cer, and GM$_3$. All of the free forms were lacking in these glycolipids, but did contain GgOse$_3$Cer (asialo GM$_2$), GgOse$_4$Cer (asialo GM$_1$), and FucGgOse$_4$Cer (fucosyl asialo GM$_1$). The results clearly indicate that the highly malignant free forms lack enzymatic synthesis of GM$_3$ and of globo-series glycolipids (see Table IX). In fact, a defect of the enzymes in free form has been demonstrated.[197,198] Mouse 3T3 cells transformed with MSV (3T3KiMSV) had a high level of GgOse$_3$Cer.[176] The level and the crypticity of this glycolipid have been correlated to the metastatic property of various clones of murine sarcoma cell lines[234] (see Table X).

4.5.4. Chemical Carcinogenesis and Glycolipids

A typical type 2 change was observed in chemical carcinogen-induced hepatomas *in vitro*[152] and *in vivo*[153–157,160–162] and in methylcholanthrene-induced mammary carcinoma.[158] The change of glycolipids and their enzymatic basis have been extensively studied during the process of liver carcinogenesis with *N*-2-fluorenylacetamide. The disialosyl ganglioside pathway (GM$_3$→GD$_3$→GD$_{1b}$→GT) was greatly impaired, whereas the monosialosyl

Table X. Metastatic Capability of Various Clones of Murine Sarcoma Cell Lines as Correlated to the Quantity and the Exposure of Ganglio-N-triaosylceramide (GgOse$_3$Cer)[a]

Clone	Incidence of spontaneous lung metastasis (%)	Composition of neutral glycolipids (% of total label from [^{14}C]-Gal)			Surface labeling activity of asialo GM$_2$ (cpm/mg nonlipid residue)[b]
		CM	CD	Asialo GM$_2$	
3T3	—	74	12	14	165
MuSV85C13	4.7	35	55	10	410
KB521	37.5	12	32	56	1005
KA31	100.0	9	34	56	5615
KA31-L1-4	100.0	6	20	74	2320
K234	100.0	10	30	60	4970
K234-L1-6	100.0	10	43	47	3215

[a] From Ref. 234.
[b] The activity represents ^{3}H label after cells were treated with galactose oxidase and NaB[^{3}H]$_4$ followed by separation of asialo GM$_2$.

ganglioside pathway (GM$_3$→GM$_1$→GD$_{1a}$) was enhanced. As a result, GD$_3$, GD$_{1b}$, and GT were decreased greatly or deleted, whereas GM$_1$ and GD$_{1a}$ accumulated to a large extent.[160,161] This accumulation was obvious soon after administration of the carcinogens and in precancerous nodules. The plasma membrane of Morris hepatoma was characterized by the absence of GT and the accumulation of GM$_1$ and GD$_{1a}$. In addition, accumulation of neutral glycolipids, GgOse$_4$Cer and GbOse$_3$Cer, was remarkable in plasma membranes isolated from Morris hepatoma 5123.[155] A similar type 2 change with a large accumulation of GM$_1$ and GD (GD$_{1a}$ and GD$_{1b}$) was observed in rat hepatoma 27. Regenerating rat liver contained GT, and the general pattern was similar to that of normal liver.[156] A remarkable accumulation of these gangliosides due to the type 2 change (Table VII) may reflect the enhanced ganglioside concentration in serum of rats bearing Morris hepatoma[199] and mice and humans bearing mammary carcinoma.[200] Rat embryo cells transformed with 3-methylcholanthrene and combined 3-methylcholanthrene–Rauscher leukemia virus showed a change similar to type 2. A loss of density-dependent glycolipid response was observed in two chemically transformed cell lines.[108]

4.5.5. Studies of Human Cancer

Some of the earlier studies have already been reviewed[191,201]; therefore, only supplemental data are presented here. An increase of GlcCer and LacCer in glioma as a result of type 1 change (Table VII) was further studied in many different types of glioma and meningioma.[174] Type 3 change (Table VII) was found for gangliosides of chronic lymphocyte leukemia cells, namely, deletion of higher ganglioside and accumulation of GM$_2$.[165] The major gangliosides of leukemic polymorphonuclear leukocytes were identified as NeuAcLacOse$_4$Cer (sialosylparagloboside) and its disialosyl derivative.[202] In addition to these gangliosides, human neutrophils contain a relatively large amount of LacOse$_3$Cer and LacOse$_4$Cer.[203] The accumulation of LacOse$_3$Cer in some human gastrointestinal adenocarcinomas[99] and human melanoma[172] may partially reflect neutrophil reaction in human tumor tissue. However, the change may also result from the precursor accumulation of lacto-series glycolipids due to the blocked synthesis of blood group antigens. Impaired synthesis of blood group ABH antigen has been found in various gastrointestinal tumors.[166,204–206] A hexaosylceramide with Le^b activity appeared to be consistently present in all human adenocarcinomas, irrespective of their Lewis status.[206,207] A significant increase of LacCer was also found.[206] A remarkable accumulation of various fucolipids occurred in some gastrointestinal adenocarcinomas, and a few cases showed the accumulation of a new type of fucolipid, fucosylceramide (see Section 5.9).[181]

4.5.6. Search for Glycolipid Tumor Antigens or Cell-Surface Markers

This classical problem, once described as "cytolipin" antigens,[208,209] has been recurring in recent years with newer ideas and technology. A claim that LacCer is a tumor-specific antigen[208] has been conflicting with the presence of LacCer as a common membrane component in almost all kinds of cells and tissues, although antibodies that react with LacCer are present in sera of normal human subjects[210] as well as in human cancer and pregnant women.[211] Recently, anti-LacCer antibody titer measured by radioimmunoassay was reported to be significantly higher in various malignancies.[212] This result suggested that LacCer in the tumor altered its organization and stimulated an immune response. An increased quantity of LacCer in human cancer[174,206] was noticed, while LacCer at the normal cell surface could be cryptic. Further extensive studies are needed to determine the validity of this explanation. The NILpy tumor grown in hamsters seems to stimulate antibody production directed to paragloboside (LcnOse$_4$Cer) because tumorigenic NILpy cells contain this glycolipid, whereas low-tumorigenic NIL cells contain sialylparagloboside (NeuAcLcnOse$_4$Cer).[177] Mouse BALB/c–3T3 cells transformed with KiMSV and the tumor grown in BALB/c mice and L5178Y lymphoma cells grown in DBA mice contained high levels of GgOse$_3$Cer, whereas various tissues and organs of these mice did not contain chemically detectable amounts of the same glycolipid.[176,178] This glycolipid could be a specific marker of these tumors, although the host does not produce antibodies directed to this glycolipid while the tumors are growing. Vesicular stomatitis virus (VSV) obtained from SV40-transformed hamster cell lines acquired strong SV40-TSTA activity, causing SV40 tumor rejection. A glycolipid fraction (Folch's upper phase) prepared from VSV incorporated into liposomes was sufficiently immunogenic to suppress tumor growth.[213] Antiserum prepared to a liposome containing the polar glycolipids of SV40-transformed hamster tumor cells after being absorbed on normal hamster tissue specifically stained the SV40-transformed cells.[214] The results implied that a specific glycolipid was an SV40-induced antigen. Forssman antigen (GbOse$_5$Cer) is absent in the majority of gastrointestinal mucosae in a normal human population examined, whereas gastrointestinal tumor often contains this glycolipid.[183] On the other hand, about 20% of the population contain Forssman that disappeared in tumors derived from Forssman-positive tissue.[183] Blood group A-like antigen appeared in tumors of blood group B or O individuals.[184] Some of them have been identified as glycolipids,[166,264] and some of their structures were determined as difucosylated heptasaccharide bearing A determinants.[188] The gastric cancer case of a patient (Mrs. J.) with blood group p (genotype *pp*) contained incompatible P_1 antigen, which is now identified as a ceramide pentasaccharide.[185,186] One of the distinct antigens found in the culture medium of human malignant melanoma cell lines contained glycolipid antigen that reacted with human melanoma antibodies.[215] Recent studies with monoclonal antibodies, directed to various human cancers, revealed the presence of various glycolipid antigens associated with human cancer (see Chapter 5). Certainly a search for glycolipid specific markers for tumor cells is important for practical purposes. Such knowledge could be directly applied to immunotherapy or targeting of drugs by antibodies.[216] Glycolipid markers have obvious advantages over other chemically ill-defined antigens. Chemically well-defined monoclonal antiglycolipid antibodies have been produced[217,218]; thus, immunotherapy or immunochemotherapy directed to glycolipid markers is now close to reality (see Chapter 5).

4.5.7. Biological Significance of Tumor-Associated Glycolipid Changes

Why do tumor cells change their carbohydrate chain in glycolipids and glycoproteins? The question is closely related to the general basic function of carbohydrate chains at the cell surface. Our knowledge in this area is extremely immature and fragmentary despite great efforts paid to this question. However, various types of glycolipid changes (Table VII) and their diversity can be explained by a retrogenetic expression of membrane phenotype, since a large variety of carbohydrate changes have been observed at the early stage of ontogenesis and at the later phase of differentiation. Many glycolipid molecules can be expressed in the early embryo, suppressed with the progress of differentiation, and reappear after morphogenesis is completed; expression of some glycolipids remains suppressed after morphogenesis is completed. Those glycolipids, whose expression was suppressed, can appear on oncogenic transformation, whereas those glycolipids, whose synthesis was activated after morphogenesis was completed, can be suppressed and disappear on oncogenic transformation. The change of glycolipid patterns associated with transformation may well be related to the degree of retrogenesis and the status of suppression or repression of gene expression (see Fig. 15). Further systematic studies in relation to ontogenic changes and regulation of glycolipid expression may provide some general rule.

One trend of studies is to find a possible role of carbohydrates in cell recognition and to try to understand how aberrant carbohydrate structures, including their anachronistic expression, may cause failure of cell-to-cell recognition among cells. The absence of contact response in glycolipid synthesis[55–59,106–109] may reflect the failure of cell recognition.

An aberrant profile of cell surface glycolipids, which occurs as a conse-

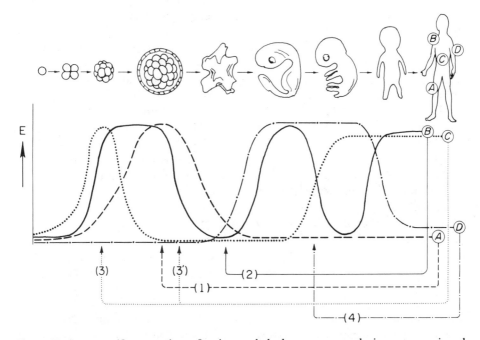

Figure 15. Stage-specific expressions of various carbohydrate structures during ontogenesis and retrogenetic expression of these structures in individual tumors. A large variety of specific carbohydrate structures bound to lipid (glycolipids) or to proteins (glycoproteins) are expressed at different stages and different loci of embryonic cells and tissues. They are maximally expressed at specific stages of ontogenesis, and they disappear, reappear during ontogenesis, or are continuously absent even after development. Each specific carbohydrate chain, including isogeneic or allogeneic antigens (e.g., blood group or heterophil antigens), behaves in the same way. Therefore, each carbohydrate chain at the cell surface in adult tissue has its own history of ontogenesis. In this figure, ontogenetic stage-specific expression of four arbitrary carbohydrate chains *(A, B, C, D)* is demonstrated. Chain A [(1)- - -] is expressed maximally at the early stage of blastocyte and disappears later and is continuously suppressed in adult tissue (this is similar to "F-9" or "SSEA-1"). Chain B [(2)—] is expressed maximally from morulae to blastocyte stage, disappears quickly and reappears in certain tissues at midstage embryo, but is suppressed when morphogenesis progresses. It may reappear after morphogenesis is complete (such a complex change can be seen in blood group A antigen). Chain C [(3,3'). . . . .] is expressed at the early embryo, is suppressed during ontogenesis and appears at the later stage of embryo, and is continuously expressed in various tissues in adult (an example of such a chain is globoside). Chain D [(4)—·—·] is expressed only at a certain stage of embryo in a specific cell and disappears after birth (such an example is deduced for i antigen). When cells are oncogenically transformed and eventually grow tumors, many carbohydrate chains display retrogenetic expression. Tumor A would display carbohydrate A through its retrogenesis (route 1). Tumor B would not display carbohydrate B because of its retrogenesis to a certain point of development where chain B was not expressed (route 2). Carbohydrate C would express continuously in tumor C if its retrogenetic process operates to the developmental stage where chain C was fully expressed (route 3), whereas it would be deleted if the retrogenetic process stops at the stage where it was suppressed (route 3'). A large variation of carbohydrate expression in various tumors can be well explained by this hypothetical scheme.

quence of blocked synthesis or neosynthesis, may cause a failure of cellular interaction through a carbohydrate recognizing protein (cognin) at the surface of counterpart cells; but cognin is unchanged upon transformation. The cognin could be a glycosyltransferase,[244,245] hydrolase,[246,247] or animal lectin.[248] Although startling discoveries of specific cell-surface receptors in liver cells for serum glycoproteins and in certain fibroblasts for lyzosomal enzymes have been made, and the mechanism for receptor-mediated endocytosis has been greatly advanced in recent years (for review see Neufeld and Ashwell),[249] the knowledge has not been applied to a possible role of carbohydrate-cognin interaction in oncogenesis. One phase of cell recognition involves cell attachment through an adhesion matrix,[250] an adhesion plaque,[251] or a detergent-insoluble matrix.[252] Recently, much attention has been focused on molecular mechanisms associated with these structures since a loss of cell adhesion is associated with oncogenic transformation, accumulation of the src gene product (P60^{src}) at the cytoplasmic loci of focal adhesion plaques,[253] and tyrosine-phosphorylation of vinculin at the adhesion site.[254] A set of membrane proteins (140K, 170K, 250K), fibronectin and cytoskeletal components, and vinculin and actinin, are cooperating in a highly organized fashion.[252,255] Although gangliosides are not considered as functional components of adhesion plaques, they could play important roles in cell adhesion since (i) gangliosides are present in substrate attachment matrixes, and in detergent insoluble matrixes,[256] and (ii) gangliosides can inhibit cell adhesions on fibronectin coated substrates,[257] and nonspecifically on other substrates.[258]

On the other hand, certain gangliosides, such as GM_3, which is ubiquitously distributed in a large variety of cells and which is deleted or reduced greatly on transformation, may have a basic function in the regulation of cell growth. Recent studies on cell growth regulation in chemically-defined medium revealed that GM_3 may regulate the function of the receptor for fibroblast growth factor.[237] Similarly, GM_1 in other types of cells may regulate receptor function for platelet-derived growth factor (PDGF) and perhaps other growth factor receptors as well (Bremer, Bowen-Pope, Ross, and Hakomori, unpublished data).

Possible involvement of gangliosides in two basic cellular functions (i) cell adhesion and recognition, and (ii) regulation and maintenance of growth factor receptor function, would immediately suggest that defective synthesis of these gangliosides may cause loss of growth control and anchorage-dependent cell proliferation, both of which are the most common denominators of oncogenesis.

4.6. Glycolipids as Possible Mediators of Immune Response

4.6.1. Modulation of B-Cell or T-Cell Response by Gangliosides

When spleen cells are incubated with sheep red blood cells (SRBC) and thymocytes, an anti-SRBC response is stimulated that can be measured by Jerne's hemolytic plaque-forming colony (PFC) assay. GM_1 and GD_{1b}, but not

GD_{1a} in lecithin–cholesterol liposomes, added in the afore described system could inhibit the PFC response. The ganglioside liposome added as late as the 4th day of culture could inhibit the PFC response, and the inhibition seemed to be specific for B cells.[219] Furthermore, a factor shed from suppressor T cells delayed the PFC response by several days. Anti-GM_1 antibodies obviated the inhibitory effect, and a GM_1-like fraction isolated from the medium of suppressor T cells inhibited the PFC response. Thus, GM_1 or GD_{1b} either present on the cell surface or released into the medium may modulate the primary immune response of B cells.[220] These results were also interpreted to support GM_1 or GD_{1a} as a specific murine T-cell marker[221] through which B and T cells could cooperate or interact with each other.[220] The B-cell modulation by gangliosides was supported by Ryan and Shinitsky,[222] who observed that gangliosides in aqueous solution could suppress the lipopolysaccharide-stimulated B-cell mitogenesis, but failed to suppress *Phaseolus* lectin-stimulated T-cell mitogenesis. A ganglioside peptide complex was, in contrast, mitogenic to B cells. On the other hand, Con A-stimulated murine T-cell mitogenesis was inhibited by gangliosides, most effectively by GT_1. In this case, Con A-stimulated DNA synthesis was inhibited, but not Con A-stimulated lactate production.[222] Ganglioside modulation of T-cell mitogenesis was also observed in human lymphocytes, stimulated by Con A, pokeweek mitogen, and *Phaseolus* lectin, and in the mixed allogeneic lymphocyte reaction.[223] These results suggested that the enhanced concentration of gangliosides in serum from tumor-bearing hosts[199,200] could be responsible for immune suppression.

4.6.2. Gangliosides as Mediators of Immune Cell Recognition

When inactivated spleen cells (mitomycin C-treated to prevent splenic cell DNA synthesis) are mixed with allogeneic thymocytes, the thymocytes respond to the antigens on the spleen cell surfaces; the thymocyte response can be measured by the enhanced uptake of [^{3}H]thymidine. On mixing syngeneic or autologous thymocytes, the response was usually not observable. However, if spleen cells were preincubated with calf or mouse brain gangliosides, a remarkable response of autologous thymocytes was observed. A greatly enhanced allogeneic thymocyte response was also noted.[224] Among gangliosides, GD_{1a} was the most effective. This provocative result suggests that ganglioside(s) on the cell surface under a certain organization may strongly mediate immune cell response, whereas a large excess of gangliosides in the medium may block the response.

4.6.3. Glycolipids as Lymphokine Receptors

Antigen- or mitogen-stimulated lymphocytes release a variety of soluble mediators known as lymphokines that affect the functional properties of macrophages, neutrophils, and other cells. One of the most extensively studied lymphokines is the "macrophage migration-inhibitory factor" (MIF). There is some evidence that the initial interaction involves the binding of MIF on the macrophage receptor. MIF may be related, if not identical, to "macrophage-activation factor" (MAF), which renders macrophages cytotoxic for tumor cells (e.g., Pick[225] and Higgins *et al.*[226]). The elimination of MIF activity can be effected by a ganglioside–Sepharose column or by liposomes containing gangliosides. The elimination was not effected by the purified known gangliosides such as GM_1, GD_{1a}, GM_3, GD_{1b}, and GT_{1b}. On the other hand, addition of L-fucose or treatment of macrophages with fucosidase eliminated the MIF; therefore, the fucosyl residue must be important for the MIF receptor on macrophages.[226,227] Of various oligosaccharides both synthetic and natural, only α-L-fucosyl oligosaccharide inhibited MIF activity.[228] A fucose-binding lectin such as the lectins of *Lotus tetragonolobus* and *Ulex europeus* agglutinin I, but not other lectins, interfered with the response of macrophages to MIF.[228] Treatment of guinea pig macrophages with liposomes containing glycolipid extracted from guinea pig macrophages greatly enhanced their responsiveness to MIF. In contrast, incubation of macrophages with liposomes containing glycolipid isolated from guinea pig neutrophils or brain tissue had no effect.[226] Recently, a new fucolipid with the structure $Gal\alpha 1 \rightarrow 3(Fuc\alpha 1 \rightarrow 2)Gal\beta 1 \rightarrow 3GalNAc\beta 1 \rightarrow 3Gal\beta 1 \rightarrow 4Glc \rightarrow Cer$ was isolated from rat granuloma and peritoneal macrophages. This glycolipid inhibited MIF activity; however, a fucolipid with lactoglycosyl structure did not inhibit the activity. The results indicate an important point: that MIF interacts with the fucosylated ganglio-series structure $(Fuc \rightarrow Gal\beta 1 \rightarrow 3GalNAc\beta 1 \rightarrow R)$, but not with the fucosylated lacto-series structure $(Fuc \rightarrow Gal\beta 1 \rightarrow 4GlcNAc\beta 1 \rightarrow R)$, although it may not distinguish the variation of fucosylated linkage whether $1 \rightarrow 2$, $1 \rightarrow 3$, or $1 \rightarrow 6$.[229] All of these findings suggest that MIF will bind to a specific ganglio-series structure that may well be a fucoganglioside, or to its analog, which could be present in glycoprotein, because MIF interaction can also be destroyed or greatly diminished by protease treatment of macrophages.

4.7. Postscript: Enigmas Concerning Glycolipid Functions

Despite all of the positive functions of glycolipids in cellular interaction, differentiation, growth control, and oncogenesis that have been described in this chapter, the author could not close without referring to a few enigmas concerning glycolipid functions. One is the existence of normal individuals who, by some genetic alteration, lack major glycolipids or glycoproteins in specific cells, tissues, or organs. For example, the absence of globoside in erythrocytes of blood group p individuals,[230] the total absence of ABH determinants in the blood group Bombay population,[231] the essential absence of a branched I structure in blood group i individuals,[51] and the absence of glycophorin in En^{a-} erythrocytes[232] have been well documented. These individuals have no recorded symptoms and their life is normal, although their susceptibility to diseases is not known. Obviously, those carbohydrate chains the absence of which does not threaten the function of cells could be regarded as nonessential. Another feature of some glycolipids is their well-documented species differences[13,233]; i.e., the major glycolipid of erythrocytes is clearly distinctive for each species, yet the functions of erythrocytes do not differ among different species. This again suggests that the major glycolipid and probably the major carbohydrate chain in glycoprotein of erythrocytes may have no specific function.

Two possibilities are considered: (1) Some glycolipids such as ganglioside or simple glycolipid (e.g., LacCer, GalCer, GlcCer) that are present in essentially all animal cells may have basic functions. Gangliosides the deficiency of which leads to an aberrant cell social behavior such as malignancy have an obvious function, as discussed in this chapter. However, many neutral glycolipids with complex carbohydrate chains show extensive variation among species and even among individuals, and these carbohydrate chains may have no essential or vital function although they are important allo- or isoantigens. Nature provides a number of nonessential structures that have been generated through a history of selection and mutation during phylogenetic development. These luxury structures are analogous to the great variety of bacterial polysaccharide structures that are important strain-specific or group-specific antigens, but that may have not vital function to bacteria. (2) Many of these "luxury" glycolipids that have no apparent function after development could have some important function during ontogenetic development, as discussed in this chapter. Theoretically, their synthesis should have been turned off after cells or tissues were differentiated. For some unknown reason, their synthesis is kept continuously on.

Another enigma concerns the presence of the same carbohydrate sequence in both glycolipids and glycoproteins, although the majority of the carbohydrate structure of glycolipids is essentially different from that of glycoproteins. Obviously, glycolipids are better suited than glycoproteins to transducing signals intracellularly, since they are directly located at the lipid bilayer; on the other hand, glycoprotein carbohydrates are better suited than glycolipids to reacting with outside signals and ligands. It is natural to imagine the two-step interaction from glycoprotein to glycolipid, as has been discussed.

4.8. References

1. Yamakawa, T., and Suzuki, S. *J. Biochem (Tokyo)* **38,** 199 (1951).
2. Klenk, E., and Lauenstein, K. *Z. Physiol. Chem.* **288,** 220 (1951).
3. Dod, B. J., and Gray, G. M. *Biochim. Biophys. Acta 150,* 397 (1968).
4. Weinstein, D. B., Marsh, J. B., Glick, M. C., and Warren, L. *J. Biol. Chem.* **245,** 3928 (1970).
5. Keenan, T. W., Huang, C. M., and Morré, D. J. *Biochem. Biophys. Res. Commun.* **47,** 1277 (1972).
6. Critchley, D. R., Graham, J. M., and Macpherson, I. *FEBS Lett.* **32,** 37 (1973).
7. Klenk, H. D., and Choppin, P. W. *Proc. Natl. Acad. Sci. U. S. A.* **66,** 57 (1970).
8. Renkonen, O., Gahmberg, C. G., Simons, K., and Kaarianen, L. *Acta Chem. Scand.* **24,** 733 (1970).
9. Gahmberg, C. G., and Hakomori, S. *J. Biol. Chem.* **248,** 4311 (1973).
10. Steck, T. L., and Dawson, G. *J. Biol. Chem.* **249,** 2135 (1974).
11. Abrahamsson, S., Dahlén, B., Löfgren, H., Pascher, I., and Sundell, S. In: *Structure of Biological Membranes* (Abrahamsson, S., and Pascher, I, eds.), Plenum Press, New York and London (1977), pp. 1–23.
12. Sharom, F. J., and Grant, C. W. M. *J. Supramol. Struct.* **6,** 249 (1977).

13. Yamakawa, T., and Nagai, Y. *Trends Biochem. Sci.* **3,** 128 (1978).
14. Kinsky, S. C. *Biochim. Biophys. Acta* **265,** 1 (1972).
15. Watanabe, K., Powell, M. E., and Hakomori, S. *J. Biol. Chem.* **253,** 8962 (1978).
16. Watanabe, K., Powell, M. E., and Hakomori, S. *J. Biol. Chem.* **254,** 8223 (1979).
17. Watanabe, K., and Hakomori, S. *Biochemistry* **24,** 5502 (1979).
18. Koscielak, J., Miller-Podraza, H., Krauze, R., and Piasek, A. *Eur. J. Biochem.* **71,** 9 (1976).
19. Koscielak, J., Zdaebska, E., Wilcynska, Z., Miller-Podraza, H., and Dzierzkowa-Borodej, W. *Eur. J. Biochem.* **96,** 331 (1979).
20. Marcus, D. M., and Schwarting, G. A. *Adv. Immunol.* **23,** 203 (1976).
21. Hakomori, S., and Young, W. W., Jr. *Scand. J. Immunol.* **6,** 97 (1978).
22. Gahmberg, C. G., and Hakomori, S. *Biochem. Biophys. Res. Commun.* **59,** 283 (1974).
23. Gahmberg, C. G., and Hakomori, S. *J. Biol. Chem.* **250,** 2438 (1975).
24. Lingwood, C. A., and Hakomori, S. *Exp. Cell Res.* **108,** 385 (1977).
25. Kishimoto, Y., Yahara, S., and Podulso, J. In: *Cell Surface Glycolipids* (Sweeley, C. C., ed.), American Chemical Society Monograph, American Chemical Society, Washington, D.C. (1980), pp. 10–23.
26. Karlsson K.-A. In: *Structure of Biological Membranes* (Abrahamsson, S., and Pascher, I., eds.), Plenum Press, New York (1976), pp. 245–274.
27. Svennerholm, L. In: *Advances in Experimental Biology and Medicine,* Vol. **125,** *Structure and Function of Gangliosides* (Svennerholm, L., Mandel, P., Dreyfus, H., and Urban, P.-F., eds.), Plenum Press, New York (1980), pp. 533–544.
28. Caputto, R., Maccioni, A. H. R., and Caputto, H. L. *Biochem. Biophys. Res. Commun.* **74,** 1046 (1977).
29. Partington, C. R., and Daly, J. W. *Mol. Pharmacol.* **15,** 484 (1979).
30. Hu, V. W., and Wisnieski, B. J. *Proc. Natl. Acad. Sci. U. S. A.* **76,** 5460 (1979).
31. Lingwood, C. A., Hakomori, S., and Ji, T. H. *FEBS Lett.* **112,** 265 (1980).
32. Watanabe, K., Hakomori, S., Powell, M. E., and Yokota, M. *Biochem. Biophys. Res. Commun.* **92,** 638 (1980).
33. Marcus, D. M., and Cass, L. *Science* **164,** 553 (1969).
34. Laine, R. A., and Hakomori, S. *Biochem. Biophys. Res. Commun.* **54,** 1039 (1973).
35. Keenan, T. W., Schmid, E., Franke, W. W., and Wiegandt, H. *Exp. Cell Res.* **92,** 259 (1975).
36. Krishnaraj, R., Saat, Y. A., and Kemp, R. G. *Cancer Res.* **40,** 2808 (1980).
37. Sharom, F. J., and Grant, C. W. M. *Biochem. Biophys. Res. Commun.* **74,** 1039 (1977).
38. Sharom, F. J., and Grant, C. W. M. *Biochim. Biophys. Acta* **507,** 280 (1978).
39. Lee, P. M., Ketis, N. V., Barber, K. R., and Grant, C. W. M. *Biochim. Biophys. Acta* **601,** 302 (1980).
40. Uchida, T., Nagai, Y., Kawasaki, Y., and Wakayama, M. *Biochemistry* **20,** 162 (1981).
41. Barenholz, Y., Ceastarp, B., Lichtenberg, D., Freire, E., Thompson, T. E., and Gott, S. In: *Advances in Experimental Medicine and Biology,* Vol. 125, *Structure and Function of Gangliosides* (Svennerholm, L, Mandel, P., Dreyfus, H., and Urban, P. F., eds.), Plenum Press, New York (1980), pp. 105–135.
42. Révész, T., and Greaves, M. J. *Nature (London)* **257,** 103 (1975).
43. Sedlacek, H. H., Stärk, J., Seiler, F. R., Ziegler, W., and Wiegandt, H. *FEBS Lett.* **61,** 272 (1976).
44. Sela, B.-A., Raz, A., and Geiger, B. *Eur. J. Immunol.* **8,** 268 (1978).
45. Bretscher, M. S. *Nature (London)* **260,** 21 (1976).
46. Tonegawa, Y., and Hakomori, S. *Biochem. Biophys. Res. Commun.* **76,** 9 (1977).
47. Hakomori, S. *Semin. Hematol.* **18,** 39 (1981).
48. Watanabe, K., Laine, R. A., and Hakomori, S. *Biochemistry* **14,** 2725 (1975).
49. Hakomori, S., Stellner, K., and Watanabe, K. *Biochem. Biophys. Res. Commun.* **49,** 1061 (1972).
50. Hakomori, S., Watanabe, K., and Laine, R. A. *Pure Appl. Chem.* **49,** 1215 (1977).
51. Fukuda, M., Fukuda, M. N., and Hakomori, S. *J. Biol. Chem.* **254,** 3700 (1979).
52. Fukuda, M. N., Fukuda, M., and Hakomori, S. *J. Biol. Chem.* **254,** 5458 (1979).
53. Finne, J., Krusius, T., Rauvala, H., and Hemminski, K. *Eur. J. Biochem.* **77,** 319 (1977).
54. Järnefelt J., Finne, J., Krusius, T., and Rauvala, H. *Trends Biochem. Sci.* **3,** 110 (1978).
55. Hakomori, S. *Proc. Natl. Acad. Sci. U. S. A.* **67,** 1741 (1970).

56. Sakiyama, H., Gross, S. K., and Robbins, P. W. *Proc. Natl. Acad. Sci. U. S. A.* **69,** 872 (1972).
57. Critchley, D. R., and Macpherson, I. A. *Biochim. Biophys. Acta* **296,** 145 (1973).
58. Yogeeswaran, G., and Hakomori, S. *Biochemistry* **14,** 2151 (1975).
59. Mandel, P., Dreyfus, H., Yusufi, A. N. K., Sarlieve, L., Robert, J., Neskovic, N., Harth, S., and Rebel, G. In: *Advances in Experimental Medicine and Biology,* Vol. 125, *Structure and Function of Gangliosides* (Svennerholm, L., Mandel, P., Dreyfus, H., and Urban P.-F., eds.), Plenum Press, New York (1980), pp. 515–531.
60. Marchase, R. B. *J. Cell Biol.* **75,** 237 (1977).
61. Piearce, M. Department of Biology, University of Pennsylvania, Philadelphia, Ph.D. thesis, (1980, Mentor, S. Roth).
62. Obata, K., Oide, M., and Handa, S. *Nature (London)* **266,** 369 (1977).
63. Steinberg, M. S. *Science* **141,** 401 (1963).
64. Barondes, S. H., and Rosen, S. D. In: *Neuronal Recognition* Barondes, S.H., ed. Plenum Press, New York (1976), pp. 331–358; Barondes, S. H. *Ann. Rev. Biochem.* **50,** 207 (1981).
65. Roseman, S. *Chem. Phys. Lipids* **5,** 270 (1970).
66. Marchase, R. B., Vosbeck, K., and Roth, S. *Biochim. Biophys. Acta* **457,** 385 (1976).
67. Keenan, T. W., and Morré, D. J. *FEBS Lett.* **55,** 8 (1975).
68. Huang, R. T. C. *Nature (London)* **276,** 624 (1978).
69. Frazier, W., and Glazer, L. *Annu. Rev. Biochem.* **48,** 491 (1979).
70. Schengrund, C. L., Rosenberg, A., and Repman, M. A. *J. Cell Biol.* **70,** 555 (1976).
71. Von Figura, K., and Voss, B. *Exp. Cell Res.* **121,** 267 (1979).
72. Rauvala, H., Carter, W. G., and Hakomori, S. *J. Cell Biol.* **88,** 127 (1981).
73. Carter, W. G., Rauvala, H., and Hakomori, S. *J. Cell Biol.* **88,** 138 (1981).
74. Rauvala, H., and Hakomori, S. *J. Cell Biol.* **88,** 149 (1981).
75. Hakomori, S., Young, W. W., Jr., Patt, L. M., Yoshino, T., Halfpap, L., and Lingwood, C. A. In: *Advances in Experimental Medicine and Biology,* Vol. 125, *Structure and Function of Gangliosides* (Svennerholm, L., Mandel, P., Dreyfus, H., and Urban, P.-F., eds.), Plenum Press, New York (1980), pp. 247–261.
76. Carter, W. G., Fukuda, M., Lingwood, C. A., and Hakomori, S. *Ann. N. Y. Acad. Sci.* **312,** 160 (1978).
77. Hood, L., Huang, H. V., and Dryer, W. J. *J. Supramol. Struct.* **7,** 531 (1977).
78. Szulman, A. E. *J. Exp. Med.* **111,** 785 (1960).
79. Willison, K. R., and Stern, P. L. *Cell* **14,** 785 (1978).
80. Stern, P. L., Willison, N. R., Lennox, E., Galfré, G., Milstein, C., Secher, D., and Ziegler, A. *Cell* **14,** 775 (1978).
81. Marsh, W. L. *Br. J. Haematol.* **7,** 200 (1961).
82. Kapadia, A., Feizi, T., and Evans, M. J. *Exp. Cell Res.* **131,** 185 (1981).
83. Artzt, K., DuBois, P., Bennett, D., Condamine, H., Babinet, C., and Jacob, F. *Proc. Natl. Acad. Sci. U. S. A.* **70,** 2988 (1973).
84. Muramatsu, T., Gachelin, G., Damonneville, M., Delarbre, C., and Jacob, F. *Cell* **18,** 183 (1979).
85. Larraga, V., and Edidin, M. *Proc. Natl. Acad. Sci. U. S. A.* **76,** 2912 (1979).
86. Solter, D., and Knowles, B. B. *Proc. Natl. Acad. Sci. U. S. A.* **75,** 5565 (1978).
87. Kemler, R., Babinet, C., Eisen, H., and Jacob, F. *Proc. Natl. Acad. Sci. U. S. A.* **74,** 4449 (1977).
88. Nudelman, E., Hakomori, S., Knowles, B. B., Solter, D., Nowinski, R. C., Tam, M. R., and Young, W. W., Jr. *Biochem. Biophys. Res. Commun.* **97,** 443 (1980).
89. Coulen-Morelec, M., and Buc-Caron, M. *Dev. Biol.* **83,** 278 (1981).
90. Glickman, R. M., and Bouhours, J. F. *Biochim. Biophys. Acta* **424,** 17 (1976).
91. Bouhours, J. F., and Glickman, R. M. *Biochim. Biophys. Acta* **441,** 123 (1976).
92. Hakomori, S. *Vox Sang.* **16,** 478 (1969).
93. Niemann, H., Watanabe, K., Hakomori, S., Childs, R. A., and Feizi, T. *Biochem. Biophys. Res. Commun.* **81,** 1286 (1978).
94. Watanabe, K., Hakomori, S., Childs, R. A., and Feizi, T. *J. Biol. Chem.* **254,** 3221 (1979).
95. Feizi, T., Childs, R. A., Watanabe, K., and Hakomori, S. *J. Exp. Med.* **149,** 975 (1979).
96. Maniatis, A., Papayannopoulou, T., and Bertles, J. F. *Blood* **54,** 159 (1979).

97. Fukuda, M. N., and Matsumura, G. *J. Biol. Chem.* **251,** 6218 (1976).

98. Fukuda, M. N., Watanabe, K., and Hakomori, S. *J. Biol. Chem.* **253,** 6814 (1978).

99. Watanabe, K., and Hakomori, S. *J. Exp. Med.* **144,** 644 (1976).

100. Romans, D. G., Tilley, C. A., and Dorrington, K. J. *J. Immunol.* **124,** 2807 (1980).

101. Yeung, K.-K., Moskal, J. R., Chien, J.-L., Gardner, D. A., and Basu, S. *Biochem. Biophys. Res. Commun.* **59,** 252 (1974).

102. Moskal, J. R., Gardner, D. A., and Basu, S. *Biochem. Biophys. Res. Commun.* **61,** 751 (1974).

103. Whatley, R., Ng, S. K.-C., Roger, J., McMurray, W. C., and Sanwal, B. *Biochem. Biophys. Res. Commun.* **70,** 180 (1976).

104. McKay, J. Glycolipids in myogenesis, Ph.D. thesis, Department of Biological Structure, University of Washington, Seattle (1980).

105. Schwarting, G., and Summers, A. *J. Immunol.* **124,** 1691 (1980).

106. Kijimoto, S., and Hakomori, S. *Biochem. Biophys. Res. Commun.* **44,** 557 (1971).

107. Chandrabose, K. A., Graham, J. M., and Macpherson, I. A. *Biochim. Biophys. Acta* **429,** 112 (1976).

108. Langenbach, R., and Kennedy, S. *Exp. Cell Res.* **112,** 361 (1978).

109. Sakiyama, H., Robbins, P. W. *Fed. Proc. Fed. Am. Soc. Exp. Biol.* **32,** 86 (1973).

110. Prokazova, N. V., Kocharov, S. L., Zvezdina, N. D., Buznikov, G. A., Shaposhnikova, G. I., and Bergelson, L. D. *Biokhimiya* **43,** 1805 (1978).

111. Chatterjee, S., Sweeley, C. C., and Velicer, L. F. *Biochem. Biophys. Res. Commun.* **54,** 585 (1973).

112. Chatterjee, S., Sweeley, C. C., and Velicer, L. F. *J. Biol. Chem.* **250,** 61 (1975).

113. Narishimhan, R., Hay, J. B., Greaves, M. F., and Murray, R. K. *Biochim. Biophys. Acta* **431,** 578 (1976).

114. Inouye, Y., Handa, S., and Osawa, T. *J. Biochem. (Tokyo)* **76,** 791 (1974).

115. Rosenfelder, G., Van Eijk, R. V. W., Monner, D. A., and Mühlradt, P. F. *Eur. J. Biochem.* **83,** 571 (1978).

116. Rosenfelder, G., Von Eijk, R. V. W., and Mühlradt, P. F. *Eur. J. Biochem.* **97,** 229 (1979).

117. Callies, R., Schwarzmann, G., Radsak, K., Siegert, R., and Wiegandt, H. *Eur. J. Biochem.* **80,** 425 (1977).

118. Yamada, K. M., and Olden, K. *Nature (London)* **275,** 179 (1978).

119. Hynes, R. O., Ali, I. V., Destree, A. T., Mautner, J., Perkins, M. E., Sanger, D. R., Wagner, P. D., and Smith, K. K. *Ann. N. Y. Acad. Sci.* **312,** 317 (1978).

120. Mannino, R. J., and Burger, M. M. *Nature (London)* **256,** 19 (1975).

121. Laine, R. A., Yogeeswaran, G., and Hakomori, S. *J. Biol. Chem.* **249,** 4460 (1974).

122. Lingwood, C. A., Ng, A., and Hakomori, S. *Proc. Natl. Acad. Sci. U. S. A.* **75,** 6049 (1978).

123. Ginsburg, E., Salomon, D., Sreevalson, T., and Freese, E. *Proc. Natl. Acad. Sci. U. S. A.* **70,** 2457 (1973).

124. Fishman, P. H., Simmons, J. L., Brady, R. O., and Freese, E. *Biochem. Biophys. Res. Commun.* **59,** 292 (1974).

125. Simmons, J. L., Fishman, P. H., Freese, E., and Brady, R. O. *J. Cell Biol.* **66,** 414 (1975).

126. Fishman, P. H., and Atikkan, E. E. *J. Biol. Chem.* **254,** 4342 (1979).

127. Leder, A., and Leder, P. *Cell* **5,** 319 (1975).

128. Huberman, E., Heckman, C., and Langenbach, R. *Cancer Res.* **39,** 2618 (1979).

129. Lotan, R., and Nicolson, G. L. *J. Natl. Cancer Inst.* **59,** 1717 (1977).

130. Sporn, M. B., Dunlop, N. M., Newton, D. L., and Smith, J. M. *Fed. Proc. Fed. Am. Soc. Exp. Biol.* **35,** 1332 (1976).

131. Patt, L. M., Itaya, K., and Hakomori, S. *Nature (London)* **273,** 379 (1978).

132. Hakomori, S., and Murakami, W. T. *Proc. Natl. Acad. Sci. U. S. A.* **59,** 254 (1968).

133. Hakomori, S., Teather, C., and Andrews, H. D. *Biochem. Biophys. Res. Commun.* **33,** 563 (1968).

134. Mora, P. T., Brady, R. O., Bradley, R. M., and McFarland, V. W. *Proc. Natl. Acad. Sci. U. S. A.* **63,** 1290 (1969).

135. Brady, R. O., and Mora, P. T. *Biochim. Biophys. Acta* **218,** 308 (1970).

136. Robbins, P. W., and Macpherson, I. *Nature (London)* **229,** 569 (1971).

137. Nigam, V. N., Lallier, R., and Brairlorsky, C. *J. Cell Biol.* **58,** 307 (1973).

138. Diringer, H., Strobel, G., and Koch, M. *Z. Physiol. Chem.* **353,** 1769 (1972).
139. Yogeeswaran, G., Sheinin, R., Wherrett, J., and Murray, R. K. *J. Biol. Chem.* **247,** 5146 (1972).
140. Hakomori, S., Saito, T., and Vogt, P. K. *Virology* **44,** 609 (1971).
141. Mora, P. T., Fishman, P. H., Bassin, R. H., Brady, R. O., and McFarland, V. W. *Nature (London)* **245,** 226 (1973).
142. Fishman, P. H., Brady, R. O., Bradley, R. M., Aaronson, A. S., and Todaro, G. J. *Proc. Natl. Acad. Sci. U. S. A.* **71,** 298 (1974).
143. Mora, P. T., Cumar, F. A., and Brady, R. O. *Virology* **44,** 609 (1971).
144. Den, H., Sela, B.-A., Roseman, S., and Sachs, L. *J. Biol. Chem.* **249,** 659 (1974).
145. Hammarström, S., and Bjursell, G. *FEBS Lett.* **32,** 69 (1973).
146. Gahmberg, C. G., Kiehn, D., and Hakomori, S. *Nature (London)* **248,** 413 (1974).
147. Warren, L., Critchley, D., and Macpherson, I. *Nature (London)* **235,** 275 (1972).
148. Hakomori, S., Wyke, J., and Vogt, P. K. *Virology* **76,** 485 (1977).
149. Onodera, K., Yamaguchi, N., Kuchino, T., and Aoi, J. *Proc. Natl. Acad. Sci. U. S. A.* **73,** 4090 (1976).
150. Itaya, K., and Hakomori, S. *FEBS Lett.* **66,** 65 (1976).
151. Buehler, R., and Moolten, F. *Biochem. Biophys. Res. Commun.* **67,** 91 (1975).
152. Brady, R. O., Borek, C., and Bradley, R. M. *J. Biol. Chem.* **244,** 6552 (1969).
153. Siddiqui, B., and Hakomori, S. *Cancer Res.* **30,** 2930 (1970).
154. Cheema, P., Yogeeswaran, G., Morris, M. P., and Murray, R.K. *FEBS Lett.* **11,** 181 (1970).
155. Dnistrian, A. M., Skipski, V. P., Barclay, M., and Stock, C. C. *Cancer Res.* **37,** 2182 (1977).
156. Dyatlovitskaya, E. V., Novikov, A. M., Gorkova, N. P., and Bergelson, L. D. *Eur. J. Biochem.* **63,** 357 (1976).
157. Van Hoeven, R. P., and Emmelot, P. In: *Tumor Lipids: Biochemistry and Metabolism* (Wood, E., ed.), American Oil Chemists' Society Press, Champaign, Illinois (1973), pp. 126–138.
158. Keenan, T. W., and Morré, D. J. *Science* **182,** 935 (1973).
159. Langenbach, R. *Biochim. Biophys. Acta* **388,** 231 (1975).
160. Merritt, W. D., Richardson, C. L., Keenan, T. W., and Morré, D. J. *J. Natl. Cancer Inst.* **60,** 1313 (1978).
161. Merritt, W. D., Morré, D. J., and Keenan, T. W. *J. Natl. Cancer Inst.* **60,** 1329 (1978).
162. Morré, D. J., Kloppel, T. M., Merritt, W. D., and Keenan, T. W. *J. Supramol. Struct.* **9,** 151 (1978).
163. Baumann, H., Nudelman, E., Watanabe, K., and Hakomori, S. *Cancer Res.* **39,** 2637 (1979).
164. Hildebrand, J., Stryckmans, P. A., and Stoffyn, P. *J. Lipid Res.* **12,** 361 (1971).
165. Hildebrand, J., Stryckmans, P. A., and Vanhoud, J. *Biochim. Biophys. Acta* **260,** 272 (1972).
166. Hakomori, S., Koscielak, J., Bloch, H., and Jeanloz, R. W. *J. Immunol.* **98,** 31 (1967).
167. Yang, H.-J., and Hakomori, S. *J. Biol. Chem.* **246,** 1192 (1971).
168. Hakomori, S. In: *Ciba Foundation Symposium on Fetal Antigens and Cancer.* (Bodmer, W., ed.) (1983) Ciba Foundation, London, in press.
169. Seifert, H., and Uhlenbruck, G. *Naturwissenschaften* **52,** 190 (1965).
170. Kostic, D., and Buchheit, F. *Life Sci.* **9,** 589 (1970).
171. Karlsson, K.-A., Samuelsson, B. E., Schersten, G. O., Steen, T., and Wahlquist, L. *Biochim. Biophys. Acta* **337,** 349 (1974).
172. Karlsson, K-A. In: *Structure and Function of Biological Membranes* (Abramson, S., and Pascher, I., eds.), Plenum Press, New York (1976), pp. 245–274.
173. Yoda, Y., Ishibashi, T., and Makita, A., *J. Biochem (Tokyo)* **88,** 1887 (1980).
174. Kanazawa, I., and Yamakawa, T. *Jpn. J. Exp. Med.* **44,** 379 (1974).
175. Ishizuka, I., and Yamakawa, T. *J. Biochem. (Tokyo)* **76,** 221, (1974).
176. Rosenfelder, G., Young, W. W., Jr., and Hakomori, S. *Cancer Res.* **37,** 1333 (1977).
177. Sundsmo, J., and Hakomori, S. *Biochem. Biophys. Res. Commun.* **68,** 799 (1976).
178. Young, W. W., Jr., Durdik, J. M., Urdal, D., Hakomori, S., and Henney, C. S. *J. Immunol.* **126,** 1 (1981).
179. Nakahara, K., Ohashi, T., Oda, T., Hirano, T., Kasai, M., Okumura, K., and Tada, T. *N. Engl. J. Med.* **302,** 674 (1980).
180. Itaya, K., Hakomori, S., and Klein, G. *Proc. Natl. Acad. Sci. U. S. A.* **73,** 1568 (1976).

181. Watanabe, K., Matsubara, T., and Hakomori, S. *J. Biol. Chem.* **251,** 2385 (1976).
182. Holmes, E., and Hakomori, S. *J. Biol. Chem.* **257,** 7698 (1982).
183. Hakomori, S., Wang, S.-H., and Young, W. W., Jr. *Proc. Natl. Acad. Sci. U. S. A.* **74,** 3023 (1977).
184. Häkkinen, I. *J. Natl. Cancer Inst.* **44,** 1183 (1970).
185. Levine, P., Bobbitt, O. B., Waller, R. K., and Kuhmichel, A. *Proc. Soc. Exp. Biol. Med.* **77,** 403 (1951).
186. Levine, P. *Semin. Oncol.* **5,** 28 (1978).
187. Kawanami, J. *J. Biochem. (Tokyo)* **72,** 783 (1972).
188. Breimer, M.E. *Cancer Res.* **40,** 897 (1980).
189. Hakomori, S. *Adv. Cancer Res.* **18,** 265 (1973).
190. Murray, R. K., Yogeeswaran, G., Sheinan, R., and Schimmer, B. F. In: *Tumor Lipids* (Wood, R., ed.), American Oil Chemists' Society Press, Champaign, Illinois (1973), pp. 285–302.
191. Hakomori, S. *Biochim. Biophys. Acta* **417,** 55 (1975).
192. Brady, R. O., and Fishman, P. *Biochim. Biophys. Acta* **335,** 121 (1975).
193. Richardson, C. L., Keenen, T. W., and Morré, D. J. *Biochim. Biophys. Acta* **488,** 88 (1976).
194. Critchley, D. R., and Vicker, M. G. In: *Dynamic Aspects of Cell Surface Organization* (Poste, G., and Nicolson, G.L., eds.), Elsevier/North-Holland, Amsterdam (1977), pp. 308–370.
195. Matsumoto, M., and Taki, T. *Biochem. Biophys. Res. Commun.* **71,** 472 (1975).
196. Taki, T., Hirabayashi, Y., Suzuki, Y., Matsumoto, M., and Kojima, K. *J. Biochem. (Tokyo)* **83,** 1517 (1978).
197. Hirabayashi, Y., Taki, T., Matsumoto, M., and Kojima, K. *Biochim. Biophys. Acta* **529,** 96 (1978).
198. Taki, T., Hirabayashi, Y., Ishiwata, Y., Matsumoto, M., and Kojima, K. *Biochim. Biophys. Acta* **572,** 105 (1979).
199. Skipski, V. P., Katapodis, N., Prendergast, J. S., and Stock, C. C. *Biochem. Biophys. Res. Commun.* **67,** 1122 (1975).
200. Kloppel, T. M., Keenan, T. W., Freeman, J. J., and Morré, D. J. *Proc. Natl. Acad. Sci. U. S. A.* **74,** 3011 (1977).
201. Hakomori, S. *Prog. Biochem. Pharmacol.* **10,** 167 (1975).
202. Dacremont, G., and Hildebrand, J. *Biochim. Biophys. Acta* **424,** 315 (1976).
203. Macher, B. A., and Klock, J. C. *J. Biol. Chem.* **255,** 2092 (1980).
204. Stellner, K., Hakomori, S., and Warner, G. A. *Biochim. Biophys. Res. Commun.* **55,** 439 (1973).
205. Dabelsteen, E., Roed-Peterson, B., and Pindborg, J. J. *Acta Pathol. Microbiol. Scand. Sect. A* **83,** 292 (1975).
206. Siddiqui, B., Whitehead, J., and Kim, Y. S. *J. Biol. Chem.* **253,** 2168 (1978).
207. Hakomori, S., and Andrews, H. D. *Biochim. Biophys. Acta* **202,** 225 (1970).
208. Rapport, M. M., and Graf, L. *Cancer Res.* **21,** 1225 (1961).
209. Rapport, M. M., and Graf, L. *Prog. Allergy* **13,** 271 (1961).
210. Taketomi, T. *J. Biochem. (Tokyo)* **65,** 239 (1969).
211. Tal, C., and Halperin, M. *Isr. J. Med. Sci.* **6,** 708 (1970).
212. Jóźwiak, W. and Kościelak, J. *Eur. Cancer Clin. Oncol.* **18,** 617 (1982).
213. Huet, C., and Ansel, S. *Int. J. Cancer* **20,** 61 (1977).
214. Ansel, S., and Huet, C. *Int. J. Cancer* **25,** 797 (1980).
215. Gupta, R. K., Irie, R. F., Chee, D. O., Kern, D. H., and Morton, D. L. *J. Natl. Cancer Inst.* **63,** 347 (1979).
216. Urdal, D., and Hakomori, S. *J. Biol. Chem.* **255,** 10509 (1980).
217. Young, W. W., Jr., MacDonald, E. M. S., Nowinski, R., and Hakomori, S. *J. Exp. Med.* **150,** 1008 (1979).
218. Nowinski, R., Berglund, C., Lane, J., Lostrom, M., Bernstein, I., Young, W. W., Jr., Hakomori, S., Hill, L., and Cooney, M. *Science* **210,** 537 (1980).
219. Wang, T. J., Freimuth, W. W., Miller, H. C., and Esselman, W. J. *J. Immunol.* **121,** 1361 (1978).
220. Esselman, W. J., and Miller, H. C. *J. Immunol.* **119,** 1994 (1977).
221. Esselman, W. J., and Miller, H. C. *J. Exp. Med.* **139,** 445 (1974).
222. Ryan, J. L., and Shinitsky, M. *Eur. J. Immunol.* **9,** 171 (1979).

223. Lengle, E. E., Krishnaraj, R., and Kemp, R. G. *Cancer Res.* **39,** 817 (1979).
224. Sela, B.-A. *Cell Immunol.* **49,** 196 (1980).
225. Pick, E. In: *Immunopharmacology* (Hadden, J.W., Coffey, R.G., and Spreafico, F., eds.), Plenum Press, New York (1977), pp. 163–182.
226. Higgins, T. J., Sabatino, A. P., Remold, H. G., and David, J. R. *J. Immunol.* **121,** 880 (1978).
227. Poste, G., Kirsh, R., and Fidler, I. J. *Cell. Immunol.* **44,** 71 (1979).
228. Poste, G., Allen, H., and Matta, K. L. *Cell. Immunol.* **44,** 89 (1979).
229. Miura, T., Handa, S., and Yamakawa, T. *J. Biochem. (Tokyo)* **86,** 773 (1979).
230. Marcus, D. M., Naiki, M., and Kundu, S. K. *Proc. Natl. Acad. Sci. U. S. A.* **73,** 3263 (1976).
231. Bhende, Y. M., Deshpanda, C. K., Bhatia, H. M., Sanger, R., Race, R. R., Morgan, W. T. J., and Watkins, W.M. *Lancet* **1,** 903 (1952).
232. Tanner, M. J. A., and Anstee, D. *Biochem. J.* **153,** 271 (1976).
233. Yamakawa, T. In: *Lipoide* (16 Colloquium Mosbach/Baden) (Schutte, E., ed.), Springer-Verlag, Berlin and New York (1966), pp. 87–111.
234. Yogeeswaran, G., and Stein, B. S. *J. Natl. Cancer Inst.* **65,** 1980 (1980).
235. Saito, M., Nojiri, H., and Yamada, M. *Biochem. Biophys. Res. Commun.* **97,** 452 (1980).
236. Terasima, T., and Tolmach, L. J. *Exp. Cell Res.* **30,** 344 (1963).
237. Bremer, E. G., and Hakomori, S. *Biochem. Biophys. Res. Commun.* **106,** 711 (1982).
238. Hakomori, S., Nudelman, E., Levery, S., Solter, D., and Knowles, B. B. *Biochem. Biophys. Res. Commun.* **100,** 1578 (1981).
239. Hakomori, S., Nudelman, E., Kannagi, R., and Levery, S. B. *Biochem. Biophys. Res. Commun.,* **109,** 36 (1982).
240. Kannagi, R., Nudelman, E., Levery, S. B., and Hakomori, S. *J. Biol. Chem.,* **257,** 14865 (1982).
241. Gooi, H. C., Feizi, T., Kapadia, A., Knowles, B. B., Solter, D., and Evans, J. M. *Nature* **292,** 156 (1981).
242. Hounsell, E. F., Gooi, H. C., and Feizi, T. *FEBS Lett.* **131,** 279 (1981).
243. Dabelsteen, E., Vedtofte, P., Hakomori, S., and Young, W. W. *J. Invest. Dermatol.* **79,** 3 (1982).
244. Roth, S., McGuire, E. J., and Roseman, S. *J. Cell Biol.* **51,** 536 (1971).
245. Shur, B. D. *J. Biol. Chem.* **257,** 6871 (1982).
246. Rauvala, H., Carter, W. G., and Hakomori, S. *J. Cell Biol.* **88,** 127 (1981).
247. Rauvala, H., and Hakomori, S. *J. Cell Biol.* **88,** 149 (1981).
248. Barondes, S. *Ann. Rev. Biochem.* **50,** 207 (1981).
249. Neufeld, E. F., and Ashwell, G. In: *The Biochemistry of Glycoproteins and Proteoglycans* (Lennarz, W. J., ed.), Plenum Press, New York (1980), p. 241–262.
250. Culp, L. A. *Current Topics in Membranes and Transport* **11,** 327 (1978).
251. Geiger, B. *Cell* **18,** 193 (1979).
252. Carter, W. G., and Hakomori, S. *J. Biol. Chem.* **256,** 6953 (1981).
253. Rohrschneider, L. R. *Proc. Natl. Acad. Sci. U. S. A.* **77,** 3514 (1980).
254. Sefton, B. M., Hunter, T., Ball, B. H., and Singer, S. J. *Cell* **24,** 165 (1981).
255. Chen, W., and Singer, S. J. *J. Cell Biol.* **93,** 205 (1982).
256. Bremer, E., Mungai, G., Okada, Y., and Hakomori, S. *Fed. Proc.* **41,** 1171 (1982).
257. Kleinman, H. K., Martin, G. R., and Fishman, P. H. *Proc. Natl. Acad. Sci. U. S. A.* **76,** 3367 (1979).
258. Rauvala, H., Carter, W. G., and Hakomori, S. *J. Cell Biol.* **88,** 127 (1981).
259. Kannagi, R., Levery, S., and Hakomori, S. *Proc. Natl. Aca. Sci. U. S. A.,* in press (1983).
260. Young, W. W., Jr., Portoukalian, J., and Hakomori, S. *J. Biol. Chem.* **265,** 10967 (1981).
261. Nudelman, E., Kannagi, R., Hakomori, S., Lipinski, M., Wiels, J., Parsons, M., Fellous, M., and Tursz, T. *Science,* in press (1983).
262. Shevinsky, L. H., Knowles, B. B., Damjanov, I., and Solter, D. *Cell* **30,** 697 (1982).
263. Willison, K. R., Karol, R. A., Suzuki, A., Kundu, S. K., and Marcus, D. M. *J. Immunol.* **129,** 603 (1982).
264. Yokota, M., Warner, G., and Hakomori, S. *Cancer Res.* **41,** 4185 (1981).
265. Kannagi, R., Levine, P., Watanabe, K., and Hakomori, S. *Cancer Res.* **42,** 5249 (1982).

Chapter 5

Glycolipid Antigens and Genetic Markers

Sen-itiroh Hakomori and William W. Young, Jr.

5.1. Introduction

The classical immunological phenomena described as species specificity and tissue specificity were intensively studied in the early part of this century. Some of these antigens were identified as "lipoid," namely, those extractable with ethanol or ethanol–ether mixtures, and had extremely complex chemical composition containing phosphorus, nitrogen, sugars, and other constituents.[1,2] These properties were considered to be equivalent to those of type-specific antigens of bacteria that are currently known as "lipopolysaccharides." In these classical studies, heterologous antisera were prepared against whole tissue homogenates, and specific reactions were observed between the antisera and the alcohol-soluble fraction of cells or tissues by complement-fixation tests (for reviews, see Landsteiner,[3] Day,[4] and Rapport and Graf[12]). These "alcohol-soluble antigens" were not immunogenic unless they were mixed and co-injected with carrier proteins ("Schlepper").[5] Thus, the serological concept of "lipid hapten" was well established during the Landsteiner era, although the chemical properties and structures of these haptens were far from being known.

The earliest and best-characterized lipid hapten was Pangborn's cardiolipid,[6] the phospholipid hapten of the syphilis antigen, the structure of which was established in 1942. The concept of glycolipid hapten was not developed in those early days. Perhaps the earliest but clearest demonstration of glycolipid antigen was the doctoral dissertation of Brunius[7] in 1936 on the chemical nature of the Forssman hapten, which described the hapten as a lipid–carbohydrate complex the activity of which was due to a specific carbohydrate structure containing hexose and hexosamine. In the early 1950s, it became clear that the "lipoid" fraction of erythrocytes and tissues, even after extensive fractionation with various organic solvents, contained hexoses and hexosamines and carried blood group A and B specificities.[8–10] The lipid fraction with Forssman activity as originally described by Brunius was further purified and proved to be a glycosphingolipid containing glucose, galactose, galactosamine, and ceramide.[11] During the latter half of the 1950s, a series of tissue-specific lipid haptens called "cytolipins" were separated and identified as glycosphingolipids by Rapport and associates (for a review, see Rapport and Graf.[12] The complement fixation of "cytolipin H" with its antibody was inhibited by lactose; thus, the immunodominant group of a glycosphingolipid hapten "cytolipin H" was established as residing on the carbohydrate residue.[41,210,207]

These early studies on Forssman, blood group AB, and tissue-specific "cytolipins" indicated that glycosphingolipids are important cellular antigens and that specificities depend on carbohydrates. During the past ten years, many studies have agreed with the concept that the majority (80–90%) of glycosphingolipids are present in plasma membranes. This is an important basis for understanding the function of glycolipids as cellular antigens and immunogens as well as receptors for cellular recognition (see Chapter 6). Important developments in glycolipid antigen studies are listed in Table I.

Table I. Important Developments in the Establishment of the Chemical Nature of Glycolipid Antigens and Cell Surface Markers

Demonstration of Forssman hapten as a lipocarbohydrate complex	Brunius (1936)[7]
Glycolipid fraction of erythrocytes and tissues showed blood group ABH activity	Hamasato (1950)[9]
	Tokura (1952)[8]
	Yamakawa and Iida (1953)[10]
Identification of glycolipids as haptens of cellular antigens	
Cytolipin H as a human cancer hapten	Rapport et al. (1959)[41]
Cytolipin R as a rat sarcoma hapten	Rapport et al. (1964)[32]
Galactocerebroside as the brain-specific lipid hapten	Joffe et al. (1963)[13]
Isolation of a pure blood group ABH glycolipid from human erythrocytes and its identification as a fucose-containing glycosphingolipid	Hakomori and Strycharz (1968)[14]
	Stellner et al. (1973)[57]
	Ando and Yamakawa (1973)[138]
	Koscielak et al. (1973)[130]
Identification of Le^a and Le^b glycolipids that are taken up onto erythrocytes from serium	Marcus and Cass (1969)[15]
Structures of Forssman antigen and its relation to globoside	Hakomori et al. (1971)[16]
	Siddiqui and Hakomori (1971)[17]
Isolation and characterization of P$_1$ antigen and identification of globoside and ceramide trihexoside as P and P^k antigens	Naiki and Marcus (1975)[18]
	Marcus et al. (1976)[19]
Structural characterization of Ii glycolipid antigen	Niemann et al. (1978)[20]
	Watanabe et al. (1979)[21]
	Feizi et al. (1979)[22]
Glycolipid as markers for lymphoid cell subsets GM$_1$ and asialo GM$_1$ for T-cell lineage, asialo GM$_1$ for natural killer cells	Stein-Douglas et al. (1976)[23]
	Stein et al. (1978)[24]
	Young et al. (1980)[271]
	Kasai et al. (1980)[272]
Tumor-associated glycolipid markers	Rosenfelder et al. (1977)[44]
	Sundsmo and Hakomori (1976)[43]
	Hakomori et al. (1977)[25]
	Ansell and Huet (1980)[289]
	Koprowski et al. (1981)[294]
	Magnani et al. (1981)[295]

The glycosphingolipid antigens, including the heterogenetic antigen of Forssman type, various blood group antigens, tissue-specific antigens, and tumor-distinctive antigens, are listed in Table II. Recently, antigens related to a certain type of autoimmune process have been suggested to be glycolipids. The list of antigens may certainly increase within a few years. Chemical and immunological properties of each antigen will be detailed in the following sections. Readers should also refer to the excellent review by Marcus and Schwarting.[45]

5.2. General Properties of Glycolipid Antigens

5.2.1. Unique Immunogenicity and Antigenicity of Glycolipids

Immunization of animals with glycoproteins produces antibodies specific for the protein portion, but in general the carbohydrate moiety is not immunogenic. In contrast, antibodies can be produced against the carbohydrate portion of glycolipids when animals are immunized with the glycolipid in association with a foreign protein or in a lipid bilayer or cell membrane. This difference in immunogenicity may be attributed to the fact that glycoprotein carbohydrate structures are limited to only a few basic types,[46] and the common structures shared with glycolipids (see Section 4.24) could be too flexible to be immunogenic.

Glycolipid antigens possess several unique immunochemical properties as compared to other antigens:

1. Although glycolipids associated with other lipid classes and foreign proteins are immunogenic, purified glycolipids *per se* are not immunogenic. The copresence of carrier molecules as originally described by Landsteiner and Simms[5] is required for immunogenicity. Practically, to obtain efficient antibody response, glycolipids must be either mixed with foreign proteins,[47-52] covalently linked to foreign proteins or synthetic polypeptides,[53-55] included in liposomes,[56] or presented in natural cell membranes.[31,32,43,44]

2. Individual glycolipid molecules are univalent and amphipathic. In aqueous solution, they exist as micelles that, although they represent multivalent structures, display minimal reactivity with antibodies. However, the addition of "auxiliary lipids" (most often a phospholipid and cholesterol) can greatly increase this reactivity.[12,47,48]

3. Glycolipid antigenicity is masked by other glycolipids. The antigenic activity of glycolipids often cannot be detected in the presence of a large excess of other glycolipids. While this phenomenon seems especially true for glycolipids with shorter carbohydrate chains, it has also been found to apply to blood group H glycolipid[57] and Forssman hapten.[25] For many years, no H activity could be demonstrated in the glycolipid fraction, because the activity of a relatively small quantity of H glycolipids was masked by a predominance of globoside (see Section 5.4). Similarly, Ii-antigen activity was not detectable when the antigen was present in a large excess of ganglioside,[58] and Forssman

Table II. Glycosphingolipid Antigens

Antigen	Structure identified	References
1. Heterogenetic Forssman antigen	GalNAcα1→3GalNAcβ1→3Galα1→4Galβ1→4Glc→Cer (see Table III)	17,26,27
2. Blood group antigens		
ABH antigens	See Tables IV–VI	
Lewis antigens	See Table VII	
I antigen	Galβ1→4GlcNAcβ1 ↘⁶Galβ1→4GlcNAcβ1→3Galβ1→4Glcβ1→1Cer Galβ1→4GlcNAcβ1 ↗³	21,22
i antigen	Galβ1→4GlcNAcβ1→3Galβ1→4GlcNAcβ1→3Galβ1→4Glcβ1→1Cer	20
P antigen	globoside: GalNAcβ1→3Galα1→4Galβ1→4Glc→Cer	18,19
P₁ antigen	Galα1→3Galβ1→4GlcNAcβ1→3(or 4)Galβ1→4Glcβ1→1Cer	18,19,29
Pᵏ antigen	Galα1→4Galβ1→4Glc→Cer	18,19,29
J antigen	Not identified	30
3. Tissue-specific antigens		
Myelin specific antigen	Gal→Cer	31
"Cytolipin K"	Same as globoside	32

4. Antigens related to autoimmune processes

Part of experimental allergic encephalomyelitis antigen	Gal→Cer	33,34
Paroxysmal cold hemoglobinuria	Same as globoside	35,36
Multiple sclerosis	Digalactosyl diglyceride and other unidentified	37
"Experimental ganglioside syndrome"	GD_{1a} ganglioside	38
Experimental allergic neuritis	Gal→Cer	39
A-specific ovarian glycolipid in female mice inducing anti-A antibody	A-specific glycolipids	40

5. Tumor-distinctive antigens

"Cytolipin H"	Galβ1→4Glc→Cer	41
"Cytolipin R"	GalNAcβ1→3Galα1→3Galβ1→4Glc→Cer	42
Paragloboside in NILpy tumor	Galβ1→4GlcNAcβ1→3Galβ1→4Glc→Cer	26,43
Gangliotriaosylceramide (asialo GM_2) in Kirsten tumor and lymphoma L5178	GalNAcβ1→4Galβ1→4Glc→Cer	44,292,293
Forssman antigen in F⁻ individuals	Same as above	25
Globotriaosylceramide (P^k antigen) in Burkitt lymphoma	Galα1→4Galβ1→4Glcβ1→1Cer	307
GD_3 ganglioside in human melanoma	NeuAc2→8NeuAc2→3Galβ1→4Glcβ1→1Cer	267,268
Sialosyl-Le^a in human gastric, colonic, and pancreatic cancer	See Table IX	
Di- and polyfucosylated type 2 chain glycolipids	See Table IX	

antigen activity was masked in the presence of a large amount of other lipid components.[25]

5.2.2. Immune Response to Glycolipid Antigens

Antibodies directed to glycolipids are produced by natural immune stimulation, autoimmunization, and by active immunization. Monoclonal antibody technology[64,65] enables us to study the degree of immunogenicity of each glycolipid at the cell surface.

5.2.2.1. Natural Antibodies against Carbohydrate Structures of Glycolipids

Most animals including humans possess so-called "natural antibodies," some of which react with various glycolipids. In most cases, these antibodies are believed to be produced against cross-reactive exogenous structures present in microorganisms that the animal has encountered. The strongest support for this so-called "immunogenetic hypothesis"[59] was provided by the elegant experiments of Springer *et al.*[60] (reviewed in Springer[61]). They found that whereas the majority of ordinary chicks had detectable anti-blood group B agglutinins by the age of 30 days, none could be demonstrated in germ-free chicks up to the age of 60 days. The presence of serum antibodies reactive with a given determinant has generally been interpreted as indicating that the host lacks that respective antigen. For example, normal human sera contain antibodies specific for blood group antigens of the ABH system and also for the Forssman antigen.[62] However, in contrast to this prevailing view, in female mice it has been shown that natural anti-A antibodies are produced against autoantigenic ovarian glycolipids during the onset of puberty.[40] The naturally occurring anti-carbohydrate antibodies in sera are IgM class with a few exceptions.

An additional type of spontaneous anticarbohydrate antibody is produced in patients with Waldenström's macroglobulinemia. Some of these sera contain high levels of monoclonal antibodies specific for determinants of the blood group I–i system[20–22] and sialylglycolipids.[63]

5.2.2.2. Experimental Production of Polyclonal and Monoclonal Antibodies to Glycolipids

In general, rabbits develop good antibody responses against glycolipids. In many cases, these antibodies are dominantly directed to the nonreducing terminal sugar(s). In addition, there are specificity differences depending upon the immunoglobulin class; e.g., the IgM fraction of anti-asialo GM_1 showed a great deal of cross-reaction with GM_1 ganglioside, whereas the IgG fraction contained antibodies of stricter specificity.[47]

Using the methods that produce good antiglycolipid responses in rabbits, it has been relatively difficult to raise similar responses in mice.[44] However, by adsorption of glycolipids onto an acid-treated mutant of *Salmonella minnesota*,[216] an effective immunogen has been produced that stimulates low levels of antibodies in mice against several glycolipids.[66] By fusion of spleen cells from such immunized mice with myeloma cells according to the hybridoma technique developed by Köhler and Milstein,[64,65] stable hybrid cell lines have been derived that produce high levels of monoclonal antiglycolipid antibodies.[66] With this procedure, it will be possible to produce a battery of specific, high-titer antiglycolipid antibodies that may eventually replace plant lectins as probes of cell-surface carbohydrate structure and function. Interestingly, these mouse antibodies appear to be restricted to either the IgM class or the IgG$_3$ subclass, in agreement with the subclass restriction previously defined for other mouse anticarbohydrate antibodies.[67] A number of monoclonal antibodies, directed to cell surface glycolipids, have been established and characterized when whole cells or membranes were used as immunogens; but only a few instances of successful isolation of hybridoma-producing monoclonal anti-glycolipid antibodies are known when isolated glycolipids were used as immunogens.

5.2.2.3. Production of Cellular Lymphoid Responses Specific for Glycolipids

To date, immunization protocols have not been devised for generating cytolytic T cells specific for glycolipids, even though such hapten-specific responses have been generated against other haptenic groups.[68] However, hyperimmunization of rabbits with either gangliosides[38] or galactocerebroside[39] results in an autoimmune condition that likely has a cellular cytotoxic component. Glycolipids can serve as the target for antibody-dependent cellular cytotoxicity using human peripheral blood lymphocytes (killer-cell activity) as the source of the cytotoxic cells.[69]

Recently, Sela[70] reported that mouse splenocytes incorporated with mixed brain gangliosides triggered an efficient mixed-lymphocyte reaction in autologous thymocytes. While these results suggest a possible role of glycolipids in cellular recognition, it should be noted that individual purified gangliosides were much less effective than the mixture. This raises the possibility that the stimulation may have been caused by a nonglycolipid contaminant of the ganglioside preparation, such as the amphipathic membrane proteins that have been recently described.[71] Further studies indicate that the component that induces autologous reaction could be the tetrasialo ganglioside GQ$_{1b}$.[296]

5.2.3. Methods for Detecting Antiglycolipid Antibodies

For detection of specific antibodies directed to glycolipid antigens, the methods that were established for detection of protein antigens are not always useful. Special techniques and precaution are needed that are adapted for detection of glycolipid antigens.

5.2.3.1. Complement-Fixation Reaction

The reaction requires "auxiliary lipid," namely, cholesterol and lecithin[12,45]; such complexes of glycolipids and "auxiliary lipids" are now termed

ANTIGEN TITRATION

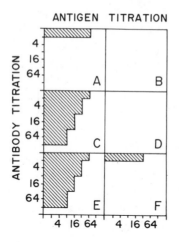

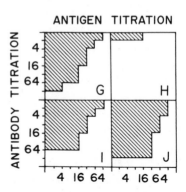

ANTIGEN TITRATION

Figure 1. *Left:* Complement-fixation patterns produced by antigloboside and various glycolipids. The determination was performed on "microtiter plates" according to the modified procedure by Lennette[73] using sodium barbiturate 6.7 mM buffer, pH 7.0, containing 0.1% gelatin, 0.15 mM calcium, and 0.5 mM magnesium. *Ordinate:* Reciprocals of antibody dilution ("dilution 1" contained 3 µg of antibody protein/well). *Abscissa:* Reciprocals of glycolipid antigen dilution ["dilution 1" contained 63 ng of glycolipid and 315 ng each of cholesterol and lecithin as auxiliary lipids (see the text)]. Antibody and antigen controls were used in each determination. After the reaction was carried out, the contents of the wells that showed inhibited hemolysis were transferred to small test tubes and centrifuged, and the optical density of the supernatant fraction was read against the content of wells that showed complete hemolysis. Thus, a series of fixation curves can be depicted. Hatched areas indicate only 100% fixation. A, Forssman glycolipid; B, ceramide lactoside; C, globoside; D, lacto-*N*-*neo*tetraosylceramide; E, cytolipin R; F, CTH. *Right:* Complement fixation of anti-*N*-glycolylhematoside antibody as tested against *N*-glycolylhematoside (G), *N*-acetylhematoside (H), *N*-glycolylhematoside methyl ester (I), and *N*-glycolyl-*O*-acetylhematoside (J). Hatched areas indicate 100% complement fixation, using sheep red cells and guinea pig complement. (From Ref. 51.)

liposomes.[72] The method has been widely used in classical studies of glycolipid antigens.[31,32,41,42] The complement-fixation test can be conveniently carried out in a microtiter plate according to the method originally described by Lennette.[73] The reaction is carried out with excess complement and therefore is only semiquantitative (Fig. 1).

The micro-complement-fixation technique of Wasserman and Levine[74] is carried out with limited complement concentration and is highly sensitive, but is also sensitive to a nonspecific anticomplementary effect of glycolipid. Nonpurified glycolipids often display an unexplainable anticomplementary effect. A typical complement-fixation curve is presented in Fig. 2.

5.2.3.2. Passive Hemagglutination and Inhibition of Passive Hemagglutination

Glycolipids have a characteristic property of being easily incorporated into erythrocyte membranes by incubating an aqueous micellar solution of glycolipid with trypsinized erythrocytes.[15] Hemagglutination of glycolipid-

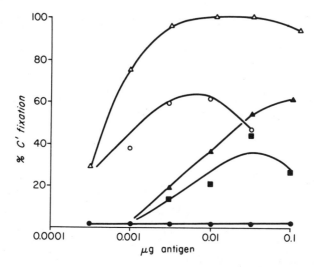

Figure 2. Micro-complement-fixation curve demonstrated between antisulfatide antiserum and sulfatide and analogs. △, Fixation curve with 1 : 50 diluted serum; anti-bovine serum albumin (BSA) was removed by BSA–cellulose; reaction with natural sulfatide. ▲, Fixation curve with the same serum as above (△); reaction with synthetic sulfatide β-galactopyranosyl(6-sulfate)ceramide. ○, Fixation curve with 1 : 200 diluted serum; anti-BSA was precipitated by BSA and centrifuged; reaction with natural sulfatide. ■, Fixation curve with the same serum as above (○); reaction with synthetic sulfatide. ●, Fixation curve with the same serum as above (○); reaction with cerebroside; the same fixation curve was obtained by cerebroside as with the 1 : 50 diluted serum (△). (From Ref. 50.)

coated red blood cells is the sensitive way to detect antiglycolipid antibodies, particularly of the IgM class. Additional information on the specificity can be further confirmed by inhibition of hemagglutination by glycolipid liposomes.[47–49] In those instances in which a glycolipid occurs naturally in a given erythrocyte, direct hemagglutination and hemagglutination inhibition can be readily carried out; e.g., anti-asialo GM_2 antibodies were detected by hemagglutination of guinea pig erythrocytes.[44]

5.2.3.3. Classical Precipitin Reaction in Agarose

This method is least sensitive but most convenient. However, it is not applicable in all cases; e.g., anti-N-acetylhematoside is not usually a precipitating antibody. Depending on the state of micellar aggregation, free glycolipids sometimes show a remarkable precipitin line. However, the reactivity is usually enhanced greatly by addition of auxiliary lipids.[47,48] A pure glycolipid sometimes gives more than two precipitin bands, and the number of precipitin lines depends on the state of micellar aggregation[51]; therefore, precipitin reactions of glycolipids may not be useful for qualitative identification of the nature of the precipitin line.

5.2.3.4. *Immune Release Assay (Complement-Dependent Liposome Lysis)*

Kinsky and associates (for a review, see Kinsky[72]) developed the immune release or liposome lysis assay. In this method, a lipid antigen (glycolipid) is incorporated into the bilayers of synthetic lipid vesicles. A variety of markers can be trapped in the aqueous spaces between the bilayers. Upon incubation with the appropriate antibody plus complement, the permeability of the lipid bilayers is altered, causing the trapped marker to be released. The trapped markers that have been utilized include glucose,[72] the fluorogenic substrate umbelliferone phosphate,[75] fluorescent compounds plus quenchers,[76] and spin-label compounds.[77] This immune release assay can be used not only to determine the specificity of antiglycolipid antibodies (Fig. 3) but also to detect the presence of glycolipid antigens in cell extracts.[25]

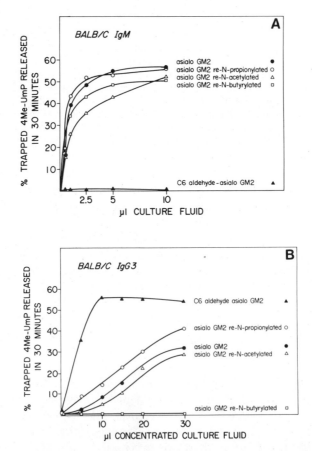

Figure 3. Effect of monoclonal BALB/c anti-asialo GM_2 on release of 4-methylumbelliferone phosphate from multicompartment glycolipid liposomes. Liposomes were prepared with the indicated structural analogs of asialo GM_2 and tested for susceptibility to antibody–complement attack using a fluorimetric assay. A, IgM antibody; B, IgG_3 antibody. See Young *et al.*[66] for assay details.

5.2.3.5. Liposome Lysis Inhibition

A lysis of multicompartment liposomes by antibodies and complement can be inhibited in the presence of single compartment liposomes. The method was found by Uemura et al.[298] to be sensitive and stoichiometric. This can overcome some drawbacks inherent to direct liposome lysis assay such as variability in liposomes and background value. The method was successfully applied in analysis of the GD_3 ganglioside of melanoma.[299]

5.2.3.6. Liposome Agglutination and Radioimmunoprecipitation Tests

Fry et al.[78] described two assays for detecting antibody to galactocerebroside. One utilized [^{3}H]cholesterol incorporated into glycolipid liposomes in a radioimmunoprecipitation test. The other involved simple visual detection of agglutination of glycolipid liposomes in the presence of appropriate antibody. Hamers et al.[79] quantitated this agglutination reaction by monitoring light-scattering changes at 600 nm; with this assay, it is possible to determine the optimal antibody/antigen ratio in 1–2 hr.

5.2.3.7. Radioimmunoassay with Polyacrylic Glycolipid Polymers

Although radioimmunoassays (RIAs) are commonly used for detecting many substances, this method had been difficult to apply to glycolipid antigens primarily because of mixed micelle formation between the radiolabled glycolipid antigen and the unlabeled antigen pool. However, we have recently developed an RIA for measuring picomolar quantities of Forssman antigen.[80] This assay is based on competition for rabbit anti-Forssman antibodies between Forssman glycolipid and a radiolabeled Forssman polyacrylic hydrazide polymer (Fig. 4). The competition curve is shown in Fig. 5.

Glycolipid-containing liposomes can also be used to inhibit the reaction of a radioactive glycoprotein and its antiserum. The presence of blood group Ii-active ganglioside was successfully demonstrated by such an RIA.[20–22,58]

5.2.3.8. Iodinated Protein A Plate Binding Assay (Solid Phase Radioimmunoassay)

Holmgren[81] demonstrated that when an aqueous suspension of glycolipid is incubated in a plastic tube, a small portion of the glycolipid becomes adsorbed to the plastic surface. We utilized this adsorption to screen for antiglycolipid hybridoma antibodies as shown in Fig. 6.[66] We have improved the procedure by using a soft vinyl strip. A lipid solution is prepared by dissolving 1 μg of glycolipid antigen, 5 μg of lecithin, and 3 μg of cholesterol in 1 ml of ethanol. Ten–15 μl of the solution is added to each well of the vinyl strip and ethanol is allowed to evaporate at room temperature. The glycolipid–phospholipid–cholesterol film is now treated by the same procedure as previously described.[66] The method is useful and sensitive for detection of small amounts of antigen and antibody (Kannagi, R., Cochran, N., and Hakomori, S., unpublished observations).

Forssman glycolipid

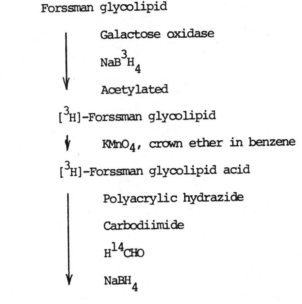

$$[^{3}H]GalNAc\alpha1{\rightarrow}3GalNAc\beta1{\rightarrow}3Gal\alpha1{\rightarrow}4Gal\beta1{\rightarrow}4Glu{\rightarrow}OCH_2$$

Figure 4. Synthesis of Forssman polyacrylic hydrazide. (From Ref. 80.)

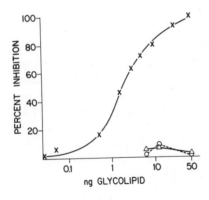

Figure 5. Inhibition of precipitation of $^{14}C/^{3}H$-labeled Forssman polyacrylic hydrazide by unlabeled glycolipid antigens. Lipid vesicles of the indicated composition were incubated with anti-Forssman IgG (50 μl, 1 : 5000 dilution) for 90 min at 4°C in a total volume of 100 μl. [$^{14}C/^{3}H$]Forssman polyacrylic hydrazide (10,000 cpm, 100 μl) was added, and incubation was continued for 30 min at 4°C. *Staphylococcus aureus* (7 mg, 700 μl) was added and after 10 min removed by centrifugation at 1000*g* for 10 min. A 450-μl aliquot of supernatant was counted. ×, Forssman/SM/CHOL; ○, globoside/SM/CHOL; △, A[b] glycolipid/SM/CHOL. Molar ratio glycolipid/SM/CHOL: 1 : 250 : 200. (From Ref. 80.)

Figure 6. Plate binding assay for detecting hybridoma cells secreting anti-asialo GM_2 immunoglobulins. Asialo GM_2 (20 μg/ml) was adsorbed to the wells of Microtest II plates by overnight incubation at 37°C. The nonspecific binding capacity of the plates was then blocked by incubation with 5% bovine serum albumin (BSA) in phosphate-buffered saline (PBS) for 2 hr at 37°C. The assay was performed in three steps: (1) Culture fluids (50 μl) from the hybrid cell lines were replicate-plated onto the glycolipid-adsorbed wells; following a 45-min incubation at 37°C, the wells were washed with PBS containing 1% BSA. (2) Rabbit anti-mouse immunoglobulin serum (50 μl, 1 : 2000 diluted) was added to each well; following a 45-min incubation at 37°C, the wells were washed with PBS containing 1% BSA. (3) ^{125}I-labeled protein A (10^5 cpm) was added to each well in 50 μl of PBS; following a 45-min incubation at 37°C, the wells were washed with PBS, dried, and used for autoradiography. Autoradiography was enhanced by the use of X-ray intensifying screens. This figure shows the autoradiograms of the assays that were used for the isolation of the hybrid cell line 2D4, an IgM producer. The *top left* well of each glycolipid-adsorbed plate was used as positive control; instead of incubation with culture fluids, this well was treated with a 1 : 100 dilution of rabbit anti-asialo GM_2 serum. Reactions in the INITIAL PLATE autoradiogram were from the original culture fluids. Cells from the positive well (C11) were minicloned into 94 wells. Reactions in the MINICLONES autoradiogram were from culture fluids of these miniclones. Cells from

INITIAL PLATE

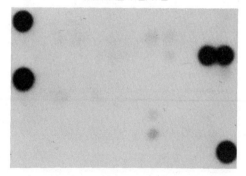

MINICLONES

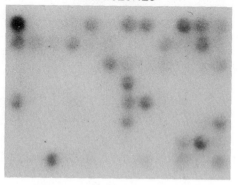

1°CLONE

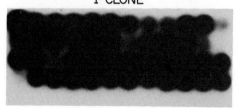

one positive miniclone (well H3) were then closed. At 3 weeks later, the clones were coalesced into a single plate and retested by the plate binding assay. Reactions in the 1°CLONE autoradiogram were from culture fluids of these clones. Culture fluids from each of these clones were positive in the plate binding assay. The negative reactions on this plate were from wells that were not seeded with hybrid cells. (From Ref. 66.)

5.2.3.9. Fluorescence-Activated Cell Sorter

Parks *et al.*[82] have shown that the fluorescence-activated cell sorter (FACS) can be used to clone hybridoma cells. This technique is based on the fact that the hybridoma cells not only secrete antibody but also have surface immunoglobulin of the same specificity that can bind fluoresceinated antigen and thus be sorted by the FACS. Leserman *et al.*[83] showed that lipid vesicles containing a lipid antigen in the bilayer and 6-carboxyfluorescein trapped in

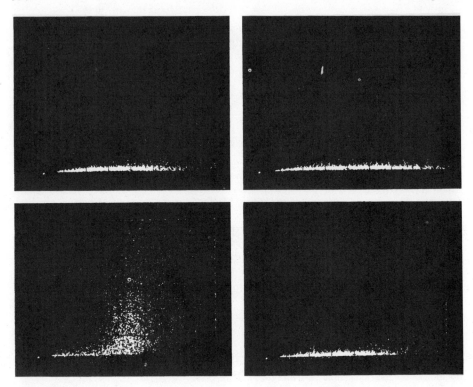

Figure 7. Binding of hybridoma cells to glycolipid vesicles. Hybridoma cells, cloned by the technique shown in Fig. 6 and secreting IgM specific for asialo GM_2, were incubated with fluorescent asialo GM_2 vesicles according to the procedures of Leserman *et al.*[83]; control cells were the mutant myeloma cell line NSI-1 used in the hybridoma fusion procedure. FACS analysis of extent of vesicle binding to cells. *Abscissa:* Cell size determined by light scatter; *ordinate:* intensity of fluorescence. *Top left,* NSI cells incubated with asialo GM_2 vesicles; *top right,* NSI cells incubated with control vesicles containing GM_1 ganglioside instead of asialo GM_2; *bottom left,* anti-asialo GM_2 hybridoma cells incubated with asialo GM_2 vesicles; *bottom right,* with GM_1 ganglioside vesicles.

the aqueous compartments would bind to myeloma cells having surface immunoglobulin specific for the lipid antigen. Figure 7 indicates that it may be feasible to clone hybridoma cells that are producing antiglycolipid antibodies using fluorescein-containing glycolipid vesicles and the FACS.

5.3 Heterophil Antigens

5.3.1. Forssman Antigen

A common antigen among various species was first noticed in 1911 by Forssman[84] and was called a heterophil or heterogenetic antigen. At that time, there was a strong belief among immunologists that antigens of tissues or cells were species-specific. The presence of a common antigen distributed over various species was against the established idea at that time. According to our present knowledge, most carbohydrate structures, including blood group ABH, Ii, and PP^k, are common for various species and therefore they are essentially all heterogenetic antigens. However, distribution of Forssman antigen has been most extensively studied and represents the most typical heterophil antigen. Since animals can be classified by the presence or absence of Forssman antigen in their organs and tissues, Forssman-positive animals such as guinea pig, horse, cat, dog, sheep, goat, mouse, pigeon, chicken, and turtle should have the capability to synthesize the glycolipids. Some of these "Forssman-positive" animals such as guinea pigs do not express the antigen in erythrocytes, but do contain it in various organs and tissues. On the other hand, Forssman-negative animals such as rabbit, rat, cow, pig, human, monkey, and frog are assumed to lack the capability to synthesize this glycolipid. These "Forssman-negative" species have been considered to lack antigen in both erythrocytes and organs (for a classical review see Buchbinder).[300] Recent study indicates that there are exceptions of Forssman expression in human tissue (see below).[25] The structure of Forssman antigen is extremely interesting in that the distinction between positive and negative animals depends on the presence or absence of a specific carbohydrate transferase for synthesis of a specific structure.

A great deal of effort has been paid to the isolation and structural characterization of this antigen. Brunius[7] purified the antigen, and it was identified as a hexosamine-containing, water-soluble glycolipid, and this was the first evidence for the presence of glycolipid antigens. Papiermeister and Mallette[11] succeeded in isolating the antigen and identified it as a water-soluble, galactosamine-containing glycosphingolipid, similar to that of globoside. Yamakawa et al.[85] purified the Forssman active fraction by silicic acid chromatography and finally obtained a pure glycolipid fraction showing a strong dextrorotary property with a carbohydrate composition similar to that of globoside. The active glycolipid was further characterized as having a carbohydrate composition and methylation pattern similar to that of globoside but showing different optical rotation.[86] The anomeric structure of globoside was determined by enzymatic degradation and nuclear magnetic resonance spectroscopy as GalNAcβ1$\rightarrow$3Galα1$\rightarrow$4Galβ1$\rightarrow$4Glcβ1$\rightarrow$1Cer in which the subterminal Gal was established to be α.[16] Consequently, Forssman structure was determined by successive degradation by enzymes and by methylation analysis. It was found that degradation of Forssman glycolipid by hog liver α-N-acetylgalactosaminidase resulted in the formation of globoside.[17] Therefore, it was made clear that globoside was the precursor of Forssman glycolipid. Further studies with methylation analysis identified the terminal structure, GalNAcα1$\rightarrow$3GalNAcβ1$\rightarrow$R, as the determinant of the Forssman hapten, as shown in Table III.[27] This structure was essentially verified by various investigators using Forssman glycolipid isolated from dog intestinal mucosa,[88] goat erythrocytes,[89] sheep erythrocytes,[90] and horse kidney.[91] It is interesting that Forssman glycolipid, irrespective of the source of the material, showed an identical terminal structure (Table III).

Hamster fibroblast NIL cells contain two kinds of Forssman glycolipid;

Table III. Heterophil Glycolipid Antigens

A. Forssman antigen

GalNAcα1→3GalNAcβ1→3Galα1→4*Galβ1→4Glc→Cer

Horse spleen	Siddiqui and Hakomori,[17] Stellner et al.[27]
Dog intestine	Sung et al.,[88] Karlsson et al.,[91] Smith et al.[128]
Goat erythrocytes	Taketomi et al.[89]
Sheep erythrocytes	Ziolkowski et al.[90]
Hamster NIL cells	Gahmberg and Hakomori[26]

GalNAcα1→3GalNAcβ1→3Galα1→4Gal1→1Cer

Hamster NIL cells	Gahmberg and Hakomori[26]

GalNAcα1→3GalNAcβ1

$\searrow$

$\overset{3}{\underset{4}{}}$Galα1→4Galβ1→4Glc→Cer

$\nearrow$

Galβ1→3GalNAcβ1

Dog gastric mucosa	Slomiany and Slomiany[92]

GalNAcα1→3GalNAcβ1→R(polysaccharide)

Streptococcus group C	Coligan et al.[93]

B. Hanganutzin–Deicher (H-D) antigen

Gangliosides containing N-glycolyl neuraminic acid	Higashi et al.,[99] Merrick et al.[100]

one of these is a ceramide tetrasaccharide having the structure Gal-NAcα1→3GalNAcβ1→3Galα1→4Galβ→Cer.[26] A branched chain glycosylceramide carrying the Forssman determinant was claimed to be present in dog gastric mucosa.[92] The same Forssman determinant, composed of two GalNAc residues at the terminus, was found in a Forssman-active polysaccharide isolated from Streptococcus group C.[93] It therefore considered to be a common determinant irrespective of its origin.

Biosynthesis of Forssman glycolipid from globoside was described using hamster and rat embryo enzymes.[94] Recently, it was demonstrated that Forssman glycolipid was more efficiently synthesized from ceramide trihexoside than from globoside, and a specific organization of the transferase complex for Forssman synthesis was suggested.[87]

Forssman glycolipid is present in certain tissues of some human populations, although the human had been considered to be a Forssman-negative species. In a study of Taiwanese, about 20–30% of cases contained Forssman glycolipid in gastrointestinal mucosa (F$^+$ group), whereas about 70–80% of cases lacked Forssman glycolipid (F$^-$ group).[25] In this study, the Forssman glycolipid was examined by thin-layer chromatography (TLC) as acetylated derivative and by the complement-dependent liposome lysis assay.[25] The sera of the majority of humans (70–80%) contained anti-Forssman IgM antibodies, whereas the sera of the minority had no detectable level of anti-Forssman antibodies.[287] Thus, Forssman antigen could be an isoantigen similar to, but independent from, ABH antigen. Forssman glycolipid was present in tumor tissue derived from F$^-$ mucosa and was therefore regarded as a potential human tumor-associated antigen (see Section 5.9).

5.3.2. Hanganutziu–Deicher Antigen

The presence of heterophil antibodies in sera of patients who received a therapeutic injection of antitoxin serum (e.g., antidiphtheria toxin and antitetanus toxin horse antisera) was described by Hanganutziu[95] and Deicher.[96] These heterophil antibodies (H-D antibodies) agglutinated not only sheep erythrocytes but also erythrocytes of bovine, horse, pig, rabbit, and guinea pig, but not human (for a review, see Kano and Milgrom[97]). Thus, clearly, H-D antibodies were distinguishable from Forssman antibodies, since they react with erythrocytes of Forssman-negative species, such as bovine and rabbit. The antibodies were also distinguishable from heterophil Paul–Bunnell antibodies in that H-D antibodies were completely absorbed by rabbit erythrocytes.[97] Early studies indicated that the antigen was heat-stable and extractable with ethanol.[98] Recent studies by Higashi *et al.*[99] and Merrick *et al.*[100] clearly indicate that the antibodies react specifically with gangliosides containing *N*-glycolylneuraminic acid, namely, hematoside and sialosylparagloboside. The H-D hemagglutination was inhibited by these gangliosides, and the sera with H-D antibodies gave a precipitin line by double-diffusion agar with these gangliosides having *N*-glycolylneuraminic acid.

5.4. Glycolipids with Blood Group ABH Specificities

The lipoid nature of blood group A and B antigens in human erythrocyte membranes was noticed in the early investigations by Witebsky,[1] Hirzfeld and Halber,[2] and Landsteiner.[3] On the other hand, polysaccharides or glycoproteins with blood group A and B activity were isolated from pepsins, hog gastric mucins, human saliva, and human gastric secretions and from human urine by a number of investigators, and the idea of blood group antigens associated with "polysaccharides" was well established.[101] Therefore, the lipoid nature of blood group antigens in human erythrocytes was suspected to be due to contamination by various "polysaccharides" present in the lipid fraction; this suspicion was based on the observation that carbohydrates are in fact soluble in the presence of large amounts of lipids and phospholipids (for a review, see Hakomori and Kobata[102]). The idea of "contamination" was further supported by a few experiments described by Hallauer[103] and Kossjakow and Tribulew.[104] These authors were not able to extract the blood group A and B antigen from human red blood cell membranes if the membranes were completely dried in a desiccator and delipidated by dried ether. They were able to extract the activities only with aqueous ethanol. On the basis of this observation, it was concluded that blood group activities were of a polysaccharide or glycoprotein nature.

In 1950, Hamasato[105] compared the extractability of A and B activities of human erythrocyte membranes under various conditions and followed the activities by treating with various organic solvents and was able to confirm the earlier finding of Witebsky and Landsteiner that blood group A and B activities were indeed associated with the lipid fraction. Furthermore, his final preparation contained carbohydrates showing the Elson-Morgan reaction for aminosugars. Masamune *et al.*[106] and Tokura[8] tried to purify the active lipid from liver, pancreas, and gastrointestinal tissues with rigorous standards employing chromatography on aluminum oxide column, electrophoresis, and counter-current partition. The samples isolated from pancreas and gastric mucosae showed a high ABO activity and contained hexosamine and hesoses.[8,106] These early studies in 1950–1952 provided a possibility that blood group determinants were carried by glycolipids. Concurrently, Yamakawa and Suzuki[108] isolated a new type of glycolipid, termed globoside, from human erythrocyte membranes containing glucose, galactose, and N-acetyl-galactosamine. A strong blood group ABO activity was found in globoside fraction by a consequent study of Yamakawa and Iida.[109] The chemical composition and optical rotation of the "globoside fraction" prepared from blood group A, B, and O erythrocyte membranes was compared, but they were almost identical irrespective of blood group sources.[110] This is not surprising, since subsequent studies indicated that the blood group active glycolipids were only the minor component present in "globoside fractions" that were separated from globoside by silicic acid chromatography or cellulose chromatography or both.[111–113] In these studies, more than two active fractions were separated from the major globoside peak. One of these fractions was further purified by precipitation with anti-A serum, and the precipitate was further extracted with chloroform–methanol.[112] A glycosphingolipid fraction with relatively high blood group A activity was isolated by silicic acid chromatography by Handa[114] and by Koscielak.[115] These fractions were characterized by having a high content of sialic acid, glucosamine, galactosamine, galactose, and glucose, but also a measurable quantity of fucose. Obviously, these fractions were grossly contaminated by ganglioside as judged by our present knowledge of the chemical composition of blood group active glycolipids.

At that time, the basic structural feature of blood group A and B active determinants was already established through a series of studies carried out by Morgan and Watkins[116,117] at the Lister Institute, London, and by Kabat and associates[118] at Presbyterian Hospital, Columbia University, New York. Through these studies, it has been apparent that blood group A and B activities depend greatly on fucosyl residues in addition to α-N-acetyl-galactosaminyl1→3galactose for the A determinant and the α-galactosyl1→3βgalactosyl residue for the B determinant.[119,120] In 1964, a fucose-containing glycolipid was isolated from human adenocarcinoma that was characterized by a high content of fucose, and the ratio of fucose, galactose, glucose, and N-acetylglucosamine was 1 : 2 : 1 : 1.[121] The glycolipid was Le^a-active, and was assumed to be lacto-N-fucopentaose(II) or (III) linked to ceramide.[122] These findings stimulated further studies of fucose-containing glycolipids in human erythrocytes, and a further systematic fractionation has been carried out starting from a large quantity of human erythrocyte membranes. Consequently, Hakomori and Strycharz[14] were able to isolate blood group active glycosphingolipids with much higher purity as indicated by a high fucose content. They reported three types of blood group A-active glycolipids and two types of blood group B-active glycolipids. The simplest type with blood group A-active glycolipid was identified as a ceramide hexasaccharide characterized as having the structure shown in Table V (A^a glycolipid). The backbone structure of this glycolipid seemed to be identical with that of the glycolipid previously isolated from human adenocarcinoma, namely, lacto-N-fucopentaosyl(III)ceramide.[123] The apparent polymorphism of blood group A and B antigens was clearly demonstrated, and the pattern of polymorphism between the glycolipids isolated from blood group A_1 and A_2 erythrocytes was described to be different; i.e.,one of the polymorphic components was lacking in the glycolipid fraction isolated from blood group A_2 erythrocytes.[14,124]

The H-active glycolipid was isolated, showing a precipitin reaction with Marcus's anti-H precipitating antiserum[125] that showed a narrow specificity with an exclusive reaction to fucosyl-lacto-N-neotetraosyl structure. The fraction was also reactive to H lectin (*Ulex europeus* and *Lotus tetragonolobus*).[14] In contrast to these findings, however, Koscielak and associates[126,127] were unable to demonstrate blood group H activity in any of their glycolipids isolated from O-erythrocyte membranes; this is now considered to be due to a masking of H activity by contaminating globoside.[57] After a few steps of chromatography, one of the H-active glycolipids was finally purified (H_1 glycolipid) as fucosyl-lacto-N-neotetraosylceramide[57] (also see Table IV). The second and third components of H-active glycolipid (H_2 and H_3 glycolipid) were eventually purified by successive chromatography on diethylaminoethyl (DEAE)–cellulose, on a long column of "Biosil A" and, finally, as acetylated compounds through TLC. The structures of the second and third H-active components have been identified[129] (see Table IV). In these studies, sequential enzymatic degradation by exoglycosidases and methylation analysis of the enzyme-degraded compound were used; however, the branching point of the H-active determinant was determined by direct probe mass spectrometry of the enzyme-degraded, permethylated–reduced glycolipid (see Chapter 1).

Koscielak and associates finally isolated and characterized two H-active components identical to H_1 and H_2 glycolipid,[130] although they had failed to detect these glycolipids in earlier studies. It is noteworthy that all these A- and H-active glycolipids of erythrocytes contained type 2 chain structure,[57,129–132] rather than type 1 structure, which is predominantly present in secretory epithelial mucosa and in secretions.[116,117] However, minor components of erythrocyte glycolipids with Le^a and Le^b specificities[15] should have the type 1 chain (see Section 5.5).

Similar to the status of H-active glycolipids, four kinds of A-active glycolipids were isolated from A-erythrocyte membranes and were termed A^a, A^b, A^c, and A^d glycolipids.[131,132,138] The structures of these glycolipids are shown in Table V. The A^a glycolipid was identified as having α-GalNAc attached to the subterminal galactose of H_1 glycolipid by 1→3 linkage. The same structure was assigned for the A-active glycolipid of human erythrocyte membranes by Ando and Yamakawa[138] and for the A-active glycolipids of human intestine by Smith *et al.*[140] and that of hog gastric mucosa by Slomiany and associates,[143,144] although the type 1 chain was predominant in human

Table IV. Blood Group H Glycolipids

From human erythrocytes

H_1 Fucα1→2Galβ1→4GlcNAcβ1→3Galβ1→4Glc→Cer   Stellner et al.,[57]
                                                 Koscielak et al[130]

H_2 Fucα1→2Galβ1→4GlcNAcβ1→3Galβ1→4GlcNAcβ1→3Galβ1→4Glc→Cer   Watanabe et al., [129]
                                                                  Koscielak et al. [130]

H_3 Fucα1→2Galβ1→4GlcNAcβ1→3   Watanabe et al. [129]
                                   ↘
                                     Galβ1→4GlcNAcβ1→3Galβ1→4Glc→Cer
                                   ↗
       Fucα1→2Galβ1→4GlcNAcβ1→6

From hog stomach

Fucα1→2Galβ1→4Galβ1→4Glc→Cer   Slomiany et al.[133]

From dog small intestine

Identical to H_1 of human erythrocytes   Smith et al.[134]

From human spleen

Fucα1→2Galβ1→3GalNAcβ1→4Galβ1→4Glc→Cer   Wiegandt[135]

        3
        ↑
        2
      NeuAc

From boar testis

Same as the fucoganglioside of human spleen   Suzuki et al.[136]

From rat ascites hepatoma

Fucα1→2Galβ1→3GalNAcβ1→3Gal1→4Glc→Cer   Matsumoto and Taki[137] intestinal glycolipid[140] in agreement with the structure of B-active pancreas glycolipid described by Wherrett and Hakomori[150] (see below). The same A-active glycolipid with type 2 chain as A^a glycolipid was isolated and characterized from dog intestine by Smith *et al.*[141]

A^b, A^c, and A^d glycolipid were identified, respectively, as shown in Table V.[107,124,132] Interestingly, A^b and A^c glycolipids are related to H_2 and H_3 glycolipids, respectively. Although H_4 glycolipid has not been purified and characterized at present, A^d-glycolipid[107] is probably related to the structure of H_4 glycolipid.

In fact, A^a, A^b, and A^c glycolipids were degraded by α-N-acetylgalactosaminidase of hog liver to H_1, H_2, and H_3 glycolipid, respectively. H_1, H_2, and H_3 glycolipids were converted by serum N-acetylgalactosaminyltransferase to A^a, A^b, and A^c glycolipids, respectively[132] (also Watanabe and Hakomori, unpublished observation). H_4 glycolipid is still a mixture, and further characterization remains to be made.

Two types of B-active glycolipids were isolated from human blood group erythrocytes and were termed B_I and B_{II} glycolipids by Koscielak *et al.*[130] The structure of B_I glycolipid was characterized as having an α-galactosyl residue attached to the subterminal galactosyl residue of H_1 glycolipid through 1→3 linkage (Table VI). More recently, Hanfland and Egge[139,151,152] identified the structure of H_1, H_2, B_I, and B_{II} glycolipids. B_{II} glycolipid was characterized as having an α-Gal residue attached to the penultimate β-Gal residue of H_2 glycolipid, and the structures of H_1 and H_2 glycolipids as described by Stellner *et al.*[57] and Watanabe *et al.*[129] were confirmed.

All of these glycolipids from human erythrocyte membranes contain type 2 chains as previously mentioned. In striking contrast, the glycolipid of human gastrointestinal tract seems to be characterized by a predominance of type 1 chain, although further detailed study is required. Wherrett and Hakomori[150] observed an accumulation of B-active glycolipid in the pancreas of a patient with Fabry's disease whose blood group status was type B. The structure was identified as shown in Table VI; namely, the majority contained type 1 chain (80%) and only the minority contained type 2 chain.

There has been considerable variation of internal carbohydrate chains that carry A and H determinants depending on the source of glycolipids. An active glycolipid of hog gastric mucosa isolated by Slomiany and Slomiany[147] contained type 2 chain with another fucosyl residue presumably attached to GlcNAc, thus forming a difucosyl-A determinant. A similar difucosyl-A glycolipid was isolated and characterized from dog intestine by Smith *et al.*[142] More recently, four types of A glycolipids having a common branched structure [Galβ1→3(Galβ1→6)Gal→R] were isolated and characterized from hog gastric mucosa.[148,149] The H-active glycolipid isolated by Wiegandt[135] from human spleen has a unique structure that contains N-acetylgalactosamine and sialic acid, and an H-active glycolipid recently isolated from rat ascites hepatoma by Matsumoto and Taki[137] was shown to have N-acetylgalactosamine as well. The fucoganglioside isolated from boar testis by Suzuki *et al.*[136] had the same structure as Wiegandt's hexasaccharide, fucosylsialosylganglio-N-tetraose [Fuc2→3Galβ1→3GalNAc(NeuAc2→3)Galβ1→4Glc]. Furthermore,

Table V. Blood Group A Glycolipids

Human erythrocytes		
A^a	GalNAcα1→3Galβ1→4GlcNAcβ1→3Galβ1→4Glc→Cer 2 ↑ L-Fucα1	Hakomori and Strycharz,[14] Hakomori *et al.*,[124,131,132] Ando and Yamakawa,[138]
A^b	GalNAcα1→3Galβ1→4GlcNAcβ1→3Galβ1→4GlcNAcβ1→3Galβ1→4Glc→Cer 2 ↑ L-Fucα1	Hakomori *et al.*[124,131,132]
A^c	L-Fucα1 ↓ 2 GalNAcα1→3Galβ1→4GlcNAcβ1 ↘ $^3_6$Galβ1→4GlcNAcβ1→3Galβ1→4Glc→Cer ↗ GalNAcα1→3Galβ1→4GlcNAcβ1 2 ↑ L-Fucα1	Hakomori *et al.*,[124,132] Fukunda and Hakomori[107]
A^d	Fucα1 ↑ 2 GalNAcα1→3Galβ1→4GlcNAcβ1 ↘ $^6_3$Galβ1→4GlcNAcβ1→3Galβ1→4Glcβ1→1Cer ↗ GalNAcα1→3Galβ1→4GlcNAcβ1→3Galβ1→GlcNAcβ1 2 ↑ L-Fucα1	Fukuda and Hakomori[107]

Human intestine
Similar to A^a glycolipid, but contains type 1 chain as the major component — Smith et al.[140]

Dog intestine
Same as A^a glycolipid — Smith et al.[141]

$$\text{GalNAc}\alpha1\to3(\text{or }4)\text{GlcNAc}\beta1\to3\text{Gal}\beta1\to4\text{Glc}\to\text{Cer}$$
with L-Fucα1 at positions 2 and (4 or 3) — Smith et al.[142]

Hog gastric mucosa
Same as A^a glycolipid of human erythrocytes, but contains some type 2 chain — Slomiany and Horowitz,[143] Slomiany et al.[144,145]

$$\text{GalNAc}\alpha1\to3\text{Gal}\beta1\to3\text{Gal}\beta1\to4\text{Glc}\to\text{Cer}\quad *$$
with L-Fucα1 at position 2 — Slomiany et al.[146]

Same as above, but one less Gal — Slomiany et al.[145]
Same as A glycolipid of dog intestine with difucosyl residues — Slomiany and Slomiany[147]

$$\text{GalNAc}\alpha1\to3\text{Gal}\beta1\to4\text{Gal}\beta1 \searrow$$
$$\overset{3}{\underset{6}{\text{Gal}\beta1}}\to4\text{Glc}\to\text{Cer}$$
$$\text{GalNAc}\alpha1\to3\text{Gal}\beta1 \nearrow$$
with L-Fucα1 at position 2 — Slomiany and Slomiany[148]

$$\text{GalNAc}\alpha1\to3\text{Gal}\beta1 \searrow$$
$$\overset{3}{\underset{6}{\text{Gal}\beta1}}\to4\text{Gal}\beta1\to4\text{Glc}\to\text{Cer}$$
$$\text{GalNAc}\alpha1\to3\text{Gal}\beta1 \nearrow$$
(or GlcNAc)
Same structure without the terminal GalNAc residue — Slomiany and Slomiany[149]

Table VI. Blood Group B Glycolipids

Human erythrocytes		
B^I	Galα1→3Galβ1→4GlcNAcβ1→3Galβ1→4Glc→Cer	Koscielak *et*
	2	*al.*,[130]Hanfland
	↑	and Egge,[139,151]
	1	Hanfland[152]
	L-Fuc	
B^{II}	Galα1→3Galβ1→4GlcNAcβ1→3Galβ1→4GlcNAcβ1→3Galβ1→4Glc→Cer	Hanfland and
	2	Egge,[139,151]
	↑	Hanfland[152]
	1	
	L-Fuc	
Human pancreas of Fabry's disease		
Same as B^I, but contains type 1 chain as the major component (80%)		Wherrett and Hakomori[150]

a unique H-active glycolipid with the structure Fucα1→2Galα1→4Galβ1→4Glc→Cer was isolated by Slomiany *et al.*[133]

Thus, it is apparent that the internal carbohydrate structure that carries the ABH determinant varies greatly depending on species and organs. The variation of the internal structure of carbohydrate for the blood group determinant could also be genetically controlled. The presence of a unique A determinant having two fucosyl residues[142,147] raises some question concerning the established specificity of A-synthetase, i.e., to transfer α-GalNAc to Gal of the H structure without the second Fuc residue linked to GlcNAc. The second fucosyl residue could be added to GlcNAc after the α-GalNAc residue was transferred to the Gal residue of the H structure.

Recently, Tilley *et al.*[153] demonstrated the presence of specific A and B blood group glycolipids in sera of A or B individuals. These glycolipids are absorbable on erythrocytes, which become agglutinable by anti-A or anti-B IgM, in a fashion similar to that by which Le^a or Le^b glycolipid can be absorbed on erythrocytes (see Section 5.5). The secretors seem to have more A or B glycolipids in serum than nonsecretors,[153] and it is probable that the *Se* gene may control the synthesis of H and release of A and B from unknown cells (or tissues). A recent study by Greaves and associates[297] indicated that blood group A antigen was already detected together with I/i antigen at the stage of burst-forming unit (BFU) of erythroblasts, before glycophorin and band 3 glycoprotein were synthesized. This clearly suggests these carbohydrate determinants are synthesized in the blood cells and may not be simply transferred from the serum.

5.5. Glycolipids with Blood Group Lewis (Le^a, Le^b, Le^c, and Le^d) Specificities

A current view on the inheritance and biochemistry of Le^a and Le^b antigens is based on classical works by Grubb, Ceppelini, Watkins, and Morgans, and by Ginsburg and associates (for a review see Watkins.[117]). The fucosyl-transferases coded by the *Le* and *Se*(H) genes are capable of making specific additions to an oligosaccharide precursor. Both Le^a and Le^b antigens are lacking (Le^{a-b-}) in erythrocytes of individuals who are lacking Le^a fucosyltransferase. Only Le^a antigen is present (Le^{a+b-}) on erythrocytes in individuals who have Le^a fucosyltransferase, but lack the fucosyltransferase coded by the *Se(H)* gene. These individuals are secretors. The individuals who have both Le^a fucosyltransferase and the *Se(H)* fucosyltransferase have Le^b antigen at the cell surface (Le^{o-b+}). Various antisera were prepared by Iseki *et al.*, Potapov, Gunson and Latham by immunization with glycoproteins secreted in saliva of Le^{a-b-} individuals or with Le^{a-b-} erythrocytes (see Graham *et al.*).[302] By the reactivity of erythrocytes with these antisera, Le^{a-b-} individuals can be further classified as Le^c ($Le^{a-b-c+d-}$) or Le^d ($Le^{a-b-c-d+}$). Le^c is associated with nonsecretors and Le^d is found in secretors.[302]

5.5.1. *Le^a* and *Le^b* Antigens

It has long been known that Le^a and Le^b antigens of human erythrocyte membranes were not synthesized *in situ*, but were acquired exogenously from serum.[154,155] A glycolipid inhibiting Le^a hemagglutination was first isolated from human adenocarcinoma.[122] A subsequent study indicated the presence of a similar glycolipid in human erythrocyte membranes that showed inhibition of Le^a and Le^b hemagglutination and displayed precipitin reactions with anti-Le^a and anti-Le^b goat antisera.[14] Consequently, an important study was made by Marcus and Cass,[15] who demonstrated clearly that Lewis glycolipid antigens are indeed acquired from serum. They isolated fucose-containing glycolipids from human serum that reacted strongly with anti-Le^a and anti-Le^b goat antisera. These glycolipids were efficiently absorbed on human erythrocytes when aqueous solutions of Le^a and Le^b glycolipids were incubated with human erythrocytes, whereas Le^a and Le^b active glycoproteins isolated from secretions were not absorbed on human erythrocyte membranes by a similar incubation. It has been well established that lacto-*N*-fucopentaose II represents the Le^a hapten[156] and that lacto-difucohexaose represents the Le^b hapten[157]; therefore, Le^a- and Le^b-specific glycolipids must have these structures. More recently, many glycolipid fractions with Le^a and Le^b specificities have been separated from serum, but their structures remain to be elucidated.[158]

The Le^a and Le^b glycolipids of human adenocarcinoma have been studied, and two Le^b-active glycolipids were isolated and their structures identified (Table VII). It was noteworthy that irrespective of the Le^a or Le^b blood group status of the host, human tumors contained both Le^a and Le^b antigens in equal quantity. However, Hattori *et al.* found that normal human gastric mucosa contain both Le^a and Le^b glycolipids[301] therefore co-presence of Le^a and Le^b antigen is not a specific phenomenon for gastric cancer. The glycolipid with Le^a structure and that with Le^b structure were isolated and characterized from dog and human intestine, respectively[142,160] (see Table VII). Recently, Wherrett and Crookston and associates[153,161] observed that the "A_1Le^b"-active glycolipid absorbed on O-erythrocytes became agglutinable with "anti-

Table VII. Lewis-Active Glycolipids and Their Analogs

Le^a	Galβ1→3GlcNAc1→3Gal1→4Glc→Cer 4 ↑ 1 Fuc	Suggested structure (exact structure not determined): Hakomori and Strycharz,[14] Marcus and Cass[15]
	Same structure as above, from human small intestine	Smith *et al.*[160]
Le^b	Galβ1→3GlcNAcβ1→3Galβ1→4Glc→Cer 2 4 ↑ ↑ 1 1 Fuc Fuc	Marcus and Cass[15]
	Same structure as above, from dog intestine	Smith *et al.*[142]
From human tumor		
Le^b	Fucα1→2Galβ1→3GlcNAc1→3Gal1→4Glc→Cer 2 ↑ 1 Fuc	Hakomori and Andrews[159]
Le^b	Fucα1→2Galβ1→3GlcNAc1→3Gal1→4GlcNAc1→Gal1→4Glc→Cer 2 ↑ 1 Fuc	Hakomori and Andrews[159]
X or Le^x	Galβ1→4GlcNAcβ1→3Galβ1→4Glc→Cer 2 ↑ 1 Fuc	Yang and Hakomori[123]
	Same structure isolated from hog gastric mucosa	Slomiany *et al.*[145]
	NeuAcα2→3GalβGlcNAcβ1→3Galβ1→4Glc→Cer 2 ↑ 1 Fuc	Rauvala[165,166]

A$_1$Le$^{b''}$ antibody. A hybrid molecule can be predicted that could have an A determinant on one branched carbohydrate chain and an Le^b determinant on another carbohydrate chain.

5.5.2. Le^c and Le^d Antigens

Graham *et al.*[302] observed that none of 98 cases tested showed more than one antigen (Le^a, Le^b, Le^c, or Le^d) detectable on erythrocytes of any individual. They predicted, therefore, that Le^c and Le^d are not derived from type 2 chain, but are derived from the common type 1 chain as in Le^a and Le^b. They also predicted type 1 chain H (Fucα1→2Galβ1→ 3GlcNAcβ1→ 3Galβ1→R) is Le^d since it is associated with secretor of Le^{a-b-}. Their prediction turned out to be indeed justified when Lemieux and associates were able to produce monospecific antibodies directed to the synthetic carbohydrate chains such as Fucα1→2Galβ1→3GlcNAcβ1→3Galβ→R   and   Galβ1→3GlcNAcβ1→

3Galβ1→R (R: protein residue). Thus Le^d and Le^c determinants were identified, respectively, as type 1 chain H,[303] and its precursor lacto-*N*-tetraosyl sequence.[304] Although glycolipids with these carbohydrate structures have not been isolated from Le^d and Le^c erythrocytes, the presence of glycolipid with such structures in Le^d and Le^c erythrocytes is predicted. Lacto-*N*-tetraosyl ceramide was isolated from meconium.[311]

5.5.3. Le^x Antigen and X-Hapten Glycolipids

The presence of another antigen, Le^x, in newborn erythrocytes (cord erythrocytes) that is independent from Le^a and/or Le^b in adult erythrocytes, has been reported. Le^x positive neonates eventually developed into Le^a or Le^b as adults. Archilla and Sturgeon suggested that Le^x antigen is produced by the action of Le-fucosyltransferase. Of 35 cord erythrocytes, 24 were Le$^{a-b-c-d-x+}$ and 11 were Le$^{a-b-c-d-x-}$. [305] It is difficult to assign the structure Galβ1→4[Fucα1→2]GlcNAcβ1→R, called X-hapten,[123,309] as a true Le^x hapten. Whereas the quantity of one of the X-hapten glycolipids (Z$_1$ in Table IV, Chapter 4) is two to three times higher in cord erythrocytes than in adult erythrocytes, the overall expression of X-hapten in cord erythrocytes by immunofluorescence with SSEA-1 antibody (see below) was similar to that of adult erythrocytes. A possibility for other types of fucosyl structures such as Fucα1→3/or 4GlcNAc or Fucα1→3GlcNAcβ1→3Galβ1→4[Fucα1→ 3]-GlcNAc→R should be tested. Although the presence of Le^x antigen was clearly demonstrated by Archilla and Sturgeon,[305] its chemical structure still remains to be elucidated. On the other hand, our knowledge of the structure and distribution of various glycolipids bearing X-hapten structure has been greatly enriched during the past few years: (i) A series of glycolipids with X-hapten determinants has been isolated and characterized from human erythrocytes[309] and tumors[28,310] (see Table IX, Chapter 4). (ii) Those isolates from tumor had an unusual fucosyl residue at the internal GlcNAc of the type 2 chain.[28] (iii) All glycolipids are unbranched in contrast to ABHI antigens that contain branched species.[309] (iv) X-hapten is identical to SSEA-1 determinants that are expressed maximally at morulae and decline after the later blastocyte (see Chapter 4). (v) Many of them increase and accumulate in a large variety of human cancer[28] (see Section 5.8.2).

5.6. Glycolipids with Blood Group P, P$_1$, and P^k Specificities

The human blood group P system expressed on erythrocytes[167] consists of three antigens, P, P$_1$ and P^k, and five phenotypes, as shown in Table VIII. Similar phenotypes and distribution of antigens have been found in cultured skin fibroblasts and lymphocytes,[168,169] as shown in Table VIII. The genetic and immunochemical relationship between these antigens is unclear, but it was suggested that the products of the P$_2$ and P$_1$ genes act in sequence to convert a precursor substance to P and P$_1$ antigens.[19,169] Immunochemical studies of a cross-reacting glycoprotein obtained from hydatid cyst fluid[170]

Table VIII. Blood Group P Phenotypes and Antigen Distribution and Their Structures

Phenotype	Frequency	Antigen	Antibody in sera	Structures
Erythrocytes[a,c]				
P_1	40%	P_1 P (P^k)[d]	—	P_1 glycolipid Galα1→4Galβ1→4GlcNAcβ1→3Galβ1→4Glc→Cer
P_2	60%	P (P^k)[d]	Anti-P_1	
P^k_1	Very rare	P_1 p^k	Anti-P	P glycolipid = globoside GalNAcβ1→3Galα1→4Galβ1→4Glc→Cer
P^k_2	Very rare	p^k	—	P^k glycolipid Galα1→3Galβ1→4Glc→Cer
P	Very rare	None	Anti-P_1 Anti-P Anti-P^k	
Skin fibro-blasts[b,c]				
P_1	40%	P_1 P p^k		
p_2	60%	P p^k		
P^k_2	Very rare	p^k		
P	Very rare	None		

[a] From Naiki and Marcus,[29] whose figures were based on Race and Sanger.[167]
[b] From Fellous et al.[168]
[c] Peripheral lymphocytes showed phenotype similar to erythrocytes, whereas cultured lymphocytes showed phenotype similar to fibroblasts.
[d] P^k antigen in P_1 and P_2 erythrocytes is chemically detectable but immunologically cryptic.

demonstrated that both P_1 and P^k antigens have carbohydrate determinants with an immunodominant structure, α-galactosyl residue.[171] About 40% of Caucasian peoples are in the P_1 group due to the presence of a unique P_1 antigen, whereas essentially all of this population, except rare P^k and p individuals, contain P antigen. In 1971, Marcus[172] isolated a P_1-active substance from P_1 erythrocyte stroma and identified it as being a glycosphingolipid. The active glycolipid was eventually isolated in pure form, and the structure was identified as a ceramide pentasaccharide having a unique structure, αGal1→4βGal, at the terminal, as shown in Table VIII.[18,29] Erythrocytes of the rare P^k phenotype lack blood group P antigen, and the very rare p phenotype erythrocytes lack both P and P^k antigens. Hemagglutination caused by anti-P (sera of P^k individual) was inhibited by globoside, and hemagglutination caused by anti-P^k (sera of p individual) was inhibited by a ceramide trihexoside (CTH: αGal1→4βGal1→4Glc→Cer). Therefore, P and P^k antigens were assumed to be globoside and ceramide trihexoside, respectively.[18,29] Chemical analysis of erythrocyte glycosphingolipids of the rare p, P^k phenotype as compared to the normal P population was carried out and revealed that P^k erythrocytes contain only a trace amount of globoside and a marked excess of ceramide trihexoside in comparison with normal P erythrocytes. The p erythrocytes lack globoside and ceramide trihexoside and contain a large excess of lactosylceramide and sialosylparagloboside.[19] Hemagglutination caused by anti-p sera preferentially agglutinate p, to a lesser extent P_2, and weak P_1 or P^k erythrocytes, was inhibited by and precipitated with sialosylparagloboside.[173] Thus, P and P^k antigens are clearly identified as globoside and ceramide trihexoside, and sialosylparagloboside or a related ganglioside could be the p antigen. On the basis of these findings, Marcus and associates[19,173] proposed a biosynthesis of glycosphingolipids in P_1, P_2, P^k, and p erythrocytes (see Fig. 8). In both P_1 and P_2 erythrocytes, synthesis of globoside through trihexosylceramide may proceed at a similar rate. Similarly, synthesis of paragloboside in P_1 and P_2 erythrocytes must be very similar, but the conversion of paragloboside to P_1 by αGal1→4 transferase is present only in P_1 erythrocytes. In p erythrocytes, synthesis of trihexosylceramide is limited; therefore, synthesis of ABH glycolipid and sialosylparagloboside through paragloboside must be greatly enhanced.

5.7. Glycolipids with Blood Group I and i Specificities

5.7.1. I and i Antigens and Antibodies

It has been known for many years that an agglutinin is present in the sera of patients suffering from acquired hemolytic anemia. The agglutinin is a "cold" antibody that reacts with erythrocytes from most of the human population to strongly agglutinate the cells at 4°C but not at 37°C.[174] In 1956, Wiener *et al.*[175] discovered that the cells of 5 individuals out of 22,000 New York inhabitants were not agglutinated by such cold agglutinins obtained from patients with acquired hemolytic anemia. Prokop and Uhlenbruch[176] also

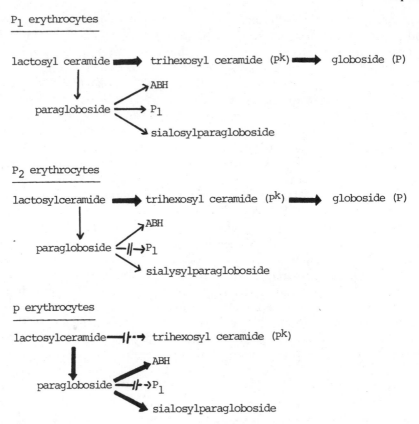

Figure 8. Proposal for the biosynthesis of glycosphingolipids in P_1, P_2, and p erythrocytes according to Marcus *et al.*[19]

found that the serum of a patient with Waldenström's macroglobulinemia named Step showed a similar agglutination. The majority of humans whose cells were agglutinated by these cold agglutinins were called blood group I and these antibodies were called anti-I, while the very rare individual whose erythrocytes were not agglutinated were called blood group i.[176] Later, umbilical cord cells and fetal erythrocytes were found not to be agglutinated by anti-I, in contrast to adult cells, which were agglutinated. This suggested that the I antigen develops during postnatal differentiation.[177] Some sera of patients with chronic hemolytic disorders and some with Waldenström's macroglobulinemia preferentially agglutinated cord and fetal cells. These antisera contained monoclonal IgM proteins and were defined as anti-i antibodies.[178,179]

The majority of humans whose erythrocytes are strongly agglutinated by anti-I antibodies but not by anti-i antibodies, are called blood group I, whereas rare individuals whose erythrocytes are not agglutinated by anti-I but are agglutinated by anti-i are called blood group i. Fetal erythrocytes are similar to blood group i adult cells, but develop into blood group I during differentiation.

To date, a number of specificities for I and i can be distinguished by different monoclonal IgM proteins, and usually the specificities are named after patients. For example, Ma, Step, Grady, Da, Phi, Sch, and Low are known to have I specificity, and Den, McD, McC, Tho, Hog, and Galli are known to have i specificity.

Antibodies with anti-I specificity are found not only in certain myeloma patients but also in patients with acquired hemolytic anemia, some chronic and acute infectious diseases such as pneumonia caused by *Mycoplasma pneumoniae*, and infectious mononucleosis. The basis of this phenomenon is not known.

The numbers of erythrocytes with I antigen decrease and those with i antigen increase in patients with bone marrow grafts,[180,181] sickle cell anemia,[182] and some cases of dysplastic bone marrow function such as Hereditary Erythroblastic Multinuclearity; Positive Acidified Serum test (HEMPAS).[184]

5.7.2. Chemical Basis of I and i Specificities

Early studies: The chemical structure of I and i antigen is therefore of great interest because the conversion of i to I or I to i occurs in conjunction with erythrocyte differentiation and hematological disorders. The chemical properties of I and i antigen have been investigated by Kabat, Feizi, and their associates. However, the structural relationship between Ii specificity and specific antigens has not been well known until recently. In an early study, Marcus *et al.*[185] found that I antigen on erythrocytes can be destroyed by β-galactosidase and β-*N*-acetylglucosaminidase. In 1971, I reactivity was found to be inhibited by the oligosaccharide preparation from human milk and by precursor glycoproteins for blood group ABH substances of human ovarian cyst and hog gastric mucin that were prepared through the one-step Smith degradation of the blood group ABH glycoproteins.[186] In a subsequent study, anti-I serum with "Ma" specificity was best inhibited by the oligosaccharide having a terminal Galβ1→4GlcNAcβ1→6Gal, which is present in a precursor blood group substance.[187] The quantitative precipitin curves by anti-I antibodies with various I specificities of various oligosaccharide fractions indicate a significant variation in their specificities, but are all directed toward precursor carbohydrate chains of blood group ABH substances.[187]

Glycolipid Ii antigens were investigated during the course of our studies on H_2 and H_3 antigens. The H_3 glycolipid, identified as ceramide decasaccharide with a branched structure containing the Galβ1→4GlcNAcβ1→6Gal structure, showed a weak but definite inhibition of hemagglutination caused by anti-I (Ma) serum.[129] Some fractions associated with ceramide octasaccharides also showed a precipitin reaction with anti-i (Dench) serum[188] (also unpublished observation). A number of subpopulations of gangliosides of human erythrocytes have been isolated by chromatography on "Iatrobeads" and on DEAE–Sephadex and some of them showed various I and i specificities by RIA.[58] Gardas[189] isolated a glycosylceramide with I activity having as many as 22 carbohydrate residues and a branched structure.

Current status: A systematic study for the isolation and characterization of i- and I-active glycolipids has been carried out recently in our laboratory in collaboration with T. Feizi of the Medical Research Council, London. A ceramide hexasaccharide having two repeating units, $Gal\beta1{\rightarrow}4GlcNAc$ and $GlcNAc\beta1{\rightarrow}3Gal$, showed a strong i activity with various anti-i sera such as Dench, Hog, McC, McDon, and Tho; only the anti-i serum Gall failed to react with this glycolipid. The same structure was not reactive with various antisera containing monoclonal anti-I antibodies Phi, Ma, Gra, and Da, but showed some reaction with anti-I Step.[20] A blood group active, branched ganglioside was isolated from bovine erythrocyte membranes, and its structure was determined as follows[21]:

$$Gal\alpha1 \longrightarrow 3Gal\beta1 \longrightarrow 4GlcNAc\beta1$$
$$\begin{array}{c} 6 \\ 3 \end{array} Gal\beta1 \longrightarrow 3GlcNAc\beta1 \longrightarrow 3Gal\beta1 \longrightarrow 4Glc\beta1 \longrightarrow 1Cer$$
$$NeuAc\alpha2 \longrightarrow 3Gal\beta1 \longrightarrow 4GlcNAc\beta1$$

Since this ganglioside contains α-Gal on one carbohydrate chain and sialic acid on the other carbohydrate chain attached to paragloboside, various structures with different side chains have been prepared by degradation of the ganglioside with sequential reactions of exoglycosidases starting from the α-Gal or the sialic acid terminus. The inhibitory activities of these structures with 11 anti-I antibodies and 6 anti-i antibodies have been determined; all anti-I except Zg react with various branched structures, and the various I specificities are located on various domains within the following branched octaglycosylceramide structure[21,22]:

$$Gal\beta1 \longrightarrow 4GlcNAc\beta1$$
$$\begin{array}{c} 6 \\ 3 \end{array} Gal\beta1 \longrightarrow 4GlcNAc\beta1 \longrightarrow 3Gal\beta1 \longrightarrow 4Glc\beta1 \longrightarrow 1Cer$$
$$Gal\beta1 \longrightarrow 4GlcNAc\beta1$$

A straight-chain structure with repeating $Gal\beta1{\rightarrow}4GlcNAc\beta1{\rightarrow}3Gal$ units again inhibited the reactivities with anti-i antibodies.[21] Furthermore, three groups of anti-I antibodies can be classified according to the inhibition pattern[22]: (1) antibodies Phi, Da, Sch, and Lo react with a structure having the complete branched structure with $1{\rightarrow}4$, $1{\rightarrow}6$ and $1{\rightarrow}4$, $1{\rightarrow}3$ carbohydrate chains; (2) anti-I antibodies Ma and Woj require one branch with $Gal1{\rightarrow}4GlcNAc1{\rightarrow}6$; and (3) anti-I Step, Gra, Ver, and Ful antibodies require one branch with $Gal\beta1{\rightarrow}4GlcNAc1{\rightarrow}3Gal$. From these results, crucial information has been obtained on the molecular basis of I and i specificities, and certain generalizations can be made: (1) nonreducing terminal β-Gal is an important part of both I and i antigenic determinants; (2) an intact straight-chain oligosaccharide with a repeating $Gal1{\rightarrow}4GlcNAc1{\rightarrow}3$ sequence expresses the majority of i and part of exceptional I antigenic determinants; and (3) the antigenic determinants recognized by the majority of I antibodies are expressed on intact 1→4, 1→6 and/or 1→4, 1→3 branched structures within the lacto-*N-isooc*taosylceramide structure. The structural basis for Ii specificities is shown in Fig. 9.

Endo-β-galactosidase from *Escherichia freundii* has been shown to directly modify specific cell-surface carbohydrate chains by hydrolyzing the β-galactosyl linkage* of a common structure, R→GlcNAcβ1→3Galβ1→4GlcNAc (or Glc)→R.[190,191] The β-galactoside linkage adjacent to the branched structure as shown below by the asterisk was not easily hydrolyzed under the same conditions: R→Galβ1→4GlcNAcβ1→3(R$_2$→Galβ1→4GlcNAcβ1→6)Galβ1→ 4GlcNAc→R$_3$. Ii antigens of human erythrocytes were found to be preferentially abolished by this enzyme with an increase of the reactivity to the antibodies directed to lacto-*N*-triaosylceramide. This structure could be the same hapten created by *Candida fragilis* infection and termed T^k antigen.[192] Accompanying these antigenic changes, surface-labeled carbohydrates showed the following changes by the endo-β-galactosidase treatment: (1) The carbohydrate moieties of Bands 3 and 4.5 glycoproteins were hydrolyzed, whereas sialoglycoproteins (PAS-1, -2, and -3) were not affected. (2) The change in glycolipids occurred mainly with long-chain carbohydrates such as H$_2$, H$_3$, and H$_4$ glycolipids and in higher gangliosides. (3) The oligosaccharide released from adult I-active erythrocytes contained higher-molecular-weight components with greater heterogeneity than that released from adult i or umbilical i erythrocytes, which contained relatively homogeneous lower-molecular-weight components (di- or trisaccharides). These results again indicate that I-active erythrocytes contain carbohydrate chains with more branching than i-active

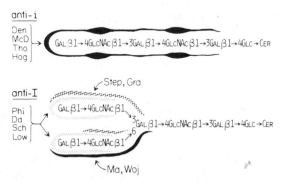

Figure 9. Minimum essential structure for expression of I and i specificities. Anti-i monoclonal antibodies, such as Den, McD, Tho, and Hog, recognize two repeating *N*-acetyllactosamine structures without branch. Elimination of the terminal β-Gal residue or *N*-acetyl group of two GlcNAc's results in complete loss of the activity. Since paragloboside (Galβ1→4GlcNAcβ1→3Galβ1→4 Glc→Cer) did not show any activity, two repeating Galβ1→4GlcNAc structures are essential to express anti-i specificity. Anti-I Phi, Da, Sch, and Low react only to a branched structure having two Galβ1→4GlcNAc chains and recognize this binary carbohydrate chain, whereas antiI Ma and Woj react to the carbohydrate chain having the Galβ1→4GlcNAcβ1→6Gal sequence and show no requirement of branching. In contrast, anti-I Step and Gra require Galβ1→4GlcNAcβ1→3Gal and the Galβ1→4GlcNAcβ1→6Gal branching structure as shown above. These results are a summary of those of Niemann *et al.*,[20] Watanabe *et al.*,[21] and Feizi *et al.*[22]

erythrocytes. The carriers of I and i antigens on erythrocytes are both gly-cosphingolipids and Band 3 glycoprotein.[191]

To further confirm the structural change of carbohydrates in Band 3 glycoprotein associated with fetal to adult differentiation, the chemical structures of Band 3 glycopeptides, prepared from erythrocytes of normal adult, umbilical cord vessel, and an i-adult variant have been studied. The Band 3 glycopeptide of cord erythrocytes gave, on permethylation analysis, predominantly 2,4,6-tri-O-methylgalactose and 3,6-di-O-methyl-2-N-methylacetam-ido-2-deoxyglucose, whereas the same glycopeptide of normal adult erythrocytes gave a much higher quantity of 2,3,4,6-tetra-O-methylgalactose and 2,4-di-O-methylgalactose as compared with that of cord erythrocytes.[191] Thus, a structural change of Band 3 carbohydrate associated with fetal to adult differentiation in analogous to that occurring in glycolipid, i.e., the conversion of a linear repeating Galβ1→4GlcNAcβ1→3Gal to a branched Galβ1→4GlcNAcβ1→3(R→6)Gal structure (see Section 4.3.2.3).

5.8. Glycolipids with Blood Group J-Antigen Specificity

Blood group J antigen, which occurs in the serum and erythrocytes of cattle, originates primarily in endothelia or allied tissues, is secreted into the serum and is then absorbed onto erythrocytes.[193–195] This situation is similar to Lewis,[196] R,[197] and A[198] antigens in man, sheep, and pig, respectively. Cattle are classified into three phenotypic classes of J-antigen status: (1) animals with J substance on the cell and in serum (J^{cs}), (2) animals with J substance in the serum but not on the cells (J^s), and (3) animals without any detectable J substance (J^a). The J hapten was extracted from J-positive sera, but could not be extracted from J-negative sera. The hapten was found in the low-density lipoproteins of serum and was easily taken up by erythrocytes. The activity of J hapten was readily destroyed by periodate oxidation, but was not destroyed by deacylation with alkali, suggesting that the hapten could be a glycosphingolipid.[199] J substance was solubilized from J-positive cells by organic solvents, but not by hypertonic salt solution, suggesting that the once-absorbed J substance is strongly associated and integrated into membrane components and cannot be readily solubilized.[200,201]

Horowitz and Slomiany[202] extensively studied the J^{cs} antigen by successive chromatography on silica gel and on thin-layer plates and finally isolated a hybrid lipid containing sphingosine sugars, glycerol, and phosphate. Further studies by Slomiany and Horowitz[30] utilizing galactose oxidase showed that galactose in intact J substance was not oxidized. It is noteworthy that the structure suggested, for the first time, the presence of a hybrid molecule between sphingolipid and glycerol-phospholipid. The chemistry of pig A antigen[198] and sheep R antigen[197] is not known, but there is a possibility that these antigens are glycolipids, in that they are synthesized, secreted into serum, and then transferred to erythrocytes.

5.9. Tissue-Specific and Tumor-Associated Glycolipid Antigens

5.9.1. Classical Studies

The presence of lipid antigens, claimed as specific to human tumor, was described by Witebsky,[1] Hirszfeld and Halber,[2] and Lehman-Facius.[203] These classical studies were followed beginning some 25 years later by those of Kobayashi[204–206] and Rapport and associates.[12,41,42,207] A lipid fraction was isolated from the ethanol extract of gastric cancer tissue from blood group 0 patients through fractionation with various solvents and by countercurrent distribution. This lipid fraction reacted strongly with rabbit antisera against gastric cancer homogenates. The final product was a water-soluble lipid containing carbohydrate and phosphorus, and its immunological reactivity determined by the complement-fixation reaction was distinguishable from that of a similar fraction isolated from normal gastric mucosa, which was blood group H-active. The property of Kobayashi's lipid hapten was similar to that of a fraction reported by Hirszfeld and Halber[2] that showed blood group activity.

In 1955, rabbit antibodies were prepared against the transplantable Murphy–Strum rat lymphosarcoma that reacted by complement fixation with a lipid hapten extracted from the lymphosarcoma. The lipid hapten was species-specific for the rat and quite limited in its distribution within the organs of the rat.[207] The hapten, later termed "cytolipin R," has been chemically characterized[42,208] (see Section 5.11).

Rabbit antisera directed against the transplantable human epidermoid carcinoma Hp40 were found to react with lipid extracts of various human tumors, but not with lipid extracts of normal tissues.[12,41] The active hapten was finally isolated from epidermoid carcinoma grown in rat. The complement-fixation reaction displayed by this lipid hapten was strongly inhibited by lactose, but not by other monosaccharides and oligosaccharides; thus, the lactosylceramide structure was predicted. In fact, the synthetic lactosylceramide[209] was found equally active as the natural hapten.[210] Graf and Rapport[211] noted that antisera to cancer tissue homogenates or extracts contained antibodies to lactosylceramide more frequently than did the antisera to normal tissue. The molecular basis of this finding is not exactly known, since lactosylceramide has been well known to be a component of normal tissue as well as tumors. It is possible that lactosylceramide could be present in higher quantity in some human tumors or that the accessibility of lactosylceramide on the tumor cell surface is greater. In this regard, Joźwiak and Kościelak *et al.*[212] recently found that the sera of cancer patients contain elevated levels of antibodies reacting with ceramide dihexoside.

5.9.2. Current Studies with Experimental Cancer and Human Cancer

Two types of glycolipid changes associated with oncogenic transformation have been known: (1) blocked synthesis of complex glycolipids, which is often accompanied by an accumulation of precursor glycolipid(s), and (2) synthesis of a new glycolipid peculiar for transformed cells. Typical examples of the former phenomenon are: (1) increased level of lacto-*N*-*neo*tetraosylceramide in hamster NIL cells transformed by polyoma virus[26,43] and (2) the presence of ganglio-*N*-triaosylceramide (asialo GM_2) in 3T3 cells transformed by the Kirsten strain of murine sarcoma virus (3T3KiMSV).[44] Typical of the latter case is the appearance of Forssman glycolipid in human tumor[25,213] and "illegimate" blood group pp_1 antigens.[214,215] In either change of glycolipid, the resulting glycolipid pattern includes a new glycolipid that was absent or markedly lower in nontransformed cells; thus, the unique glycolipid displays or behaves similar to tumor-associated antigen.

In the following sections, many instances of potential glycolipid tumor antigens are discussed, some of them defined by monoclonal antibodies directed to human cancer: (1) lacto-*N*-*neo*tetraosylceramide (paragloboside) in NILpy tumors of hamster, (2) ganglio-*N*-triaosylceramide (asialo GM_2) in 3T3KiMSV tumors and lymphoma L-5178, (3) Forssman glycolipid in human tumors of F^- humans, (4) the unknown glycolipid in simian virus 40 (SV40)-transformed hamster tumor cells (TSV_5-cl_2, EHSVi-cl_1, and EHB), (5) human melanoma-associated glycolipids, (6) Burkitt lymphoma associated antigens, (7) human colon carcinoma antigen, and (8) polyfucosylated type 2 chain glycolipid in various human malignancy.

5.9.3. Lacto-N-neotetraosylceramide in NILpy Tumor

Lacto-*N*-*neo*tetraosylceramide (paragloboside) was found as a specific surface-labeled component of NILpy cells.[26] The sera of NILpy tumor-bearing hamster contain an antibody-like material reacting with paraglobosides and auxiliary lipids (lecithin and cholesterol) but not with globosides or Forssman glycolipids. The amount of antibody administered against paraglobosides in sera of NILpy tumor-bearing hamster was parallel to the size of tumors.[43]

5.9.4. Ganglio-N-triaosylceramide in Mice Sarcoma and Lymphoma

3T3KiMSV cells and tumors in BALB/c mice accumulate a unique glycolipid that was isolated and characterized as ganglio-*N*-triaosylceramide [asialo GM_2 (GalNAcβ1→4Galβ1→4Glc→Cer)].[44] It is assumed that the glycolipid accumulates as a consequence of blocked synthesis of GM_1 ganglioside. Asialo GM_2 was barely detectable in 3T3 cells by chemical analysis or by cell-surface labeling with galactose oxidase and NaB[^{3}H]$_4$. Distribution of various neutral glycolipids in brain, thymus, lung, liver, kidney, spleen, and erythrocytes of BALB/c mice indicated that none of the organs contains significant level of asialo GM_2 as compared to a large accumulation of this glycolipid in 3T3KiMSV cells (69 μg/100 mg dry residue).[44] Antibodies specific to asialo GM_2 were produced in rabbit injected with 3T3KiMSV cells, but not in those injected with 3T3 cells. The rabbit antibody against ganglio-*N*-triaosylceramide stained only 3T3KiMSV tumor cells by indirect immunofluorescence.[44] Immunization of BALB/c mice with ganglio-*N*-triaosylceramide absorbed on acid-treated *Salmonella minnesota* (Galanos procedure[216]) resulted in the production of antibody against ganglio-*N*-triaosylceramide; however, such mice having the antiglycolipid antibody do not inhibit KiMSV tumor growth. On the other hand, passive immunization with rabbit antisera suppressed KiMSV tumor growth[69] (also Young and Hakomori, unpublished observation).

The same glycolipid was also found in mice lymphoma L5178, but not in various tissues and organs of the host DBA mice. The passive immunization of DBA mice with monoclonal anti-GgOse$_3$Cer IgG$_3$ antibodies, but not with monoclonal anti-GgOse$_4$Cer IgM antibodies, after inoculation of tumor cells completely suppressed the lymphoma cell growth, and the animals inoculated with the lymphoma were perfectly normal.[288] In contrast to the lymphoma bearing the gangliotriaosylceramide, tumor growth of some clones from the same lymphoma that did not contain gangliotriosylceramide was not cured by the passive immunization (see Fig. 10). These findings strongly indicate that gangliotriaosylceramide is the lymphoma-associated antigen or cell-surface marker.

5.9.5. Forssman Antigen in Human Cancer

The presence of Forssman glycolipid antigen in 1 case of human biliary carcinoma[213] and in 16 cases of Taiwanese gastrointestinal tumors derived from F$^-$ mucosa[25] was described. The Forssman glycolipid was identified chemically on TLC on a ceramide tetra- and pentasaccharide fraction isolated by silica gel chromatography of total neutral glycolipids and by an immunological method based on the complement-dependent lysis of liposomes containing the glycolipid fraction. Tumors derived from F$^-$ mucosa contained Forssman glycolipid characterized by hemolysis inhibition and liposome lysis. On the other hand, tumors derived from F$^+$ mucosa did not contain Forssman glycolipid; therefore, Forssman glycolipid in gastrointestinal tumors of F$^-$ mucosa behaves like a tumor-associated antigen (TAA).

5.9.6. Glycolipid Antigen Specific for SV40-Transformed Hamster Tumors

The specific antigenicity of tumor cell plasma membranes is concentrated in membranes of budding viruses using vesicular stomatitis virus (VSV); thus, the purified preparation of VSV grown on SV40-transformed hamster cells acquires a strong TAA activity. The specific tumor rejection was observed in hamsters preimmunized with VSV carrying SV40-TAA.[217] The TAA activity was associated with the lipid core of the VSV envelope rather than the peripheral protein (or glycoprotein digestable with protease).[218] A subsequent study by Huet and Ansel[219] demonstrated that TAA activity was associated with the upper phase of the Folch extract and partitioning, representing either ganglioside or long-chain neutral glycolipid. The TAA activity in this fraction was not digestable by pronase, and no protein band was demonstrated on sodium dodecyl sulfate–polyacrylamide gel electrophoresis followed by Coomassie blue staining. The vesicles containing VSV glycolipid grown on SV40-

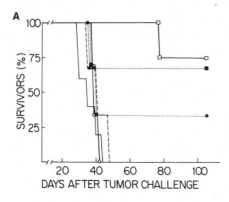

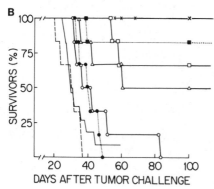

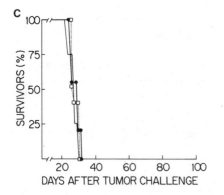

Figure 10. Serotherapy of DBA/2 mouse L5178Y lymphoma cells with monoclonal antibodies to GgOse₃Cer. A, DBA/2 males were inoculated intraperitoneally with 10^6 subclone 1A1 cells on day 0. Group 1 (———), 5 untreated control mice; group 2 (- - - -), 3 mice treated with guinea pig serum (as a complement source) alone; group 3 (○—○), 3 mice treated with IgM anti-GgOse₃Cer ascites fluid 2D4; group 4 (●. . . . . .●), 3 mice treated with 2D4 ascites fluid plus guinea pig serum; group 5 (□—□), 4 mice treated with IgG₃ anti-GgOse₃Cer ascites fluid D11G10; group 6 (■. . . . . .■), 3 mice treated with D11 G10 ascites plus guinea pig serum. Treatment consisted of intraperitoneal injections of 100 μl of ascites fluid or guinea pig serum, or both, on days 1, 3, 7, and 10. B, Mice were inoculated intraperitoneally with 10^6 subclone 1A1 cells on day 0. Group 1 (———), untreated control mice; group 2 (- - - -), 6 mice treated with nonspecific IgG₃ ascites fluid C3; group 3 (○—○), 6 mice treated with 2D4 ascites fluid; group 4 (●. . . . . .●), 6 mice treated with 2D4 ascites fluid plus guinea pig serum; group 5 (□———□), 6 mice treated with D11G10 ascites fluid; group 6 (■. . . . . .■), 6 mice treated with D11G10 ascites fluid plus guinea pig serum; group 7 (△—△), 6 mice treated with D11G10 ascites fluid injected intravenously; group 8 (×—×), 5 mice treated with purified IgG₃ (anti-GgOse₃Cer). The treatment protocol followed that in A, except for group 7, which received intravenous instead of intraperitoneal injections of ascites fluid. C, Mice were inoculated intraperitoneally with 10^5 clone 27AV cells (a variant that had no GgOse₃Cer) on day 0. Group 1 (———), 4 untreated control mice; group 2 (●—●), 5 mice treated with 2D4 ascites plus guinea pig serum; group 3 (□. . . . . .□), 5 mice treated with D11G10 ascites fluid. The treatment protocol was identical to that in A. (From Ref. 288.)

transformed hamster cells inhibited the subsequent tumor growth; however, there was an optimal range of glycolipid amount injected (700–1200 μg VSV protein equivalent). The immune sera prepared by injection of liposomes containing the polar glycolipids of the SV40-transformed hamster tumor cells, after being absorbed on normal hamster tissues, specifically stained the SV40-transformed cells.[289]

5.9.7. Human Melanoma Antigen Defined by Monoclonal Antibody

Several studies suggest that certain gangliosides may be distinctive markers of human malignant melanoma. Portoukalian *et al.*[220] found that a ganglioside fraction isolated from a melanoma liver metastasis reacted by Ochterlony double immunodiffusion with the serum of a melanoma patient, but not with normal human serum. These workers also detected higher levels of gangliosides in both plasma and erythrocytes in melanoma patients as compared to normal controls[221]; in particular, GM_3 (hematoside) was elevated in the red cells and GD_3 (disialosyllactoceramide) in the plasma. The authors did not comment on an intriguing combination of the results of these two studies, namely, that there might be specific ganglioside–antiganglioside immune complexes in the sera of melanoma patients. Such tumor antigen–IgG complexes have been detected in human cancer sera using the Raji cell RIA.[222]

Portoukalian *et al.*[223] also analyzed the ganglioside patterns of several melanoma tumors in which it was possible to estimate the proportion of malignant melanocytes in the total cell population. GM_3, GM_2, and GD_3 were found to be the major components, but no correlation was found between the malignant level and the ganglioside content of the tumors.

Gupta *et al.*[224] isolated a chloroform–methanol-soluble material from the spent tissue culture medium of cultured human malignant melanoma cells that appeared to have TAA activity. Recently, many monoclonal mouse antibodies were selected that are directed to the cell-surface antigens of the immunizing human melanoma cell line (SK-MEL-28). The antigen system, defined by a group of monoclonal antibodies, "R24," was tentatively identified as glycolipid, which was found only in melanoma and astrocytoma and was absent in various fibroblasts, epithelial cells, and cells of hematopoietic origin.[291] The glycolipid antigen defined by R24 antibody was finally isolated and identified as GD_3 ganglioside (see Table II).[267] A similar independent study on monoclonal antibodies directed to human melanoma established an IgM antibody designated "4.2".[183] The antigen was isolated and characterized as GD_3 ganglioside with a ceramide having C22 to C24 fatty acids.[268] Specific recognition of GD_3 as a human melanoma-associated antigen is due to its unusually high concentration in melanoma and its unusual immunogenicity for its high content of long chain fatty acids.[268]

5.9.8. Burkitt Lymphoma-Associated Antigen Defined by Monoclonal Antibody

A rat monoclonal IgM antibody directed to Burkitt lymphoma was generated by the hybridoma technique. This antibody defines an antigen specifically expressed on most malignant B cell lines derived from Burkitt lymphoma, irrespective of whether the lymphoma cells contain Epstein-Barr virus (EBV) genome (Central-East African endemic form), or does not contain EBV genome (European-North American form). The antigen was not detectable on EBV-positive lymphoblastoid cell lines, or malignant cells from patients affected by various non-Burkitt lymphoproliferative disorders.[306] Thus, it was concluded that the antibody defines a specific antigen associated with Burkitt lymphoma. The antigen was finally identified as globotriaosylceramide (Galα1→4Galβ1→4Glcβ1→1Ceramide) and the antibody did not react with globoisotriaosylceramide (Galα1→3Galβ1→4Glcβ1→1Ceramide).[307] Although this antigen has been found in moderate quantities in human erythrocytes (1.6–2.0 μg/100 mg dried cell residue), the quantity of the antigen in Burkitt lymphoma was 100 times higher than in normal human erythrocytes. The antibody was not reactive to normal erythrocytes, but was reactive to erythrocytes of a rare blood group P^k individual.

5.9.9. Monosialo Ganglioside of Human Colon Carcinoma Defined by a Specific Monoclonal Antibody

A monoclonal antibody specific for colon carcinoma cells has been described.[294] The binding of this antibody to antigen is inhibited by the sera of most patients with advanced colorectal carcinoma, but not by the sera of normal patients with inflammatory bowel diseases or most patients with other malignancies. This indicated that the antigen is highly specific for human colonic carcinoma. Recently, Magnani *et al.*[295] found that the binding of the antibody to colorectal carcinoma cells was unaffected by treatment of the cells with protease, but was abolished by treatment of the cells with neuraminidase. A direct binding of the antibody to an unidentified monosialo ganglioside extracted and partially purified from the colorectal carcinoma cell line was demonstrated on TLC reacted by the antibody, followed by detection with ^{125}I-labeled Fab of rabbit IgG antibodies to mouse immunoglobulin. More recently, the antigen was isolated and characterized as having the structure, sialosyl-Le^a (see Table IX).[308]

5.9.10. Polyfucosylated Lactosaminolipids in Human Gastrointestinal, Lung, and Liver Adenocarcinoma

The study of the glycolipid antigen defined by the monoclonal antibody SSEA-1 that is directed to mouse F9 embryonal carcinoma cells, disclosed the presence of a series of glycolipid having X-determinant (Galβ1→4[Fucα1→3]GlcNAc).[309] The structures were determined as described in Table IV in Chapter 4. Interestingly, the glycolipids with difucosylated lacto-N-*nor*hexaosylceramide and trifucosylated lacto-N-*nor*octaosylceramide were found to accumulate in various types of human cancers and are suspected to be common human tumor antigens.[28] The specificity and clinical significance of these unique markers remain to be studied. The structure of these antigens are listed in Table IX.

5.10. Modification of Blood Group Antigens

Chemical and metabolic changes of membrane glycolipids associated with oncogenic transformation have been well established (see Chapter 4). In var-

Table IX. Important Glycolipid Antigens in Human Cancer

Antigen	Structure	Reference
A-like antigen in tumors of blood group O	GalNAcα1→3Galβ1→HexN→(Hex)$_n$→Hex→Cer	Yokota *et al.*[290]
P-like antigen in tumor of a rare blood group p individual	GalNAcβ1→3Galβ1→4GlcNAcβ1→3Galβ1→4Glcβ1→1Cer	Levine *et al.*,[214,215] Kannagi *et al.*[242]
Forssman antigen in F⁻ (fsfs) tumor	GalNAcα1→3GalNAcβ1→3Galα1→4Galβ1→4Glcβ1→1Cer	Hakomori *et al.*[25]
	GalNAcα1→3GalNAcβ1→Hex→Cer	Yokota *et al.*[290]
Sialosyl Le^a	NeuAcα2→3Galβ1→3GlcNAcβ1→3Galβ1→4Glcβ1→1Cer with Fucα1→4 on GlcNAc	Manani *et al.*[295,308]
Difucosyl X	Galβ1→4GlcNAcβ1→3Galβ1→4GlcNAcβ1→3Galβ1→4Glcβ1→1Cer with Fucα1→3 on each GlcNAc	Hakomori *et al.*[28]
Poly-X (Tri-X)	Galβ1→4GlcNAcβ1→3Galβ1→4GlcNAcβ1→3Galβ1→4GlcNAcβ1→3Galβ1→4Glcβ1→1Cer with Fucα1→3 on each GlcNAc	Hakomori *et al.*[28,309]

Table X. Modification of Blood Group Antigens in Human Tumors

1. Deletion of A and B determinants	Oh-Uti,[225] Masamune et al.[226]
2. Accumulation of precursors	
Ii accumulation	Simmons and Pearlman,[229] Feizi et al.[230]
GlcNAcβ1→3Galβ1→4Glc accumulation	Watanabe and Hakomori[231]
T, Tn appearance	Springer et al.[232]
Le^x accumulation	Yang and Hakomori[123]
3. Appearance of "incompatible" antigens	Hakomori et al.[122]
O → A conversion	Häkkinen[233]
P → P conversion	Levine et al.,[214] Levine[215]
Le^a and Le^b copresence	Hakomori and Andrews[159]
Forssman appearance in tumors of F$^-$ individuals	Hakomori et al.[25]

ious human tumors, the glycolipid changes can be expressed as a modification of blood group activity. The following changes have been described: (1) deletion of A and B determinants, (2) precursor accumulation, and (3) the presence of incompatible or illegitimate blood group antigens (see Table X). These changes are discussed in the following sections.

5.10.1. Deletion of A and B Determinants

The deletion of blood group A and B antigens in human gastric cancer was first observed by Oh-Uti[225] when blood group substance isolated from the cancer tissue of group A patients failed to inhibit A-specific hemagglutination. The optical rotation of the blood group substance of cancer tissue was different from that of normal blood group substance. Subsequent studies comparing the blood group substance of gastric cancer, normal gastric mucosa, and metastatic lesion to liver confirmed the previous finding.[234] Further studies were carried out by the same research group, and at least six glycoproteins were separated. Some glycoproteins were isolated in a homogeneous state on electrophoresis and were noticed to carry blood group haptens. Blood group activities and chemical and physical properties were examined in greater detail.[226,235] The decrease of both A and B activities and corresponding chemical changes were noticed, whereas H activity determined by eel sera and by anti-human O saliva chicken sera increased significantly.[226,235] Unfortunately, these data were based on analysis of pooled tumor tissue and compared to pooled normal mucosa; therefore, possible variations of individual tumor tissue were ignored. Iseki et al.[236] compared blood group substances isolated from individual tumors and individual gastric mucosa. Significant variations in immunological activities were noticed. The blood group active fraction in some tumors showed high H activity, whereas others showed high Le^a activity; some tumors showed low H and low Le^a activities, but the deletion of A and B activities was invariably observed.

A mixed hemagglutination technique was applied by Kay and Wallace[237] and demonstrated that both A and B antigens were not detected in bladder cancer cells, but were detected in normal urinary tract epithelial cells. Similarly, Nairn et al.[238] showed a deletion of blood group A and B antigens in human adenocarcinoma cells. Further extensive studies using immunohistological methods such as adherence of erythrocytes on tissue sections were carried out by Davidsohn *et al.*[227,228] and revealed that ABH antigens were detectable in normal tissues and were reduced or absent in the course of transformation of carcinoma. Such changes were observed in carcinoma of gastrointestinal tract and ovary and squamous cell carcinoma of skin, tongue, larynx, and urinary bladder.[227] The immunoadherent reactions applied on histological samples from 82 cases of benign and malignant cervical cancer patients indicated that ABH isoantigens were present in all benign lesions, but were deleted in all metastatic lesions. In 18 out of 21 cases with early infiltrating carcinomas without clinical evidence of metastatis, the isoantigens were decreased or absent.[228] In these studies, isoantigens were always demonstrated on noncancerous metaplastic or dysplastic epithelium, whereas in the early infiltrative region isoantigens were no longer detectable. The deletion of ABH antigens in neoplastic oral epithelium was reported by Prendergast *et al.*[239] A study by Dabelsteen and Fulling[240] showed that A and B antigens were lost in premalignant dysplasia of oral epithelial tissue.

Because of sensitive changes of A and B determinants associated with malignant transformation of human epithelial and endodermal cells, enzymatic studies have been carried out for synthesis of A and B determinants. A much higher rate of conversion of H glycolipids to A glycolipids took place when catalyzed by the enzyme of normal epithelial mucosa as compared to the same reaction catalyzed by the enzyme of adenocarcinoma tissue derived from epithelial carcinoma of the same individual.[241] The difference was also clear when the yield of radioactive A^a glycolipids was compared by increasing the incubation time. The hydrolase activity for A^a glycolipids of normal mucosal tissue was nearly identical to that of adenocarcinoma.

Thus, the deletion of A and B determinants in epithelial tumors is due to a deficiency of glycosyltransferases for synthesis of A and B determinants, not to enhanced hydrolase activity. The deficient A or B enzyme for glycolipid could be the same enzyme for synthesis of A and B glycoproteins as well, because blood group glycoproteins of tumors were also shown to be deficient.

5.10.2. Lewis Fucolipids in Tumors

Lewis antigen activities of blood group glycoproteins isolated from human gastric adenocarcinomas showed some inconsistency, i.e., enhanced Le^a activity in some cases and enhanced H–Le^b activity in other cases.[236]

Accumulation of a fucose-containing glycosphingolipid having the carbohydrate composition fucose/glucose/galactose/GlcNAc 1 : 1 : 2 : 1[121] was noticed in some human adenocarcinomas. Subsequent studies showed that this glycolipid fraction inhibited Le^a hemagglutination and showed A-like immunogenicity, although the fraction was heterogeneous.[122] By further fractionation of the fucose-containing glycolipids as acetylated compounds on TLC, several components have been isolated. A major component, besides the Le^a-active glycolipid, was finally identified as Galβ1→4(Fucα1→3) GlcNAcβ1→3Galβ1→4Glcβ1→1Cer, i.e., lacto-*N*-fucopentaosyl(III)ceramide. The ceramide moiety of this lipid was characterized as having 4-hydroxyoctadecasphinganine as the major base component.[123]

In five cases of gastrointestinal adenocarcinoma, the copresence of Le^a and Le^b glycolipids has been noticed irrespective of the Lewis status of the hosts. Two Le^b-active glycolipids were separated and their structures determined.[159]

In some cases of human adenocarcinoma of gastric mucosa, pancreas, bronchogenic lung tumor, and cecal tumor, accumulation of two isomeric fucose-containing glycolipids was especially noticeable: Le^a and its isomer, Le^x. Di- or tri-fucosylated polylactosaminolipids and sialosyl-Le^a (see Table IX), the most characteristic glycolipids present in various human cancers, are both derivatives of Le^{x-} or Le^a-pentaglycosylceramides. The aberrant metabolism of the fucolipids is an important basis for human cancer.

5.10.3. Precursor Accumulation

Recently, precursor glycolipid with the structure GlcNAcβ1$\rightarrow$ 3Galβ1$\rightarrow$ 4Glc$\beta$$\rightarrow$Cer was found in some cancer tissues and is considered as a precursor accumulation of various blood group glycolipids.[231] The I antigen is also present in colonic cancers and is regarded as the precursor of blood group antigen.[230] The I activity was detected in carcinoembryonic antigen (CEA), and one of the CEA specificities was assumed to be identical to I (Ma).[229] The presence of T and Tn antigens in breast cancer has been reported[232]; they are usually sialylated in various normal glycoproteins.

A lung cancer case was reported in which the tumor was regressed in a patient with an incidental "gammopathy" which secreted a warm-reactive IgMλ. The antibody was directed to Galβ1$\rightarrow$3GlcNAcβ1$\rightarrow$3Gal sequence, the type 1 chain precursor of the blood group determinant.[164] It is assumed that the tumor was inhibited due to its accumulation of blood group precursor.

5.10.4. Blood Group Determinants in Human Tumors Foreign to the Host (Illegitimate Blood Group Antigens)

It has long been known that some human cancers can have a blood group determinant incompatible with the host's blood group. In the classical literature, Witebsky[1] described a rabbit antiserum immunized by a case of uterine carcinoma homogenate that showed a positive complement-fixation reaction with a lipid extract of blood group A erythrocytes, but not with O erythrocytes. In 1967, rabbit antisera prepared against a fucose-containing glycolipid fraction isolated from gastric adenocarcinoma of a blood group O host showed a property of anti-A specificity and agglutinated A erythrocytes preferentially.[122] In a subsequent study by Häkkinen,[233] the appearance of an A-like antigen in gastric cancer of patients with blood group O or B was noticed by immunofluorescence with anti-A antibody and was called "neo A" antigen. Similarly, Denk et al.[243] found blood group A and B incompatible reactions in adenocarcinoma of gastric and colonic mucosa. Recently, Breimer[244] obtained mass spectrometric evidence for the presence of blood group A-similar glycolipid in a gastric adenocarcinoma from a blood group B individual. The chemical properties of these A-like antigens that appeared in gastric cancer of patients of blood group O or B are of particular interest, since the A-like antigen could be recognized as foreign and the tumor growth could be affected; epidemiological studies have demonstrated a statistically significant higher incidence of gastric cancer in A individuals as compared to B or O (for a review, see Mourant *et al.*[245]). It was suggested that the growth of gastric cancer in O or B individuals could be inhibited at an early stage of growth by anti-A antibody present in O or B individuals.[122,233] Forssman glycolipid antigen has been isolated and characterized from several cases of gastrointestinal tumors.[25,213] Some of the A-like antigens in human tumors could be Forssman antigen. However, some of the A-like antigens react only with anti-A but not with anti-Forssman antibodies; therefore, these A-like antigens must be independent of Forssman antigen. Recently, a glycoprotein with A activity as well as Forssman activity was isolated from the tumor of a blood group O individual. Mass spectrometry of the methylated glycolipid and enzymatic hydrolysis indicated that the A-like antigen may have the GalNAcα1$\rightarrow$Gal structure without fucose.[290]

Another version of the presence of illegitimate blood group antigens in gastric cancer came from the work of Levine and associates.[214,215] In 1951, Levine *et al.*[214] reported the case described in Table XI. The serum of a 66-year-old woman (Mrs. Ja) who had gastric adenocarcinoma was found to agglutinate erythrocytes of the majority of the human population except her own erythrocytes, and the antibody was called "anti-Tja." On surgical operation, her tumor was subtotally extirpated. Prior to surgery, a small amount of incompatible blood was transfused, resulting in a severe reaction for incompatible blood type. Following subtotal gastrectomy, her anti-Tja titer was raised to 1 : 525; more surprisingly, the tumor did not reappear thereafter. The patient died of cerebral hemorrhage 25 years later!

Table XI. Possible Relationship between P Antigen and Tumor-Associated Antigen: The Case of Mrs. D. S. Ja[a]

Patient
   Mrs. D. S. Ja, 66 years old in 1951
Diagnosis
   Gastric adeoncarcinoma
Blood type
   O, rare *pp* (indicence 1 : 150,000)[e]r serum contained anti-P_1/P/P^k (called anti-Tja)
Cancer tissue
   Lyophilized powder absorbed anti-P, P^k; 20 mg of the powder absorbed anti-Tja 16–32
     units. Glycolipid of the tumor powder was examined in 1976–1981. P,P_1-like glycolipids
     were detected in neutral glycolipid fraction,[242] and P-like activity was detected in
     ganglioside.[b]
Clinical treatment
   Subtotal resection; blood transfusion with 25 ml of incompatible blood, severe reaction
Results
   (1) Severe reaction for incompatible blood transfusion, titer up to 1 : 512; (2) tumor
     spontaneously cured
Conclusion
   Genetic mutation for illegitimate antigen foreign to host

[a] From Ref. 214.
[b] Watanabe, Hakomori and Levine (unpublished).

The anti-Tja is now identified as a mixture of anti-P, anti-P_1, and anti-P^k and Mrs. Ja obviously belonged to a rare genotype "*pp*" (see Table XI). Her tissue after lyophilization was found to absorb anti-Tja (5 mg, about 16–32 units). Recently, glycolipids of Mrs. Ja's tumor tissue, which were lyophilized and kept in a deep freezer for 30 years, were analyzed, and the major neutral glycolipid was found to be a P-like glycolipid with the structure GalNAcβ1→3Galβ1→4GlcNAcβ1→3Galβ1→4Glc→Cer that cross-reacts with globoside. P_1-like glycolipid, that was susceptible to α-galactosidase, was also present in the tumor.[242]

In conclusion, these results suggest that Mrs. Ja, who belonged to blood type pp, developed a tumor that contained P,P_1-like glycolipid; therefore, she elicited anti-P or anti-P_1 antibodies as her tumor developed. After the incompatible blood transfusion, her tumor regressed because of the high anti-P,P_1 titer.

5.11. Tissue-Specific or Organ-Specific Glycolipid Antigens

5.11.1. Tissue-Specific or Organ-Specific Glycolipids

The heterologous antibody response directed against immunization of tissue or organ homogenates is characteristic of the particular tissue or organs rather than depending on the species difference. This indicates the possible presence of a tissue- or organ-specific antigen irrespective of species. Such specific antibodies often react with the lipid fraction, particularly to glycolipids of the original tissue used as the immunogen.

Brandt *et al.*[246] described a brain homogenate that, when injected into rabbit, elicited a specific antibody that reacted with brain extracts from a variety of species, but did not react with other organs. Joffe *et al.*[13] studied the nature of the antigen(s) using antisera directed against bovine brain homogenate and found a high degree of reactivity in a complement-fixation test with brain lipid common to ten different mammalian species, whereas no reactivity was found with lipids of ox heart, liver, lung, kidney, and spleen. Among the brain lipids, the activity was found in cerebroside in the presence of auxiliary lipids, and the synthetic *N*-lignoceryl-cerebroside (phrenosine) had approximately the same activity as the natural cerebroside. The brain antigenicity may depend on the galactocerebroside residue in myelin[31]; anticerebroside antibodies react with suspensions of pure myelin prepared according to Autilio *et al.*[247] Myelin membranes are extremely rich in galactocerebroside as compared to other membrane structures in brain or extraneural cells. Raff *et al.*[248] demonstrated that galactocerebroside represented a cell-surface antigenic marker for oligodendrocytes in culture. Although antigalactocerebroside can be regarded as a brain-specific antibody, galactocerebroside is not a specific product of brain or myelin. It is widely distributed in various tissues. The organ specificity of glycolipids based on immunogenicity is therefore not strictly related to the chemical quantity of glycolipid present in cells and tissues (Section 5.11 for further discussion of galactocerebroside and experimental allergic neuritis).

A glycosphingolipid called "cytolipin K" was assigned as a kidney-specific hapten.[32] The hapten was identified as a ceramide tetrasaccharide indistinguishable from the globoside of Yamakawa *et al.*[249] Chemically, globoside is the major glycolipid of human erythrocytes and is now established as the P antigen.[19] Obviously, cytolipin K or globoside is not a specific product of kidney.

A ceramide tetrasaccharide, originally described as a specific hapten of rat lymphosarcoma, was detected by the complement-fixation reaction between rabbit anti-rat lymphosarcoma antisera and the lipid fraction of lymphosarcoma.[207] The pure hapten, cytolipin R, has been isolated[250] and chemically characterized.[208] However, a glycolipid with the same structure was isolated from normal rat kidney.[251] More recently, cytolipin R was isolated and characterized from rat spleen [252]; therefore, cytolipin R is not a rat lymphosarcoma-specific but a species-specific glycolipid for rats.

It has become increasingly apparent that some membrane-bound glycolipids are immunologically dominant when cells or tissue homogenates are injected into heterologous animals and that the specificity depends on the carbohydrate structure of the glycosphingolipids. The tissue specificity or organ specificity of glycolipid antigens is difficult to claim, as outlined above. The antiglycolipid response in heterologous animals against tissue homogenates occurs rather irregularly. For example, only 8 out of 70 rabbits showed an obvious antibody response against "cytolipin K,"[32] although antiglycolipid responses were consistent when glycolipids with carrier proteins were injected.[253,254] Therefore, the degree of tissue or organ specificity is difficult to evaluate in many cases.

Sulfogalactoglycerolipid [(SGG) 1-*O*-alkyl-2-*O*-acyl-3-(3'-sulfo-β-galactosyl)-glycerol][257,258] is the major glycolipid of the mature rat testis and has also been found in the testis of humans,[255] rabbit, bull, guinea pig,[256] and boar.[257] The conversion of galactosyl diglyceride to SGG by a sulfotransferase apparently occurs during the early spermatocyte stages[258,259]; therefore, SGG represents a differentiation marker during spermatogenesis in the rat. Antisera specific for SGG have recently been raised in female rabbits.[260]

5.11.2. Lymphoid-Cell-Type-Specific Glycolipids

The origin of this relatively new aspect of glycolipid research lies in attempts to define the chemical nature of theta antigens (Thy-1), which have been widely used as markers for thymocytes and peripheral T cells in mice. Early reports suggested that theta antigen might be lipoprotein or glycolipid.[261,262] Esselman and Miller[263] reported that the cytotoxicity of anti-Thy-1.2 serum was inhibited by GM_1 ganglioside and that GD_{1b} ganglioside inhibited rabbit anti-brain-associated theta antiserum (BA Θ)[264]; they proposed that GM_1 might be the Thy-1.2 antigen and GD_{1b} the Thy-1.1 antigen. Thiele *et al.*[265] reported that CBA thymocyte receptors for choleragen and anti-Thy-1.2 antibodies cocapped. However, more recent studies indicated that anti-GM$_1$ reacted equally well with thymocytes of both Thy-1.1 and Thy-1.2 phenotypes[266] and that the capping of receptors for anti-Thy 1.2 and anti-GM$_1$ occurred independently.[269] At present, theta antigenic determinants are claimed to reside not only on a glycoprotein but also on minor ganglioside components.[269,270]

The clearest indication that glycolipid determinants might serve as useful markers for defining lymphoid subpopulations came from the immunofluorescence studies of Stein-Douglas *et al.*[23] and Stein *et al.*[24] They found that anti-GM$_1$ antibodies reacted with murine thymocytes and peripheral T cells independent of their Thy-1 phenotypes.[23] In contrast, antibodies specific for asialo GM$_1$ failed to react with most thymocytes, but did stain about 30% of mature T (immunoglobulin-negative) cells in several mouse strains.[24] Recently, mouse natural killer cell activity was found to be depleted by anti-asialo GM$_1$ plus complement[271,272]; under the same conditions, mouse alloimmune cytotoxic T cells were not destroyed. Finally, in a separate study, Nakahara *et al.*[273] reported that antibodies against asialo GM$_1$ reacted with leukemic cells from patients with acute lymphoblastic leukemia, but not with human lymphoid cells from normal individuals. However, this finding was not supported by other studies.[312,313]

A separate area of research has suggested that specific macrophage glycolipids may act as receptors for migration-inhibitory factor (MIF). Such studies were prompted by reports that L-fucose, but not other monosaccharides, blocked the effect of MIF on macrophages and that L-fucosidase treatment of macrophages made them unresponsive to MIF.[274-276] Moreover, other studies suggested that macrophage glycolipids may be receptors for MIF.[277,278] Hanada *et al.*[279,280] found that a fucolipid [Galα1→3(Fucα1→2)Galβ1 → 3GalNAcβ1→3Galβ1→4Glcβ1→Cer] was a major constituent of rat granuloma cells and macrophages. This glycolipid specifically blocked MIF activity, whereas a structurally related B-active glycolipid [Galα1→3(Fucα1→2) Galβ1→4GlcNAcβ1→3Galβ1→4Glcβ1→Cer] did not.[281]

5.12. Glycolipid Antigens Associated with Autoimmune Processes

Experimental allergic neuritis (EAN) and experimental allergic encephalomyelitis (EAE) are autoimmune, demyelinating diseases of the peripheral and central nervous system (CNS), respectively. They are classically produced in animals by injection of homogenates of nervous system tissue. Niedieck and associates[282,283] found that the serum of a rabbit developing EAE formed a precipitin line with either a lipid extract of brain or pure cerebroside plus auxiliary lipid. Furthermore, rabbit antisera to galactocerebroside demyelinated organotypic CNS cultures and inhibited the myelination and sulfatide synthesis in immature CNS cultures.[284-286] Although immunization with one or two injections of galactocerebroside was not encephalitogenic,[283] repeated boosting produced EAN in 13 out of 31 rabbits.[39] Similarly, intensive immunization of rabbits with either total bovine brain gangliosides or particular gangliosides such as GD$_{1a}$ or GM$_1$ produced a "ganglioside syndrome" that closely resembled EAN.[38]

Myelination-inhibiting or demyelinating antibodies are also found in patients with multiple sclerosis (MS). Using the liposome immune release assay described in Section 5.2.3.4, Hirsch and Parks[37] found that antibodies against digalactosyl diglyceride occurred more frequently in MS sera than in normal sera or those with other neurlogical diseases. Sela and associates found that MS-like disease can be induced in rabbits by immunization with brain gangliosides,[314] and peripheral lymphocytes of MS patients were susceptible to be stimulated by MS brain gangliosides, particularly GQ_{1b}.[315] These findings suggest a possible role for glycolipids in the MS disease process.

5.13. References

1. Witebsky, E. *Z. Immunitätsforsch.* **62,** 35–73 (1929).
2. Hirszfeld, L., and Halber, W. *Z. Immunitätsforsch.* **67,** 286–318, (1930).
3. Landsteiner, K. *The Specificity of Serological Reactions,* 2nd ed., Harvard University Press, Cambridge, Massachusetts (1945).
4. Day, E. D. In: *The Immunochemistry of Cancer* (Kugelmass, T. N., ed.), Charles C. Thomas, Springfield, Illinois (1965), pp. 5–19.
5. Landsteiner, K., and Simms, S. *J. Exp. Med.* **38,** 127 (1923).
6. Pangborn, M. C. *J. Biol. Chem.* **143,** 247 (1942).
7. Brunius, E. *Ark. Kemi Mineral. Geol.* **12B,** 1–3 (1936).
8. Tokura, M. *Tohoku J. Exp. Med.* **56,** 299, 307 (1957).
9. Hamasato, Y. *Tohoku J. Exp. Med.* **52,** 17, 29, 35 (1950).
10. Yamakawa, T., and Iida, T. *Jpn. J. Exp. Med.* **23,** 327–331 (1953).
11. Papiermeister, B., and Mallette, M. F. *Arch. Biochim. Biophys.* **57,** 94–106 (1955).
12. Rapport, M. M., and Graf, L. *Prog. Allergy* **13,** 273–331 (1969).
13. Joffe, S., Rapport, M. M., and Graf, L. *Nature (London)* **197,** 60 (1963).
14. Hakomori, S., and Strycharz, G. D. *Biochemistry* **7,** 1285–1286 (1968).
15. Marcus, D. M., and Cass, L. *Science* **164,** 553–555 (1969).
16. Hakomori, S., Siddiqui, B., Li, Y.-T., Li, S.-C., and Hellerqvist, C. G. *J. Biol. Chem.* **246,** 2271–2277 (1971).
17. Siddiqui, B., and Hakomori, S. *J. Biol. Chem.* **246,** 5766–5769 (1971).
18. Naiki, M., and Marcus, D. M. *Biochemistry* **14,** 4837–4841 (1975).
19. Marcus, D. M., Naiki, M., and Kundu, S. K. *Proc. Natl. Acad. Sci. U.S.A.* **73,** 3263–3267 (1976).
20. Niemann, H., Watanabe, K., Hakomori, S., Childs, R. A., and Feizi, T. *Biochem. Biophys. Res. Commun.* **81,** 1286–1293 (1978).
21. Watanabe, K., Hakomori, S., Childs, R. A., and Feizi, T. *J. Biol. Chem.* **254,** 3221–3228 (1979).
22. Feizi, T., Childs, R. A., Watanabe, K., and Hakomori, S. *J. Exp. Med.* **149,** 975–980 (1979).
23. Stein-Douglas, K., Schwarting, G. A., Naiki, M., and Marcus, D. M. *J. Exp. Med.* **143,** 822–832 (1976).
24. Stein, K. N., Schwarting, G. A., and Marcus, D. M. *J. Immunol.* **120,** 676–679 (1978).
25. Hakomori, S., Wang, S.-H., and Young, W. W., Jr. *Proc. Natl. Acad. Sci. U.S.A.* **74,** 3023–3027 (1977).
26. Gahmberg, C. G., and Hakomori, S. *J. Biol. Chem.* **250,** 2438–2446 (1975).
27. Stellner, K., Saito, H., and Hakomori, S. *Arch. Biochem. Biophys.* **155,** 464–472 (1973).
28. Hakomori, S., Nudelman, E., Kannagi, R., and Levery, S. B. *Biochem. Biophys. Res. Commun.* **109,** 36 (1982).
29. Naiki, M., and Marcus, D. M. *Biochem. Biophys. Res. Commun.* **60,** 1105–1111 (1974).
30. Slomiany, B. L., and Horowitz, M. I. *Immunochemistry* **9,** 1067 (1972).
31. Rapport, M. M., Graf, L., Autilio, L. A., and Norton, W. L. *J. Neurochem.* **11,** 855 (1964).
32. Rapport, M. M., Graf, L., and Schneider, H. *Arch. Biochem. Biophys.* **105,** 431 (1964).
33. Bornstein, M. B., and Apple, S. H. *J. Neuropathol. Exp. Neurol.* **20,** 141 (1961).

34. Fry, J. M., Weissbarth, S., Lehrer, G. M., and Bornstein, M. B. *Science* **183**, 540 (1974).
35. Levine, P., Celano, M. J., and Falkowski, F. *Ann. N. Y. Acad. Sci.* **124**, 456 (1965).
36. Schwarting, G. A., Kundu, S. K., and Marcus, D. M. *Blood* **53**, 186–192 (1979).
37. Hirsch, H. E., and Parks, M. E. *Nature (London)* **264**, 785–787 (1976).
38. Nagai, Y., Momoï, T., Saito, M., Mitsuzawa, E., and Ohtani, S. *Neurosci. Lett.* **2**, 107 (1976).
39. Saida, T., Saida, K., Dorfman, S. H., Silberberg, D. H., Sumner, A. J., Manning, M. C., Lisak, R. P., and Brown, M. J. *Science* **204**, 1103–1106 (1979).
40. Arend, P., and Nijssen, J. *Nature (London)* **269**, 255–257 (1977).
41. Rapport, M. M., Graf, L., Skipski, V. P., and Alonzo, N. F. *Cancer* **12**, 438–445 (1959).
42. Rapport, M. M., Schneider, H., and Graf, L. *Biochim. Biophys. Acta* **137**, 409 (1967).
43. Sundsmo, J., and Hakomori, S. *Biochem. Biophys. Res. Commun.* **68**, 799–806 (1976).
44. Rosenfelder, G., Young, W. W., Jr., and Hakomori, S. *Cancer Res.* **37**, 1333–1339 (1977).
45. Marcus, D. M., and Schwarting, G. A. *Adv. Immunol.* **23**, 203–240 (1976).
46. Kornfeld, R., and Kornfeld, S. *Annu. Rev. Biochem.* **45**, 217 (1976).
47. Naiki, M., Marcus, D. M., and Ledeen, R. J. *Immunol.* **113**, 84–93 (1974).
48. Koscielak, J., Hakomori, S., and Jeanloz, R. W. *Immunochemistry* **5**, 441–455 (1968).
49. Marcus, D. M., and Janis, R. J. *Immunol.* **104**, 1530–1539 (1970).
50. Hakomori, S. J. *Immunol.* **112**, 424–426 (1974).
51. Laine, R. A., Yogeeswaran, G., and Hakomori, S. J. *Biol. Chem.* **249**, 4460–4466 (1974).
52. Razin, S., Prescott, B., and Chanock, R. M. *Proc. Natl. Acad. Sci. U.S.A.* **67**, 590 (1970).
53. Taketomi, T., and Yamakawa, T. *Lipids* **1**, 31 (1966).
54. Taketomi, T., and Yamakawa, T. J. *Biochem. (Tokyo)* **54**, 444 (1963).
55. Arnon, R., Sela, M., Rachaman, E. S., and Shapiro, D. *Eur. J. Biochem.* **2**, 79 (1967).
56. Nagai, Y., and Ohsawa, T. *Jpn. J. Exp. Med.* **44**, 451–464 (1974).
57. Stellner, K., Watanabe, K., and Hakomori, S. *Biochemistry* **12**, 656–661 (1973).
58. Feizi, T., Childs, R. A., Hakomori, S., and Powell, M. E. *Biochem. J.* **173**, 245 (1978).
59. Dupont, M. *Arch. Intern. Med* **9**, 133–167 (1934).
60. Springer, G. F., Horton, R. E., and Forbes, M. J. *Exp. Med.* **110**, 221–244 (1959).
61. Springer, G. F. *Prog. Allergy* **15**, 9–77 (1971).
62. Young, W. W., Jr., Hakomori, S., and Levine, P. J. *Immunol.* **123**, 92–96 (1979).
63. Tsai, C.-M., Zopf, D. A., Yu, R. K., Wistar, R., and Ginsburg, V. *Proc. Natl. Acad. Sci. U.S.A.* **74**, 4591–4594 (1977).
64. Köhler, G., and Milstein, C. *Nature (London)* **256**, 495 (1975).
65. Köhler, G., and Milstein, C. *Eur. J. Immunol.* **6**, 511 (1976).
66. Young, W. W., Jr., MacDonald, E. M. S., Nowinski, R. C., and Hakomori, S. J. *Exp. Med.* **150**, 1008–1019 (1979).
67. Perlmutter, R., Hansburg, D., Briles, D. E., Nicolotti, R. A., and Davie, J. M. J. *Immunol.* **121**, 566 (1978).
68. Schmitt-Verhulst, A.-M., Pettinelli, C. B., Henkart, P. A., Lunney, J. K., and Shearer, G. M. J. *Exp. Med.* **147**, 352 (1978).
69. Young, W. W., Jr., and Hakomori, S. *Fed. Proc. Fed. Am. Soc. Exp. Biol.* **38**, 5247 (1979) (abstract).
70. Sela, B.-A. *Cell. Immunol.* **49**, 196–201 (1980).
71. Watanabe, K., Hakomori, S., Powell, M. E., and Yokota, M. *Biochem, Biophys. Res. Commun.* **92**, 638–646 (1980).
72. Kinsky, S. C. *Biochim. Biophys. Acta* **265**, 1–23 (1972).
73. Lennette, E. H. In: *Diagnostic Procedures for Viral and Rickettsial Infections* (Lennette, E. H., and Schmidt, N. J., eds.), American Public Health Association, New York (1969), pp. 52–54.
74. Wasserman, E., and Levine, L. J. *Immunol.* **87**, 290 (1960).
75. Six, H. R., Young, W. W., Jr., Uemura, K., and Kinsky, S. C. *Biochemistry* **13**, 4050 (1974).
76. Geiger, B., and Smolarsky, M. J. *Immunol. Methods* **17**, 7–19 (1977).
77. Wei, R., Alving, C. R., Richards, R. L., and Copeland, E. S. J. *Immunol. Methods* **9**, 165–170 (1975).
78. Fry, J. M., Lisak, R. P., Manning, M. C., and Silberberg, D. H. J. *Immunol. Methods* **11**, 185–193 (1976).
79. Hamers, M. N., Donker-Icoopman, W. E., Reijngoud, D., Schram, A. W., and Tager, J. M. *Immunochemistry* **15**, 97–105 (1978).

80. Young, W. W., Jr., Regimbal, J. W., and Hakomori, S. *J. Immunol. Methods* **28,** 59–69 (1979).
81. Holmgren, J. *Infect. Immun.* **8,** 851 (1973).
82. Parks, D. R., Bryan, V. M., Oi, V. T., and Herzenberg, L. A. *Proc. Natl. Acad. Sci. U.S.A.* **76,** 1962–1966 (1979).
83. Leserman, L. D., Weinstein, J. N., Blumenthal, R., Sharrow, S. O., and Terry, W. D. *J. Immunol.* **122,** 585–591 (1979).
84. Forssman, J. *Biochem. Z.* **37,** 78 (1911).
85. Yamakawa, T., Irie, R., and Iwanaga, M. *J. Biochem.* **48,** 490–507 (1960).
86. Makita, A., Suzuki, C., and Yosizawa, Z. *J. Biochem. (Tokyo)* **60,** 502 (1966).
87. Kijimoto-Ochiai, S., Yokosawa, N., and Makita, A. *J. Biol. Chem.* **255,** 9719–9723 (1980).
88. Sung, S.-J., Esselman, W. J., and Sweeley, C. C. *J. Biol. Chem.* **248,** 6528–6533 (1973).
89. Taketomi, T., Hara, A., Kawamura, N., and Hayashi, M. *J. Biochem. (Tokyo)* **75,** 197 (1974).
90. Ziolkowski, C. H. J., Fraser, B. A., and Mallette, M. F. *Immunochemistry* **12,** 297–302 (1975).
91. Karlsson, K.-A., Pascher, I., Pimlott, W., and Samuelson, B. E. *Biomed. Mass Spectrometry* **1,** 49–56 (1974).
92. Slomiany, A., and Slomiany, B. L. *Eur. J. Biochem.* **76,** 491–498 (1977).
93. Coligan, J. E., Fraser, B. A., and Kindt, T. J. *J. Immunol.* **118,** 6–11 (1977).
94. Kijimoto, S., Ishibashi, T., and Makita, A. *Biochem. Biophys. Res. Commun.* **56,** 177 (1974).
95. Hanganutziu, M. *C. R. Soc. Biol. (Paris)* **91,** 1457–1459 (1924).
96. Deicher, H. *Z. Hyg. Infekttions kr.* **106,** 561–579 (1926).
97. Kano, K., and Milgrom, F. *Curr. Top. Microbiol. Immunol.* **77,** 43–69 (1977).
98. Schiff, F. *J. Immunol.* **33,** 305–313 (1937).
99. Higashi, H., Naiki, M., Matsuo, S., and Okouchi, K. *Biochem. Biophys. Res. Commun.* **79,** 388–395 (1977).
100. Merrick, J. M., Zadarlik, K., and Milgrom, F. *Int. Arch. Allergy Appl. Immunol.* **57,** 477–480 (1978).
101. Kabat, E. A. *Blood Group Substances,* Academic Press, New York (1956).
102. Hakomori, S., and Kobata, A. In: *The Antigens,* Vol. 2, (Sela, M., ed.), Academic Press, New York (1974) pp. 79–140.
103. Hallauer, C. *Z. Immunitätsforsch. Exp. Ther.* **83,** 114 (1934).
104. Kossjakow, P. N., and Tribulew, G. P. *Immunitätsforsch. Exp. Ther.* **98,** 261 (1940).
105. Hamasoto, Y. *Tohoku J. Exp. Med.* **52,** 17, 29, 35 (1950).
106. Masamune, H., and Siojima, S. *Tohoku J. Exp. Med.* **54,** 319 (1951).
107. Fukuda, M. N., and Hakomori, S. *J. Biol. Chem.* **257,** 446 (1982).
108. Yamakawa, T., and Suzuki, S. *J. Biochem.* **39,** 393–402 (1952).
109. Yamakawa, T., and Iida, T. *Jpn. J. Exp. Med.* **23,** 327 (1953).
110. Yamakawa, T., Matsumoto, M., and Suzuki, S. *J. Biochem. (Tokyo)* **43,** 63–72 (1956).
111. Radin, N. S. *Fed. Proc. Fed. Am. Soc. Exp. Biol.* **16,** 825 (1957).
112. Yamakawa, T., Irie, R., and Iwanaga, M. *J. Biochem. (Tokyo)* **48,** 490–507 (1960).
113. Hakomori, S., and Jeanloz, R. W. *J. Biol. Chem.* **236,** 2827–2834 (1961).
114. Handa, S. *Jpn. J. Exp. Med.* **33,** 347 (1963).
115. Koscielak, J. *Biochim. Biophys. Acta* **78,** 313 (1963).
116. Watkins, W. M. In: *Glycoproteins* (Gottschalk, A. ed.) Elsevier, Amsterdam (1966) pp. 462–515.
117. Watkins, W. M. In: *Advances in Human Genetics,* Vol. 10 (Harris, H., and Hirschorn, K., eds.) Plenum Press, New York (1980) pp. 1–136.
118. Lloyd, K. O., and Kabat, E. A. *Proc. Natl. Acad. Sci. U.S.A.* **61,** 1470 (1968).
119. Schiffman, G., Kabat, E. A., and Thompson, W. *Biochemistry* **3,** 113, 587 (1964).
120. Lloyd, K. O., and Kabat, E. A. *Biochem. Biophys. Res. Commun.* **16,** 385 (1964).
121. Hakomori, S., and Jeanloz, R. W. *J. Biol. Chem.* **239,** 3606–3607 (1964).
122. Hakomori, S., Kóscielak, J., Bloch, K. J., and Jeanloz, R. W. *J. Immunol.* **98,** 31–38 (1967).
123. Yang, H.-J., and Hakomori, S. *J. Biol. Chem.* **246,** 1192–1200 (1971).
124. Hakomori, S., Watanabe, K., and Laine, R. A. In: *Human Blood Groups, Proceedings of the 5th International Convocation on Immunology* (Mohn, J. F., Plunkett, R. W., Cunningham, R. K., and Lambert, R. M., eds.), S. Karger, Basel (1977), pp. 150–163.
125. Marcus, D. M., and Cass, L. *J. Immunol.* **99,** 987–993 (1967).
126. Kóscielak, J., Piasek, A., and Gorniak, H. In: *Blood and Tissue Antigens* (Aminoff, D., ed.), Academic Press, New York, (1970), pp. 163–183.

127. Gardas, A., and Kóscielak, J. *Vox Sang.* **20,** 137–149 (1971).
128. Smith, E. L., McKibbin, J. M., Karlsson, K.-A., Pascher, I., and Samuelsson, B. E. *Biochim. Biophys. Acta* **388,** 171–179 (1975).
129. Watanabe, K., Laine, R. A., and Hakomori, S. *Biochemistry* **14,** 2725–2733 (1975).
130. Kóscielak, J., Piasek, A., Gorniak, H., and Gregor, A. *Eur. J. Biochem.* **37,** 214 (1973).
131. Hakomori, S., Stellner, K., and Watanabe, K. *Biochem. Biophys. Res. Commun.* **49,** 1061–1068 (1972).
132. Hakomori, S., Watanabe, K., and Laine, R. A. *Pure Appl. Chem.* **49,** 1215–1227 (1977).
133. Slomiany, B. L., Slomiany, A., and Horowitz, M. I. *Eur. J. Biochem.* **43,** 161–165 (1974).
134. Smith, E. L., McKibbin, J. M., Karlsson, K. A., Pascher, I., Samuelsson, B. E., and Li, S.-C. *Biochemistry* **14,** 3370–3376 (1975).
135. Wiegandt, H. *Hoppe-Seyler's Z. Physiol. Chem.* **354,** 1049–1056 (1973).
136. Suzuki, A., Ishizuka, I., and Yamakawa, T. *J. Biochem. (Tokyo)* **78,** 947–954 (1975).
137. Matsumoto, M., and Taki, T. *Biochem. Biophys. Res. Commun.* **71,** 472–475 (1975).
138. Ando, S., and Yamakawa, T. *J. Biochem. (Tokyo)* **73,** 387–396 (1973).
139. Hanfland, P., and Egge, H. *Chem. Phys. Lipids* **16,** 201–214 (1976).
140. McKibbin, J. M., Smith, E. L., Månsson, J.-E., and Li, Y.-T. *Biochemistry* **16,** 1223–1228 (1977).
141. Smith, E. L., McKibbin, J. M., Karlsson, K.-A., Pascher, I., and Samuelsson, B. E. *Biochemistry* **14,** 2120–2124 (1975).
142. Smith, E. L., McKibbin, J. M., Breimer, M. E., Karlsson, K.-A., Pascher, I., and Samuelsson, B. E. *Biochim. Biophys. Acta* **398,** 84–91 (1975).
143. Slomiany, B. L., and Horowitz, M. I. *J. Biol. Chem.* **248,** 6232–6238 (1973).
144. Slomiany, A., Slomiany, B. L., and Horowitz, M. I., *J. Biol. Chem.* **249,** 1225–1230 (1974).
145. Slomiany, B. L., Slomiany, A., and Horowitz, M. I. *Eur. J. Biochem.* **56,** 353–358 (1975).
146. Slomiany, B. L., Slomiany, A., and Horowitz, M. I. *Biochim. Biophys. Acta* **326,** 224–231 (1973).
147. Slomiany, A., and Slomiany, B. L. *Biochim. Biophys. Acta* **388,** 135–145 (1975).
148. Slomiany, B. L., and Slomiany, A. *Biochim. Biophys. Acta* **486,** 531–540 (1977).
149. Slomiany, B. L., and Slomiany, A. *Chem. Phys. Lipids* **20,** 57–69 (1977).
150. Wherrett, J. R., and Hakomori, S. *J. Biol. Chem.* **248,** 3046–3051 (1973).
151. Hanfland, P., and Egge, H. *Chem. Phys. Lipids* **15,** 243–247 (1975).
152. Hanfland, P. *Chem. Phys. Lipids* **15,** 105–124 (1975).
153. Tilley, C. A., Crookston, M. C., Brown, B. L., and Wherrett, J. R. *Vox Sang.* **28,** 25–33 (1975).
154. Sneath, J. S., and Sneath, P. H. A. *Nature (London)* **176,** 172 (1955).
155. Mäkela, O., Mäkela, P. H., and Kortekangas, A. *Ann. Med. Exp. Biol. Fenn.* **45,** 159–164 (1967).
156. Watkins, W. M., and Morgan, W. T. J. *Nature (London)* **180,** 1038–1040 (1957).
157. Watkins, W. M., and Morgan, W. T. J. *Vox Sang.* **7,** 129 (1962).
158. Hanfland, P. *Eur. J. Biochem.* **87,** 161 (1978).
159. Hakomori, S., and Andrews, H. D. *Biochim. Biophys. Acta* **202,** 225–228 (1970).
160. Smith, E. L., McKibbin, J. M., Karlsson, K-A., Pascher, I., Samuelsson, B. E., Li, Y.-T., and Li, S.-C. *J. Biol. Chem.* **250,** 6059–6064 (1975).
161. Wherrett, J. R., Brown, B. L., Tilley, C. A., and Crookston, M. C. *Clin. Res.* **29,** 784 (1971).
162. Jordal, K. *Acta Pathol. Microbiol. Scand.* **42,** 269–284 (1958).
163. Sturgeon, P. H., and Archilla, M. B. *Vox Sang.* **18,** 301–322 (1970).
164. Kabat, E. A., Liao, J., Shyong, J., and Osserman, E. F. *J. Immunol.* **128,** 540 (1982).
165. Rauvala, H. *Biochim. Biophys. Acta* **424,** 284–295 (1976).
166. Rauvala, H. *J. Biol. Chem.* **251,** 7517–7520 (1976).
167. Race, R. R., and Sanger, R. *Blood Groups in Man,* 5th ed., Blackwell, Oxford, (1968), pp. 136–170.
168. Fellous, M., Gerbal, A., Tessier, C., Frezal, J., Dausset, J., and Salmon, C. *Vox Sang.* **26,** 518–538 (1974).
169. Fellous, M., Couillin, P., Neauport-Santes, C., Frezal, J., Bilardon, C., and Dausset, J. *Eur. J. Immunol.* **3,** 543 (1973).

170. Cameron, G. L., and Stavely, J. M. *Nature (London)* **179**, 147 (1957).
171. Watkins, W. M., and Morgan, W. T. In: *Proceedings of the 9th Congress International Society of Blood Transfusion, Mexico, 1962.* Karger, Basel, (1964) pp. 230–234.
172. Marcus, D. M. *Transfusion* **11**, 16 (1971).
173. Schwarting, G. A., Marcus, D. M., and Metaxas, M. *Vox Sang.* **32**, 257–261 (1977).
174. Race, R. R., and Sanger, R. *Blood Groups in Man*, 4th ed. Blackwell, Oxford (1962), pp. 17–74.
175. Wiener, A. S., Unger, L. T., Cohen, L., and Feldman, J. *Ann. Intern. Med.* **44**, 221 (1956).
176. Prokop, O., and Uhlenbruch, G. *Human Blood and Serum Groups*, English ed. (Raven J. L., transl.), Maclaren and Sons, London (1965) pp. 302–305.
177. Marsh, W. L. *Br. J. Haematol.* **7**, 200–209 (1961).
178. Christenson, W. N., and Dacie, J. V. *Br. J. Haematol.* **3**, 153–164 (1957).
179. Feizi, T. *Nature (London)* **215**, 540–542 (1967).
180. Giblett, E. F., and Crookston, M. C. *Nature (London)* **201**, 1138 (1964).
181. Dacie, J. V., Lewis, S. M., and Tills, D. *Br. J. Haematol.* **6**, 362 (1960).
182. Maniatis, A., Frieman, B., and Bertles, J. F. *Vox Sang.* **33**, 29–36 (1977).
183. Yeh, M.-Y., Hellström, I., Abe, K., Hakomori, S., and Hellström, K. E. *Int. J. Cancer* **29**, 269 (1982).
184. Crookston, J. H., Crookston, M. C., Burnie, K. L., Francombe, W. H., Dacie, J. V., Davis, J. A., and Lewis, S. M. *Br. J. Haematol.* **17**, 11 (1969).
185. Marcus, D. M., Kabat, E. A., and Rosenfield, R. E. *J. Exp. Med.* **118**, 175 (1963).
186. Feizi, T., Kabat, E. A., Vicai, G., Anderson, B., and Marsh, W. L. *J. Exp. Med* **133**, 39–52 (1971).
187. Feizi, T., Kabat, E. A., Vicai, G., Anderson, B., and Marsh, W. L. *J. Immunol.* **106**, 1578–1592 (1971).
188. Hakomori, S., and Watanabe, K. in: *Glycolipid Methodology* (Witting, L. A. ed.), American Oil Chemists' Society Press, Champaign, Illinois, (1976), pp. 13–47.
189. Gardas, A. *Eur. J. Biochem.* **68**, 185–191 (1976).
190. Fukuda, M. N., Fukuda, M., and Hakomori, S. *J. Biol. Chem.* **254**, 5458–5465 (1979).
191. Fukuda, M., Fukuda, M. N., and Hakomori, S. *J. Biol. Chem.* **254**, 3700–3703 (1979).
192. Bird, G. W. G., and Wingham, J. *Br. J. Haematol.* **23**, 759–762 (1972).
193. Stormont, C. *Proc Natl. Acad. Sci. U.S.A.* **35**, 323 (1949).
194. Stone, W. H. *Ann. N. Y. Acad. Sci.* **97**, 269 (1962).
195. Schmid, D. O. Z. *Immunitätsforsch. Exp. Ther.* **124**, 101 (1962).
196. Mäkela, O., and Mäkela, P. *Ann. Med. Exp. Biol. Fenn.* **34**, 157 (1956).
197. Rasmussen, A. *Ann. N. Y. Acad. Sci.* **97**, 306 (1962).
198. Andresen, E. *Ann. N. Y. Acad. Sci.* **97**, 205 (1962).
199. Thiele, O. W., and Koch, J. *Eur. J. Biochem.* **14**, 379 (1970).
200. Schröffel, J., Thiele, O. W., and Koch, J. *Eur. J. Biochem.* **22**, 294 (1971).
201. Schröffel, J., Radas, A., Thiele, O. W., and Koch, J. *Eur. J. Biochem.* **22**, 396 (1971).
202. Horowitz, M. I., and Slomiany, B. L. In: *Blood and Tissue Antigens* (Aminoff, D., ed.), Academic Press, New York, (1970), p. 131.
203. Lehman-Facius, M. Z. *Immunitätsforsch.* **56**, 464–515 (1928).
204. Kobayashi, K. *Tohoku J. Exp. Med.* **63**, 185–189 (1956).
205. Kobayashi, K. *Tohoku J. Exp. Med.* **63**, 191–202 (1956).
206. Kobayashi, K. *Tohoku J. Exp. Med.* **63**, 203–220 (1956).
207. Rapport, M. M., Graf, L., and Alonzo, N. *Cancer* **8**, 546 (1955).
208. Laine, R. A., Sweeley, C. C., Li, Y.-T., Kisic, A., and Rapport, M. M. *J. Lipid Res.* **13**, 519–524 (1972).
209. Shapiro, D., and Rachman, E. S. *Nature (London)* **201**, 878 (1964).
210. Rapport, M. M., and Graf, L. *Nature (London)* **201**, 879 (1964).
211. Graf, L., and Rapport, M. M., *Cancer Res.* **20**, 546 (1960).
212. Jóźwiak, W., and Kościelak, J. *Eur. J. Cancer Clin. Oncol.* **18**, 617 (1982).
213. Kawanami, J. *J. Biochem. (Tokyo)* **72**, 783–785 (1972).
214. Levine, P., Bobitt, O. B., Waller, R. K., and Kuhmichel, A. *Proc. Soc. Exp. Biol. Med.* **77**, 403–405 (1951).

215. Levine, P. *Ann. N. Y. Acad. Sci.* **277,** 428–435 (1976).
216. Galanos, C., Lüderitz, O., and Westphal, O. *Eur. J. Biochem.* **24,** 116–122 (1971).
217. Ansel, S. *Int. J. Cancer* **13,** 773–784 (1974).
218. Ansel, S., Huet, C., and Tournier, P. *Int. J. Cancer* **20,** *51–60 (1977).*
219. Huet, C., and Ansel, S. *Int. J. Cancer* **20,** 61–66 (1977).
220. Portoukalian, J., Zwingelstein, G., Dore, J.-F., and Bonrgoin, J.-J. *Biochimie* **58,** 1285–1287 (1976).
221. Portoukalian, J., Zwingelstein, G., Abdul-Malak, N., and Dore, J. F. *Biochem. Biophys. Res. Commun.* **85,** 916–920 (1978).
222. Theofilopoulos, A. N., Andrews, B. S., Urist, M. M., Morton, D. L., and Dixon, F. J. *J. Immunol.* **119,** 657–663 (1977).
223. Portoukalian, J., Zwingelstein, G., and Dore, J.-F. *Eur. J. Biochem.* **94,** 19–23 (1979).
224. Gupta, R. K., Irie, R. F., Chee, D. O., Kern, D. H., and Morton, D. L. *J. Natl. Cancer Inst.* **63,** 347–356 (1979).
225. Oh-Uti, K. *Tohoku J. Exp. Med.* **51,** 297–304 (1949).
226. Masamune, H., Kawasaki, H., Abe, S., Oyama, K., and Yamaguchi, Y. *Tohoku J. Exp. Med.* **68,** 81–91 (1958).
227. Davidsohn, I., Kovarik, S., and Lee, C. L. *Arch. Pathol.* **81,** 381–390 (1966).
228. Davidsohn, I., Kovarik, S., and Ni, Y. *Arch. Pathol.* **87,** 306–314 (1969).
229. Simmons, D. A. R., and Pearlman, P. *Cancer Res.* **33,** 313–322 (1973).
230. Feizi, T., Turberville, C., and Westwood, J. H. *Lancet* **2,** 391–393 (1975).
231. Watanabe, K., and Hakomori, S. *J. Exp. Med.* **144,** 644–653 (1976).
232. Springer, G. F., Desai, P. R., and Banatwala, I. *J. Natl. Cancer Inst.* **54,** 335–338 (1975).
233. Häkkinen, I. *J. Natl. Cancer Inst.* **44,** 1183–1188 (1970).
234. Masamune H., Yosizawa, Z., Oh-Uti, T., Matsuda, J., and Masukawa, A. *Tohoku J. Exp. Med.* **56,** 37–42 (1952).
235. Kawasaki, H. *Tohoku J. Exp. Med.* **68,** 119–132 (1958).
236. Iseki, S., Furukawa, K., and Ishihara, K. *Proc Jpn. Acad. Sci.* **38,** 556–566 (1962).
237. Kay. H. E. H., and Wallace, B. H. *J. Natl. Cancer Inst.* **26,** 1349–1366 (1961).
238. Nairn, R. C., Fothergill, J., and McEntegart, H. *Br. Med. J.* **1,** 1791–1793 (1962).
239. Prendergast, R. C., Tot, P. D., and Gargiulo, A. W. *J. Dent. Res.* **47,** 306–310 (1968).
240. Dabelsteen, E., and Fulling, J. H. *Scand. J. Dent. Res.* **79,** 387–393 (1971).
241. Stellner, K., Hakomori, S., and Warner, G. A. *Biochem. Biophys. Res. Commun.* **55,** 439–455 (1973).
242. Kannagi, R., Levine, P., Watanabe, K., and Hakomori, S. *Cancer Res.* **42,** 5249 (1982).
243. Denk, H., Tappeiner, G., Davidovits, A., Eckerstorfer, R., and Holzner, J. H. *J. Natl. Cancer Inst.* **53,** 933–938 (1974).
244. Breimer, M. E. *Cancer Res.* **40,** 897–908 (1980).
245. Mourant, A. E., Kopec, A. C., and Domaniewska-Sobezak, K. In: *Blood Groups and Diseases: A Study of Association of Diseases with Blood Groups and Other Polymorphisms,* Oxford University Press, New York (1978), pp. 13–15.
246. Brandt, R., Guth, H., and Muller, R. *Klin. Wochenschr.* **5,** 655 (1926).
247. Autilio, L. A., Norton, W., Kay, T., and Terry, R. D. *J. Neurochem.* **11,** 17 (1964).
248. Raff, M. C., Mirsky, R., Fields, K. L., Lisak, R. P., Dorfman, S. H., Silberberg, D. H., Gregson, N. A., Leibowitz, S., and Kennedy, M. C. *Nature (London)* **272,** 813 (1979).
249. Yamakawa, T., Yokayama, S., and Handa, N. *J. Biochem. (Tokyo)* **53,** 28 (1963).
250. Rapport, M. M., Schneider, H., and Graf, L. *Biochim. Biophys. Acta* **137,** 409 (1967).
251. Siddiqui, B., Kawanami, J., Li, Y.-T., and Hakomori, S. *J. Lipid Res.* **13,** 657–662 (1972).
252. Arita, H., and Kawanami, J. *J. Biochem. (Tokyo)* **81,** 1661–1664 (1977).
253. Hakomori, S. *Vox Sang.* **16,** 478–484 (1969).
254. Koscielak, J., Hakomori, S., and Jeanloz, R. W. *Immunochemistry* **5,** 441–455 (1968).
255. Levine, M., Bain, J., Narasimhan, R., Palmer, B., Yates, A. J., and Murray, R. K. *Biochim. Biophys. Acta* **441,** 134 (1976).
256. Suzuki, A., Ishizuka, I., Ueta, N., and Yamakawa, T. *Jpn. J. Exp. Med.* **43,** 435 (1973).
257. Ishizuka, I., Suzuki, M., and Yamakawa, T. *J. Biochem. (Tokyo)* **73,** 77 (1973).
258. Kornblatt, M. J., Knapp, A., Levine, M., Schachter, H., and Murray, R. K., *Can. J. Biochem.* **52,** 689 (1974).

259. Letts, P. J., Hunt, R. C., Shirley, M. A., Pinterie, L., and Schachter, H. *Biochim. Biophys. Acta* **541,** 59 (1978).
260. Lingwood, C. A., Murray, R. K., and Schachter, H. *J. Immunol.* **124,** 769–774 (1980).
261. Reif, A. E., and Allen, J. M. *Nature (London)* **209,** 521 (1966).
262. Vitetta, E. S., Boyse, E. A., and Uhr, J. W. *Eur. J. Immunol.* **3,** 446 (1973).
263. Esselman, W. J., and Miller, H. C. *Fed. Proc. Fed. Am. Soc. Exp. Biol.* **33,** 771 (1974).
264. Miller, H. C., and Esselman, W. J. *Ann. N. Y. Acad. Sci.* **249,** 54 (1975).
265. Thiele, H.-G., Arndt, R., and Stark, R. *Immunology* **32,** 767 (1977).
266. Milewicz, C., Miller, H. C., and Esselman, W. J. *J. Immunol.* **117,** 1774–1780 (1976).
267. Pukel, C. S., Lloyd, K. O., Trabassos, L. R., Dippold, W. G., Oettgen, H. F., and Old, L. J. *J. Exp. Med.* **155,** 1133 (1982).
268. Nudelman, E., Hakomori, S., Kannagi, R., Levery, S., Yeh, M.-Y., Hellström, K. E., and Hellström, I. *J. Biol. Chem.* **257,** 12752 (1982).
269. Wang, T. J., Freimuth, W. W., Miller, H. C., and Esselman, W. J. *J. Immunol.* **121,** 1361–1365 (1978).
270. Kato, K. P., Wang, T. J., and Esselman, W. J., *J. Immunol.* **123,** 1977–1984 (1979).
271. Young, W. W., Jr., Hakomori, S., Durdik, J. M., and Henney, C. S. *J. Immunol.* **124,** 199–201 (1980).
272. Kasai, M., Iwamori, M., Nagai, Y., Okumura, K., and Tada, T. *Eur. J. Immunol.* **10,** 175 (1980).
273. Nakahara, K., Ohashi, T., Oda, T., Hirano, T., Kasai, M., Okumura, K., and Tada, T. *N. Engl. J. Med.* **302,** 674–677 (1980).
274. Remold, H. G. *J. Exp. Med.* **138,** 1065 (1973).
275. Fox, R. A., Gregory, D. S., and Feldman, J. D. *J. Immunol.* **112,** 1867–1872 (1974).
276. Poste, G., Allen, H., and Matta, K. L. *Cell. Immunol.* **44,** 89–98 (1979).
277. Higgins, T. J., Sabatino, A. P., Remold, H. G., and David, J. R. *J. Immunol.* **121,** 880–886 (1978).
278. Poste, G., Kirsh, R., and Fidler, I. J. *Cell. Immunol.* **44,** 71–88 (1979).
279. Hanada, E., Konno, K., Handa, S., and Yamakawa, T. *J. Biochem. (Tokyo)* **81,** 1079–1083 (1977).
280. Handa, E., Hanada, S., Konno, K., and Yamakawa, T. *J. Biochem. (Tokyo)* **83,** 85–90 (1978).
281. Miura, T., Handa, S., and Yamakawa, T. *J. Biochem. (Tokyo)* **86,** 773–776 (1979).
282. Niedieck, B., and Pette, E. *Klin. Wochenschr.* **41,** 773 (1963).
283. Niedieck, B., Kuwert, E., Palacios, O., and Drees, O. *Ann. N. Y. Acad. Sci.* **122,** 266 (1965).
284. Dubois-Daloq, M., Niedieck, B., and Buyse, M. *Pathol. Eur.* **5,** 331 (1970).
285. Fry, J. M., Weissbarth, S., Lehrer, G. M., and Bornstein, M. B. *Science* **183,** 540 (1974).
286. Hruby, S., Alvord, E. C., Jr., and Seil, F. J. *Science* **195,** 173–175 (1977).
287. Young, W. W., Jr., Hakomori, S., and Levine, P. *J. Immunol.* **123,** 92–96 (1979).
288. Young, W. W., Jr., and Hakomori, S. *Science* **211,** 487 (1981).
289. Ansel, S., and Huet, C. *Int. J. Cancer* **25,** 797 (1980).
290. Yokota, M., Warner, G., and Hakomori, S. *Cancer Res.* **41,** 4185 (1981).
291. Dippold, W. G., Lloyd, K. O., Li, L. T. C., Ikeda, H., Oettgen, H. F., and Old. L. J. *Proc. Natl. Acad. Sci. U.S.A.* **77,** 6114–6118 (1980).
292. Young, W. W., Jr., Durdik, J. M., Urdal, D., Hakomori, S., and Henney, C. S. *J. Immunol.* **126,** 1 (1981).
293. Urdal, D. L., and Hakomori, S. *J. Biol. Chem.* **255,** 10,509 (1980).
294. Koprowski, H., Herlyn, M., Steplewski, Z., and Sears, H. F. *Science* **212,** 53 (1981).
295. Magnani, J. L., Brockhaus, M., Smith, D. F., Ginsburg, V., Blaszczyk, M., Mitchell, K. F., Steplewski, Z., and Koprowski, H. *Science* **212,** 55–56 (1981).
296. Sela, B. A. *Eur. J. Immunol.* **11,** 347–349 (1981).
297. Sieff, C., Bicknell, D., Robinson, J., Lam, G., and Greaves, M. F. *Blood* **60,** 703 (1982).
298. Uemura, K., Hattori, H., and Taketomi, T. In: *Glycoconjugates, Proceedings of the 6th International Symposium on Glycoconjugates* (Yamakawa, T., Osawa, T., and Handa, S., eds.) Japan Scientific Soc. Press, Tokyo (1981) pp. 124–125.
299. Nudelman, E., Hakomori, S., Kannagi, R., Levery, S., Yeh, M.-Y., Hellström, I., and Hellström, K.-E. *J. Biol. Chem.* **257,** 12752 (1982).
300. Buchbinder, L. *Arch. Path. (Chicago)* **19,** 841 (1935).

301. Hattori, H., Uemura, K., and Taketomi, T. *Biochim. Biophys. Acta* **666**, 361 (1981).
302. Graham, H. A., Hirsch, H. F., and Davies, O. M. In: *Human Blood Groups, 5th International Convocation Immunology* (Mohn, J. F., ed.) (1976) pp. 257–267.
303. Lemieux, R. U., Baker, D. A., Weinstein, W. M., and Switzer, C. M. *Biochemistry* **20**, 199 (1981).
304. Pendu, J. L., Lemieux, R. U., and Oriol, R. *Vox Sang.* **43**, 188 (1982).
305. Archilla, M. B., and Sturgeon, P. *Vox Sang.* **26**, 425 (1974).
306. Wiels, J., Fellows, M., and Tursz, T. *Proc. Natl. Acad. Sci. U.S.A.* **78**, 6485 (1981).
307. Nudelman, E., Kannagi, R., Hakomori, S., Lipinski, M., Wiels, J., Parsons, M., Fellous, M., and Tursz, T. *Science* (in press).
308. Magnani, J., Nilsson, B., Brockhaus, M., Zopf, D., Steplewski, Z., Koprowski, H., and Ginsburg, V. *J. Biol. Chem.* **257**, 14365 (1982).
309. Kannagi, R., Nudelman, E., Hakomori, S., and Levery, S. B. *J. Biol. Chem.* **257**, 14865 (1982).
310. Hakomori, S., Nudelman, E., Levery, S., Solter, D., and Knowles, B. B. *Biochem. Biophys. Res. Commun.* **100**, 1578 (1981).
311. Karlsson, K.-A., and Larsson, G. *J. Biol. Chem.* **254**, 9311 (1979).
312. Schwarting, G. A., Parkinson, D. R., Manson, D., and Zielinski, C. *New England J. Med.* **304**, 300 (1981).
313. Kiguchi, K., Iwamori, M., Nagai, Y., Eto, Y., and Akatsuka, J. *Gann (Jap. J. Cancer Res.)* (in press).
314. Cohen, O., Sela, B.-A., Schwartz, M., Eshkar, N., and Cohen, I. R. *Israel J. Med. Sci.* **17**, 711 (1981).
315. Sela, B., Konat, G., and Offner, H. *Israel J. Med. Sci.* **17**, 38 (1981).

Chapter 6

Glycosphingolipids as Receptors

Julian N. Kanfer

A suggestion was offered a number of years ago that the glycosphingolipids may function as receptors for biological materials, and much of the early attention concerning this possibility was focused upon the gangliosides and the sulfatides, which are the glycosphingolipids possessing a charge at physiological pH. An attractive reason for attempting to assign roles as neurotransmitter receptors for these compounds is the high glycosphingolipid concentration in the nervous system.

6.1. Gangliosides and the Serotonin Receptor

The studies initiated by Woolley and Gommi provided circumstantial evidence for the possibility that ganglioside may act as the naturally occurring receptor molecule for serotonin. These workers found that the serotonin-stimulated contraction of isolated strips of rat stomach was abolished by neuraminidase pretreatment of tissues.[1] The inactivated tissue sample would slowly recover to a level of 10% of that seen with control tissues. Reactivation could also be obtained by exposing the inactivated strips to either a crude lipid extract from normal stomach or a commercial sample of purified beef brain gangliosides. Subsequent studies attempted to identify the ganglioside species that was the most effective in reestablishing serotonin susceptibility to neuraminidase-inactivated strips. These workers concluded that a ganglioside termed "new G" was superior to others tested. This ganglioside could have been disialosyl lactosylceramide (G' lact, or GD_3); however, unequivocal structural evidence was not provided.[2] Subsequent work from another laboratory clarified the structure[3] and demonstrated the ability of serotonin to bind to this ganglioside species.[4] An examination was undertaken of the effects of inhibitors of sialic acid biosynthesis on the serotonin-induced contraction of neuraminidase-treated stomach strips. The results led the investigators to conclude that sialic acid is a component of the serotonin receptor.[5]

Uterine muscle membrane was treated with neuraminidase, and free sialic acid was found to be released only from the nonlipid sialic acid-containing compounds with no loss of lipid-bound sialic acid being detected. These observations led to the suggestion that gangliosides are not serotonin receptors.[6] Cat brain nerve-ending membranes were isolated and shown to be active in [^{14}C]serotonin binding. It was found, after a particular extraction procedure, that the radioactivity was recovered in the butanol phase while the gangliosides appeared in the aqueous phase. There was no effect of or increased uptake by the addition of gangliosides to these incubation mixtures. These workers concluded that a "proteolipid" rather than a ganglioside was responsible for the binding of serotonin.[7] This was confirmed by others who proposed that gangliosides are not responsible for serotonin uptake but rather that a sialoprotein or a sialolipid is the receptor.[8] The serotonin binding to gangliosides in model systems is diminished when gangliosides are present in a complex with cerebrosides.[9]

Platelets actively bind and store serotonin and these cells have gangliosides, 92% of which is GM_3 with small quantities of GM_1 and GD_{1a}. Both of the purified monosialogangliosides bind small amounts of serotonin. Platelet lipid mixtures, which contain 8% gangliosides, 66% phospholipid, and 14% protein, as well as "crude" mixtures obtained from other tissues irreversibly bind labeled serotonin. These investigators do not regard gangliosides as the receptor but suggest that they may function merely as a "recognition" process for hydroxyindole compounds.[10]

Brief treatment of human platelets with neuraminidase has been shown to result in enhanced serotonin uptake; however, more extensive treatment caused a decreased uptake by these cells. This led to the conclusion that sialic acid present on the outer cell surface is not involved with serotonin uptake but that intermembranous sialic acid-containing proteins or lipids may be involved in the transport process.[10a]

The evidence available in the literature suggests that some sort of sialoconjugate may be involved in the serotonin receptor mechanism. This may or may not be a glycosphingolipid such as a ganglioside. The native receptor may be present only in trace quantities and undetectable by conventional analytical techniques. Thus, the cited study[7] on the butanol distribution of serotonin in the absence of detectable gangliosides should be taken with the reservation that a minute amount of the "correct" ganglioside could have been associated with the transmitter. It should also be remembered that commercial preparations of this enzyme have been shown to often contain other enzymes and chemical stabilizers, which can affect tissues and tissue extracts. Therefore, the reduced responsiveness of neuraminidase-treated tissue may not have been due solely to sialic acid release. Neuraminidase treatment of biological material is a common general approach for implicating a biological role for sialic acid in a variety of situations. It should be emphasized, however, that sialic acid release by such treatment is rarely documented.

6.2. Gangliosides and Acetylcholine

A correlation was suggested between the enrichment of both acetylcholine and gangliosides in the synaptic vesicles. This observation led to the hypothesis that the gangliosides may be responsible for the binding and subsequent

Table I. *Effect of Calcium Ion Concentration on Gangliosides and Proteins in Synaptic Vesicles and Membrane Fractions*[a,b]

Fraction	1×10^{-6} M Ca^{2+}			1×10^{-3} M Ca^{2+}		
	Gangliosides	Proteins	RSA[c]	Gangliosides	Proteins	RSA
SM2A	13%	38%	0.3	5%	12%	0.4
M2A	54%	34%	1.6	9%	33%	0.3
SM2B	14%	3%	4.9	7%	3%	2.5
M2B	19%	25%	0.8	79%	52%	1.5

[a] Modified from Ref. 13.
[b] The synaptic vesicles and membrane fractions were prepared except that part of the M2 pellet was fractionated in the presence of calcium ions at 1×10^{-6} M and part at 1×10^{-3} M.
[c] RSA, relative specific activity.

release of acetylcholine by synaptic vesicles.[11] Subsequently, it was demonstrated that more rigorously prepared synaptic vesicles are practically devoid of gangliosides.[12] This discrepancy may be explained by a methodological difference involving the quantity of Ca^{2+} present in the media used for preparation of these subcellular fractions. The distribution of gangliosides between the several synaptosomal membranous components can be manipulated by suitable adjustments of the Ca^{2+} content of the media as shown in Table I.[13] At 1×10^{-6} M Ca^{2+}, 19% of the gangliosides are in the M2B fraction whereas 79% are in this fraction at 1×10^{-3} M Ca^{2+}. A polysialoganglioside abolished the binding of an antiserum specific to cholinergic terminal to membranes.[13a]

6.3. Gangliosides and Interferon

Interferon binding to mouse L cells was shown to be blocked by several plant lectins including phytohemagglutinin and this interference could be reversed by the addition of fetuin. This observation suggested that the membrane binding sites for interferon contained carbohydrates. Exposure of interferon–Sepharose beads to the gangliosides GM_2 and GT_1 blocked its usual antiviral activity, in addition, gangliosides covalently bound to Sepharose beads could be shown to bind interferon. However, addition of gangliosides to the L cells did not enhance the antiviral activity of interferon as might be expected due to increased uptake as a result of producing a greater number of binding sites.[14] Although GM_3 was less effective than GM_2 or GT_1, it was suggested that the sialosyllactose unit of the ganglioside is involved in the interferon binding (Table II).[15] Preincubation of gangliosides with certain strains of interferon-insensitive cells rendered them responsive to interferon. In contrast to the work on L cells, there did not appear to be any particular ganglioside that was more effective than the commercial sample of sphingolipids employed.[16] Both GM_3 and GM_2 had the capacity to block the antiviral effect of interferon on transformed human fibroblasts. However, the addition of gangliosides to cultures of these cells could not prevent the anticellular effects of interferon. This may suggest that the growth suppression effects of inter-

Table II. Inhibition of Antiviral Activity of Interferon–Sepharose by Glycolipids[a]

Interferon–Sepharose preincubated with	Virus yield (% of control) after treatment of L cells with interferon–Sepharose preincubated with glycolipids in	
	Water	Ethanol
—	0.7	2.5
GM_2	51	100
GM_3	1	16
Ganglio Tri-Cer	1	5
Globo Tri-Cer	0.8	4.7
Globo Tet-Cer	1.2	5

[a] Modified from Ref. 15.

feron may not involve cell surface gangliosides.[16a] A glycoprotein that binds interferon has been obtained from both mouse Ly cells and human K-B3 cells. Surface labeling of K-B3-cell lipid-bound sialic acid or galactose was slightly reduced by the presence of interferon, suggesting a possible ganglioside protection from the labeling reagents, In addition, there was decreased galactose labeling of at least four membrane glycoproteins and certain glycolipids.[16b] These results suggest that perhaps the natural acceptor is a glycoprotein and that certain gangliosides with similar oligosaccharide chain merely "cross-react."

Type I and type II interferons differ with respect to molecular weight, antigenic determinants, stability at low pH, and sources. However, both types seem to have similar effects on the same target cell. Gangliosides prevent the antiviral and antigrowth effects of type I but not type II interferon. Type I but not type II interferon binds to ganglioside affinity columns.[16c] These differences between the interactions of type I and II interferons have been observed independently.[16d] There were no significant differences in the sensitivity to interferon between wild-type L929 mouse fibroblasts and its ganglioside-deficient NCTC 2071 clone.[16e] The ganglioside composition was identical in L1210 cells either resistant or sensitive to interferon.[16f]

6.4. Gangliosides and Bacterial Toxins

There is a fairly extensive literature describing the interactions between gangliosides and toxins produced by a variety of bacterial sources.

6.4.1. Botulinum Toxin

Compounds occurring in brain tissue were evaluated for their ability to inactivate botulinum toxin using both an *in vivo* and an *in vitro* assay system. Evidence was obtained indicating that a trisialoganglioside, GT_1, was the most

Figure 1. Effects of different gangliosides on botulinum toxin-induced paralysis of the isolated phrenic nerve–diaphragm preparation. The straight line plots the dose–response curve obtained by poisoning neuromuscular preparations with various amounts of botulinum toxin (abscissa). Arrows indicate the paralysis times (ordinate) of neuromuscular preparations (five in each group), receiving 5.0 μg of toxin plus 5.0 μg of ganglioside. Increased time until paralysis means decreased toxin potency. (From Ref. 17.)

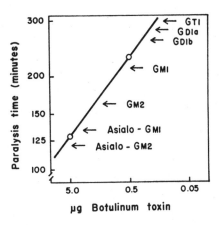

effective compound tested as shown in Fig. 1.[17] The other gangliosides, several sialic acid-free glycosphingolipids, steroids, and fatty acids did not result in toxin inactivation employing these tests.[18] Under "physiological" conditions, phospholipids do not prevent the toxin activity in the test systems. However, it was found that acidic lipids such as phosphatidic acid, cardiolipid, or certain bile acids could inactivate the toxin, whereas phosphatidylcholine enhanced the toxin's activity. It was concluded that phospholipids were not the natural toxin receptor.[19] Proteolipid prepared from brain had no effect on toxin activity.

The observation on the ability of ganglioside to inactivate botulinum toxin was not confirmed by one group.[20] The binding of labeled *C. botulinum* to monkey brain synaptosomes was prevented by GT_{1b}.[20a]

6.4.2. Tetanus Toxin

Tetanus toxin acts on the central nervous system by suppressing synaptic transmission, and early workers demonstrated that premixing the toxin with an emulsion of brain tissue diminished the toxin's activity.[21] A hot ethanol extract of brain tissue termed "protagon," which is enriched in sphingolipids, was shown to have toxin receptor activity.[22] The cerebroside phrenosine was thought to be the receptor molecule; however, this lipid was found to be enriched in white matter while the toxin receptor activity was found enriched in gray matter.

A reproducible assay was developed for the toxin inactivation using lipoidal preparations obtained from mammalian brain tissues. These studies confirmed the earlier observations and showed that a "protagon" fraction prepared from gray matter was superior to that prepared from either whole brain or white matter and that crude but not purified "phrenosine" was quite effective in toxin inactivation. Therefore, it appeared that a gray matter component associated with, but different than cerebrosides was responsible for the effectiveness of "protagon."[23] A spectrophotometric assay was developed for quantitating the protein of tetanus toxin absorbed by the "receptor." It was shown that protagon did not absorb serum globulins, albumins, lysozyme, pepsin, trypsin, *C. welchii* toxins, diphtheria toxin, or dysentery toxins.[24] Using the *in vitro* assay developed for toxin absorption, attempts were undertaken to identify the material present in "protagon" that was responsible for inactivation. It was found that the activity resided in at least two separable components both of which were required since neither possessed activity in the test system. The results of a series of investigations implicated cerebrosides, sphingomyelin, perhaps Ca^{2+}, and an unknown water-soluble component. The *in vitro* assay system developed depended upon binding of the toxin to the "receptor" and subsequent removal of this "insoluble" complex by centrifugation and thus "soluble" complexes would not be observable. It has previously been established that gray matter is superior to white matter for toxin inactivation. It has also been established that gangliosides or "strandin" are enriched in gray matter. Therefore, these complex sialic acid-containing glycosphingolipids were examined as the possible water-soluble component obtained from protagon and it was found that the gangliosides could substitute for the water-soluble component. These experimental observations suggested that gangliosides were responsible for the binding of toxin. The Ca^{2+}, cerebroside, and sphingomyelin interacted with this complex in order to render it insoluble and precipitable in a centrifugal field.[25]

An interaction between gangliosides and tetanus toxin was demonstrated employing both boundary electrophoresis and analytical ultracentrifugation techniques. Ganglioside specificity was suggested since sialic acid-free gangliosides, gangliosides in which the sialic acid carboxyl group was chemically methylated, free sialic acid, and a sialoprotein, ovine mucoid substance, did not bind tetanus toxin. The gangliosides present in the mixture were resolved on a silicic acid column into two fractions. These were designated as "fast" or "slow" with respect to chromatographic migration. The slower migrating ganglioside possessed a higher sialic acid content and was shown to be more effective in tetanus toxin binding activity.[26] The relationship between the relative proportions of ganglioside and cerebroside present in the test system greatly influenced tetanus toxin binding, and optimum binding occurred with a mixture containing 25% gangliosides. This complex did not fix botulinum toxin, albumin, serotonin, or strychnine. The polysialogangliosides GT_1 and GD_{1b}, which possess a disialosyl group attached to the lactosyl unit, were more effective binders than GM_1 and GD_{1a}.[27] Thus, it appears that the cerebrosides, which are less water soluble, may participate by rendering the ganglioside–toxin complexes insoluble.

Studies with subcellular organelles from guinea pig brain showed that fractions enriched in synaptosomes were more effective in toxin binding than those enriched in either mitochondria or microsomes.[28] Subsequently, it was demonstrated that synaptic vesicles possessed little ability to bind the toxin, and this activity was present in a fraction presumably containing external synaptosomal membranes.[29] This active fraction contains more gangliosides than the synaptic vesicles. Studies have been carried out using ^{125}I-labeled tetanus toxin in order to localize its binding. *In vivo* administration showed that it was preferentially found in the gray matter of spinal cord and brain stem but little if any was found associated with forebrain or cerebellum. Presumably, the blood–brain barrier restricts penetration into the central nervous system.[30] ^{125}I-labeled tetanus toxin can bind to the nerve terminals of the neuromuscular junction and spinal cord synaptic endings.[30a] Lyophilized rat brain homogenates bind the toxin more effectively than spinal cord, whereas liver is inactive. A mixture of gangliosides and cerebrosides binds to a greater extent than whole brain. Autoradiography of brain slices treated with labeled toxin shows preferential gray matter localization.[31] Four separate cultured brain cells lacking the polysialogangliosides did not absorb ^{125}I-labeled tetanus toxin, whereas homogenates and primary cultures of CNS containing polysialogangliosides were able to bind the toxin (Table III).[32] This is in accord with analytical studies of gangliosides on cultured glioma and neuroblastoma cells, which are reported to lack GD_{1b} and GT_1.[33,34] Ganglioside addition to a hybrid (neuroblastoma × glioma) cell lacking these gangliosides resulted in enhanced toxin binding.[32] Toxin binding to uncharacterized rat brain membranes was prevented by cholera toxin, gangliosides, and acidic phospholipids.[34a]

A newer *in vitro* assay has been developed to examine toxin–ganglioside interactions. Tetanus toxin binds firmly to Sephadex G-100 gel, and small columns containing this matrix have the property of removing ^{3}H-labeled gangliosides from solutions. Employing this assay system, it was observed that both GD_{1b} and GM_1 were bound to these columns, whereas the ceramide-free oligosaccharide prepared from GD_{1b} was not removed. There is no apparent reason for lack of the GM_1–tetanus toxin interactions reported by earlier investigators. Tetanus toxoid, chymotrypsin, diphtheria toxin, and bovine serum albumin inhibited the ganglioside–toxin interactions. The differences between this assay and previous ones are based upon the removal of gangliosides rather than toxin from the system and the capability to measure smaller quantities of materials. These workers also indicated that the heavy polypeptide chain exhibited the same binding as intact toxin.[35]

A tetanus toxin subfraction was obtained that was 1000 times less toxic

Table III. Correlation between Binding of ^{125}I-Labeled Tetanus Toxin and Content of Long-Chain Gangliosides[a]

Tissue tested	Binding of toxin	Content of long-chain gangliosides
Homogenate of adult rat CNS	+	+
Homogenate of embryonic rat CNS	+	+
Primary tissue culture of embryonic rat CNS	+	+
Hybrid cell line 108CC15	−[b]	−
Neuroblastoma cell line 2A	−	−
Oligodendroglial cells	−	−
Glioma C_6	−	−

[a] From Ref. 32.
[b] −, not detectable.

than the parent material but retains the antigenic cross-reactivity. This fraction, with a molecular weight of 46,000, is capable of binding to gangliosides more effectively than the native toxin. One of the procedures employed in these studies was binding measurements of ^{125}I-labeled toxin or its fragment to ganglioside affinity chromatographic columns.[36] The composition of the ganglioside mixture used to prepare the column was not described; however, some GD_{1b} was presumably present. The chemical coupling of the ganglioside presumably resulted in the abolition of the sialic acid carboxyl group,[37] which had been reported as a required site for toxin binding. This requirement of a disialosyl group for tetanus toxin binding to gangliosides has been disputed.[38] Gangliosides can be made to adhere to plastic dishes and their interactions with various ligands studied. GT_{1B}, GQ_{1b}, and GD_{1b} were shown to effectively bind tetanus toxin.[38a] This suggests that a disialosyl grouping optimizes the toxin–ganglioside interaction. The neurophysiological effects and biochemical interaction of tetanus toxin have been reviewed.[38b]

6.4.3. Cholera Toxin

There has been an increasing literature describing the interactions of cholera toxin and gangliosides. Much of this recent proliferation of activity is based upon several early observations.

The clinical consequence of cholera toxin infection is severe diarrhea due to water extrusion from the intestine. Using intestinal tissue *in vitro*, it was shown that exposure to this toxin resulted in elevated cyclic AMP levels due to enhanced adenylate cyclase activity.[39] Crude ganglioside preparations were shown to inactivate the toxin's biological activity.[40,41] Subsequently, it was demonstrated that the ability to prevent the cholera toxin-induced elevated adenylate cyclase of intestinal mucosal cells as well as the toxin-induced increased local capillary permeability of rabbit skin appeared restricted to a single ganglioside, GM_1.[42] Only GM_1 ganglioside of several glycosphingo-lipids tested was found to give a precipitation line against the toxin using an immunodiffusion assay procedure. This ganglioside also blocked cholera toxin-stimulated fluid accumulation in ileal loops. At higher concentrations, GD_{1b} and GT_1 were also found active in the intestinal tissue assay.[43] This series of physiological investigations suggested that gangliosides, especially GM_1, could interact with cholera toxin and prevent several of its biological effects on cells or tissues.

Several publications appeared in a single issue of *Biochemistry* in 1973 that acted as the catalyst for much of the activity in this area.[44–47] The binding of ^{125}I-labeled cholera toxin to isolated rat fat cells and liver membranes was demonstrated. A variety of carbohydrate-containing compounds interfered with the binding to liver membranes. The most effective material on a quantitative basis was a specific ganglioside species, GM_1. Accordingly, 0.01 nmoles of GM_1 caused a 45% inhibition of the binding of 0.01 pmole of cholera toxin.

It should be pointed out that fetuin and undiluted serum itself was reasonably effective in interfering with binding of cholera toxin to these membranes as seen in Table IV. However, the glycopeptides produced by proteo-

Table IV. Binding of Cholera Toxin to Various Glycoproteins, Lipoproteins, and Serum[a,b]

	Amount		Binding of ^{125}I-labeled cholera toxin (cpm)	
			With liver membranes	Without liver membranes
None			19,300 ± 300	900 ± 100
Fetuin,	0.5	mg/ml	1,800 ± 200	1,100 ± 200
	150	μg/ml	8,600 ± 400	800 ± 100
	50	μ/ml	12,700 ± 300	900 ± 100
Human RBC glycoprotein,	0.5	mg/ml	14,100 ± 300	1,400 ± 200
	100	μg/ml	18,700 ± 700	1,100 ± 200
Rat liver membrane glycoprotein,[c]	80	μg/ml	6,500 ± 400	900 ± 100
	20	μg/ml	14,600 ± 600	1,000 ± 200
Thyroglobulin,	15	mg/ml	26,400 ± 400	27,400 ± 1000
Horse serum glycoprotein,	20	mg/ml	25,100 ± 400	26,100 ± 800
	2	mg/ml	18,700 ± 300	6,200 ± 300
Thyrotropin,	25	mg/ml	24,600 ± 1100	23,900 ± 900
Human α-globulin,	25	mg/ml	10,300 ± 300	11,400 ± 300
	5	mg/ml	7,100 ± 200	1,200 ± 200
Bovine β-lipoprotein (IV-4),	50	mg/ml	24,500 ± 900	20,700 ± 1100
	0.5	mg/ml	18,100 ± 600	1,400 ± 200
Bovine β-lipoprotein (III-0),	50	ng/ml	23,100 ± 700	21,300 ± 700
	0.5	ng/ml	17,900 ± 900	1,800 ± 300
Rabbit serum glycoprotein,	25	mg/ml	14,200 ± 500	2,200 ± 200
Human β-globulin,	30	mg/ml	15,900 ± 600	7,700 ± 500
	0.5	mg/ml	18,700 ± 800	1,200 ± 200
Human serum glycoprotein,	25	mg/ml	20,500 ± 700	1,000 ± 100
Bovine γ-globulin,	25	mg/ml	19,500 ± 300	900 ± 200
Ovomucoid,	30	mg/ml	17,200 ± 500	1,000 ± 0
Lima bean trypsin inhibitor,	30	mg/ml	19,100 ± 900	1,100 ± 100
Egg albumin		–	19,200 ± 300	1,000 ± 100
Serum,		undiluted	1,200 ± 200	1,600 ± 600
	1:10 dilution		12,600 ± 500	1,200 ± 300

[a] From Ref. 44.

[b] ^{125}I-labeled cholera toxin (3.2×10^4 cpm, 15 ng/ml) was preincubated at 24°C for 60 min in 0.2 ml of Krebs–Ringer–bicarbonate buffer containing 0.1% albumin and protein. Samples (50 μl) of each incubation mixture were then added to 0.2 ml of the same buffer without other additions or containing 50 μg of liver membrane protein. After incubating at 24°C for 30 min, the incubation mixtures were filtered over EGWP Millipore filters. Since some proteins adsorb strongly to the filters in the absence of liver membranes, it was possible to measure directly the binding of ^{125}I-labeled cholera toxin to these proteins. The data in the presence and absence of liver membranes are therefore presented. A number of proteins thought not to be glycoproteins had no effect when tested at 0.5 mg/ml: pancreatic ribonuclease, lysozyme, serum albumin, Conc A, wheat germ agglutinin, hemoglobin.

[c] At least 80% of the inhibiting activity is lost after extracting this material with chloroform–methanol (2 : 1, vol./vol).

lysis of fetuin were not active, suggesting that the noncarbohydrate portion of this glycoprotein contributed to cholera toxin interactions. Interestingly, the toxin could also bind directly to thyroglobulin, horse serum glycoprotein, thyrotropin, and bovine β lipoprotein.[44] GM_1 ganglioside and fetuin itself partially overcame the cholera toxin-stimulated lipolysis of isolated adipocytes. Unfortunately, carefully designed quantitative comparative studies on the effectiveness of GM_1 and fetuin were not undertaken with the fat cells. Fat cells and liver membranes presoaked in gangliosides exhibited increased binding of cholera toxin. Ganglioside-treated fat cells also showed enhanced lipolysis when subsequently exposed to the toxin as seen in Fig. 2.[45] Due to the great interest in the possibility that a "natural membrane receptor" might have been discovered, virtually all subsequent studies in this area have concentrated on GM_1 ganglioside with little interest in membrane glycoproteins. Some of the general properties of the toxin-stimulated lipolysis were examined and especially the characteristic initial lag period. It was speculated that initially an inactive toxin–ganglioside complex is formed that relocates and spontaneously becomes active.[46] Choleragenoid does not stimulate lipolysis but

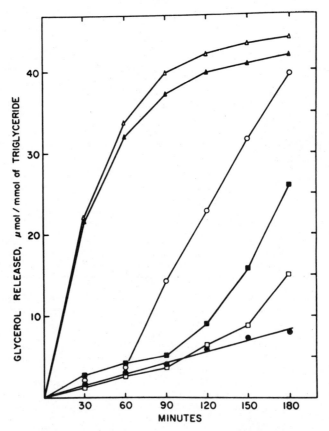

Figure 2. Effect of bovine brain gangliosides and fetuin on cholera toxin and on L-epinephrine-stimulated lipolysis in isolated fat cells. Fat cells were incubated at 37°C in Krebs–Ringer–bicarbonate buffer containing 3% (w/v) albumin, in the absence of additional compounds (●) or with 3 μg/ml of cholera toxin (○), 3 μg/ml of cholera toxin plus 0.1 mg/ml of gangliosides (□), 3 μg/ml of cholera toxin plus 0.4 mg/ml of fetuin (■), 3 μg/ml of L-epinephrine (△), and 3 μg/ml of L-epinephrine plus 0.1 mg/ml of gangliosides (▲). The cholera toxin was preincubated with gangliosides at 24°C for 20 min before addition to the cells. Fetuin was added to the cells 5 min (24°C) before the addition of cholera toxin. At the concentrations used here, fetuin and gangliosides in the absence of cholera toxin did not significantly alter lipolysis. (From Ref. 45.)

does bind to membranes and this binding is blocked by the presence of either cholera toxin or GM_1 in the incubation media. Pretreatment of membranes with GM_1 results in increased choleragenoid binding and these observations suggest a common binding site for both choleragenoid and intact cholera toxin.[47]

Cholera toxin was separated into heavy (H) and light (L) subunits and it was shown that the cell binding and ganglioside interactions occurred with the L subunit while toxicity was associated with the H subunit.[48–50] Chemical or enzymatic modification of the toxin revealed that its toxicity could be affected independently of GM_1 binding or antibody binding. The most potent reagents were those believed to affect arginine residues (Table V).[51]

The toxin reactive properties were compared for native GM_1, acetyl-sphingosine, a chemically modified GM_1 in which the fatty acid residue is replaced by an acetyl group, and the isolated oligosaccharide chain prepared from GM_1. Intact GM_1 and the acetyl analog formed precipitin lines against cholera toxin in comparative double-diffusion experiments. Although the oligosaccharide chain did not form a precipitation band, it was capable of inhibiting the reaction between GM_1 and toxin. The oligosaccharide unit was 20,000 times less effective than either of the other compounds in preventing the skin toxicity of cholera toxin.[52]

Similar observations appeared from another laboratory, namely, that GM_1, but not the oligosaccharide chain, interfered with an anticholeragen antibody reaction. Some interaction between the carbohydrate unit and choleragen did appear to occur as judged by altered electrophoretic mobility of the toxin. Lipophilic derivatives of the carbohydrate chain, however, could inhibit the antibody–antigen reaction.[53] The requirement for an available carboxyl group of the sialic acid is uncertain. There appears to be a specific requirement for the correct sialic acid since the polysialogangliosides, the asialo GM_1, and the less complex ganglioside GM_2 are much less effective than GM_1. However, amide derivatives of the sialic acid carboxyl group of GM_1 that are either soluble or insoluble effectively bind to cholera toxin.[54] These observations suggested that only a specific subunit of cholera toxin binds to gangliosides

Table V. Specificity of Reagents with Differential Effects on the Toxic, GM_1-Binding, and Antibody-Fixing Activities of Cholera Toxin[a]

Reagent	Incubation conditions	Specificity	Effect on		
			Toxicity	GM_1 binding	Antibody binding
Cyclohexanedione	1 mM in 0.05 M NaOH, 23°C, 1 hr	Arginine	+	−	−
Phenanthrenequinone	1 mM in 0.05 M NaOH, 5% ethanol, 23°C, 1 hr	Arginine	+	−	−
Butandione	1 mM in phosphate pH 7.0, 23°C, 3 hr	Arginine	+	−	(+)

[a] From Ref. 51.

and that both the ceramide (lipophilic) and the carbohydrate chain (hydrophilic) portions of gangliosides are required for the binding. Chemically modified derivatives of the oligosaccharide portion of GM_1 ganglioside were prepared in order to elucidate the structural requirements for interaction with cholera toxin. It was found that the sialic acid must have a free carboxyl group and that a terminal galactose residue is essential for reaction with the toxin.[54a] As a result of these structural requirements, these workers question the reported ability of an affinity column linking the sialic acid carboxyl residue of GM_1 ganglioside to the inert support to be effective for toxin binding. The kinetics of binding of the oligosaccharide derived from GM_1 to cholera toxin was carried out using equilibrium dialysis. A positive cooperation of sugar binding was observed that was not decreased by 2-mercaptoethanol. A total of four binding sites for the oligosaccharide was found with the toxin rather than the anticipated five corresponding to the five protomers of the B subunits.[54b]

A comparison of the physical interaction of GM_1 and structurally related synthetic compounds and cholera toxin was undertaken. Evidence obtained suggested that the ceramide portion was not involved in the binding to toxin but was involved in aggregate formation of the ganglioside–toxin complex.[54c] The direct binding of ^{125}I-labeled cholera toxin to thin-layer plates has been devised as a sensitive procedure to detect toxin-binding materials.[54d]

Three transformed mouse embryo kidney cell lines differing with respect to the presence of undetectable quantities of GM_1 were employed to examine their response to cholera toxin. The SVS AL/N cell line, devoid of both GM_1 and GM_2, bound very little of labeled toxin. Cell line TAL/N (P > 200), which had both GM_3 and GM_2 but undetectable GM_1, was about as effective in toxin binding as cell line TAL/N (P > 60), which possessed GM_1. Twice as much cholera toxin was required for the SVS AL/N cell line as the TAL/N cell line for toxin-stimulated adenylate cyclase. The SVS AL/N cells had no detectable GM_1 but was still capable of binding one-sixth the amount of labeled toxin, and was one-half as sensitive to toxin with respect to stimulated adenylate cyclase as the TAL/N (P > 60). The authors felt that this provided strong supportive evidence for GM_1 being the toxin receptor. The TAL/N (P > 200) cell line also had no detectable GM_1 and did not possess the galactosyltransferase believed responsible for converting GM_2 to GM_1 and yet was almost as active in toxin binding as the control cell line TAL/N (P > 46) (Fig. 3).[55] Contrary to the authors' biases, these studies do not provide unequivocal evidence for the role of GM_1 as the toxin receptor in these cells. Indeed, they can be interpreted as demonstrating that GM_1 may not represent the receptor.

Subsequent work demonstrated that the A subunit of cholera toxin was the adenylate cyclase activator of pigeon erythrocytes[56] and the B subunit presumably was responsible for binding to the cell surface receptor, which could be GM_1.[57,58]

"Capping" of surface structures is a phenomenon that occurs when cells are exposed to a variety of polyvalent ligands such as antibodies or lectins. The "capping" is thought to represent a clustering of diffusely or randomly distributed specific membrane receptors into a restricted area on the cell

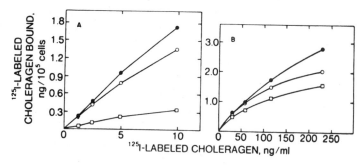

Figure 3. Binding of cholera toxin to transformed AL/N cells. ●, TAL/N, P = 46; ○, TAL/N, P = 269; □, SVS AL/N. (A) Binding to 8×10^4 cells in 0.4 ml of buffer. (B) Binding to 3.2×10^5 cells in 0.2 ml of buffer. (From Ref. 55.)

surface and is often conveniently visualized microscopically by using a fluorescent labeled ligand or its antibody. This "capping" was observed with rat mesenteric lymph node cells and fluorescent-labeled cholera toxin.[59] Human lymphocytes, monocytes, and granulocytes can bind cholera toxin and this process is inhibited by GM_1 ganglioside, and "capping" of the receptor is also observed with these cells. Cells obtained from patients with active lymphoblastic leukemia only weakly bind this toxin but this can be increased by prior exposure of the cells to GM_1 ganglioside.[60] Documentation of the ganglioside composition of circulating white blood cells is limited; however, it has been reported that the quantity of gangliosides in cells from chronic myelogenous leukemia is increased twofold over controls. In contrast, only a trace of gangliosides is present in cells obtained from patients with chronic lymphocytic leukemia. Thin-layer chromatographic analysis revealed the presence of approximately six ganglioside bands.[61] Compositional studies indicate that the major ganglioside of cells obtained from a patient with chronic myelogenous leukemia has as its proposed structure: *N*-acetylneuraminyl-galactosyl-*N*-acetylglucosaminyl-galactosyl-glucosylceramide. There appeared to be approximately six other uncharacterized gangliosides one of which cochromatographed with GM_1 standard.[62] The structure of gangliosides of cells from lymphocytic leukemia is not established, making it difficult to correlate it with "capping" by cholera toxin.

It has been proposed that there is an initial binding of cholera toxin to a surface membrane component, perhaps a ganglioside-like compound. This mobile complex then moves laterally within the membrane, permitting the "active" toxin component to interact with the adenylate cyclase system.[63,64]

Rat intestinal microvillous membranes exhibit a specific and saturable binding of labeled cholera toxin.[65] Studies on the gangliosides of intestinal tissue and their relationship to cholera toxin utilized isolated mucosal cells. Distinct species differences in both the absolute quantities and the patterns of the gangliosides were observed for human, pig, and bovine mucosal cells. GM_1 was present in all three species; however, its concentration varied 400-fold between human and bovine tissue. Exposure of cells to exogenous GM_1 in general resulted in increased binding of cholera toxin concomitant with

Table VI. Membrane Incorporation of GM_1 Ganglioside in Intestinal
Mucosal Cells and Effect on Cholera Toxin Binding[a]

Species	GM_1 (μM)[b]	Toxin molecules maximally bound per cell $\times 10^{-4}$	$[^3H]\text{-}GM_1$ molecules incorporated per cell $\times 10^{-4}$
Man	0	1.5	
	0.2	2.5	
	2.0	8.0	
Pig	0	12	
	0.07	25	18
	0.2	60	44
	0.7	110	166
Beef	0	260	
	0.2	290	4.7
	0.7	295	22
Rabbit	0	1.0 n[c]	
	0.07	1.7 n[c]	
	0.7	4.2 n[c]	

[a] Modified from Ref. 66.
[b] The cells were incubated at 37°C for 25 min with GM_1 ganglioside in minimal medium/ bovine serum albumin and then carefully washed.
[c] The numerical values could not be calculated since the experiment was performed in intestinal segments whose number of cells is unknown.

uptake of the ganglioside, except for the bovine cells (Table VI). However, the increased toxin binding was not stoichiometric to the number of GM_1 molecules incorporated. A 10-fold increase of 18 to 166 $\times 10^4$ molecules of GM_1 present on pig intestinal cells resulted in only a 4-fold increase in the quantity of toxin bound (Table VI). This could implicate GM_1 in toxin binding. The accumulation of fluid in rabbit small intestinal loops was used as a test for the biological effects of cholera toxin. It was found that low levels of the toxin resulted in fluid accumulation only with samples that had previously been exposed to cholera toxin. There was no enhanced activity by added ganglioside on fluid accumulation at higher toxin concentration. Exposure of these intestinal preparations to choleragenoid, which binds as efficiently to cells *in vitro*,[48] or to GM_1[43] followed by treatment with cholera toxin, did not result in the usual fluid accumulation produced by the toxin. This suggested that the choleragenoid occupied the toxin receptors. If the cells were first treated with GM_1 gangliosides, followed by choleragenoid and then finally exposed to toxin, there was no inhibition caused by the choleragenoid. These observations could readily be interpreted as indicating that GM_1 ganglioside may not be the natural cholera toxin receptor of intestinal cells. If this ganglioside was the receptor, then there should have been an inhibition similar to that seen in the control samples. The authors, however, concluded that this report supported the role of GM_1 as the cholera toxin receptor.[66] There is a dose-dependent inhibition of rabbit intestinal cell guanylate cyclase and repressed cGMP phosphodiesterase by cholera toxin.[67] There is no information available relating this observation to GM_1.

Gastric mucin prevents the cholera toxin-stimulated fluid secretion of rat intestine. The mucin binds labeled cholera toxin. These effects were not seen with salivary or intestinal mucin.[67a] Glycoproteins, which were glycolipid free, were isolated from rat intestinal microvillous membranes and shown to bind cholera toxin. They seem to have carbohydrate chains terminating in galactose or *N*-acetylgalactosamine and may be structurally similar to GM_1.[67b] These data reinforce the possibility that glycolipids are not the sole receptors present in intestinal tissue for cholera toxin. The group committed to GM_1 as the cholera toxin receptor has rejected glycoproteins as being quantitatively important receptors.[67c]

Synthetic analogs of the carbohydrate unit of GM_1 gangliosides containing various aliphatic hydrocarbon chains have been synthesized and examined for their ability to interact with cholera toxin. Precipitation of the toxin occurred with those compounds having an aliphatic chain of at least 16 carbon atoms.[68] A synthetic dansyl fluorescent-labeled GM_1-like compound was incorporated into lymphocyte membranes and "capping" was observed when these coated cells were later exposed to cholera toxin. This migration was accompanied by the IgG receptors, suggesting that there may not be independent movement of lipid in the membrane.[69]

In vivo exposure of experimental animals to cholera toxin resulted in increased hormone sensitivity of their liver membranes.[63] Several other observations have appeared indicating distinctions between cholera toxin and hormone stimulation of adenylate cyclase activity. Adrenal tumor cells rendered unresponsive to hormonally stimulated cAMP production were unaffected in their response to cholera toxin.[70] Treatment of thymus cells with 1-ethyl-3-(3-dimethylaminopropyl) carbodiimide reduced their cholera toxin sensitivity by 90%; however, there was no effect on cAMP production by epinephrine or prostaglandin E_1 (Table VII).[71] These authors also suggest in terms of effects of cholera toxin that "findings in one kind of cell might not have general applicability." Differences have been reported between the adenylate cyclase stimulation by cholera toxin and the hormones glucagon and epinephrine.[72] A mutant derived from Balb/c/3T3 cells, referred to as AD6, is deficient in glucosamine-6-phosphate acetyltransferase with a presumable decrease in sialoconjugate content, both as glycolipid and as glycoprotein. These cells have reduced cAMP activation upon exposure to cholera toxin without any reduction to stimulations by GTP, epinephrine, prostaglandin, and fluoride. Exposure of this cell line to GM_1 ganglioside restored the cholera toxin responsiveness.[72a] The possibility has been raised that GM_1 is not a true receptor on adrenal cells but may merely be acting as a pseudoreceptor.[73]

Semiquantitative estimates were made of molecules of cholera toxin bound as a function of GM_1 content of pigeon erythrocytes and correlated with the degree of enhanced adenylate cyclase activity. The authors concluded that approximately 90% of the binding sites in treated and untreated cells were

Table VII. Specific Inhibition of cAMP Response to Cholera Toxin by Cell Modification with 1-Ethyl-3-(3-dimethylaminopropyl)carbodiimide (EDC)[a,b]

Stimulant	cAMP (pmoles/10^7 cells)	
	EDC-treated	Control
Cholera toxin	34	300
Epinephrine	196	122
Prostaglandin E₁	389	340
No additive	6	6

[a] Modified from Ref. 71.

[b] Thymocytes (10^7 in 1 ml of tissue culture medium RPMI 1640 supplemented with 10% fetal calf serum) were incubated at 37°C for 30 min with or without EDC (1 mM) and then washed. Thereafter, the cells were suspended in fresh medium and incubated at 37°C with cholera toxin (10^{-9} M, 50 min), epinephrine (10^{-6} M, 25 min), prostaglandin E₁ (10^{-4} M, 15 min), or with no stimulant (50 min). Intracellular cAMP was then determined.

nonproductive. This suggests that the presence of GM_1 itself is not sufficient for toxin activation of adenylate cyclase.[74]

A series of studies were carried out on fat cells to reexamine the hypothesis that GM_1 was the cholera toxin receptor. Using conventional methodology for ganglioside quantitation, GM_1 was undetectable in these cells or their isolated membranes. It was suggested that GM_1 binding, previously believed capable of "creating native receptor sites,"[45] was nonspecific since it was not possible to saturate these cells even at a level of 7×10^9 molecules per cell (Table VIII). GM_2 and glucosylsphingosine appeared to bind nearly as effectively as GM_1 while several other lipids were somewhat less effective. In contrast to observations by previous workers,[45] there was no increase of absolute lipolytic activity of GM_1-coated cells but merely an increased rate upon exposure to cholera toxin was observed. 4-Methylumbelliferyl-β-D-galactoside had an effect qualitatively similar to that observed with GM_1 on the isolated adipocytes. It was suggested that GM_1 by virtue of its ability to bind both to the cells and to cholera toxin might act as a "glue" merely facilitating the saturation of preexisting non-GM_1 receptor sites.[75]

Employing a combined galactose oxidase treatment and NaB^3H_4 reduction, indirect evidence was offered for the presence of GM_1 ganglioside in a fat cell particulate fraction at a concentration of 2 to 5×10^5 molecules per fat cell. The intact ganglioside was not analyzed but rather an oligosaccharide derivative was obtained through ozonolysis, chromatography, and coprecipitation with choleragen by the toxin's antibody. These workers estimate that 94% of the fat cell ganglioside is GM_3, corroborating previous estimates.[75] When the radioactive oligosaccharide derived from a purified mixed ganglioside was treated with cholera toxin and its antibody, only 3% of the label was present in the precipitate. The remaining 96% of the radioactivity, which was nonprecipitable, could not have been derived from gangliosides since neither GM_3 or GD_{1a} are labeled using the procedure employed. GM_2 which appears to also be present in the fat cell membranes, is much more difficult to label under these conditions and represents about 6% of the total ganglio-

Table VIII. Accumulation of Various Exogenously Added Radiolabeled Sphingolipids on Fat Cells[a]

Lipid ligand	Total lipid incubated		Bound lipid		
	nmoles	cpm	cpm	nmoles/ 5 × 10^5 cells	%
[^{3}H]Asialo GM$_1$	2.0	364	4 ± 3	NS[b]	NS[b]
[^{14}C]Glucosyleramide	6.8	1,581	25 ± 6	0.11	1.6
[^{3}H]Lactosylceramide	0.8	4,021	86 ± 7	0.02	2.1
[^{35}S]Sulfatide	1.0	3,669	327 ± 9	0.09	8.9
[^{3}H]Galactosylsphingosine	0.8	9,098	1,058 ± 31	0.09	11.6
[^{14}C]-GM$_2$	7.7	1,007	146 ± 12	1.1	14.5
[^{14}C]-GM$_2$	6.3	822	126 ± 11	1.1	15.3
[^{3}H]-GM$_1$	0.8	4,716	835 ± 22	0.14	17.7
[^{3}H]-GM$_1$	0.7	4,041	788 ± 16	0.14	19.5
[^{3}H]-GM$_1$[c]	35.0[c]	4,041	679 ± 13	5.9	16.8
[^{14}C]-GM$_2$	10.0	1,299	280 ± 15	2.2	21.6
[^{3}H]-GM$_1$	0.8	4,681	1,024 ± 31	0.18	21.9
[^{3}H]-GM$_1$	7.6	44,290	10,168 ± 72	1.7	23.0
[^{3}H]-GM$_1$	0.9	5,320	1,266 ± 26	0.22	23.8
[^{14}C]Glucosylsphingosine	8.0	1,873	649 ± 21	2.8	34.7
[^{14}C]Glucosylsphingosine	9.4	2,204	815 ± 34	3.5	37.0

[a] The experimental conditions and procedure used to measure the binding of ligands to intact cells were the same as those employed by Cuatrecasas.[44] With each designated amount of ligand, about 5×10^5 cells (0.05 ml of packed cells) were incubated in a final volume of 0.2 ml of Krebs–Ringer–bicarbonate buffer, pH 7.4, containing 0.1% bovine albumin, for 30 min at 24°C with gentle shaking. Afterward, cells were immediately washed five times with chilled buffer by filtration through EAWP Millipore filters. Radioactivity associated with the cells on the filters was counted. Total lipid incubated represents the cpm measured in separate but equivalent aliquots of ligand incubated without cells. Bound lipid, that amount of radioactive ligand associated with the washed cells on the filters, was calculated as percentage of the total lipid incubated. Each value for bound lipid is the average (± S.E.) of triplicate assays from a single incubation with cells. (See Ref. 75).
[b] NS, not significant.
[c] Included an amount of unlabeled GM$_1$ 50 times greater than 0.7 nmole of [^{3}H]-GM$_1$; total cpm incubated was therefore assumed to be the same as that determined in the absence of unlabeled GM$_1$.

sides, according to this publication. This suggests that carbohydrates derived from nonganglioside components are the predominant materials incorporating tritium under these conditions. The relationship of these materials to the cholera toxin binding site was not considered. The thin-layer chromatogram areas corresponding to GM$_1$ and GD$_{1a}$ standards were eluted and when added to NCTC-2071 cells caused a cholera toxin-enhanced level of cAMP. Another area on the plate, which did not correspond to any standards employed was even more active. No comments were provided about this observation.[75a] 3T3-L1 cells can be induced to differentiate into adipocytes when cultivated in the presence of insulin. Choleragen binding is reduced 50–60%, total ganglioside content is decreased to about the same extent, and GM$_1$ is reduced to about 10–25% in the differentiated cells.[75b]

Some of the more suggestive supportive evidence for the potential GM$_1$ function as a toxin receptor comes from studies of mutant cell lines. Transformed mouse fibroblasts, designated NCTC 2071 cells, appear to be deficient in gangliosides including GM_1. These cells when exposed to these glycosphingolipids take up about 4% of the gangliosides, which is independent of the amount present in the growth medium. In the study reported, there appeared to be an all-or-nothing phenomenon. There was a fourfold increase in the cholera toxin-stimulated basal level of cAMP when 34×10^4 molecules of GM_1/cell were bound and a fivefold increase when 1700×10^4 molecules of GM_1 were bound (Table IX). These cells appeared to have an insatiable ability to bind GM_1, which increased proportionately with the amount added in the culture media and suggests that a large percentage of the "newly created" sites were nonproductive. The cholera toxin stimulation of adenylate cyclase was not GM_1 dependent with broken cell preparations.[76] These NCTC 2071 cells may lack certain glycosyltransferases presumably responsible for GM_1 production and were rendered sensitive to cholera toxin-stimulated cAMP formation both by GM_1 and by GD_{1a} (Table X). A different line of these cells contained substantial amounts of a GM_2-like ganglioside. These cells possessed only a moderate responsiveness to cholera toxin, which was increased by prior exposure to GM_1.[77] Exposure of cultured HeLa cells to butyrate resulted in increased binding of cholera toxin and this may have been correlated with GM_1 appearance.[77a] Similar increased toxin binding was seen with rat C6 glial and Friend erythroleukemic cells; however, there was no information provided about the gangliosides of these cell populations.

It has been suggested that the number of cholera toxin binding sites for maximal adenylate cyclase stimulation of KBalb/3T3, Balb/c/3T3 cells, and rat adipocytes are similar and may be in the neighborhood of 20,000. In contrast, the number of toxin molecules that can saturate these cells varies from 2×10^4 to 6×10^6, indicating that as little as 0.5% occupancy of all the potential cholera toxin binding sites of cultured cells results in maximal adenylate cyclase activity. This may suggest the presence of "spare" binding

Table IX. Uptake of $[^3H]$-GM_1 and Response to Choleragen
by NCTC 2071 Cells[a,b]

$[^3H]$-GM_1 (pmoles/dish)	Uptake of $[^3H]$-GM_1		cAMP (pmoles/mg protein)
	%/mg protein	Molecules/cell $\times 10^{-3}$	
0	—	—	3.9
3.8	4.2	34	16
10.2	3.8	81	17
16.1	3.5	120	25
38.9	3.1	250	21
61.4	4.0	520	22
171	3.2	1700	21

[a] Modified from Ref. 76.
[b] Cells were incubated for 18 hr in 7 ml of NCTC medium with $[^3H]$-GM_1 as indicated. The cells were then washed three times and incubated with choleragen, 1 µg/ml, for 3 hr before determination of cell ^3H and cAMP content.

Table X. Choleragen Responsiveness of NCTC 2071(A) Cells
following Ganglioside Binding[a]

Ganglioside added	pmoles/dish	Molecules bound/cell ($\times$ 10s-^{4})	cAMP (pmoles/mg protein)[b]
		—[c]	8.7 ± 0.5
GM$_3$	125	170	6.3 ± 0.5
GM$_2$	423	570	8.3 ± 1.3
GM$_1$	1.7	5.9	13.9 ± 0.3
GD$_{1a}$	14.9	22	12.7 ± 1.0
GD$_{1a}$	5.7	10	10.9 ± 0.9
GD$_{1a}$	3.5	5.4	11.9 ± 0.3
GD$_{1a}$	1.6	2.4	9.8 ± 0.4
GD$_{1a}$	0.9	1.1	9.1 ± 0.4

[a] NCTC 2071(A) cells were incubated for 18 hr in NCTC 135 medium containing [^{14}C]-GM$_3$ (66.1 μCi/μmole), [^{14}C]-GM$_2$ (4 μCi/μmole), [^{3}H]-GM$_1$ (4.37 mCi/μmole), or [^{3}H]-GD$_{1a}$ (4.37 mCi/μmole) as indicated. Then the cells were washed three times with PBS and incubated in NCTC 135 medium with choleragen (1 μg/ml) for 3 hr at 37°C. The medium was aspirated and the cells were washed with PBS and scraped in 2 ml 5% trichloroacetic acid. The precipitates following centrifugation were dissolved in 1 M NaOH and samples were analyzed for protein content and radioactivity. cAMP in the supernatant was purified and assayed. Basal cAMP content of cells not exposed to choleragen was 5.4 pmoles/mg protein. (See Ref. 77.)

[b] Values are the mean ± S.D. for triplicate determinations.

[c] Cells contained 4×10^6 molecules of a ganglioside presumed to be GM$_2$.

sites for cholera toxin.[77b] An alternative possibility is that there are two different binding sites for cholera toxin. One might be a "specific" receptor while the other is an unrelated "nonspecific" site such as GM$_1$ ganglioside.

Peroxidase has been coupled to cholera toxin and employed to visualize binding sites in a glioblastoma TC 593 cell line. There appeared to be an inverse relationship between surface labeling and time in culture.[78] GM$_1$ ganglioside coating of cultured neuroblastoma cells increased the uptake of cholera toxin covalently linked to peroxide.[78a]

Normal human skin fibroblasts grown in culture have GM$_3$ and GD$_3$ as their major gangliosides; GM$_1$ represents approximately 1% of the total gangliosides. These cells were reported to be deficient in the N-acetylgalactosaminyl transferase believed responsible for the conversion of GM$_3$ to GM$_2$. The presence of either [^{14}C]galactose or N-acetyl[^{3}H]mannosamine in the culture medium resulted in radioactive material being formed that chromatographed like GM$_3$ and GD$_{1a}$ but did not label material that chromatographed like GM$_1$. Labeled GM$_1$ could be taken up by the cells from a chemically defined medium but not from a serum-containing incubation mixture. In the absence of added GM$_1$, these cells were sensitive to cholera toxin stimulation of cAMP production (Table XI).[79]

A combination of galactose oxidase followed by NaB^{3}H$_4$ reduction has been shown to be an effective procedure for labeling the primary hydroxy group of galactose present in glycoconjugates including the glycosphingo-

Table XI. Effect of Choleragen on
Intracellular cAMP Accumulation in
Human Fibroblasts[a,b]

Choleragen (ng/ml)	cAMP (pmoles/mg protein)	Fold stimulation
0	10	
2	65	6.5
20	194	19
200	417	42
2000	478	48

[a] From Ref. 79.
[b] Fibroblasts were incubated for 90 min at 30°C in Hanks' medium at the indicated choleragen concentrations. The medium was aspirated and cAMP was isolated and assayed. cAMP levels represent the average of duplicate determinations.

lipids. When intact human fibroblasts are exposed to this procedure, radioactivity was found associated with materials cochromatographing with the nonsialic acid-containing tri- and tetrahexosylceramides as well as with GM_3, GM_1, and GM_2. Pretreatment of the cells with choleragen appears to result in a net reduction of GM_1 labeling and an apparent reduction of GM_3 labeling without any decrease in neutral glycosphingolipid labeling (Fig. 4). Similar types of experiments were undertaken with NCTC 2071 cells pretreated with GM_1. These studies indicate that cholera toxin protects susceptible GM_1 molecules from oxidation by galactose oxidase.[80] Unfortunately, these studies using tissue culture experiments cannot be regarded as providing definitive information about such binding specificity. The specificity, if any, could have been investigated if these workers had endeavored to examine the effects of cholera toxin on labeling of nonganglioside compounds (e.g., glycoproteins) that are also membrane components.

This technique was also applied to an examination of thyroid plasma membranes. Thin-layer chromatographic examination of the gangliosides derived from these membranes showed the presence of five resorcinol-positive bands that did not exactly correspond with GM_3, GM_1, and GD_{1a} standards. The labeling pattern with galactose oxidase–NaB^3H_4 revealed a doublet designated to be GM_1 and only traces in the GM_2 and GM_3 areas of the plate. Pretreatment with cholera toxin reduced the radioactivity associated with the two GM_1-like materials but greatly increased that associated with GM_3, GM_2, GD_{1b}, and GT_1 portions of the chromatogram. This observation is difficult to judge since there is no information about the detailed structure of the materials that are supposed to be GM_1-like. The presence of a well-separated GM_1 doublet is rare, and increased labeling of other undefined products including polysialogangliosides has not usually been seen in other systems.[81]

Neuraminidase-treated, galactose oxidase–NaB^3H_4-labeled cultured Balb/c/3T3 cells were exposed to cholera toxin. After removal of excess toxin, the cells were solubilized with detergent. The extract was incubated with antibody

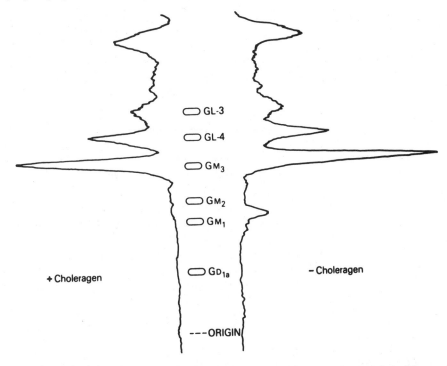

Figure 4. Effect of choleragen on labeling of GM_1 of human fibroblasts by galactose oxidase and NaB^3H_4. Cells were incubated for 2 hr with or without 10 μg/ml of choleragen. The cells were washed, incubated with galactose oxidase for 2 hr, and treated with 1 mCi NaB^3H_4 per dish. Glycolipids were extracted and chromatographed in chloroform–methanol–0.25% $CaCl_2$ (60 : 35 : 8, vol./vol.). Radioscans of chromatograms of glycolipids from cells incubated with or without choleragen are shown. (From Ref. 80.)

to cholera toxin and the receptor–antibody–toxin complex to protein A-containing bacteria. A labeled material migrating (on TLC) like GM_1 was seen. The author concludes that GM_1 acted as a receptor for cholera toxin. However, the possibility is considered that a glycoprotein may also be involved. There is an unexpectedly low yield with cells not pretreated with neuraminidase.[81a] The binding of choleragen to cultured cells decreases the ability of Triton X-100 to extract either choleragen or GM_1 ganglioside.[81b]

It should be pointed out that (1) sialic acid seems to be required for cholera toxin interactions with potential receptor molecules, (2) the sialic acid of GM_1 gangliosides is neuraminidase resistant under conventional conditions, and (3) the sialic acid of glycoproteins is readily removed with neuraminidase. Therefore, it is possible that neuraminidase treatment of cells destroys the natural receptor, perhaps a sialoprotein. The only membrane sialoconjugate remaining then could be GM_1, which has been shown to interact *in vitro* with cholera toxin.

Cholera toxin and the A protomer can hydrolyze NAD to ADP, ribose, and nicotinamide. This hydrolysis is effectively inhibited by GM_2, GD_{1a} and

GM$_1$ presumably due to their binding of the toxin. These compounds also increased the fluorescence of the A protomer, while only GM$_1$ induced a "blue shift" in the spectrum of the B protomer.[82] An interaction between tryptophanyl residues and GM$_1$ ganglioside is suggested both from specific chemical modifications of the B protomer of cholera toxin and from spectral emission investigations.[82a,b] The release of glucose from entrapped liposomes containing GM$_1$ gangliosides was evoked by cholera toxin and the B subunit but not the A subunit.[83] Choleragen caused agglutination of both erythrocytes and lysosomes possessing GM$_1$ gangliosides.[83a,b] These extend the previously established interactions of GM$_1$ and cholera toxin.

Peroxidase-conjugated cholera toxin antibodies have been used in an attempt to localize the bound toxin. Dispersed cells or pieces of tissue were incubated with the toxin, the excess removed and the tissues were then exposed to the conjugated antibodies and examined by electron microscopy. The results of these investigations largely confirm previous work since there was an approximate correlation between staining and GM$_1$ concentration and evidence for lateral mobility of the complexes as well as "capping" in both pre- and postsynaptic membranes of the synaptic terminals of rabbit and rat brain.[84]

6.4.4. Other Toxins

Gangliosides have been shown to inactivate the hemolysin of *V. parahaemolyticus* and this inactivation appears to reside in neuraminidase-sensitive polysialogangliosides rather than in GM$_1$.[85]

The anti-α hemagglutinating activity of various tissues has been speculated to be due to soluble glycolipids, presumably gangliosides.[86] *Staphylococcus* α-toxin hemolytic activity can be inhibited by a glucosamine-containing ganglioside isolated from human red blood cells that is thought to have the following structure: NeuAcα2→3Galβ1→4GlcNAcβ1→3Galβ1→4GlcCer.

Other sphingolipids such as globosides and galactosamine-containing gangliosides were less active and did not form precipitin bands.[87] The ability to confirm this observation is uncertain. Induced steroidogenesis in cultured adrenal tumors by both *E. coli* enterotoxin and cholera toxin is blocked by GM$_1$ gangliosides and competition experiments suggest a common receptor for both of these bacterial toxins.[88] The enterotoxin of *E. coli* will bind to ganglioside-coated plates.[88a] GM$_2$ inhibited the agglutination of erythrocyte suspensions.[88b]

Liposomes containing gangliosides have the capacity to bind Sendai virus and inhibit the virus' hemagglutinating activity.[89] Sendai virus binds to GT$_{1a}$, and GQ;$_{1b}$-coated plastic dishes.[89a] HeLa cell glycoproteins, but not gangliosides, incorporated into liposomes were found to bind Sendai virus.[89b] Trisialoganglioside and a disialoganglioside obtained from mouse brain inhibited the hemagglutination of goose red blood cells by Eastern equine encephalitis virus.[90] GD$_{1a}$ was reported to be the most effective ganglioside, of those tested, in preventing hemolysis by both streptolysin D and *L. monocytogenes* hemolysins.[90a] GM$_1$ gangliosides incorporated into liposomes interact with

R. communis toxin[90b] and may also be the receptor on leukocytes for *S. aureus* leukocidin.[90c]

6.5. Gangliosides and Lymphocyte Markers

Functionally distinct subpopulations of lymphocytes have been described. B cells, derived from bone marrow, are thought to be responsible for specific antibody production; a T-cell subset, derived from the thymus, is believed to be responsible for delayed hypersensitivity, antigen recognition, and allograft rejection. T cells possess a surface antigen referred to as theta or Thy-1.[91] A material identical to, or immunologically very similar to Thy-1 antigen is present in the brain tissue of rodents.[92,93]

Attempts have been made to isolate and identify the compound that possesses the Thy-1 antigenic determinant. Lysates of radioiodinated T cells were treated with Thy-1 antiserum and radioactivity was found in the resulting precipitate suggesting that the Thy-1 antigen is present on the cell surface membrane. A detergent was capable of removing an appreciable amount of radioactivity from these precipitates, and lipid extraction of the detergent-solubilized material removed 50% of the radioactivity. Similarly, if the cell lysate was directly detergent extracted, much of the precipitable material was removed. These authors feel that such results implicated a lipid as a part of the Thy-1 antigen. It should be noted that under ordinary circumstances, gangliosides would not be expected to become radioiodinated nor would the more complex gangliosides be expected to partition into the chloroform phase according to the procedures used in these studies. SDS electrophoresis of the precipitates gave a major peak of radioactivity in the 35,000-dalton range, which is similar to that reported with gangliosides.[94] The Thy-1 and Thy-2 antigens can be labeled with galactose and also slightly with amino acids (Table XII). It was also suggested that the Thy-1 antigen is present on the surface membrane, is associated with some carbohydrate-containing material, probably has a lipid component, and may be a glycolipid.[95] These properties could also describe an intrinsic membrane glycoprotein.

Several other investigators have concluded from their observations that the Thy-1 antigen is a glycoprotein. Mouse lymphocyte Thy-1 antigen can be precipitated with appropriate antiserum. If the cells are grown in the presence of [^{3}H]palmitate, no radioactivity is recovered in the antigen–antibody precipitate, which would argue against the Thy-1 antigen being a lipid.[96] A cultured cell line, S-49.1-TB 2.3, can be obtained in large amounts and contains a considerable quantity of Thy-1 antigen. Limited papain digestion of these cells liberates the antigen, which does not behave as a ganglioside using the classical partitioning procedure of Folch. Prolonged proteolysis destroys the antigenic material, suggesting that this is most likely a protein and not a lipid.[97] The Thy-1 antigen from rabbit brain and from thymocytes was purified to homogeneity as judged by SDS gel electrophoresis and the results of the properties examined indicate that the antigen is a glycoprotein and not a ganglioside.[98,99] Highly purified Thy-1 was extracted from rat brain and

Table XII. Precipitation of ^{3}H-Labeled Thy-1 Antigen from Lysate of Balb/c (Thy-1.2)
Thymocytes Using Congenic Anti-Thy-1.2^a

^{3}H-labeled precursor	Acid-precipitable radioactivity (cpm)	Lysate cleared with	Percent of immunoprecipitable radioactivity		
			Mouse serum (control)	Anti-Thy-1.1 (control)	Anti-Thy-1.2
L-Tyrosine	3.4 × 10⁷	Rabbit anti-ϕX + goat anti-	0.12	0.20	0.30
L-Leucine	6.1 × 10⁷	rabbit Ig	0.20	0.30	0.50
L-Fucose	7.4 × 10⁴		0.09	0.13	0.19
D-Galactose	1.2 × 10⁵		3.21	4.20	15.00
L-Tyrosine	3.4 × 10⁷	Rabbit anti-ϕX + goat anti-	0.10	0.11	0.10
L-Leucine	6.1 × 10⁷	rabbit Ig followed by anti-thy-	0.04	0.09	0.09
L-Fucose	7.4 × 10⁴	1.1 + goat anti-mouse Ig	—	—	—
D-Galactose	1.2 × 10⁵		—	—	—

^{a} From Ref. 95.

thymocytes with deoxycholate and purified to apparent homogeneity in a protocol that included a lentil lectin affinity column. This lectin has an affinity for glucose and mannose, suggesting that the Thy-1 antigen is not a ganglioside. A molecular weight of 27,000 was obtained by zonal centrifugation in the presence of deoxycholate, which is nearly 20-fold greater than the reported molecular weight of gangliosides. Analysis indicated that the material from both rat tissues was a glycoprotein containing approximately 30% carbohydrate.[99a,b]

Evidence was presented supporting the possibility that GD_{1a} is the brain-associated Thy-1 antigen. The ganglioside fraction from either CBA/J mouse brain or thymocytes inhibited the test system. Testing of individual thin-layer chromatographically separated bands of the mixed ganglioside fraction showed that a resorcinol-positive material migrating like GD_{1b} accounted for the inhibitory activity.

Bone marrow and thymus lymphocytes were utilized as the test system for attempts to gain information about their antigens. Individual gangliosides were obtained from the brains of both CBA and AKR strains of mice. A very peculiar observation was that GM_1 isolated from CBA mice was more effective than the GM_1 from AKR mice in absorbing the Thy-1.2 antigen, while the GD_{1b} from AKR but not from CBA mice absorbed the anti-brain-associated Thy antigen (Table XIII).[100,100a] This is an extremely perplexing observation since there are no known chemical differences between the same gangliosides from different strains of mice. Other experiments were performed suggesting that the target cell for GM_1-containing liposomes is the B lymphocyte. The antigenic material was partially purified by Sepharose 6B and isoelectric focusing from a detergent extract of thymocytes and had an apparent molecular weight of 35,000. Evidence was presented indicating that neither a mixed ganglioside fraction nor GD_{1b} could by the Thy-1 antigen.[101] Antibodies to GM_1 react with mouse T cells and thymocytes independent of the Thy-1 type. This may be due to cross-reacting materials present on the surface membrane rather than the GM_1-specific antibody. These observations do not support the proposal that GM_1 is the Thy-1 antigen.[102] Subsequently, it was shown that the Thy-1.2 antigen and GM_1 cap independently on the C3H thymocytes, suggesting that the ganglioside is not the Thy-1.2 antigen.[102a] The Thy-1 antigen activity of rat brain and thymus does not reside in the isolated gangliosides but rather in the chloroform–methanol-insoluble residue of these tissues.[102b] Cholera toxin and choleragenoid block the cytotoxic effect on bone marrow cells of anti-Thy-1 or anti-GM_1 antiserum and this would be consistent with a commonality in the receptor molecules.[103] Cocapping on thymocytes from CBA mice by cholera toxin and anti-Thy-1 antiserum has

Table XIII. Absorption of Anti-Thy-1.2 and Anti-BAO with AKR and CBA Brain Gangliosides[a]

Ganglioside[b]	Quantity used for absorption (μg)	Cytotoxicity index after absorption[c]			
		Anti-Thy-1.2		Anti-BAO	
		CBA	AKR	CBA	AKR
GM_3	1.5	87[d]	82	84	84
GM_2	1.5	77	84	80	86
GM_1	2.0	1	10	7	23
	1.0	4	22	20	49
	0.5	23	60	55	82
	0.1	65	80	75	84
GD_{1a}	2.0	89	86	74	70
	1.0	85	84	85	84
	0.5	84	83	82	84
GD_2	2.0	85	84	91	87
	1.0	88	90	83	84
	0.5	85	88	88	84
GD_{1b}	2.0	1	65	6	12
	1.0	39	90	5	13
	0.5	65	88	8	47
	0.1	85	90	5	69
GT_1	2.0	92	84	87	84
	1.0	85	89	83	82
	0.5	85	87	80	78

[a] Modified from Ref. 100.
[b] Gangliosides were formulated with cholesterol–lecithin at a ratio of 1 μg ganglioside: 5 μg cholesterol: 5 μg lecithin for each incubation with 0.05 ml serum (diluted 1 : 8, titer 1 : 64).
[c] Cytotoxicity index = [(% dead antiserum − % dead normal serum)(100 − % dead normal serum)] × 100.
[d] Pooled cytotoxic indexes from two observations.

been demonstrated. This further supports the possibility that a part of the Thy-1 material is associated with a cholera toxin receptor in these cells.[104]

An *in vitro* immune response assay was employed to investigate the Thy-1 antigens and it was found that five times less glycoprotein than GM_1 ganglioside was required for the response. The GM_1 ganglioside sample employed in this test system appeared as a single entity in two separate solvent systems. However, in a third solvent system, the Thy-1.2 antigen was separated from this glycosphingolipid but the nature of this material is currently defined.[104a] Commercial polysialoganglioside increased cAMP levels of mouse thymocytes *in vitro*. Purification of these samples by Florisil columns abolished this effect[104b] cautioning interpretation of many previous studies on the specificity of ganglioside as biological active agents.

6.5.1. Miscellaneous

A variety of somewhat isolated observations that have not been extensively pursued concerning interactions of gangliosides have appeared in the literature. Morphine and nalorphine were shown to affect Ca^{2+}–ganglioside binding in model systems.[105] The interaction of divalent cations, proteins, and gangliosides in biphasic systems was reported.[106] Basic proteins such as protamine[107] and nucleohistones[108] can bind gangliosides. In studies of Ca-dependent binding of colchicine to gangliosides, it has shown that GT_1 and GD_{1a} are more effective than GM_1.[109] Complex formation between micellar and monomeric GM_1 ganglioside and bovine serum albumin suggests that the interactions are largely hydrophobic.[109a] Spectrophotometric changes have been demonstrated to occur in the absorption spectrum of bilirubin when mixed with brain gangliosides.[110]

Monoclonal antibodies were prepared against chick embryo retinal cells and it was found that GQ ganglioside inhibited the cytotoxicity of the test system.[110a] However, the structure of the antigen is presently unknown. Mixed brain gangliosides enhanced adenylate cyclase activity of a membrane fraction prepared from rat cerebral cortex.[110b] The binding of cultured Chinese hamster ovary cells to collagen-coated dishes is mediated by fibronectin and this process is blocked by several gangliosides.[110c]

Cerebroside sulfate has been shown to bind narcotic analgesics[111] and biogenic amines,[112] and this lipid has been implicated in the sodium ion transport of the salt gland of sea birds.[113] Antisulfatide antibodies, detected by an immunofluorescent technique, were found in areas of the brain enriched in opiate receptors. Pretreatment with morphine analogs prevented binding to these areas but not to myelin. These results suggest that a sulfatide-like molecule may be a part of the opiate receptor.[113a]

6.6. Gangliosides and Glycoprotein Hormone Receptors

A considerable literature has evolved concerning the possible role of gangliosides as toxin receptors and information is becoming available concerning their potential as hormone receptors. One of the most extensive series of studies has been about their possible function in thyroid tissue as thyrotropin (TSH) receptors. This receptor, which was successfully solubilized with lithium diiodosalicylate, was trypsinized to yield materials with an approximate molecular weight of 15,000 to 30,000 that retained the ability to bind TSH. A 250-fold purification was achieved in a procedure that included affinity chromatography on TSH-Sepharose columns. The ability of the receptor to bind the hormone was decreased by both immobilized neuraminidase or immobilized Con A, suggesting that the receptor might be a sialic acid-containing glycoconjugate (Table XIV).[114] A series of studies showed that GT_1 and GD_{1b} were the most effective inhibitors of TSH binding to thyroid plasma membranes of the series of gangliosides tested. Since the same rank order of ability of the various gangliosides to effect a fluorescent change in TSH was observed, it was concluded that the gangliosides were interacting directly with the TSH rather than with the thyroid cell membrane. These observations led to the suggestion that gangliosides may be "an important structural component of TSH receptors in all tissues or a structural analog of an "active site" important for binding in these membranes."[115]

Plasma membranes from a TSH-unresponsive rat thyroid tumor have a reduced capacity for hormone binding. These cells have as their major, if not exclusive, ganglioside the relatively simple GM_3, which is in distinct contrast to the presence of higher ganglioside homologs in normal thyroid tissue. The tumors appear to be devoid of the GM_3 N-acetylgalactosaminyl transferase believed essential for the synthesis of the more complex gangliosides.[116] This might suggest that the unresponsiveness may be related to the ganglioside composition in these tumors. Cholera toxin can interfere with both TSH binding and TSH stimulation of adenylate cyclase. The nature of this inhibition is complex, displaying both noncompetitive and competitive compo-

Table XIV. *Effects of Sepharose Derivatives on [³H]-TSH Binding Activity of Lithium Diiodosalicylate-Solubilized TSH Receptor Preparation[a]*

Addition	Net cpm bound
None	11,773
TSH–Sepharose	0
Sepharose 2B or 4B	10,998
Neuraminidase–Sepharose	6,334
Con A–Sepharose	4,115
Procollagen–Sepharose	11,200
Collagen–Sepharose	11,800

[a] The standard 150-ml binding assay was carried out with 40 ml of a membrane extract that had been pretreated with the Sepharose derivatives indicated. Pretreatment consisted of incubating with shaking 100 μl of membrane extract with 100 μl of the indicated packed volume of swollen, buffer-washed Sepharose derivative beads for 30 min at 37°C. The beads were then pelleted by centrifugation, and the supernatant fluid was assayed for binding activity. (See Ref. 114.)

nents.[117] The binding of TSH to mixed brain ganglioside-containing liposomes was not an unexpected observation. The authors reported that a purified soluble glycoprotein obtained from thyroid plasma membranes when incorporated into liposomes similarly binds TSH.[118] The solubilization of thyroid plasma membranes with lithium diiodosalicylate resulted in the liberation of 80% of the protein, 79% of the TSH binding activity, and only 10% of the gangliosides. The insoluble residue contained 85% of the total ganglioside and the remainder of the binding capacity originally present (Table XV). Ganglioside-depleted membranes retained 24–34% of their TSH binding capacity. This particular observation is difficult to reconcile with their previous report that thyroid tumor membranes have a very much reduced TSH binding capacity and absence of complex gangliosides. It is possible that these cells were also devoid of the glycoprotein binding component. These studies deemphasize the relative importance of gangliosides as the receptor molecule for TSH, and the authors indicate that both the glycoproteins and the gangliosides have different functions in this phenomenon.[119]

The evidence for both glycoprotein and glycolipid interactions with TSH has gradually led to the hypothesis that both are components of the receptor complex. According to this theory, the glycoprotein serves as a high-affinity recognition site while the ganglioside is a low-affinity component. The ganglioside also induces a conformational change in the ligand and facilitates the entry of the ligand into the polar regions of the membrane.[119a]

Thyroid cells cultured from tissues taken from patients with Graves' disease, a form of hyperthyroidism, appear to be hypersensitive to TSH since cAMP levels are increased to a much larger extent than control cells. TSH binding to these cells is the same as the control cells. The authors believe that differences exist in the gangliosides from Graves' tissues; unfortunately, careful study to document this claim was not presented.[120] Large amounts of an

Table XV. *[^{125}I]-TSH Binding and Ganglioside Content of Intact Bovine Thyroid Plasma Membranes and of Fractions Obtained by "Solubilizing" the Membranes with Lithium Diiodosalicylate (LIS)^a*

	Total [^{125}I]-TSH binding activity^b		Total ganglioside content^c		Total protein	
	cpm × 10^{-6}	%	nmoles	%	mg	%
Intact membranes	73	100	175	100	30	100
LIS supernatant (soluble receptor activity)	58	79	17	10	24	80
LIS pellet	15	21	149	85	4.5	15

^{a} Modified from Ref. 119.
^{b} Total binding activity is that obtained by multiplying the [^{125}I]-TSH binding activity (in cpm) of an aliquot of the membrane, supernatant, or pellet preparations by the total volume of these preparations.
^{c} Total gangliosides were extracted and purified from intact membranes and the LIS fractions are expressed as nanomoles of sialic acid.
^{d} Total binding activity of intact membranes was the same, using assay procedures developed for either plasma membrane or soluble receptor activity.

antibody reacting against asialo GM_1 have been detected in patients with Graves' disease and Hashimoto's thyroiditis.[120a]

The possibility of ganglioside functioning as part of the TSH receptor has become clouded as a result of a recent report. A crude ganglioside preparation of bovine thyroid was more inhibitory for TSH binding than either a purified preparation or a saponified sample of the purified gangliosides. These observations suggested that ester-linked lipids may be involved. This was the case since acidic phospholipids such as cardiolipin, phosphatidylglycerol, and phosphatidylinositol all caused appreciable inhibition of TSH binding. These authors suggest that some of the inhibition that they previously reported by gangliosides might have been due to acidic phospholipids present in these preparations.[120b] Phospholipids are not thought to be receptor components but may be integral to the message transmission subsequent to TSH binding to the glycoconjugate receptor.[120c] ^{125}I-labeled tetanus toxin has been shown to bind to thyroid plasma membranes in a manner similar to cholera toxin and TSH.[121]

The β subunits of luteinizing, human chorionic gonadotropin, and follicle-stimulating hormones have some amino acid homology with the β chain of cholera toxin, and this may explain certain similarities of binding to membrane receptors.[121a]

6.6.1. Others

^{125}I-labeled luteinizing hormone binding to rat testicular membranes has been investigated. This system appears to have almost identical properties to that reported for the thyroid–TSH system and GT_1 and GD_{1b} are the most effective ganglioside inhibitors of the hormone–tissue interaction.[122] Similar studies have also been reported on the characteristics of human chorionic gonadotropin binding to the testicular membrane.[123] However, a series of experiments from another laboratory led to the conclusion that gangliosides are not the human chorionic gonadotropin receptors.[123a]

Progesterone production by isolated rat ovarian cells is stimulated by cholera toxin, human chorionic gonadotropin, and luteinizing hormones. The addition of gangliosides abolished the effect of cholera toxin on cAMP without any effect on the hormone response. These authors conclude that gangliosides are not involved as hormone receptors in this particular tissue.[123b–d] Mild proteolysis of mouse cortical brain cells liberates glycopeptides which decreases amino acid incorporation into protein by cultured BHK-21 and slightly with 1316 cells. Pretreatment of the 1316 cells with GM_1, but not with ceramide, increases their sensitivity to the glycopeptides.[123e]

6.7. Conclusions

There is a literature that can be interpreted as indicating that gangliosides may function as membrane receptors for a host of compounds. Unfortunately, none of the evidence presented can be regarded as unequivocal but merely as suggestive. Indeed, in several instances, subsequent investigations in different test systems have eliminated the probability that gangliosides have such a function. Unfortunately, occasionally the investigator's biases in favor of gangliosides have neglected an examination of parallel alterations in the non-ganglioside glycoconjugates. Indications exist that receptors may be glycoproteins that have a carbohydrate chain similar to that found in gangliosides. This possibility is supported by observations on "glanglioproteins" that cross-react to ganglioside antibodies.[124] Structural similarities between portions of the oligossaccharide of glycolipids and glycoproteins have recently been reviewed.[124a]

These concerns have received some support. For example, certain thyroid tumors have a decrease in glycosyltransferase both towards glycoproteins and glycolipids (L. Kohn, personal communication). Until the exact substrate specificities and properties of highly purified glycosyltransferases are available, all studies correlating these enzymes levels in crude biological samples should be viewed as tentative rather than being based upon unequivocal data. Those studies where data are provided only for the glycolipid or only for the glycoprotein rather than both should be regarded as a preliminary observation since only a single side of the coin is revealed. The trypsin sensitivity for cholera toxin binding in K-B3 and mouse LY cells suggests that the receptor may be a glycoprotein (L. Kohn, personal communication).

The critical micelle concentration of a mixed brain ganglioside sample was found to be approximately 10^{-8} M and that for GM_1 to be about 2×10^{-10} M or less. These values are considerably lower than previous determinations and were obtained by a combination of gel filtration, equilibrium dialysis, and boundary centrifugation techniques. Strong associations between ovalbumin, albumin, and fumerase and gangliosides were observed. If these lower values are more accurate and reliable, then all studies of interactions of gangliosides with hormones, toxins, and cells may have occurred with micelles rather than monomers.[124b] The interaction with many of these proteins may represent a generalized rather than specific phenomenon. This may also apply to reported interactions between follitropin and bilayers containing GM_1.[124c]

The accepted criteria of a tissue receptor emphasize a specificity both with respect to the target cell as well as the stimulatory agent. The general effect on several distinct cell types of the same ganglioside species to interact with diverse agents precludes their satisfying these criteria for receptor activity. A possible reservation to this inconsistency is that cells may have minute quantities of conventionally undetectable gangliosides species that are indeed tissue specific. This is illustrated by the observation that thyroid may contain at least 28 "gangliosides" some of which are present in trace quantities.[125] If this is true, then a new era of technical achievements and insight on the role of glycosphingolipids in cell biology may be in the foreseeable future. Hopefully, continual challenges to the dogma of gangliosides as receptor will provoke further investigations into this fascinating problem until the unequivocal results required for scientific truth are obtained.

6.8. References

1. Woolley, D. W., and Gommi, B. W. *Nature (London)* **202,** 1074 (1964).
2. Woolley, D. W., and Gommi, B. W. *Proc. Natl. Acad. Sci. U.S.A.* **53,** 959 (1965).
3. Gielen, W. *Z. Naturforsch.* **23b,** 117 (1968).
4. Gielen, W. *Z. Naturforsch.* **21b,** 1007 (1966).
5. Weseman, W., and Zilliken, F. *Biochem. Pharmacol.* **16,** 1773 (1967).
6. Carroll, P., and Sereda, D. D. *Nature (London)* **217,** 667 (1968).
7. Fiszer, S., and DeRobertis, E. *J. Neurochem.* **16,** 7201 (1969).
8. Weseman, W., Henkel, R., and Marx, R. *Biochem. Pharmacol.* **20,** 1961 (1971).
9. Van Heyningen, W. E. *Nature (London)* **249,** 415 (1974).
10. Marcus, A. J., Saifer, L. B., and Ullman, H. L. *Ciba Found. Symp.* **35,** 309 (1975).
10a. Gielen, W., and Vichofer, B. *Experientia* **30,** 1177 (1974).
11. Burton, R. M. *Int. J. Neuropharmacol.* **3,** 13 (1964).
12. Lapetina, E. G., Soto, E. F., and DeRobertis, E. *Biochim. Biophys. Acta* **135,** 33 (1967).
13. Burton, R. M. *Adv. Exp. Med. Biol.* **71,** 123 (1976).
13a. Richardson, P. J., Walker, N. H., James, W. T. and Whittaker, V. P. *J. Neurochem.* **38,** 1605 (1982).
14. Besancon, F., and Ankel, H. *Nature (London)* **252,** 478 (1974).
15. Besancon, F., Ankel, H., and Basu, S. *Nature (London)* **259,** 576 (1976).
16. Vengris, V. E., Reynolds, F. H., Hollenberg, M. D., and Petha, P. M. *Virology* **72,** 486 (1976).
16a. Kuwata, T., Handa, S., Fuse, A., and Morinaga, N. *Biochem. Biophys. Res. Commun.* **85,** 77 (1978).
16b. Grolman, E. F., Lee, G., Ramos, S., Lazo, P. S., Kaback, H. R., Friedman, R. M., and Kohn, L. D. *Cancer Res.* **38,** 4172 (1978).
16c. Ankel, H., Krishnamurti, C., Bescancon, F., Stefanos, S., and Falcoff, E. *Proc. Natl. Acad. Sci. U.S.A.* **77,** 2528 (1980).
16d. Aoyagi, T., Okuyama, A., Amizawa, H., Iwamori, M., Nagai, Y., Suzuki, J., Ishii, A., and Kobayashi, S. *Biochem. Int.* **2,** 187 (1981).
16e. Schiffmann, D., and Koschel, K. *Med. Microbiol. Immunol.* **169,** 281 (1981).
16f. MacDonald, H. S., Elconin, H. and Ankel, H. *FEBS Lett.* **141,** 267 (1982).
17. Simpson, L. L., and Rapporti, M. M. *J. Neurochem.* **18,** 1341 (1971).
18. Simpson, L. L., and Rapport, M. M. *J. Neurochem.* **18,** 1751 (1971).
19. Simpson, L. L., and Rapport, M. M. *J. Neurochem.* **18,** 1761 (1971).
20. Van Heyningen, W. E., and Mellanby, J. *Nauyn-Schmiedebergs Arch. Pharmakol.* **276,** 297 (1973).
20a. Kitamura, M., Iwamori, M., and Nagai, Y. *Biochim. Biophys. Acta* **628,** 328 (1980).
21. Wassermann, A., and Takaki, T. *Berl. Klin. Wochenschr.* **35,** 5 (1898).
22. Landsteiner, K., and Batteri, A. *Zentralbl. Bakteriol.* **42,** 562 (1906).
23. Van Heyningen, W. E. *J. Gen. Microbiol.* **20,** 291 (1959).
24. Van Heyningen, W. E. *J. Gen. Microbiol.* **20,** 301 (1959).
25. Van Heyningen, W. E. *J. Gen. Microbiol.* **20,** 310 (1959).
26. Van Heyningen, W. E., and Miller, P. A. *J. Gen. Microbiol.* **24,** 107 (1961).
27. Van Heyningen, W. E., and Millansky, J. *J. Gen. Microbiol.* **52,** 447 (1968).
28. Mellanby, J., van Heyningen, W. E., and Whittaker, V. P. *J. Neurochem.* **12,** 77 (1965).
29. Mellanby, J., and Whittaker, V. P. *J. Neurochem.* **15,** 205 (1968).
30. Habermann, E., and Dimpfel, W. *Nauyn-Schmiedebergs Arch. Pharmakol.* **267,** 327 (1973).
30a. Price, D. L., Griffin, J. W., and Pick, K. *Brain Res.* **121,** 379 (1977).
31. Habermann, E. *Nauyn-Schmiedebergs Arch. Pharmakol.* **267,** 341 (1973).
32. Dimpfel, W., Huang, R. T. C., and Habermann, E. *J. Neurochem.* **29,** 329 (1977).
33. Dawson, G., and Stoolmiller, A. C. *J. Neurochem.* **26,** 225 (1976).
34. Duffard, R. O., Fishman, P. H., Bradley, R. M., Lauter, C. J., Brady, R. O., and Trams, E. G. *J. Neurochem.* **28,** 1161 (1977).

34a. Lee, G., Grollman, E. F., Dyer, S., Beguinot, F., Kohn, L. D., Habig, W. H., and Hardegree, B., *J. Biol. Chem.* **254,** 3826 (1979).

35. Helting, T. R., Zwisler, O., and Weigandt, H. *J. Biol. Chem.* **252,** 194 (1977).

36. Bizzini, B., Stockel, K., and Schwab, M. *J. Neurochem.* **28,** 529 (1977).

37. Parikh, I., and Cuatrecasas, P. *Adv Enzymol.* **34b,** 610 (1974).

38. Wiegandt, H. *Adv. Exp. Med. Biol.* **83,** 259 (1977).

38a. Holmgren, J., Elwing, H., Fredman, P., Strannegard, O., and Svennerholm, L. In: *Structure and Function of Gangliosides* (Svennerholm, L., Mandel, P., Dreyfus, H., and Urban, P.-F. eds.), Plenum Press, New York (1980), p. 453.

38b. Mellanby, J., and Green, J. *Neurosciences* **6,** 281 (1981).

39. Kimbarg, D. V., Field, M., Johnson, J., Henderson, A., and Gershon, E. *J. Clin. Invest.* **50,** 1218 (1971).

40. Shafer, D. E., Lust, W. D., Sercar, B., and Goldelarg, M. D. *Proc. Natl. Acad. Sci. U.S.A.* **67,** 851 (1970).

41. Van Heyningen, W. E., Carpenter, C. C. J., Pierce, N. F., and Greenough, W. B., *J. Infect. Dis.* **124,** 415 (1971).

42. King, C. A., and van Heyningen, W. E. *J. Infect. Dis.* **127,** 639 (1973).

43. Holmgren, J., Lönnroth, I., and Svennerholm, L. *Infect. Immun.* **8,** 208 (1973).

44. Cuatrecasas, P. *Biochemistry* **12,** 3547 (1973).

45. Cuatrecasas, P. *Biochemistry* **12,** 3558 (1973).

46. Cuatrecasas, P. *Biochemistry* **12,** 3567 (1973).

47. Cuatrecasas, P. *Biochemistry* **12,** 3577 (1973).

48. Holmgren, J., Lindholm, L., and Lönnroth, I. *J. Exp. Med.* **139,** 801 (1974).

49. Van Heyningen, S. *Science* **183,** 656 (1974).

50. Holmgren, J., and Lönnroth, I. *J. Gen. Microbiol.* **86,** 49 (1975).

51. Lönnroth, I., and Holmgren, J. *FEBS Lett.* **44,** 282 (1974).

52. Holmgren, J., Mansson, J. E., and Svennerholm, L. *Med. Biol.* **52,** 229 (1974).

53. Staerk, J., Ronneberger, H. L., Wiegandt, H., and Ziegler, W. *Eur. J. Biochem.* **48,** 103 (1974).

54. Cuatrecasas, P., Parikh, I., and Hollenberg, M. D. *Biochemistry* **12,** 4253 (1973).

54a. Sattler, J., Schwarzmann, G., Staerk, J., Ziegler, W., and Wiegandt, H. *Hoppe-Seyler's Z. Physiol. Chem.* **358,** 159 (1977).

54b. Sattler, J., Schwarzmann, G., Knack, I., Rohm, K. H., and Wiegandt, H. *Hoppe-Seyler's Z. Physiol. Chem.* (in press).

54c. Schwarzmann, G., Mraz, W., Sattler, J., Schindler, R., and Wiegandt, H. *Hoppe-Seyler's Z. Physiol. Chem.* **359,** 1277 (1978).

54d. Magnani, J. L., Smith, D. F., and Ginsburg, V. *Anal. Biochem.* **109,** 399 (1980).

55. Hollenberg, M. D., Fishman, P. H., Bennett, V., and Cuatrecasas, P. *Proc. Natl. Acad. Sci. U.S.A.* **71,** 4224 (1974).

56. Van Heyningen, S., and King, C. A. *Biochem. J.* **146,** 269 (1975).

57. King, C. A. *J. Biol. Chem.* **250,** 6424 (1975).

58. Van Heyningen, S. *J. Infect. Dis.* **133,** 85 (1976).

59. Craig, S. W., and Cuatrecasas, P. *Proc. Natl. Acad. Sci. U.S.A.* **72,** 3844 (1975).

60. Revesz, T., and Greaves, M. *Nature (London)* **257,** 103 (1975).

61. Hildebrand, J., Stryckmans, P. A., and Vanhouche, J. *Biochim. Biophys. Acta* **260,** 272 (1972).

62. Dacremont, G., and Hildebrand, J. *Biochim. Biophys. Acta* **424,** 315 (1976).

63. Bennett, V., O'Keefe, E., and Cuatrecasas, P. *Proc. Natl. Acad. Sci. U.S.A.* **72,** 33 (1975).

64. Bennett, V., Craig, S., Hollenberg, M. D., O'Keefe, E., Sahyoun, N., and Cuatrecasas, P. *J. Supramol. Struct.* **4,** 99 (1976).

65. Walker, W. A., Feld, M., and Isselbacher, K. J. *Proc. Natl. Acad. Sci. U.S.A.* **71,** 320 (1974).

66. Holmgren, J., Lönnroth, I., Mansson, J. E., and Svennerholm, L. *Proc. Natl. Acad. Sci. U.S.A.* **72,** 2521 (1975).

67. Kiefer, H. C., Atlas, R., Moldan, D., and Kantor, H. S. *Biochem. Biophys. Res. Commun.* **66,** 1017 (1975).

67a. Strombeck, D. R., and Harrold, D. *Infect. Immun.* **10,** 1266 (1974).

67b. Morita, A., Tsao, D., and Kim, Y. S. *J. Biol. Chem.* **255,** 2549 (1980).

67c. Critchley, D. R., Magnani, J. L., and Fishman, P. H. *J. Biol. Chem.* **256,** 8724 (1981).

68. Wiegandt, H., Ziegler, W., Staerk, J., Kranz, T., Ronnenberger, H. J., Zilg, H., Karlsson, K. A., and Samuelsson, B. E. *Hoppe-Seyler's Z. Physiol. Chem.* **357,** 1637 (1976).

69. Sedlacek H. H., Stärk, J., Seiler, F. R., Ziegler, W., and Wiegandt, H. *FEBS Lett.* **61,** 272 (1976).

70. Wishnow, R. M., Lifrack, E., and Chen, C. C. *J. Infect. Dis.* **133,** S108 (1976).

71. Holmgren, J., and Lönnroth, I. *J. Infect Dis.* **133,** S64 (1976).

72. Berkenbile, F., and Delaney, R. *J. Infect. Dis.* **133,** S82 (1976).

72a. Anderson, W. B., Jaworski, C. J., Gallo, M., and Roston, I. *Nature (London)* **275,** 223 (1978).

73. Donata, S. T. *J. Infect. Dis.* **133,** S115 (1976).

74. King, C. A., van Heyningen, W. E., and Gascoyne, N. *J. Infect. Dis.* **133,** S75 (1976).

75. Kanfer, J. N., Carter, T. P., and Katzen, H. M. *J. Biol. Chem.* **251,** 7610 (1976).

75a. Pacuszka, T., Moss, J., and Fishman, P. H. *J. Biol. Chem.* **253,** 5703 (1978).

75b. Reed, B. C., Moss, J., Fishman, P. H., *J. Biol. Chem.* **253,** 5703 (1978).

76. Moss, J., Fishman, P. H., Manganiello, V. C., Vaughan, M., and Brady, R. O. *Proc. Natl. Acad. Sci. U.S.A.* **73,** 1034 (1976).

77. Fishman P. H., Moss, J., and Vaughan, M. *J. Biol. Chem.* **251,** 4490 (1976).

77a. Fishman, P. H., and Atikkan, E. E. *J. Biol. Chem.* **254,** 4342 (1979).

77b. O'Keefe, E., and Cuatrecasas, P. *J. Membrane Biol.* **42,** 61 (1978).

78. Manuelidis, L., and Manuelidis, E. E. *Science* **193,** 588 (1976).

78a. Joseph, K. C., Stieber, A., and Gonatas, N. K. *J. Cell Biol.* **81,** 543 (1979).

79. Fishman, P. H., Moss, J., and Manganiello, V. C. *Biochemistry* **16,** 1871 (1977).

80. Moss, J., Manganiello, V. C., and Fishman, P. H. *Biochemistry* **16,** 1876 (1977).

81. Mullins, B. R., Alaj, S. M., Fishman, P. H., Lee, G., Kohn, L. D., and Brady, R. O. *Proc. Natl. Acad. Sci. U.S.A.* **73,** 1679 (1976).

81a. Critchley, D. R., Ansell, S., Perkins, R., Dilks, S., and Ingram, J. *J. Supramol. Struct.* **12,** 273 (1979).

81b. Hagmann, J. and Fishman, P. H. *Biochem. Biophys. Acta* **720,** 181 (1982).

82. Moss, J., Osborne, J. C., Fishman, P. H., Brewer, H. B., Vaughan, M., and Brady, R. O. *Proc. Natl. Acad. Sci. U.S.A.* **74,** 74 (1977).

82a. De Wolf, M. J. S., Fridkin, M., Epstein, M., and Kohn, L. D. *J. Biol. Chem.* **256,** 5481 (1981).

82b. De Wolf, M. J. S., Fridkin, M., and Kohn, L. D. *J. Biol. Chem.* **256,** 5489 (1981).

83. Moss, J., Richards, R., Alving, C. R., and Fishman, P. H. *J. Biol. Chem.* **252,** 797 (1977).

83a. Richards, R. L., Moss, J., Alving, C. R. Fishman, P. H., and Brady, R. O. *Proc. Natl. Acad. Sci. U.S.A.* **76,** 1673 (1979).

83b. Fishman, P. H., Moss, J., Richards, R., Brady, R. O., and Alving, C. R. *Biochemistry* **18,** 2562 (1979).

84. Hansson, H. A., Holmgren, J., and Svennerholm, L. *Proc. Natl. Acad. Sci. U.S.A.* **74,** 3782 (1977).

85. Takeda, Y., Takeda, T., Honda, T., Tage, S., Sakuri, J., Ohtome, N., and Miwatani, T. *Jpn. J. Med. Sci. Biol.* **28,** 337 (1975); *Infect. Immunol.* **12,** 931 (1975).

86. Arend, P., and Nijssen, J. *J. Immunogenet.* **3,** 373 (1976).

87. Kato, V., and Naiki, M. *Infect. Immunol.* **13,** 289 (1976).

88. Donata, S. T., and Viner, J. P. *Infect. Immunol.* **11,** 982 (1975).

88a. Sack, D. A., Huda, S., Neogi, P. K. B., Daniel, R. R., and Spira, W. M. *J. Clin. Microbiol.* **11,** 35 (1980).

88b. Farris, A., Lindahl, M., and Wadstrom, T. *FEBS Lett.* **7,** 265 (1980).

89. Haywood, A. M. *J. Mol. Biol.* **83,** 427, 625 (1975).

89a. Holmgren, J., Svennerholm, L., Elwing, H., Fredman, P., and Strannegard, O. *Proc. Natl. Acad. Sci. U.S.A.* **77,** 1947 (1980).

89b. Wu, P.-S., Ledeen, R. W., Uden, S., and Isaacson, Y. A., *J. Virol.* **33,** 304 (1980).

90. Zapata, M. T., and Paglini, S. *Arch. Virusforsch.* **43,** 184 (1973).

90a. Takeda, Y., Takeda, T., Honda, T., and Miwata, T. *J. Med. Sci. Biol.* **31,** 198 (1978).

90b. Kayser, G., Goormaghtigh, M. Vandenbranden, M., and Ruysschaert, J. M. *FEBS Lett,* **127,** 207 (1981).

90c. Noda, M., Kato, I., Kirayama, T., and Matsuda, F. *Infect. Immun.* **29,** 678 (1980).

91. Ross, G. D. *Arch. Pathol. Lab. Med.* **101,** 337 (1977).
92. Reif, A. E., and Allan, J. M. *J. Exp. Med.* **120,** 413 (1964).
93. Golub, E. S. *J. Immunol.* **109,** 168 (1972).
94. Dutton, G. R., and Bacondes, S. H. *J. Neurochem.* **19,** 559 (1972).
95. Vitetta, E. S., Bayse, E. A., and Uhrs, J. W. *Eur. J. Immunol.* **3,** 446 (1973).
96. Towbridge, I. S., Weissmann, F. O., and Bevan, M. J. *Nature (London)* **256,** 652 (1975).
97. Kuckch, U. N., Bennett, J. C., and Johnson, B. J. *J. Immunol.* **115,** 626 (1975).
98. Letarte-Muirhead, M., Barclay, A. N., and Williams, A. F. *Biochem. J.* **151,** 685 (1975).
99. Barclay, A. N., Letarte-Muirhead, M., and Williams, A. F. *Biochem. J.* **151,** 699 (1975).
99a. Williams, A. E., Barclay, A. N., Letarte-Muirhead, M., and Morris, R. J., *Cold Spring Harbor Symp. Quant. Biol.* **41,** 51 (1976).
99b. Kuchel, P. W., Campbell, D. G., Barclay, N. A., and Williams, A. F. *Biochem. J.* **169,** 411 (1978).
100. Miller, H. C., and Esselman, W. J. *J. Immunol.* **115,** 839 (1975).
100a. Kato, K. P., Wang, T. J., and Esselman, W. J. *J. Immunol.* **123,** 1977 (1979).
101. Arndt, R., Stark, R., Klein, A. M., and Thiele, H. G. *Eur. J. Immunol.* **6,** 333 (1976).
102. Stein-Douglas, K. E., Schwarting, G. A., Naiki, M., and Marcus, D. M. *J. Exp. Med.* **143,** 822 (1976).
102a. Stein, K. E., Schwarting, G. A., and Marcus, D. M. *J. Immunol.* **120,** 676 (1978).
102b. Inokuchi, Y., and Nagai, Y. *Mol. Immunol.* **16,** 791 (1979).
103. Milewicz, C., Miller, H. C., and Esselman, W. J. *J. Immunol.* **117,** 1774 (1976).
104. Thiele, H. G., Arndt, R., and Stark, R. *Immunology* **37,** 767 (1977).
104a. Wang, T. J., Freimuth, W. W., Miller, C. C., and Esselman, W. J. *J. Immunol.* **121,** 1361 (1978).
104b. Krishnaraj, R. and Kemp, R. G. *Biochem. Biophys. Res. Commun.* **105,** 1453 (1982).
105. Greenberg, S., Diecke, F. P. J. and Long, J. P. *J. Pharm. Sci.* **61,** 1471 (1972).
106. Hayashi, K., and Katagiri, A. *Biochim. Biophys. Acta* **337,** 107 (1974).
107. Booth, D. A. *J. Neurochem.* **9,** 265 (1962).
108. Meisler, M. H., and McCluer, R. H. *Science* **154,** 896 (1969).
109. Rosner, H., and Sehonharting, M. *Hoppe-Seyler's Z. Physiol. Chem.* **358,** 915 (1977).
109a. Tomasi, M., Rodu, G., Ausiello, G., D'Agnolo, G., Venerando, B., Ghidoni, R., Sonnino, S., and Tettamanti, G. *Eur. J. Biochem.* **111,** 315 (1980).
110. Weil, M. L., and Menkes, J. H. *Pediatr. Res.* **9,** 791 (1975).
110a. Eisenbarth, G. S., Walsh, F. S., and Nirenberg, M. *Proc. Natl. Acad. Sci. U.S.A.* **76,** 4913 (1979).
110b. Partington, C. R., and Daly, J. W. *Mol. Pharmacol.* **15,** 484 (1979).
110c. Kleinman, H. K., Marten, G. R., and Fishman, P. H. *Proc. Natl. Acad. Sci. U.S.A.* **76,** 3367 (1979).
111. Cho, T. M., Cho, J. S., and Loh, H. H. *Life Sci.* **18,** 231 (1976).
112. Mosek, K., Bensch, K., and Felsenfeld, H. *FEBS Lett.* **9,** 337 (1970).
113. Karlsson, K., Samuelsson, B. C., and Steen, G. O. *Eur. J. Biochem.* **46,** 243 (1974).
113a. Zalc, B., Craves, F. B., Monge, M., Loh, H. H., and Baumann, N. A. *Colloq. INSERM* **86,** 423 (1979).
114. Tate, R. L., Holmes, J. A., Kohn, L. D., and Winand, R. J. *J. Biol. Chem.* **250,** 6527 (1975).
115. Mullins, B. R., Fishman, P. H., Lee, G., Aloj, S. M., Ledley, L. D., Winand, R. J., Kohn, L. D., and Brady, R. O. *Proc. Natl. Acad. Sci. U.S.A.* **73,** 842 (1976).
116. Meldolesi, M. F., Fishman, P. H., Aloj, S. M., Kohn, L. D., and Brady, R. O. *Proc. Natl. Acad. Sci. U.S.A.* **73,** 4060 (1976).
117. Mullins, B. R., Aloj, S. M., Fishman, P. H., Lee, G., Kohn, L. D., and Brady, R. O. *Proc. Natl. Acad. Sci. U.S.A.* **73,** 1679 (1976).
118. Aloj, S. M., Kohn, L. D., Lee, G., and Meldolesi, M. F. *Biochem. Biophys. Res. Commun.* **74,** 1053 (1977).
119. Meldolesi, M. F., Fishman, P., Aloj, S. M., Ledley, R. D., Lee, G., Bradley, R. M., Brady, R. O., and Kohn, L. D. *Biochem. Biophys. Res. Commun.* **75,** 581 (1977).
119a. Kohn, L. D., Consiglio, E., DeWolf, M. J. S., Grollman, E. F., Ladley, F. D., Lee, G., and Morris, N. P. In: *Structure and Function of Gangliosides* (Svennerholm, L., Mandel, P., Dreyfus, H., and Urban, P. F. eds.), Plenum Press, New York (1980), p. 487.

120. Lee, G., Grollman, E. F., Aloj, S. M., Kohn, L. D., and Winand, R. J. *Biochem. Biophys. Res. Commun.* **77,** 139 (1977).

120a. Sawada, K., Sakurami, T., Imura, H., Iwamori, M., and Nagai, Y. Lancet **11** (8187), **198** (1980).

120b. Omodeo-Sale, F., Brady, R. O., and Fishman, P. H. *Proc. Natl. Acad. Sci. U.S.A.* **75,** 5301 (1978).

120c. Aloj, S. M., Lee, G., Grollman, E. F., Beguinot, F., Consiglio, E., and Kohn, L. D. *J. Biol. Chem.* **254,** 9040 (1979).

121. Ledley, F. D., Lee, G., Kohn, L. D., Habig, W. H., and Hardegree, M. C. *J. Biol. Chem.* **252,** 4049 (1977).

121a. Ledley, F. D., Mullin, B. R., Lee, G., Aloj, S. M., Fishman, P., Hunt, L. T., Dayhoff, M. D., and Kohn, L. D. *Biochem. Biophys. Res. Commun.* **69,** 852 (1976).

122. Lee, G., Aloj, S. M., and Kohn, L. D. *Biochem. Biophys. Res. Commun.* **77,** 434 (1977).

123. Lee, G., Aloj, S. M., Brady, R. O., and Kohn, L. D. *Biochem. Biophys. Res. Commun.* **73,** 370 (1976).

123a. Pacuszka, T., Osborne, J. C., Brady, R. O., and Fishman, P. H. *Proc. Natl. Acad. Sci. U.S.A.* **75,** 764 (1978).

123b. Azhar, S., and Menon, K. M. *Biochim. Biophys. Acta* **81,** 205 (1978).

123c. Azhar, S., Fitzpatrick, P., and Menon, K. M. J. *Biochem. Biophys. Res. Commun.* **83,** 493 (1978).

123d. Azhar, S., and Menon, K. M. S. *Eur. J. Biochem.* **94,** 77 (1979).

123e. Kinders, R. J., Rintoul, D. A. and Johnson, T. C. *Biochem. Biophys. Res. Commun.* **107,** 663 (1982).

124. Tonegawa, Y., and Hakomori, S. *Biochem. Biophys. Res. Commun.* **76,** 9 (1977).

124a. Rauvala, H., and Finne, J. *FEBS Lett.* **97,** 1 (1979).

124b. Formisano, S., Johnson, M. L., Lee, G., Aloj, S. M., and Edelhoch, H. *Biochemistry* **18,** 1119 (1979).

124c. Deleers, M., Chatelain, P., Poss, A., and Ruysschaert, J. M. *Biochem. Biophys. Res. Commun.* **89,** 1102 (1979).

125. Mullins, B. R., Pacuszka, T., Lee, G., Kohn, L. D., Brady, R. O., and Fishman, P. H. *Science* **199,** 79 (1978).

Index